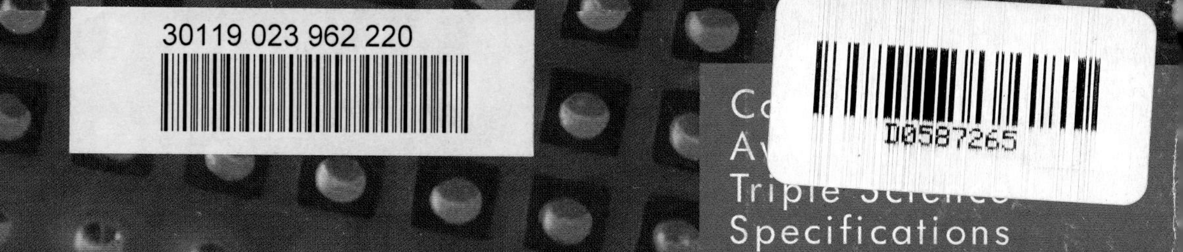

Specifications

Longman GCSE Physics

BRIAN ARNOLD
AND
STEVE WOOLLEY

Contents

Introduction

This book covers everything you would need to know for any GCSE Physics exam set by AQA, Edexcel or OCR. This means that you will not need to know everything in this book – only the parts which are relevant to the GCSE specification that you are following. It is important to find out exactly what your particular examiners are likely to ask you. To do this you will need a copy of your specification and as much other useful information as you can get.

If your teacher hasn't given you a copy of your specification, you can download it from your Awarding Body's website. Find out from your teacher which Awarding Body your school is using. If you are doing Physics as a single subject and are using OCR, find out whether you are doing Extension Block A or B.

The web addresses are:

www.aqa.org.uk

www.edexcel.org.uk

www.ocr.org.uk

Find your way to the GCSE Physics specification you want and download it. Downloading won't take more than a few minutes. The OCR Physics specification contains both Extension Blocks.

While you are on your Awarding Body's site, see what other useful things you can find. You should be able to find examples of exam papers and mark schemes. These are important so that you can see exactly what sort of questions your examiners ask, and how they expect you to answer them. You will also be able to find material written to help teachers. There's no reason why students can't make good use of this as well. Much of it will be free.

To do well in Physics at GCSE you don't need to know everything in this book, but you do need to understand exactly what your examiners want.

About this book

This book has several features to help you with GCSE Physics.

Introduction
Each chapter has a short introduction to help you start thinking about the topic and let you know what is in the chapter.

Section A: Electricity

Chapter 6: Static Electricity

Static electricity is the result of an imbalance of charge. It can be very useful, but it can also be extremely hazardous. In this chapter, you will learn about some of the uses and problems associated with static electricity.

The photograph in Figure 6.1 shows a spectacular example of **static electricity**. Bolts of lightning are seen when thunderclouds discharge their electricity. The discharging currents that flow can be as large as 20 000 A and typically take place in just 0.1 s. The flow of such a large current causes the air to heat up to temperatures of approximately 30 000°C. At such high temperatures, the air immediately around the bolt of lightning expands at supersonic speeds, causing thunder. As the sound waves travel outwards, they slow down and interact with the air and ground along their paths. The results of these interactions are the prolonged claps and rumbles we normally associate with thunder.

Figure 6.1 *Thunderclouds discharge their static electricity as lightning.*

Charges within an atom

All atoms contain small particles called **protons**, **neutrons** and **electrons**. The protons are found in the centre or **nucleus** of the atom and carry a relative charge of +1. The neutrons are also in the nucleus of the atom but carry no charge. The electrons travel around the nucleus in orbits. The electrons carry a relative charge of −1.

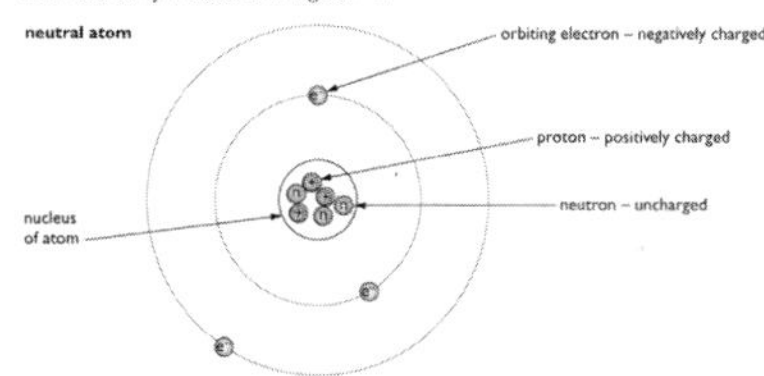

Figure 6.2 *A neutral atom has the same number of negative electrons and positive protons.*

Normally the number of protons in the nucleus is equal to the number of orbiting electrons. The atom therefore has no overall charge. It is **neutral**. If an atom gains extra electrons, it is then negatively charged. If an atom loses electrons, it becomes positively charged. An atom that becomes charged by gaining or losing electrons is called an **ion**.

Charging objects with static electricity

It is important to remember that the rubbing action does not produce or create charge. It simply separates charge – that is, it transfers some electrons from one object to another.

If an uncharged plastic rod is rubbed with an uncharged cloth, it is possible for both of them to become charged. This is sometimes called charging by friction. During the rubbing, electrons from the atoms of the rod may move onto the cloth. There is now an imbalance of charge in both objects. The rod is short of electrons and so is positively charged. The cloth has excess electrons and so is negatively charged.

48

End of chapter checklists
These lists summarise the material in the chapter. They could also help you to make revision notes because they form a list of things that you need to revise. (You need to check your specification to find out exactly what you need to know.)

End of Chapter Checklist

If you haven't got a copy of your specification, read the introduction on page iv.

ideas evidence

You will need to be able to do some or all of the following. Check your Awarding Body's specification (syllabus) to find out exactly what you need to know.

- Understand why a current-carrying conductor placed in a magnetic field experiences a force.
- Recall and use Fleming's left hand rule to predict the direction of the force and consequent movement.
- Recall that the size of the force can be increased by increasing the strength of the magnetic field or the size of the current flowing in the wire.
- Explain how electromagnetic effects are used in the simple dc motor.
- Understand that when a wire cuts through a magnetic field, or vice versa, a voltage is induced across the wire. If the wire is part of a complete circuit this induced voltage will cause a current to flow.
- Recall that the size of the induced voltage increases as the rate at which the magnetic field lines are being cut increases.
- Understand and explain how electricity can be generated by rotating a coil in a magnetic field or rotating a magnet inside a coil of wire.
- Understand the differences between ac and dc currents.
- Recall and understand the structure of a transformer.
- Recall that when an ac voltage is applied to the primary coil of a transformer an ac voltage is produced across the secondary coil.
- Recall and use the relationship $\frac{V_p}{V_s} = \frac{N_p}{N_s}$
- Describe the use of transformers in the National Grid to reduce energy losses.

Questions

More questions on electric motors and electromagnetic induction can be found at the end of Section A on page 80.

1 The diagram below shows a long wire placed between the poles of a magnet.

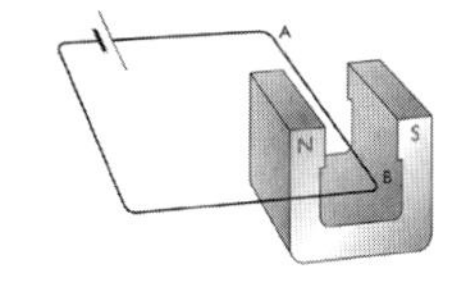

Describe what happens when:

a) current is passed through the wire flowing from A to B
b) the direction of the current is reversed – that is, it flows from B to A
c) with the current flowing from B to A, the poles of the magnet are reversed
d) a larger current is passed through the wire.

2 *a)* Draw a diagram of a simple electric motor.
b) Explain the functions of the split ring and the brushes.
c) Suggest three ways in which the speed of rotation of the motor could be increased.

46

Margin boxes
The boxes in the margin give you extra help or information. They might explain something in a little more detail or guide you to linked topics in other parts of the book.

Ideas and Evidence *ideas* evidence

This icon means that one of the Awarding Bodies has highlighted this topic as useful for learning about Ideas and Evidence. The Ideas and Evidence parts of the specifications help you to understand how scientific ideas have developed over time and how science relates to our everyday lives.

Questions
There are short questions at the end of each chapter. These help you to test your understanding of the material from the chapter. Some of them may also be research questions – you will need to use the internet and other books to answer these.
There are also questions at the end of each section. The end of section questions are written in an exam style and cover topics from all the chapters in the section.

Section A: Electricity

Chapter 1: Electric Circuits

We rely on electricity in many areas of our lives. This chapter looks at what electric current is and what makes it flow. You will learn what happens to electric current in different circuits, and what effects it has as it flows.

Look around the room you are in. If you are at home, you will probably be able to see a television, a radio or a computer. If you are in a science laboratory, you may be able to see a calculator or power supply. These and many other everyday objects need electric currents if they are to work. But what is an electric current? How are they produced and what do they do when they flow?

Figure 1.1 *Electricity is all around us.*

Conductors, insulators and electric current

An electric current is a flow of **charge**. In metal wires the charges are carried by very small particles called **electrons**. Electrons flow easily through all metals. We therefore describe metals as being good **conductors** of electricity. Electrons do not flow easily through most plastics – they are poor conductors of electricity. A very poor conductor is known as an **insulator** and is often used in situations where we want to prevent the flow of charge – for example, in the casing of a plug.

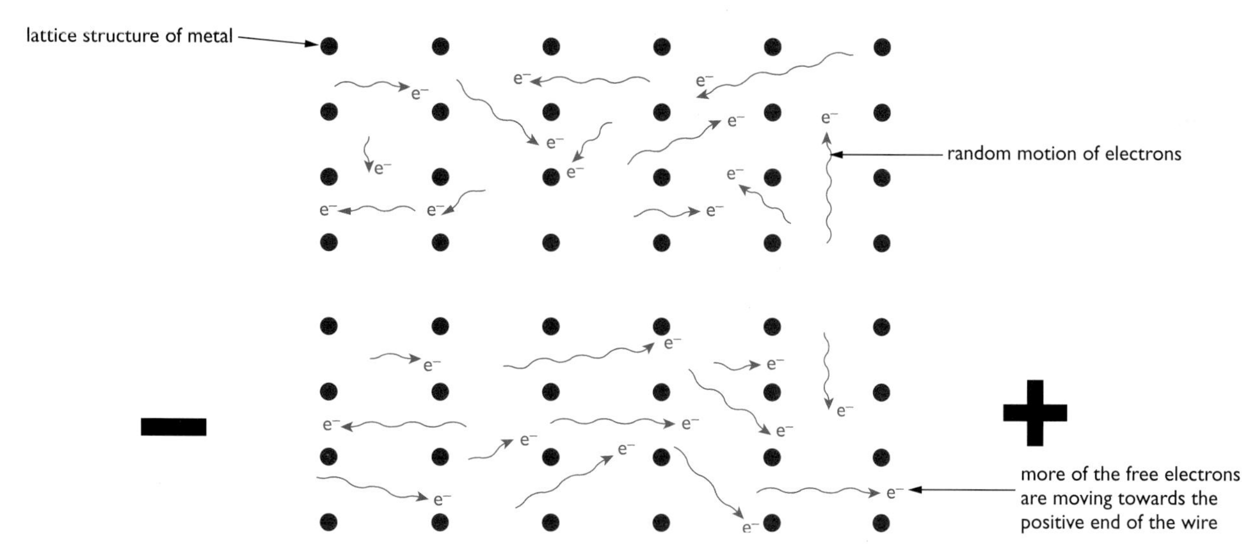

Figure 1.2 *Some electrons in a metal will flow when a voltage is applied.*

In conductors some electrons are free to drift between the atoms. Under normal circumstances this drifting is random – that is, the number of electrons flowing in any one direction is roughly equal to the number flowing in the opposite direction. There is therefore no overall flow of charge. If however a cell or battery is connected across the conductor more of the electrons now flow in the direction away from the negative terminal and towards the positive terminal than in the opposite direction – that is, there is now a net flow of charge. This flow of charge is an **electric current**.

In insulators all the electrons are held tightly in position and are unable to move from atom to atom. Charges are therefore unable to pass through insulators.

When scientists first experimented with charges flowing through wires they assumed that it was positive charges that were moving and that current therefore passes from the positive to the negative. We now know that this is incorrect and that when an electric current passes through a wire it is the negative charges or electrons that move. Nevertheless when dealing with topics such as circuits and motors it is still considered that current flows from positive to negative. This is **conventional current**.

A relatively new but very useful group of materials called **semiconductors** have some free electrons within their structure but not very many. The number of free electrons however can be increased in several ways – for example, by adding an impurity to the material, warming the material or shining light on it. Devices which contain components such as microchips, made from semiconductors, may be used in heat- or light-sensitive circuits in fire alarms, automatic street lights, and so on.

Electrolysis

Some substances such as lead bromide and sodium chloride (common salt) will not conduct electricity when they are solid but will allow charges to flow through them when they are molten or dissolved in water. Lead bromide and sodium chloride are examples of **ionic compounds**. They are made up of electrically charged particles called **ions**. In the solid state these ions are in fixed positions and therefore unable to carry charge. In the liquid state they are able to move.

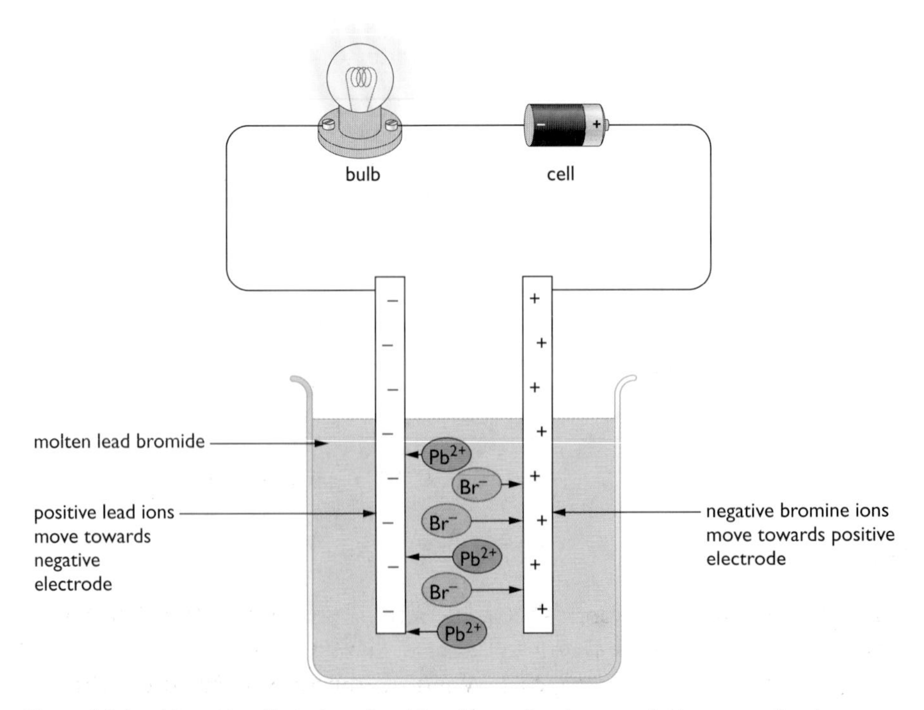

Figure 1.3 *Lead bromide will conduct electricity when molten because the ions carry the charge.*

As a result of the movement of the ions the lead bromide is broken down and simpler substances are released at the electrodes. In this example lead forms at the negative electrode and bromine forms at the positive electrode. The breaking down of a compound into simpler substances using electricity is called **electrolysis**. The mass or volume of the substances formed at the electrodes depends upon the size of the current flowing during electrolysis and the time for which this current flows.

Electrolysis is often used to coat one metal with another. For example, in silver plating of cutlery, a knife, fork or spoon is lowered into a solution containing silver and connected to the negative side of the supply. The positive silver ions are attracted over to the negatively charged cutlery, coating it with silver.

Electrical circuits

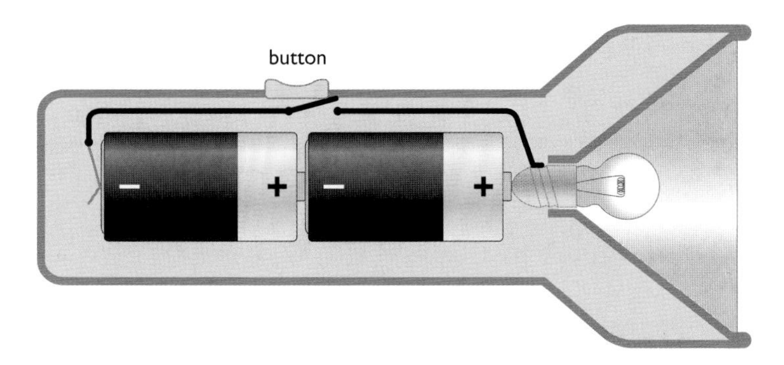

Figure 1.4 *A torch contains a simple electrical circuit – a series circuit.*

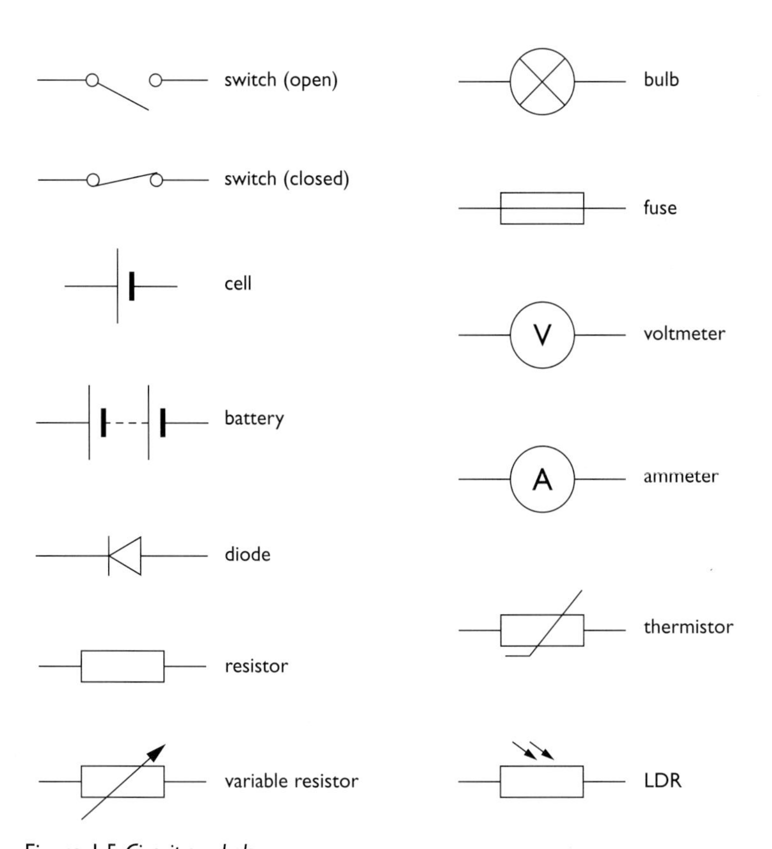

Figure 1.5 *Circuit symbols.*

Drawing diagrams of the actual components in a circuit is a very time-consuming and skilful task. It is much easier to use symbols for each of the components. Diagrams drawn in this way are called **circuit diagrams**. Here is a table of common circuit components and their symbols. You should know the common symbols but you will find a full list in Appendix B. Do not waste time memorising some of the less common ones as these will be given to you in the exam if you need them.

When the button on the torch shown in Figure 1.4 is pressed the circuit is *complete* – that is, there are no gaps. Charges are now able to flow around the circuit and the torch bulb glows. When the button is released the circuit becomes *incomplete*. Charges cease to flow and the bulb goes out.

Switches turn circuits on and off by making them complete or incomplete. In a series circuit the whole circuit is turned on and off by one switch.

In parallel circuits it is possible to turn parts of the circuit on and off using switches.

In the torch circuit in Figure 1.4 there is only one path for the charges to flow along. There are no branches or junctions. This simple "single loop" type of circuit is called a **series circuit**.

Circuits that have branches or junctions are called **parallel circuits**. There is more than one path that current can flow along.

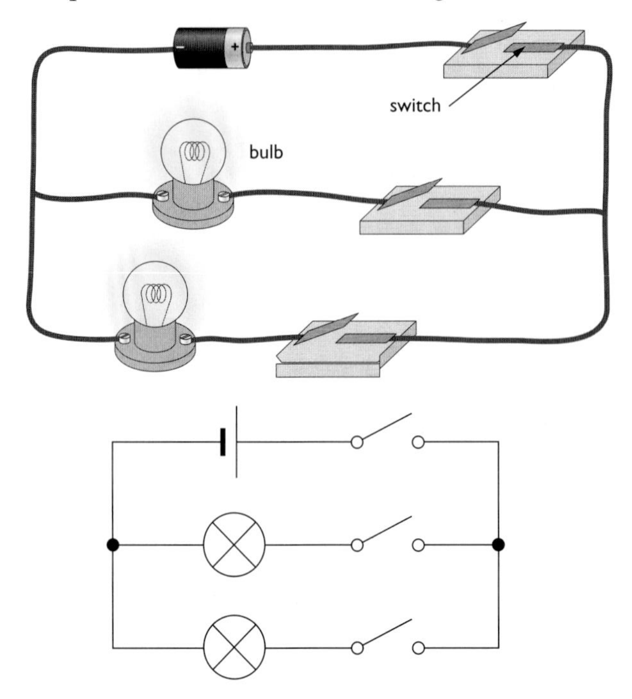

Figure 1.6 *A parallel circuit.*

Measuring current

We measure the size of the current flowing in a circuit using an **ammeter**. The ammeter is connected in series with the part of the circuit being investigated.

The size of an electric current indicates the rate at which charge flows.

We measure electric charge (Q) in units called coulombs (C). One coulomb of charge is the equivalent of the charge carried by approximately six million, million, million (6×10^{18}) electrons.

We measure electric current (I) in amperes or amps (A). If 1 C of charge flows along a wire every second the current passing through the wire is 1 A.

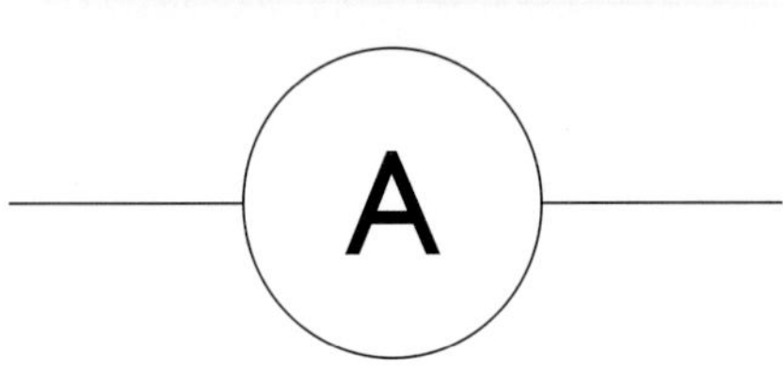

circuit symbol for an ammeter

Figure 1.7 *An ammeter is used to measure current in a circuit.*

$$1\ \text{C/s} = 1\ \text{A}$$

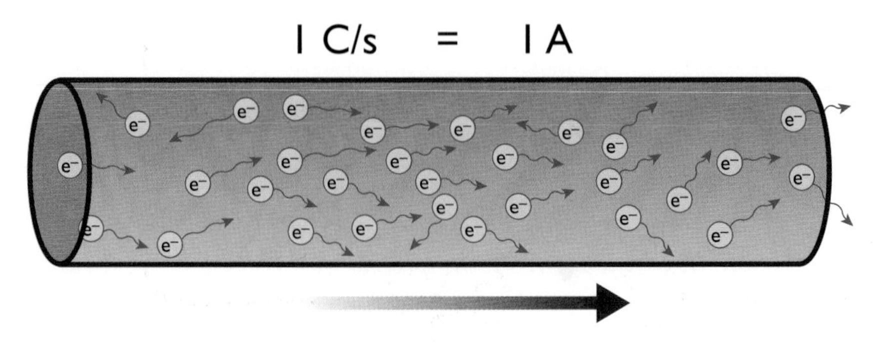

Figure 1.8 *One coulomb of charge flowing each second is one amp.*

We can calculate the current passing along a wire using the equation:

$$\text{current, I (in amps)} = \frac{\text{charge, Q (in coulombs)}}{\text{time, t (in seconds)}}$$

$$I = \frac{Q}{t}$$

Example 1

Calculate the current flowing through a wire if 15 C of charge pass along the wire in 5 s.

$I = \frac{Q}{t}$

$I = \frac{15\,C}{5\,s}$

$I = 3\,A$

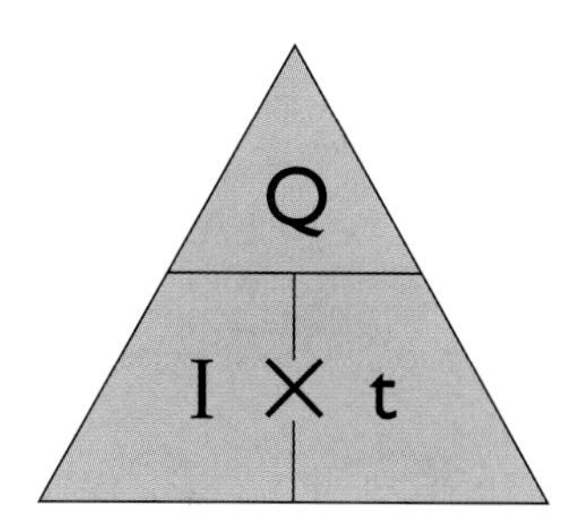

Figure 1.9 *You can use the triangle method for rearranging equations like this* $I = \frac{Q}{T}$.

Currents in series and parallel circuits

In a series circuit the current is the same in all parts. Current is not used up as it passes around a circuit.

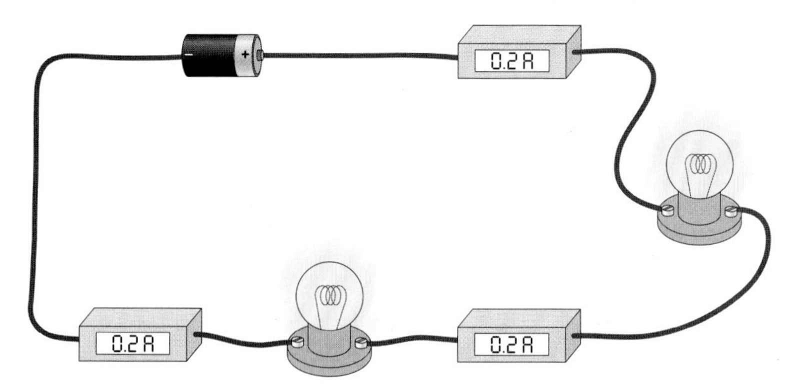

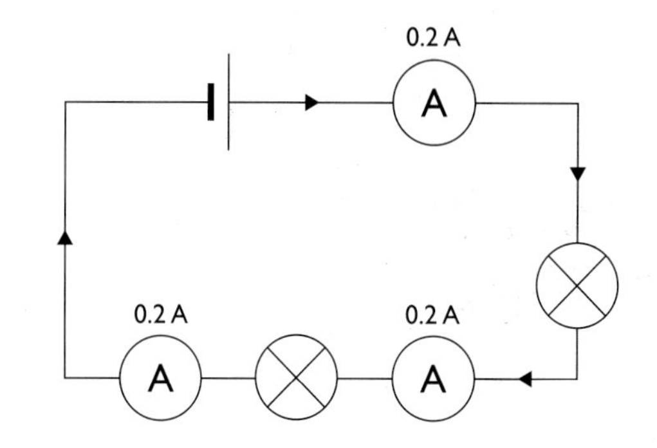

Figure 1.10 *In a series circuit the current does not vary.*

In parallel circuits the current flowing in the different branches may not be the same, but at each of the junctions the total currents flowing into the junction must be equal to the currents flowing out of the junctions.

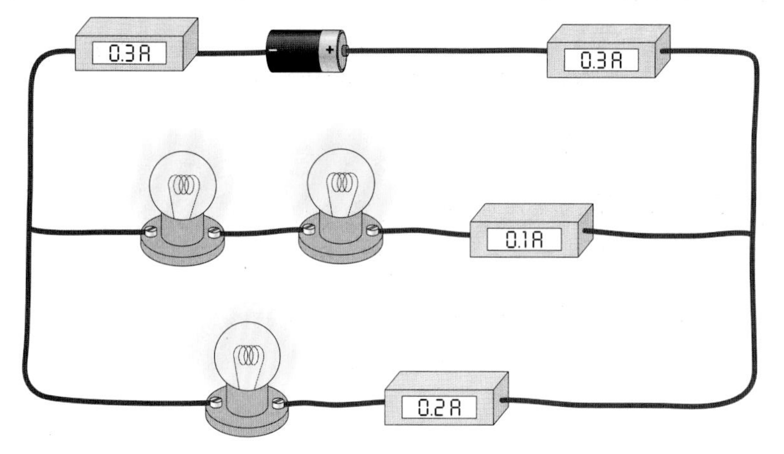

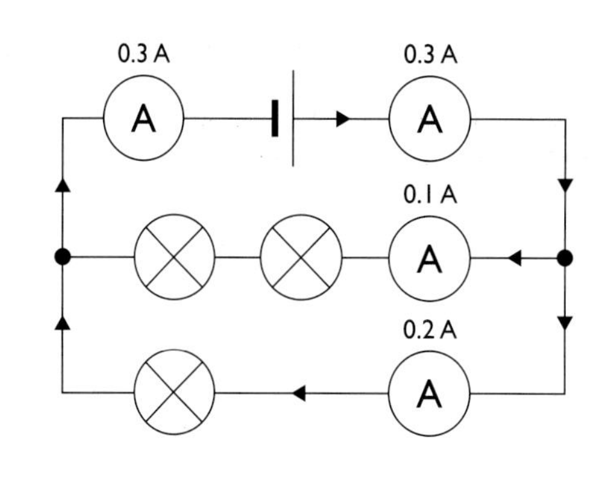

Figure 1.11 *In a parallel circuit the current can be different in different places.*

***worked* example**

You should be confident with circuit diagrams by now so we will only use these from this point on.

Example 2

Calculate the currents I_1, I_2, I_3 and I_4 in the circuits below.

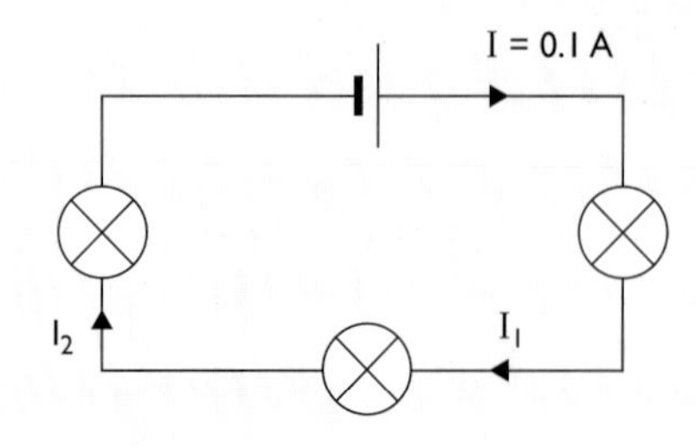

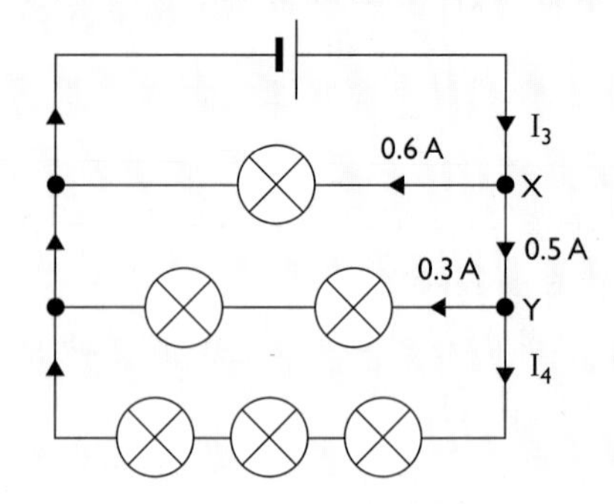

Figure 1.12.

The currents I_1 and I_2 are flowing around a series circuit. The current in a series circuit is the same everywhere. Therefore I_1 = 0.1 A and I_2 = 0.1 A.

The second circuit is a parallel circuit.

At junction X: $I_3 = 0.6\text{ A} + 0.5\text{ A}$

$I_3 = 1.1\text{ A}$

At junction Y: $0.5\text{ A} = 0.3\text{ A} + I_4$

$I_4 = 0.2\text{ A}$

Moving charges

We often use cells or batteries to move charges around circuits. We can imagine them as being "electron pumps". They transfer energy to the charges. The amount of energy given to the charges by a cell or battery is determined by its **electromotive force (emf)**. This is measured in volts (V) and is usually indicated on the side of the battery or cell.

If we connect a 1.5 V cell into a circuit and current flows, 1.5 J of energy is given to each coulomb of charge that passes through the cell.

If two 1.5 V cells are connected in series so that they are pumping in the same direction each coulomb of charge will receive 3 J of energy.

When several cells are connected together it is called a **battery**.

Cells and batteries provide current flowing in one direction. This is known as **direct current** (dc).

Figure 1.13 *When one coulomb of charge passes through this cell it gains 1.5 J of energy.*

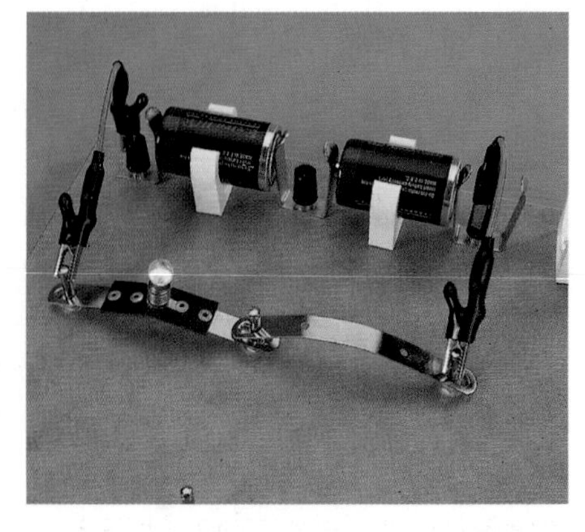

Figure 1.14 *When one coulomb of charge passes through both these cells in turn it gains 3 J of energy.*

In more general terms we can describe this energy change using the equation:

energy given to charge, E = battery voltage, V × charge, Q
(in joules) (in volts) (in coulombs)

$$E = VQ$$

where E is the total energy transferred to the charge (Q) as it flows through the cell or battery.

***worked* example**

Example 3

Calculate the total energy transferred to 100 C of charge when they pass through a battery of emf 12 V.

$E = VQ$

$E = 12\,V \times 100\,C$

$E = 1200\,J$ or $1.2\,kJ$

Voltage and potential difference (pd)

As the charges flow around a circuit the energy they carry is converted into other forms of energy by the components they pass through. The **voltage** or **potential difference** across each component tells us how much energy it is converting. The voltage, like the emf, is measured in volts (V). If the voltage across a component is 1 V this means that the component is changing 1 J of electrical energy into a different kind of energy each time 1 C of charge passes through it.

If we apply the Law of Conservation of Energy to any circuit it follows that the energy given to each coulomb of charge as it flows through a cell or battery must be equal to the energy converted into other forms as it passes around the circuit.

The Law of Conservation of Energy states that energy can neither be created nor destroyed.

Figure 1.15 illustrates this.

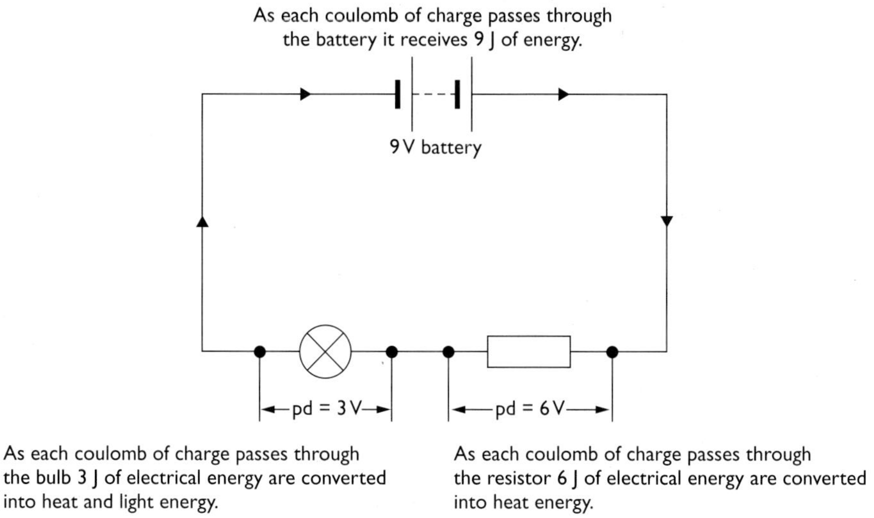

Figure 1.15 *Energy is always conserved as current flows in a circuit.*

Measuring voltages

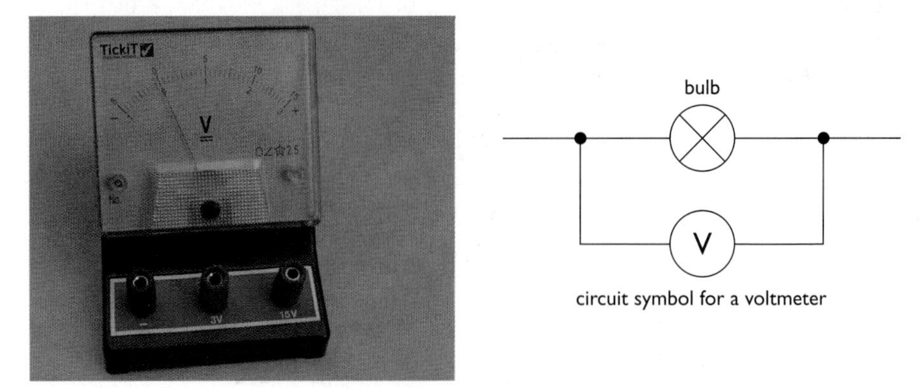

Figure 1.16 *A voltmeter measures voltages across a component.*

A voltmeter has a very high resistance, so very little current flows through it.

We measure voltages using a **voltmeter**. This is connected *across* (in parallel with) the component we are investigating. A voltmeter connected across a cell or battery will measure the energy given to each coulomb of charge that passes through it. A voltmeter connected across a component will measure the electrical energy converted into other forms when each coulomb of charge passes through it.

Voltages in series and parallel circuits

If all the energy given to the charges is converted into other forms by components in the external part of the circuit – that is, not inside the cell or battery – then the voltage across the cell or battery should be equal to the sum of the external voltages (Figure 1.17).

$$V_{cell} = V_{bulb} + V_{buzzer} + V_{res}$$

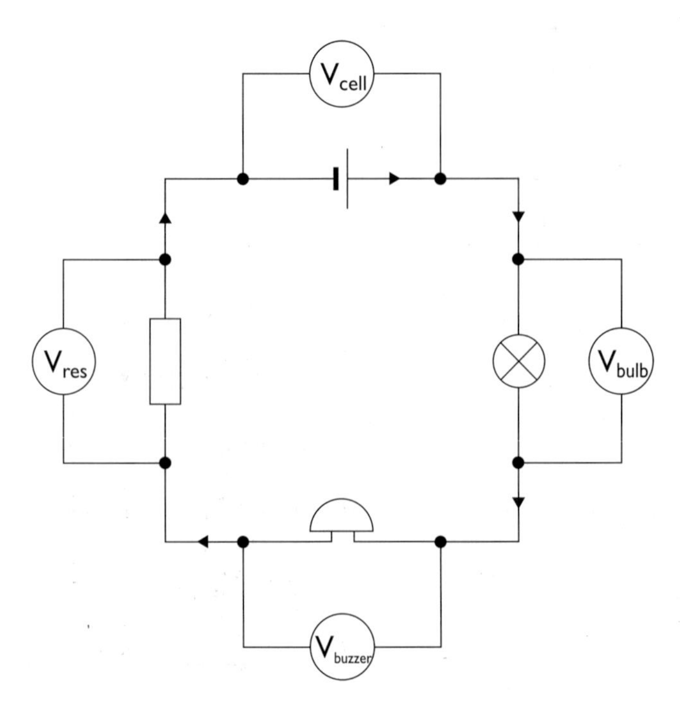

Figure 1.17 *The voltage of the cell is equal to the sum of the component voltages.*

worked example

Example 4

Calculate the electrical energy converted into heat when 1 C of charge flows through the resistor R in Figure 1.18 below.

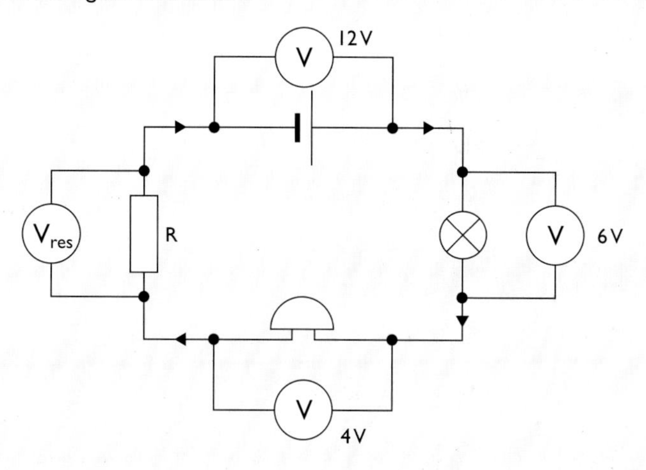

Figure 1.18.

$V_{bat} = V_{bulb} + V_{buzzer} + V_{res}$

$12\,V = 6\,V + 4\,V + V_{res}$

$V_{res} = 2\,V$

The pd across the resistor is 2 V. Therefore 2 J of electrical energy are converted into 2 J of heat energy when 1 C of charge flows through it.

In a parallel circuit the same rule is correct for any one complete loop in the circuit. For example, in Figure 1.19:

$V_{bat} = V_{bulb} + V_{resA}$

or $V_{bat} = V_{bulb} + V_{resB}$

or $V_{bat} = V_{bulb} + V_{resC}$

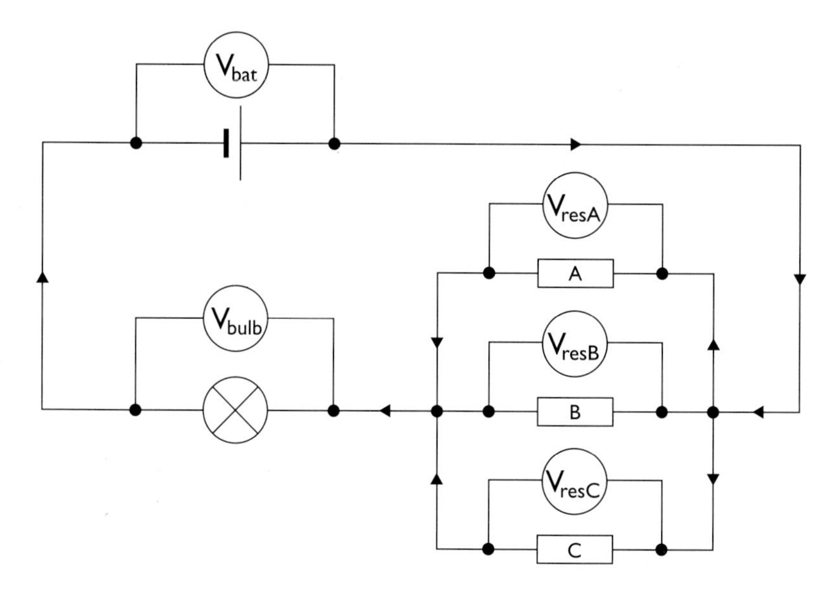

Figure 1.19.

From page 9 we can see that the voltages across the branches of a network are the same.

$$V_{resA} = V_{resB} = V_{resC}$$

worked example

Example 5

Determine the potential differences across the three resistors marks A, B and C.

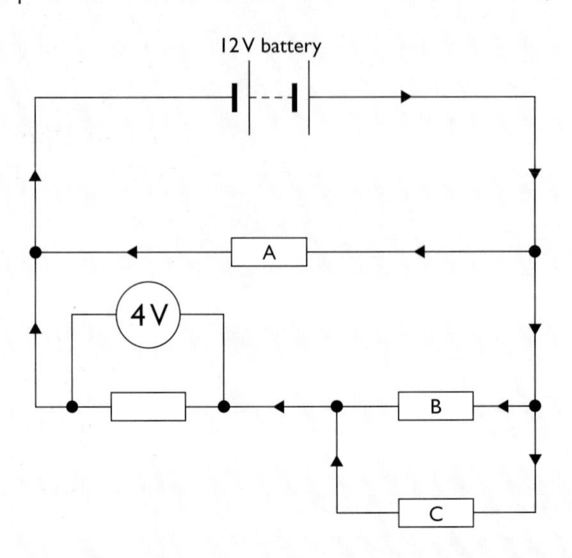

Figure 1.20.

potential difference across resistor A = potential difference of battery = 12 V

potential difference across resistor B = 12 V – 4 V

= 8 V

potential difference across resistor C = potential difference across resistor B

= 8 V

Putting a voltmeter across a connecting wire will result in a zero reading. This is because the wires are very good conductors and a negligible amount of electrical energy will be converted into other forms as the current flows through them.

End of Chapter Checklist

If you haven't got a copy of your specification, read the introduction on page iv.

You will need to be able to do some or all of the following. Check your Awarding Body's specification (syllabus) to find out exactly what you need to know.

- Draw and interpret circuit diagrams.
- Understand how the number of free charge carriers in a material determines whether a material is a conductor or an insulator.
- Recall that in series circuits ***a)*** there is only one path for the current to follow, ***b)*** the current is the same everywhere, ***c)*** the sum of the voltages is equal to the voltage across the whole circuit.
- Recall that in parallel circuits ***a)*** there is more than one path for the current to follow, ***b)*** the sum of the currents entering a junction is equal to the sum of the currents leaving that junction, ***c)*** the voltages across components that are in parallel are the same.
- Understand and use the equations $I = Q/t$ and $E = V \times Q$.
- Recall that the potential difference (pd) between two points is the number of joules of energy converted when 1 C of charge flows between them.

Questions

More questions on electrical circuits can be found at the end of Section A on page 80.

1 Current is flow of charge.

a) What are the charge carriers in metals?

b) Explain why charges are able to flow through metals but not through a plastic.

c) Explain why an ionic substance such as sodium chloride is an insulator when it is a solid but will conduct electricity when it is molten or in solution.

d) If the current flowing through a heater is 3 A, calculate the charge that flows through it in *i)* 1 s, *ii)* 10 min and *iii)* 1 hour.

2 Currents flow around circuits.

a) Explain the differences between:

i) a complete circuit and an incomplete circuit

ii) a series circuit and a parallel circuit.

b) The diagram opposite shows a junction in an electrical circuit. Calculate the size and direction of the current flowing in the wire AB.

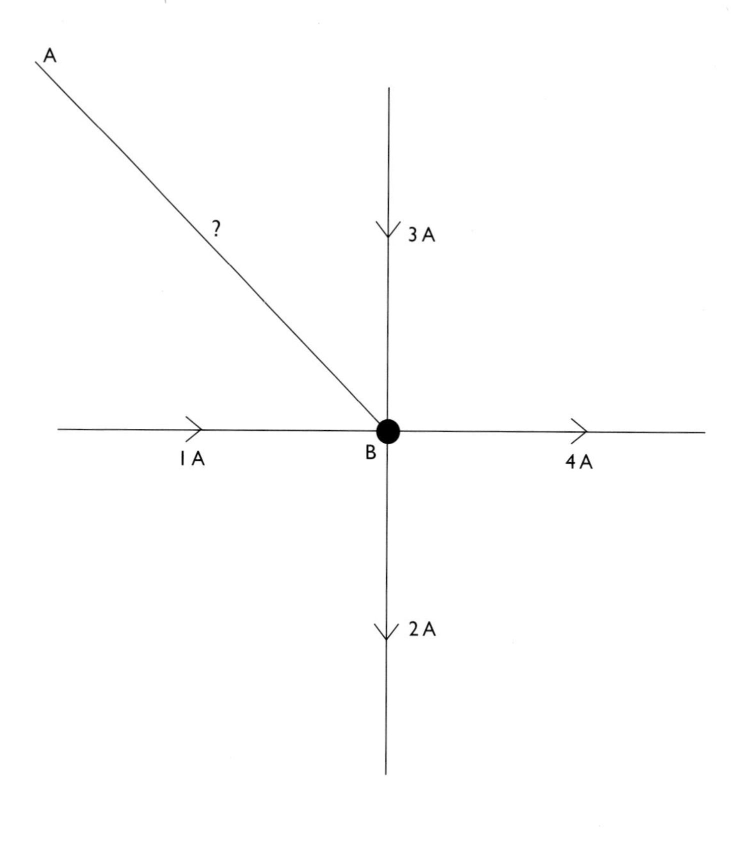

c) Look carefully at the circuits shown below. Assuming that all switches are initially closed, decide which of the bulbs go out when each of the switches in turn is opened.

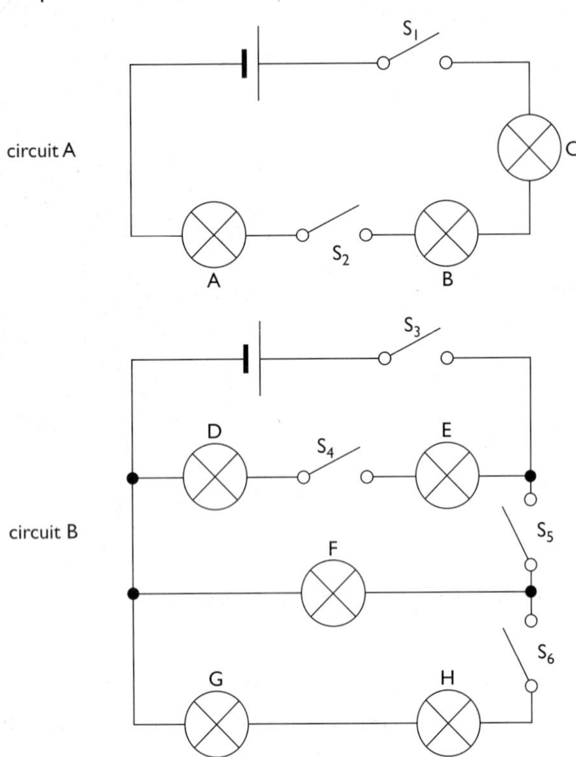

d) In circuit A, which bulb(s) glow the brightest when all the switches in all the circuits are closed?

e) Explain your answer to **d)**.

f) In circuit B, which bulb(s) glow brightest when all the switches in the circuit are closed?

g) Explain your answer to ***f)***.

3 The potential difference (pd) between two points in a circuit is measured using a voltmeter.

a) Draw a circuit diagram to show how a voltmeter should be connected to measure *i)* the pd across a bulb, and *ii)* pd of a cell.

b) Explain in your own words the phrase "a cell has an emf of 1.5 V".

c) The diagram below shows a 6 V battery connected across three bulbs. The bulbs are not identical. What is the pd across bulb C?

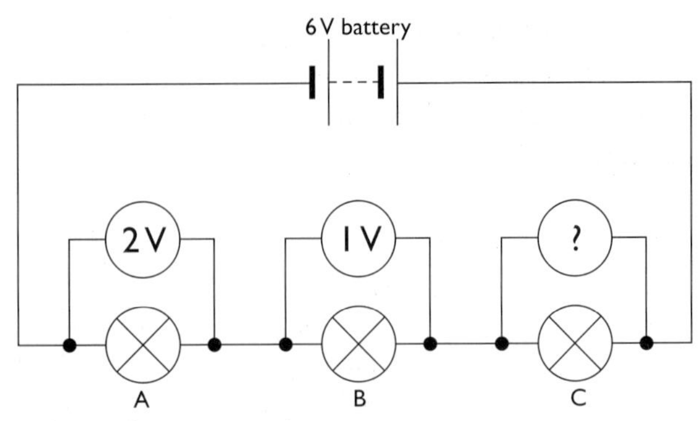

d) The pd across a bulb is 12 V. How many joules of electrical energy are converted into other forms of energy when:

i) a charge of 1 C passes through it?

ii) a charge of 50 C passes through it?

iii) a current of 0.2 A flows through it for 10 s?

iv) a current of 0.3 A flows through it for 5 min?

4 The diagram below shows a circuit containing two two-way switches.

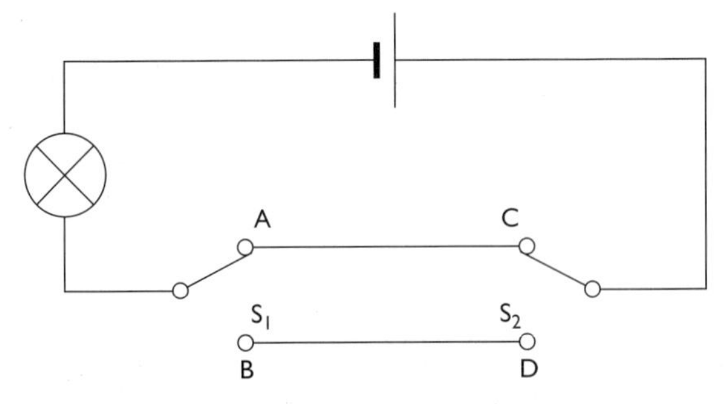

a) Explain in your own words what happens when each of the switches is moved to a new position.

b) Suggest one important application of this circuit in the home.

Chapter 2: Electrical Resistance

In this chapter you will learn what resistance is, and how it can be useful in electrical appliances. You will learn what factors affect resistance, and how to work out the resistance of a component by measuring the current flowing in it and the voltage across it (Ohm's law). You will also read about some special resistors, and their uses.

Figure 2.1 *You can control electrical appliances such as stereos by adjusting the resistance of the circuits.*

It is likely that almost every day of your life you will make some adjustments to at least one electrical appliance. You may turn up the volume of your radio or adjust the colour on your television. In each of these examples your adjustments are changing the currents and the voltages in the circuits of your appliance. You are achieving the results you require by altering the **resistance** of the circuits. This chapter will help you understand the meaning and importance of resistance and how we make use of it.

Resistance

We normally assume that connecting wires have zero resistance.

All components in a circuit offer some resistance to the flow of charge. Some (for example, connecting wires) allow charges to pass through very easily losing very little of their energy. We describe connecting wires as having very **low resistance**. The flow of current through some components is not so easy and a significant amount of energy is used to move the charges through them. This energy is converted into other forms, usually heat. Components like this are said to have a **high resistance**.

We measure the resistance (R) of a component by comparing the size of the current that flows through that component and the potential difference (voltage) applied across its ends:

$$\text{resistance, R (in ohms)} = \frac{\text{potential difference, V (in volts)}}{\text{current, I (in amps)}}$$

$$R = \frac{V}{I}$$

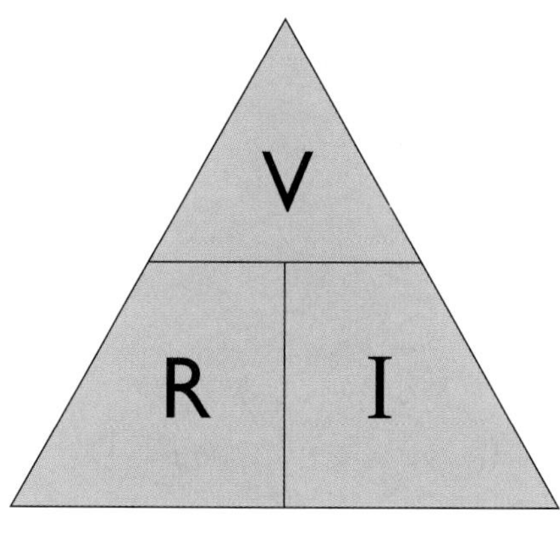

Figure 2.2 *You can use the triangle method for rearranging equations like* $R = \frac{V}{I}$.

We measure resistance in units called ohms (Ω).

worked example

Example 1

When a pd of 12 V is applied across a buzzer a current of 0.1 A flows. Calculate the resistance of the buzzer.

$$R = \frac{V}{I}$$

$$R = \frac{12\,V}{0.1\,A}$$

$$R = 120\,\Omega$$

Resistance of a piece of wire

Although we regard connecting wires as having zero resistance this is not the case for other types of wire. The resistance of a piece of wire depends upon:

1 the *length* of the wire – the longer the wire the greater its resistance; doubling the length of the wire will double its resistance

2 the *thickness* or cross-sectional area of the wire – doubling the cross-sectional area of the wire will halve its resistance

3 the *material* from which the wire is made – the more free electrons available to carry charge around a circuit the lower the resistance of the wire

4 the *temperature* of the wire – increasing the temperature of a piece of wire will increase its resistance; at higher temperatures the atoms in the wire vibrate more vigorously, impeding the flow of charge.

Using resistance

The heating effect of a current

The resistance of a component can be very useful. Next time you toast some bread in a toaster ask yourself "What is toasting the bread and how is it doing it?" The heating element in a toaster contains a wire that has electrical resistance. As current passes through the wire a lot of the energy carried by the electrons in the wire is converted into heat. It is this energy which you are using to toast the bread.

Figure 2.3 *Resistance can be useful.*

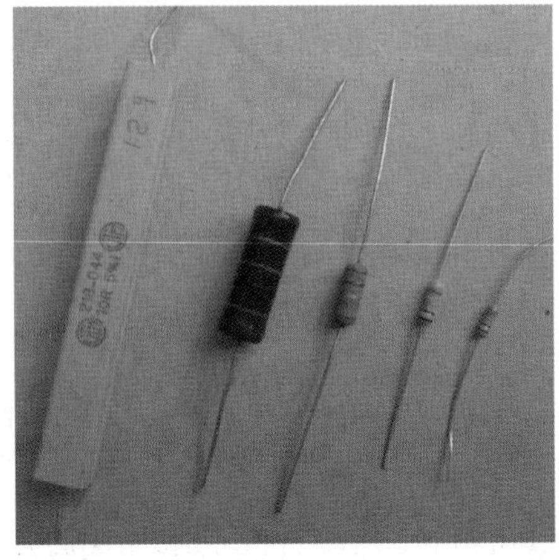

Figure 2.4 *A selection of resistors.*

Fixed resistors

In many circuits you will find components similar to those shown in Figure 2.4. They are called **fixed resistors**. They are included in circuits in order to control the sizes of currents and voltages. The resistor in the circuit in Figure 2.5 is included so that the correct voltage is applied across the bulb and the correct current flows through it. Without the resistor the voltage across the bulb may cause too large a current to flow through it and the bulb may "blow".

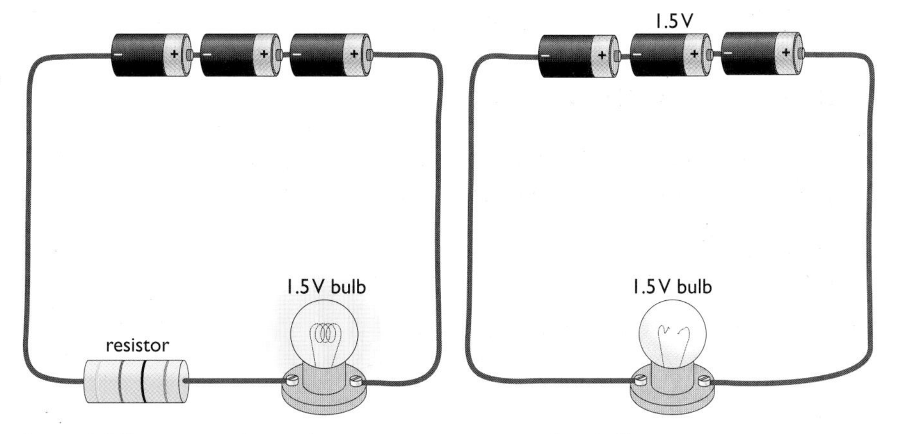

Figure 2.5 *Resistors can be used to protect components such as bulbs.*

Variable resistors

Figure 2.6 *Variable resistors.*

Figure 2.6 shows examples of a different kind of resistor. They are called **variable resistors** as it is possible to alter their resistance. When you alter the volume of your television or your radio you are using a variable resistor to do this.

In the circuit in Figure 2.7 a variable resistor is being used to control the size of the current flowing through a bulb. If the resistance is decreased a larger current flows and the bulb glows more brightly. If the resistance is increased a smaller current flows and the bulb will glow less brightly or not at all. The variable resistor is behaving in this circuit as a dimmer switch. In circuits containing electric motors variable resistors can be used to control the speed of the motor.

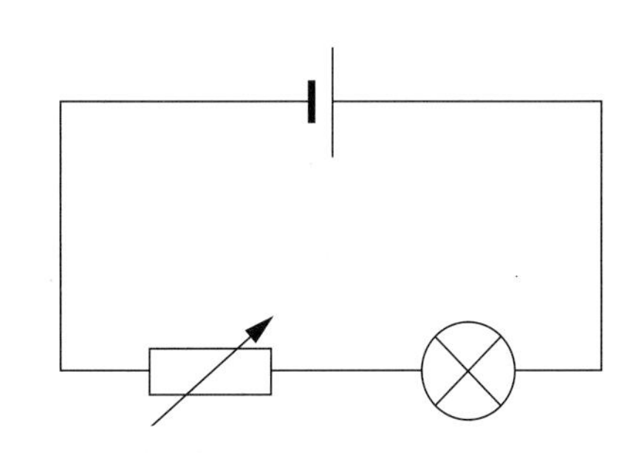

Figure 2.7 *Circuit with a variable resistor being used as a dimmer switch.*

Resistors in series and parallel

Often more than one resistor is included in a circuit. If the resistors are connected in series, their overall effect can be found by simply adding together the resistances of the individual components:

$$R_t = R_1 + R_2$$

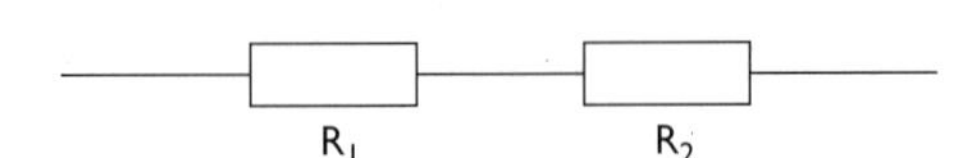

Figure 2.8 *Resistors in series.*

worked example

Example 2

Calculate the total resistance of the four resistors shown in Figure 2.9.

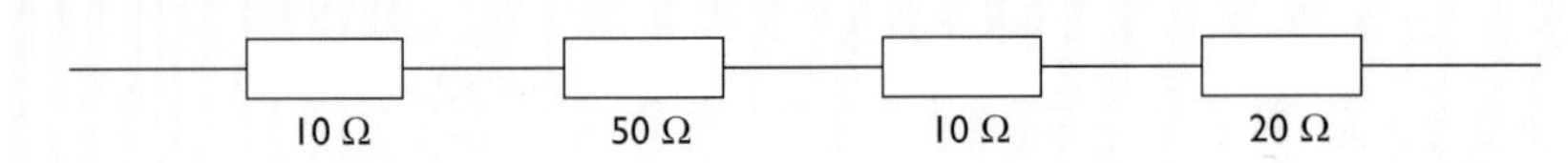

Figure 2.9.

$$R_t = R_1 + R_2 + R_3 + R_4$$
$$R_t = 10\,\Omega + 50\,\Omega + 10\,\Omega + 20\,\Omega$$
$$R_t = 90\,\Omega$$

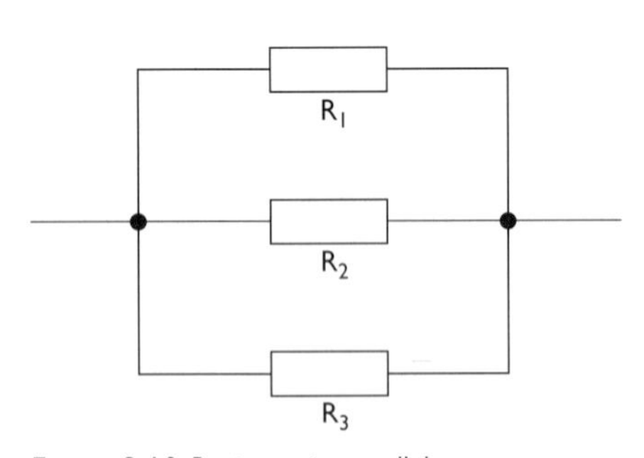

Figure 2.10 *Resistors in parallel.*

If the resistors are connected in parallel their total resistance can be calculated using the equation:

$$\frac{1}{R_t} = \frac{1}{R_1} + \frac{1}{R_2} + \frac{1}{R_3}$$

worked example

At GCSE level you can use a simpler formula for R in parallel (when there are two resistors only):

$$R_t = \frac{R_1 \times R_2}{R_1 + R_2}$$

It is worth noting that when resistors are connected in parallel the resistance of the network as a whole is always less than the resistance of any of the individual resistors. This is because the addition of each resistor increases the number of paths along which the current can flow.

Example 3

Calculate the total resistance of 4 Ω, 2 Ω and 6 Ω connected in parallel.

$$\frac{1}{R_t} = \frac{1}{R_1} + \frac{1}{R_2} + \frac{1}{R_3}$$

$$\frac{1}{R_t} = \frac{1}{4} + \frac{1}{2} + \frac{1}{6}$$

$$\frac{1}{R_t} = \frac{3}{12} + \frac{6}{12} + \frac{2}{12}$$

$$\frac{1}{R_t} = \frac{11}{12}$$

$$R_t = \frac{12}{11}$$

$$R_t = 1.1\,\Omega$$

Special resistors

Thermistors

A **thermistor** is a resistor whose resistance changes quite dramatically with temperature. It is made from a semiconducting material such as silicon or germanium. At room temperature the number of charge carriers is small and so the resistance of a thermistor is large. If however if it is warmed the number of free charge carriers increases and its resistance decreases. Thermistors are often used in temperature-sensitive circuits in devices such as fire alarms and thermostats.

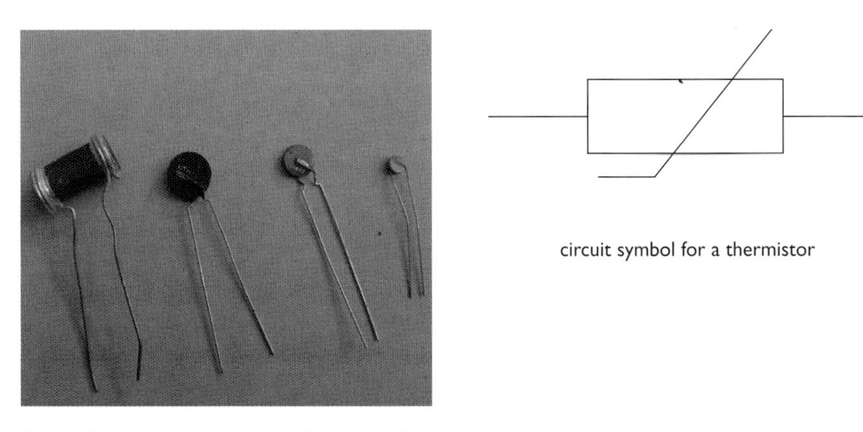

Figure 2.11 *The resistance of a thermistor decreases as the temperature rises.*

Light-dependent resistors (LDRs)

In dark conditions **light-dependent resistors** (**LDRs**) contain few "free" charge carriers and so have a high resistance. If however light is shone onto an LDR more charge carriers are freed and the resistance decreases. LDRs are often used in light-sensitive circuits in devices such as photographic equipment, automatic lighting controls and burglar alarms.

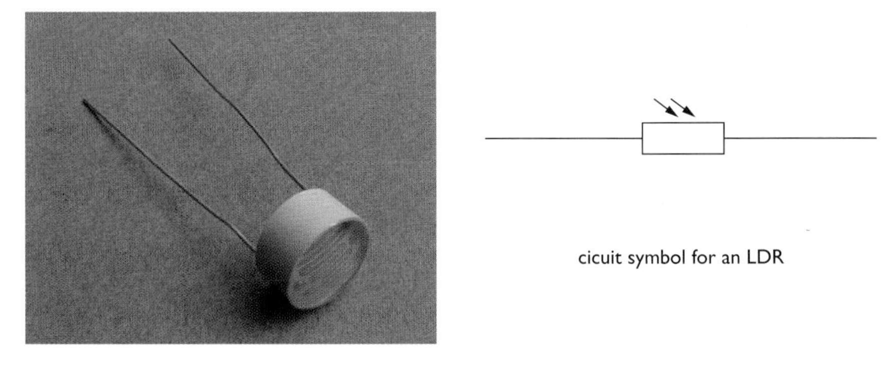

Figure 2.12 *Light-dependent resistors have a lower resistance when it is light.*

Diodes

Diodes are very special resistors that behave like one-way valves or one-way streets. When current flows through them in one direction it can do so quite easily as the diode has a low resistance. But if current tries to flow in the opposite direction the diode has a very high resistance and very little current can now flow. Diodes are often used in circuits where it is important that current flows only in one direction – for example, in rectifier circuits that convert alternating current into direct current.

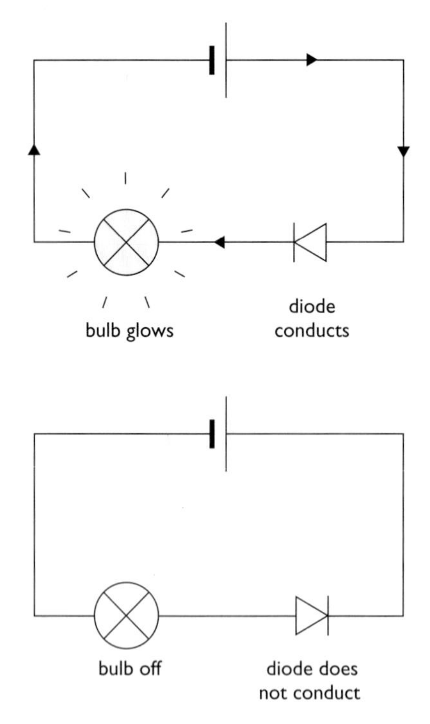

Figure 2.13 *Diodes will only let current flow one way.*

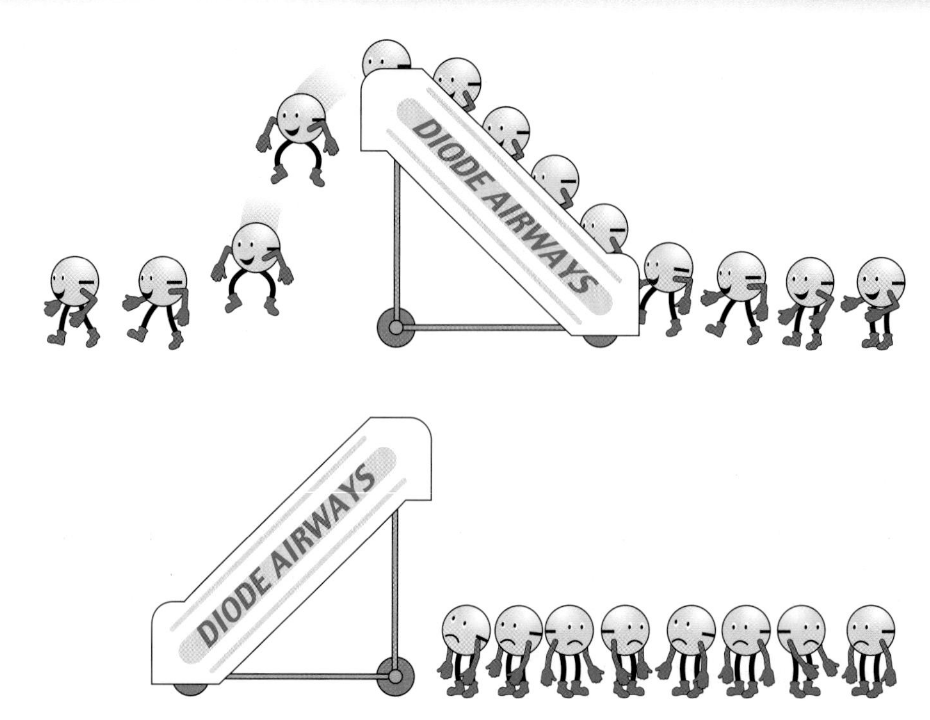

Figure 2.14 *Diodes are like aeroplane steps – from the ground, you can only climb them in one direction. In a diode the current can only flow in one direction.*

You can imagine a diode as behaving like a set of aeroplane steps which the charges have to climb over. If the charges are moving towards the right side of the steps they can "flow through it". If however they try to flow the opposite way there are no steps for them to climb so the flow in this direction is almost zero.

Light-emitting diodes (LEDs) are also "one-way" components but as current flows through them they give off light. They need very little energy to make them glow and so are often use as light indicators showing that a circuit or piece of electrical equipment – such as a TV – is turned on.

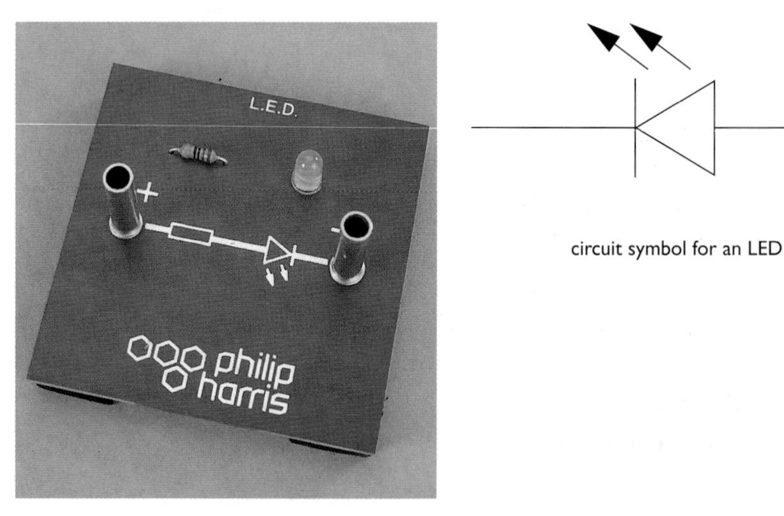

Figure 2.15 *Light-emitting diodes give out light.*

Measuring resistance – Ohm's Law

As we have already seen the resistance of a component is given by the ratio V/I where I is the current that flows through the component when a pd of V is applied across it. The circuit in Figure 2.16 can be used to investigate this relationship for a piece of resistance wire.

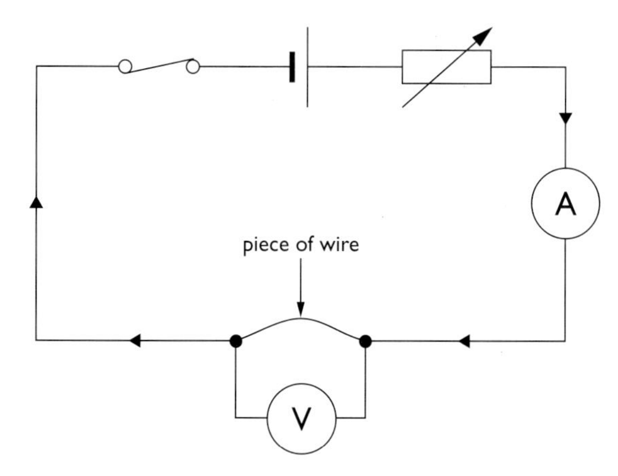

Figure 2.16 *This circuit can be used to investigate Ohm's Law.*

When the switch S is closed the readings on the ammeter and voltmeter are noted. The value of the variable resistor is then altered and a new pair of readings taken from the meters. The whole process is repeated at least six times, the results are placed into a table and a graph of current (I) against potential difference (V) is drawn.

Current, I (A)	Potential difference, V (V)
0.0	0.0
0.1	0.4
0.2	0.8
0.3	1.2
0.4	1.6
0.5	2.0

Table 2.1 *Typical results for an investigation of Ohm's Law.*

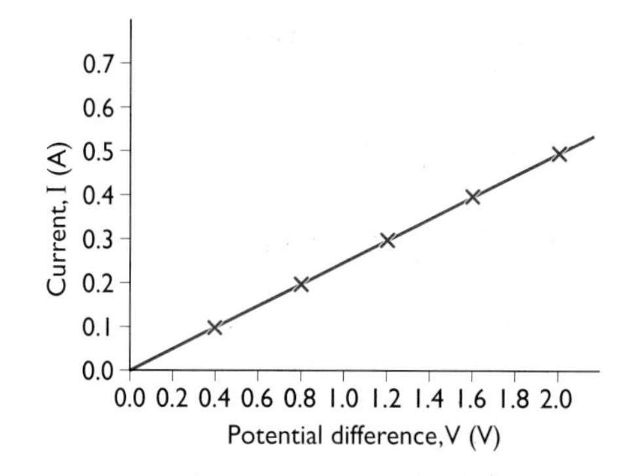

Figure 2.17 *The graph shows that current and voltage increase proportionally.*

The graph in Figure 2.17 is a straight line graph passing through the origin. This tells us that the current flowing through the wire is *directly proportional* to the pd applied across its ends – that is, if the pd across the wire is doubled the current flowing through it doubles.

The relationship between the voltage across a component and the current that flows through it is described by **Ohm's Law**, which states:

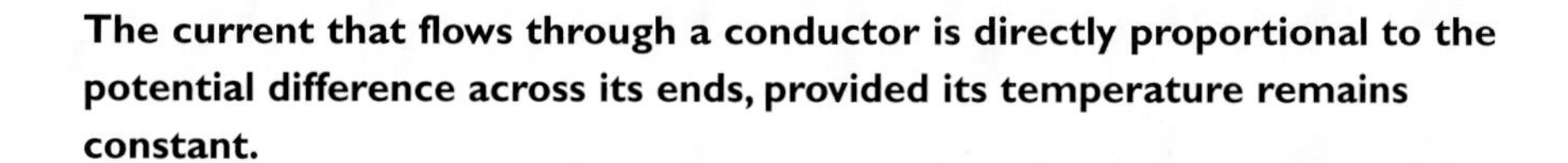
The current that flows through a conductor is directly proportional to the potential difference across its ends, provided its temperature remains constant.

The resistance of the wire can be found by selecting any value of potential difference (V), reading from the graph the current (I) that flows when this pd is applied to the wire and calculating a value for the ratio V/I.

In this case the resistance of the wire is $\frac{1.2\text{ V}}{0.3\text{ A}} = 4\ \Omega$.

If we extend the range of readings in the above experiment, such that the currents flowing cause the wire to become warm, the shape of the I/V graph changes.

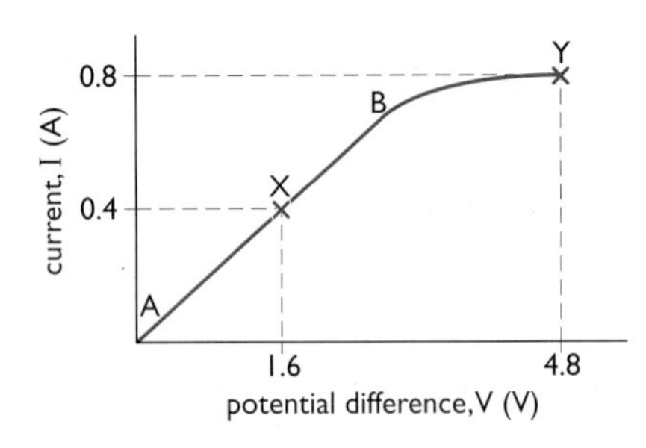

Figure 2.18 *Resistance in a wire increases as the current increases.*

Between A and B, in Figure 2.18, the relationship between current and voltage is still a direct proportionality and the wire has a constant resistance. Beyond B the current flowing through the wire is large enough to change its temperature and the resistance of the wire increases.

At point X the graph shows a current of 0.4 A is flowing through the wire so the resistance of the wire is $\frac{V}{I} = \frac{1.6\,V}{0.4\,A} = 4\,\Omega$.

At point Y the graph shows a current of 0.8 A is flowing through the wire so its resistance now is $\frac{V}{I} = \frac{4.8\,V}{0.8\,A} = 6\,\Omega$.

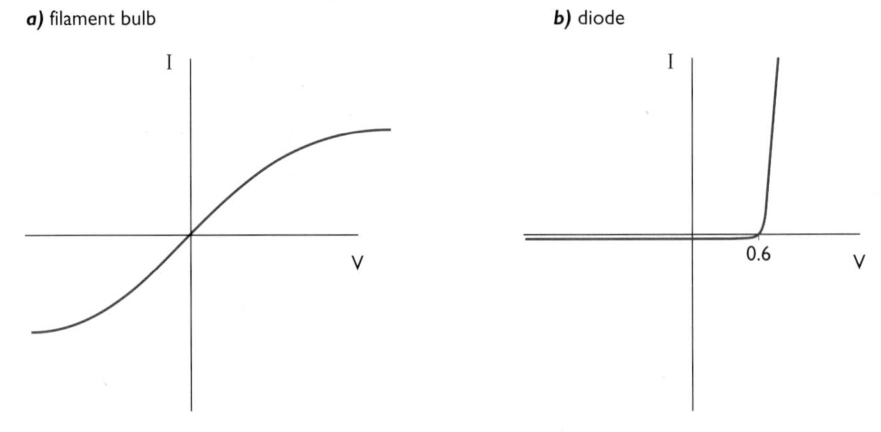

Figure 2.19 *The current/voltage graphs for a diode and a filament bulb.*

The current/voltage graph in Figure 2.19a shows that when a filament bulb is first turned on its resistance is quite low but as the filament becomes hot its resistance increases.

The graph for a diode, in Figure 2.19b, shows that a small voltage of approximately 0.6 V must be applied to the diode before it will conduct. Further increases in voltage then result in increases in current. If the voltage is applied in the opposite direction almost no current flows.

End of Chapter Checklist

If you haven't got a copy of your specification, read the introduction on page iv.

You will need to be able to do some or all of the following. Check your Awarding Body's specification (syllabus) to find out exactly what you need to know.

- Describe qualitatively the effect of changing resistance on the current flowing in a circuit.
- Understand that the size of the current that passes through a component depends upon the voltage across the component and its resistance, and use the equation $I = \frac{V}{R}$.
- Understand that when current flows through a resistor an energy transfer takes place.
- Calculate the total resistance of several resistors connected in series or in parallel.
- Recall the special properties of thermistors, light-dependent resistors and diodes.
- Describe an experiment to obtain voltage and current measurements for a number of components, construct I–V graphs and then interpret the shapes of these graphs.

Questions

More questions on electrical resistance can be found at the end of Section A on page 80.

1 The diagram below shows a series circuit containing a battery, a bulb and a piece of resistance wire.

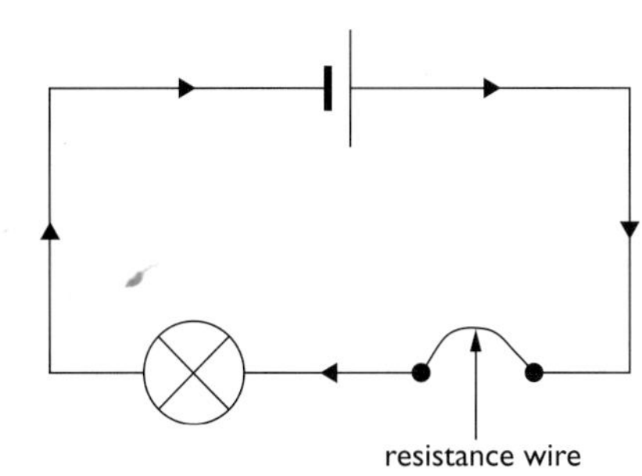

State and explain what would happen to the brightness of the bulb if each of the following changes were made.

a) The resistance wire is replaced with an identical but shorter piece of wire.

b) A second identical piece of resistance wire is connected in series with the first.

c) A second identical piece of wire is placed in parallel with the first.

d) The resistance wire is heated.

2 ***a)*** State Ohm's Law.

b) Draw a diagram of the circuit you would use to confirm Ohm's Law for a piece of wire.

c) Describe how you would use the apparatus and what readings you would take.

d) Draw a $\frac{I}{V}$ graph for:

i) a piece of wire at room temperature

ii) a filament bulb

iii) a diode.

Explain the main features of each of these graphs.

3 ***a)*** A current of 5 A flows when a pd of 20 V is applied across a resistor. Calculate the resistance of the resistor.

b) Calculate the current that flows when a pd of 12 V is applied across a piece of wire of resistance 50 Ω.

c) Calculate the pd that must be applied across a wire of resistance 10 Ω if a current of 3 A is to flow.

Remember when doing calculations like these to show all your working out and include units with your answer.

4 ***a)*** Calculate the total resistance of each of the resistor networks shown in the diagram below.

i)

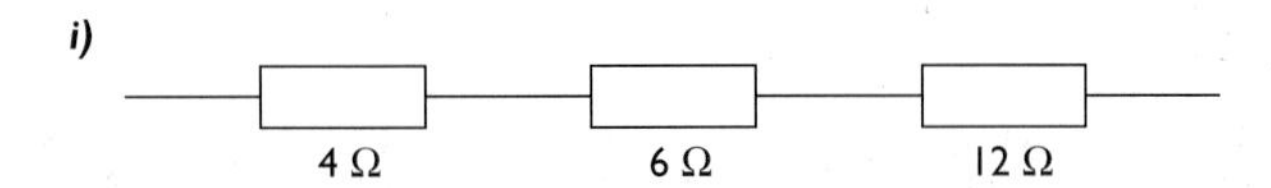

ii)

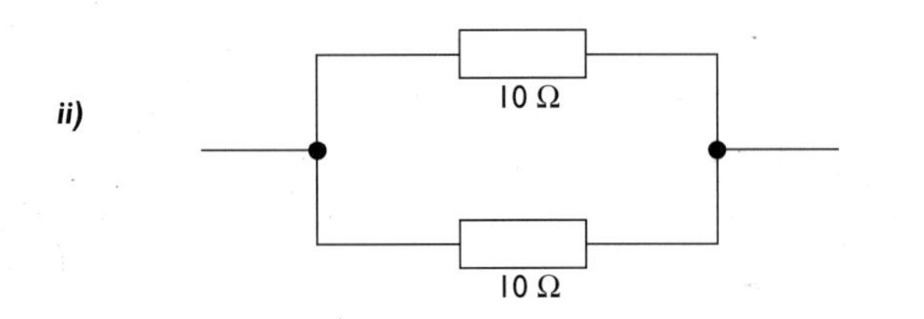

iii)

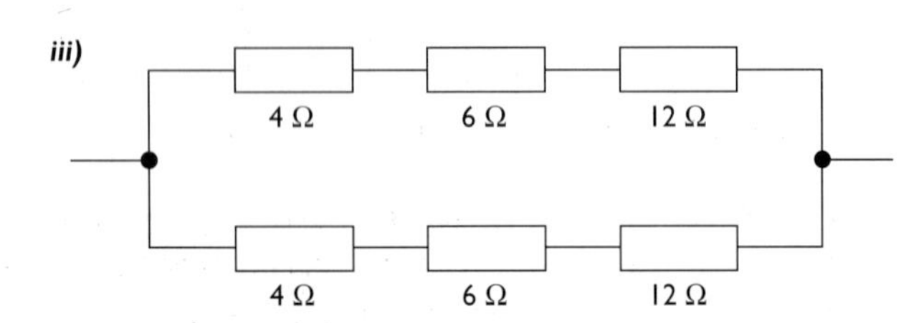

b) You are given six 10 Ω resistors. How should these resistors be connected in order to produce:

i) the maximum possible resistance

ii) the minimum possible resistance

iii) a resistance of 15 Ω.

In each case, all six resistors must be used.

5 ***a)*** Explain how the resistance of:

i) a thermistor

ii) a light-dependent resistor

can be increased.

b) Describe in detail one practical application for each of these resistors.

6 Look at the diagram.

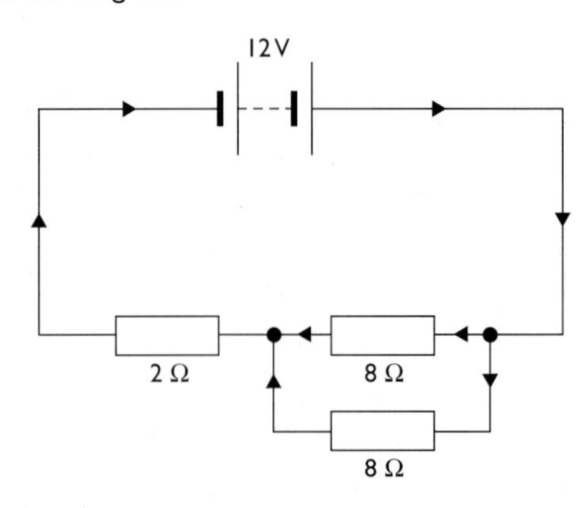

a) Calculate the total resistance of the two 8 Ω resistors connected in parallel.

b) Calculate the total resistance of the whole circuit.

c) Calculate the current that flows through the 2 Ω resistor.

d) Calculate the current that flows through each of the 8 Ω resistors.

e) Calculate the pd across the 2 Ω resistor.

f) Calculate the pd across the 8 Ω resistors in parallel.

g) Describe the energy change that takes place as 1 C of charge passes through the 12 V battery. (**Hint**: See Chapter 1 for charge formula.)

h) Describe the energy change that takes place as 1 C of charge passes through the 2 Ω resistor.

Chapter 3: Domestic Electricity

We use mains electricity, supplied by power stations, for all kinds of appliances in our homes, so it is very important to know how to use it safely. In this chapter you will learn how mains electricity is brought into our homes and supplied to appliances. You will read about devices that protect users from electric shocks, and find out how electricity suppliers work out the cost of the electricity we use.

Mains electricity

Figure 3.1 *We use a huge amount of electricity every day for heating and lighting.*

When you turn on your computer, television and most other appliances in your home the electricity you use is almost certainly going to come from the **mains supply**. This electrical energy usually enters our homes through an underground cable. The cable is connected to an **electricity meter**, which measures the amount of electrical energy used. From here, the cable is connected to a **consumer unit** or a **fuse box**, which contains fuses for the various circuits in your home. Fuses are safety devices which shut off the electricity in a circuit if the current flowing through them becomes too large (see page 24).

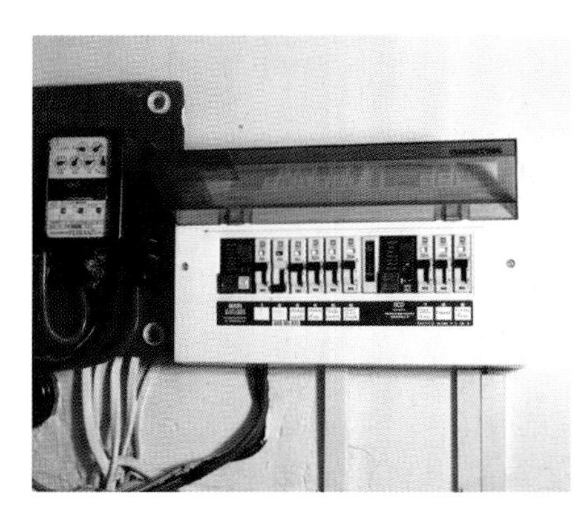

Figure 3.2 *Fuse box in a house.*

Most of the wires that leave the fuse box are connected to **ring main circuits** that are hidden in the walls or floors around each room. Individual pieces of electrical equipment are connected to these circuits using plugs.

Ring main circuits provide a way of allowing several appliances in different parts of the same room to be connected to the mains using the minimum amount of wiring. Imagine how much wire would be needed if there was just one mains socket in each room.

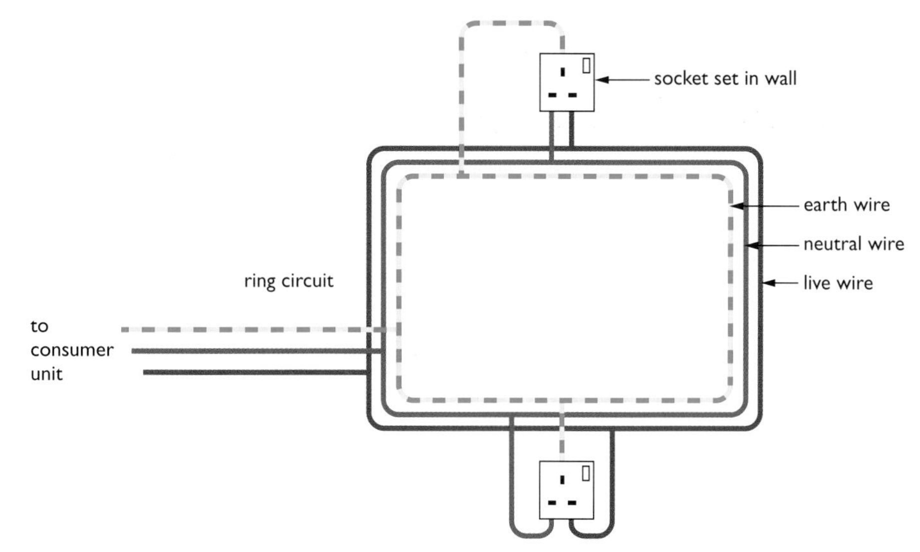

Figure 3.3 *Ring mains help to cut down on the amount of wiring needed in a house.*

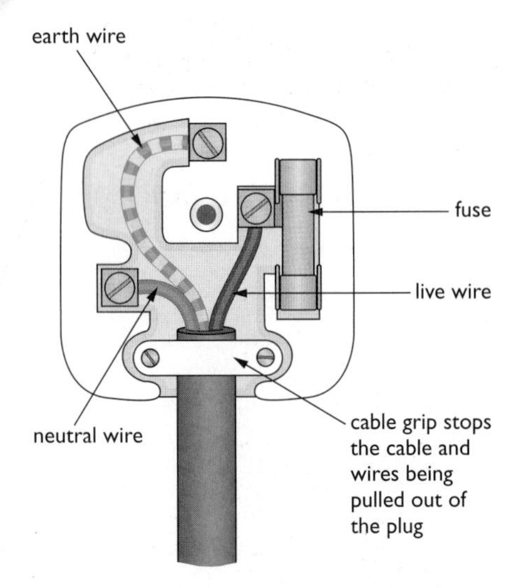

Figure 3.4 *A plug should be wired carefully so that the live wire is connected to the pin on the right, the neutral wire is connected to the pin on the left and the earth wire is connected to the pin at the top.*

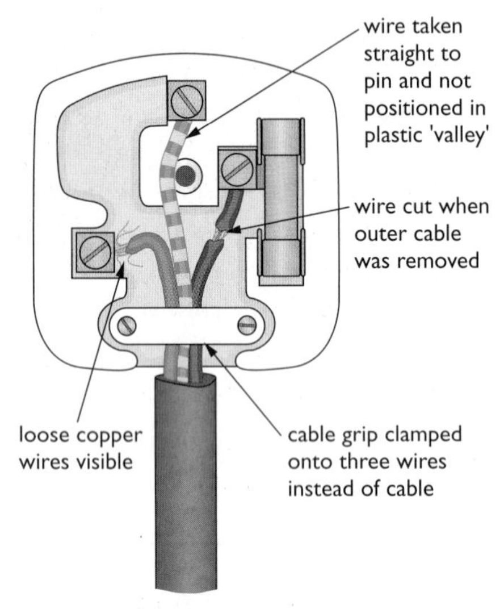

Figure 3.5 *How not to wire a plug!*

Ring circuits in the UK consist of three wires – the live wire, the neutral wire and the earth wire.

The **live wire** provides the path along which the electrical energy from the power station travels. This wire is alternately positive and negative causing **alternating current** (**ac**) to flow along it.

The **neutral wire** completes the circuit.

The **earth wire** usually has no current flowing through it. It is there to protect you if an appliance develops a fault.

The modern 13 A three-pin plug

Mains electricity is supplied to the home at about 230 V. This is a much higher voltage than the cells and batteries used in portable electrical appliances. If you come into direct contact with mains electricity you could receive a severe electric shock, which might even be fatal. The outer part of a plug, called the casing, is therefore made from plastic, which is a good insulator. Connections to the circuits are made via three brass pins, as brass is an excellent conductor of electricity.

It is important that the wires inside a plug are connected to the correct pins.

- The live wire is coloured BRown and must be connected to the pin on the Bottom Right.
- The neutral wire is BLue and must be connected to the pin on the Bottom Left.
- The earth wire is green and yellow and must be connected to the pin at the top of the plug.

Figure 3.5 shows some of the more common mistakes people make when wiring a plug. With a little care and attention it is easy to avoid making these mistakes so the plug is safe to use.

Fuses

Most plugs in the UK contain a fuse. The fuse is usually in the form of a cylinder or cartridge, which contains a thin piece of wire made from a metal that has a low melting point. If too large a current flows in the circuit the fuse wire becomes very hot and melts. The fuse "blows" shutting the circuit off. This prevents you getting a shock and reduces the possibility of an electrical fire. Once the fault causing the surge of current has been corrected, the blown fuse must be replaced with a new one before the appliance can be used again.

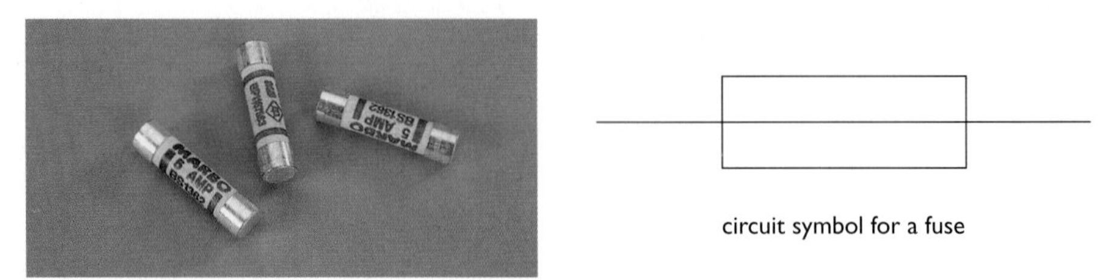

Figure 3.6 *Fuses are important safety devices in electrical appliances.*

There are several sizes of fuses. The most common in the UK are 3 A, 5 A and 13 A. The correct fuse for a circuit is the one which allows the correct current to flow but which blows if the current is a little larger. If the correct current in a circuit is 2 A then it should be protected with a 3 A fuse. If the correct current is 4 A then a 5 A fuse should be used. It is possible to calculate the correct size of fuse for an appliance but nowadays manufacturers provide this information with all appliances.

Fuses protect wiring and appliances from overheating due to excessive currents. ELCBs protect people from elecrocution.

Modern safety devices such as those you might find in your consumer unit are often in the form of **trip switches** or **circuit breakers** (see page 35). If too large a current flows in a circuit a switch opens making the circuit incomplete. Once the fault in the circuit has been corrected, the switch is reset, usually by pressing a reset button. There is no need for the switch or circuit breaker to be replaced, as there is when fuses are used.

Earth-leakage circuit breakers (ELCBs)

When you use appliances such as electric lawn mowers or hedge trimmers, you might accidentally cut through the electric cable and receive an electric shock. Using an **earth-leakage circuit breaker** (**ELCB**) like the one shown in Figure 3.7 could prevent this. ELCBs can detect small differences in the currents flowing through the live and neutral wires caused by a cable being cut. The ELCB then switches off the circuit before the user is injured.

Figure 3.7 *Earth-leakage circuit breakers cut off the current if the cable to the device is damaged.*

Double insulation

Many appliances have a metal casing. This should be connected to the earth wire so that if the live wire becomes frayed or breaks and comes into contact with the casing the earth wire provides a low-resistance path for the current. This current is likely to be large enough to blow the fuse and turn the circuit off. Without the earth wire anyone touching the casing of the faulty appliance would receive a severe electric shock as current passed through them to earth.

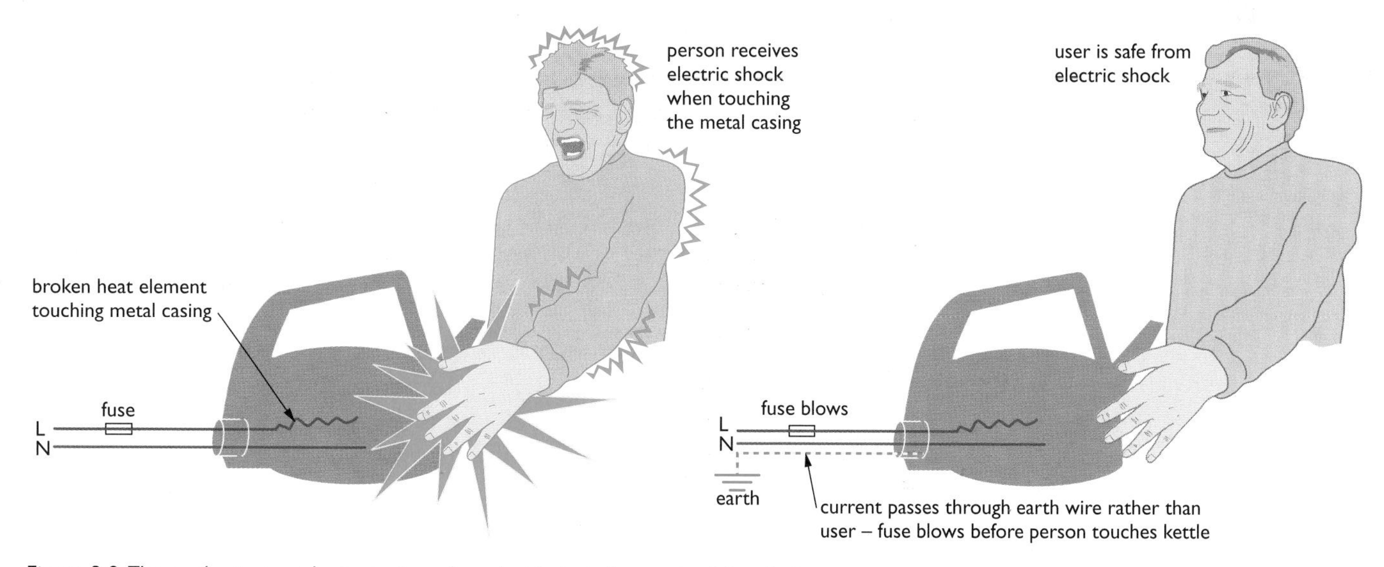

Figure 3.8 *The earth wire provides protection when electrical appliances develop a fault.*

Some modern appliances now use casings made from an insulator such as plastic rather than from metal. If all the electrical parts of an appliance are insulated in this way, so that they cannot be touched by the user, the appliance is said to have **double insulation**. Appliances that have double insulation use a two-wire flex. There is no need for an earth wire.

Figure 3.9 *This plastic kettle has double insulation which means that there is no need for an earth wire.*

Switches

Switches in mains circuits should always be included in the live wire so that when the switch is open no electrical energy can reach an appliance. If the switch is included in the neutral wire electrical energy can still enter an appliance, and could possibly cause an electric shock.

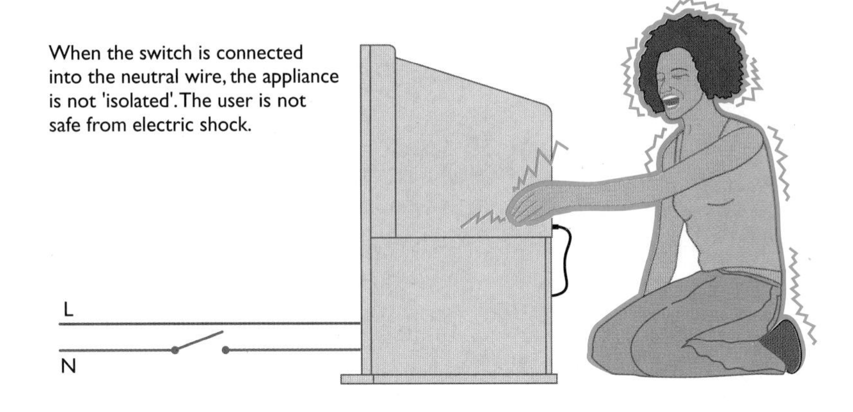

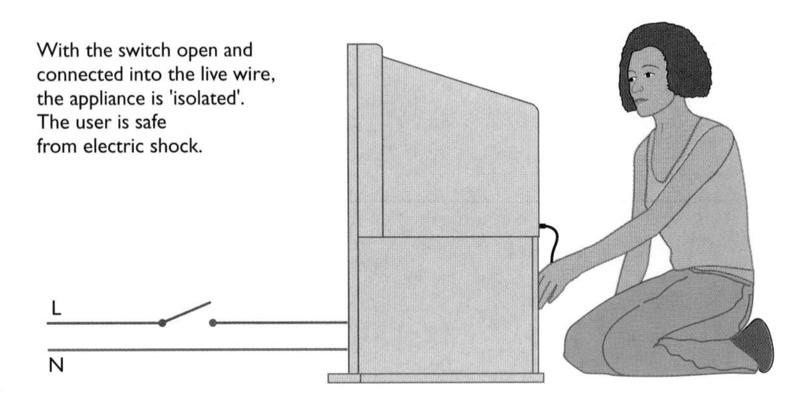

Figure 3.10 *The switch in a circuit should be in the live wire.*

Figure 3.11 *The 100 W bulb converts more electrical energy into heat and light energy every second.*

Electrical power

Figure 3.11 shows two light bulbs connected to a supply of electricity. Both bulbs are converting electrical energy into heat and light. The brighter 100 W bulb is converting 100 J of electrical energy into 100 J of heat and light energy every second. The dimmer 60 W bulb is converting 60 J of electrical energy into 60 J of heat and light energy every second. The 100 W bulb has a higher **power** rating.

Power is measured in joules per second or **watts (W)**.

The power (P) of an appliance is related to the potential difference (V) across it and the current (I) flowing through it, by the equation:

power, P (in watts) = potential difference, V (in volts) × current, I (in amps)

$$P = V \times I$$

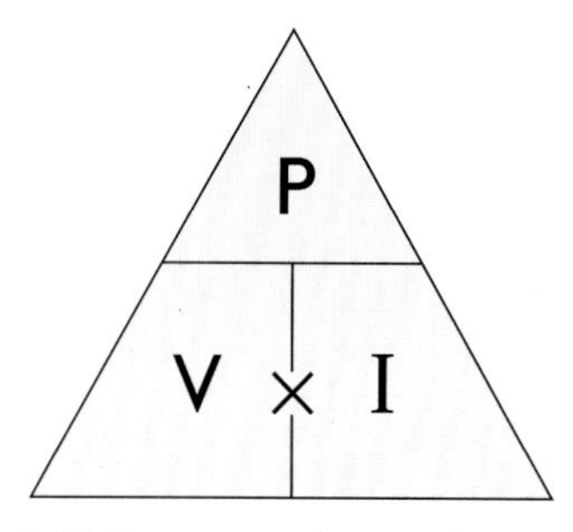

Figure 3.12 *You can use the triangle method for rearranging equations like $P = V \times I$.*

worked example

Example 1

If a 230 V television takes a current of 3 A, calculate its power.

$P = V \times I$

$P = 230\,V \times 3\,A$

$P = 690\,W$

worked example

Example 2

Calculate the correct fuse that should be used for a 230 V, 1 kW electric hairdryer (1 kW = 1000 W).

$P = V \times I$

$I = \frac{P}{V}$

$I = \frac{1000\,W}{230\,V}$

$I = 4.3\,A$

The correct fuse for this hairdrier is therefore a 5 A fuse.

As we have already seen, $V = I \times R$. Using this equation the power equation above can be written in several different forms:

$P = V \times I$

But $V = I \times R$

So $P = (I \times R) \times I$

$P = I^2 \times R$

Similarly $P = V \times I$

But $I = \frac{V}{R}$

So $P = V \times \frac{V}{R}$

$P = \frac{V^2}{R}$

worked example

Example 3

Calculate the power of a 230 V electric fire which has a heating element of resistance 50 Ω.

$P = \frac{V^2}{R}$

$P = \frac{230^2}{50}$

$P = 1058\,W$

Calculating the total energy converted by an appliance

The power of an appliance (P) tells you how much energy it converts each second. This means that the total energy (E) converted by an appliance is equal to its power multiplied by the length of time (in seconds) the appliance is being used.

energy, E (in joules) = power, P (in watts) × time, t (in seconds)

$$E = Pt$$

worked example

Example 4

Calculate the energy converted by a 60 W bulb that is turned on for ***a)*** 20 s, and ***b)*** 5 min.

a) $E = Pt$

$E = 60\,W \times 20\,s$

$E = 1200\,J$ or $1.2\,kJ$

b) $E = Pt$

$E = 60\,W \times 5 \times 60\,s$

$E = 18\,000\,J$ or $18\,kJ$

Paying for domestic electricity

Kilowatt hours and units are non-standard units of energy but they are used by all suppliers of electricity.

In most households we have lots of appliances that use and convert electrical energy. The joule is too small a unit to measure the total amount of energy we use. It would be like asking you how far you live from school in centimetres! Instead therefore we use a much larger unit of energy called a **kilowatt hour** (**kWh**) or **unit**.

If a 1 kW heater is turned on for 1 hour it will have converted 1 kWh or 1 unit of electrical energy into heat energy. If the heater is turned on for 5 hours it will have converted 5 units of electrical energy.

We can describe a more general situation using the equation:

energy, E (in units) = power, P (in kilowatts) × time, t (in hours)

$$E = Pt$$

worked example

Example 5

Calculate the number of units of electrical energy used when a 3 kW water immersion heater is turned on for 30 minutes.

$E = Pt$

$E = 3\,kW \times 0.5$ hours

$E = 1.5$ units

If we know the cost of 1 unit of electrical energy (approximately 8 p) we can calculate the cost of heating the water.

cost = 1.5 units × 8 p = 12 p

The electricity bill

Most households receive an electricity bill similar to the one shown in Figure 3.13 every three months.

The bill shows:

- the reading on the electricity meter at the start and end of the three months covered by the bill – by subtracting the first value from the second, the electricity supplier works out how many units of electricity have been used
- the cost of each unit of electrical energy
- the total cost of the electricity that has been used (that is, the number of units × the cost of each unit)
- a standing or quarterly charge paid to the electricity supplier for the use and maintenance of the equipment needed to supply a home with electricity
- the total amount due – that is, the cost of the electricity plus the standing charge (plus VAT, which is a tax charged by the government).

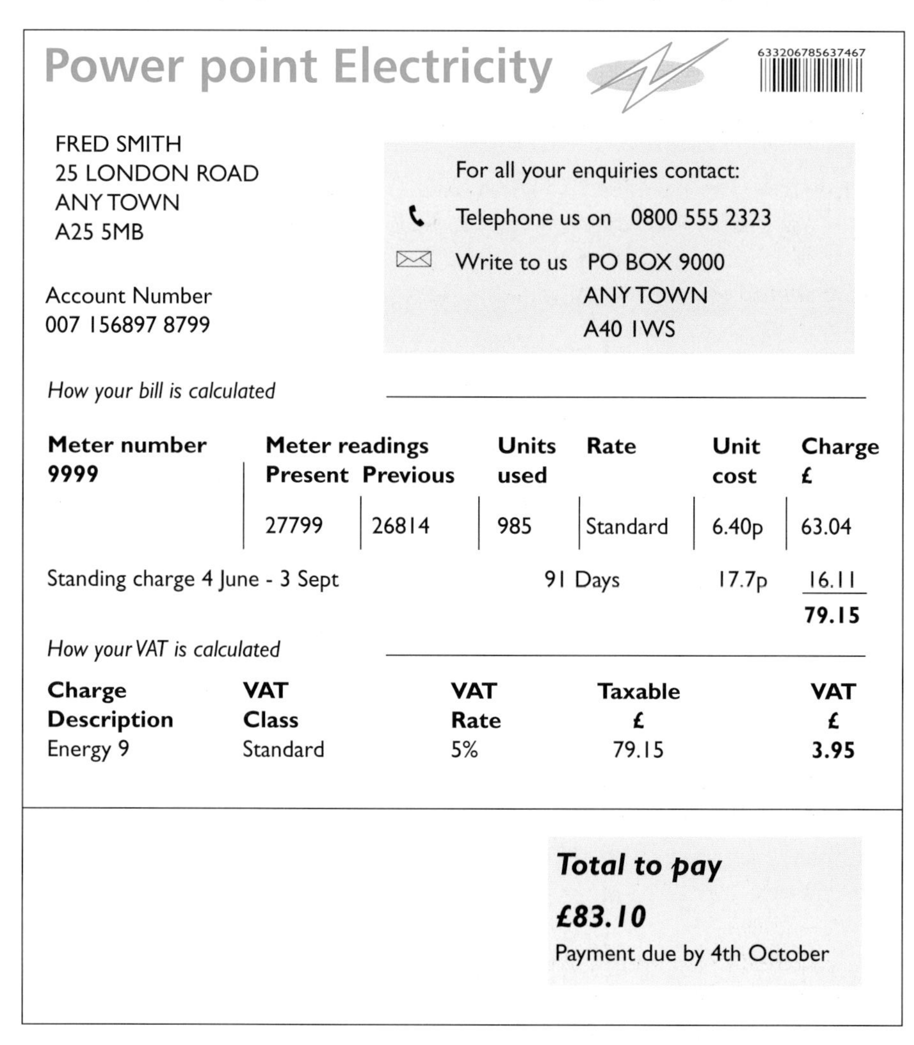

Power point Electricity

633206785637467

FRED SMITH
25 LONDON ROAD
ANY TOWN
A25 5MB

For all your enquiries contact:
Telephone us on 0800 555 2323
Write to us PO BOX 9000
ANY TOWN
A40 1WS

Account Number
007 156897 8799

How your bill is calculated

Meter number	Meter readings Present	Previous	Units used	Rate	Unit cost	Charge £
9999	27799	26814	985	Standard	6.40p	63.04
Standing charge 4 June - 3 Sept				91 Days	17.7p	16.11
						79.15

How your VAT is calculated

Charge Description	VAT Class	VAT Rate	Taxable £	VAT £
Energy 9	Standard	5%	79.15	3.95

Total to pay

£83.10

Payment due by 4th October

Figure 3.13 *Your electricity bill shows your previous meter reading, the present reading and the cost of each unit of electricity.*

End of Chapter Checklist

If you haven't got a copy of your specification, read the introduction on page iv.

You will need to be able to do some or all of the following. Check your Awarding Body's specification (syllabus) to find out exactly what you need to know.

- Identify the live, neutral and earth wires in a three-pin plug.
- Know how to wire a plug correctly and be able to identify common mistakes.
- Know the correct usage of each of these three wires: live, neutral and earth.
- Realise the dangers of using mains electricity incorrectly.
- Understand the steps taken to protect users from electric shocks – for example, use of insulators, earthing, fuses, circuit breakers and double insulation.
- Understand the phrase "power rating" when applied to an electrical appliance and be able to use the equation $\mathbf{P} = \mathbf{V} \times \mathbf{I}$ to determine the correct fuse to use with an appliance.
- Calculate the cost of electrical energy used from knowledge of power, time and cost per unit.

Questions

More questions on domestic electricity can be found at the end of Section A on page 80.

1 ***a)*** A current of 0.25 A flows through a bulb when a voltage of 12 V is applied across it. Calculate the power and resistance of the bulb.

b) Calculate the voltage that is being applied across a 10 W bulb if a current of 0.2 A flows through it.

c) Calculate the current that flows through a 60 W bulb if the pd across it is 240 V.

2 An electric kettle is marked "230 V, 1.5 kW".

a) Explain exactly what these figures mean.

b) Calculate the resistance of the heating element of the kettle.

c) Explain why a 230 V, 100 W bulb glows more brightly than a 230 V 60 W bulb when both are connected to the mains supply.

3 ***a)*** Give one advantage of using a circuit breaker rather than a wire or cartridge fuse.

b) What is an ELCB? How does it work?

c) What is meant by the sentence "The kettle has double insulation"?

4 What is the cost of running the following?

a) a 3 kW fire for 2 h

b) a 100 W bulb for 20 h

c) a 2 kW tumble drier for 30 min

d) a 650 W colour television for 5 h

Assume that the cost of a unit of electricity is 8 p.

5 The table below shows some figures taken from a typical electricity bill.

Present reading on meter	54 623
Previous reading on meter	52 881
Cost per unit	7 p
Standing charge	£12

a) How many units of electricity have been used?

b) What is the cost of these units?

c) What is the total cost of the electricity bill if VAT is charged at 17.5%?

Chapter 4: Electromagnetism

A magnet has a magnetic field around it. A current-carrying wire also has a magnetic field around it, while current is flowing. This means we can make electromagnets that can be switched on and off. In this chapter you will learn about the factors affecting the magnetic field around an electromagnet, and how electromagnets are used in several important devices.

Figure 4.1 *Electromagnets can be used to lift iron or steel objects.*

The huge electromagnet in Figure 4.1 is being used in a scrap yard to pick up large objects that contain iron or steel. When the objects have been moved to their new position the electromagnet is turned off and the objects fall. This chapter explains how magnets and electromagnets are used in everyday devices.

Magnetism and magnetic materials

Magnets are able to attract objects made from **magnetic materials** such as iron, steel, nickel and cobalt. Magnets cannot attract objects made from plastic, wood, paper, rubber and so on. These are **non-magnetic materials**.

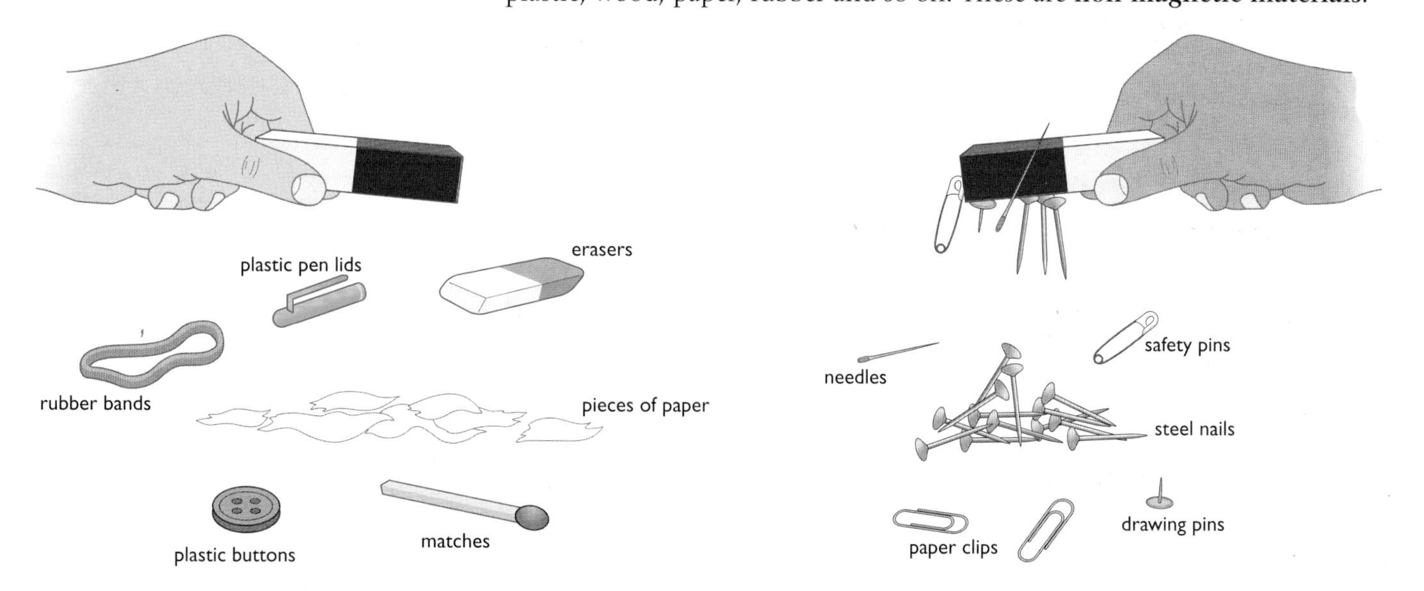

Figure 4.2 *Magnets attract some objects and not others.*

Ancient travellers would construct a crude compass by placing a splinter of magnetised rock called **lodestone** onto a piece of wood floating in water. The splinter would turn to align itself in a north–south direction.

The strongest parts of a magnet are called its **poles**. Most magnets have two poles. These are called the **north pole** and the **south pole**.

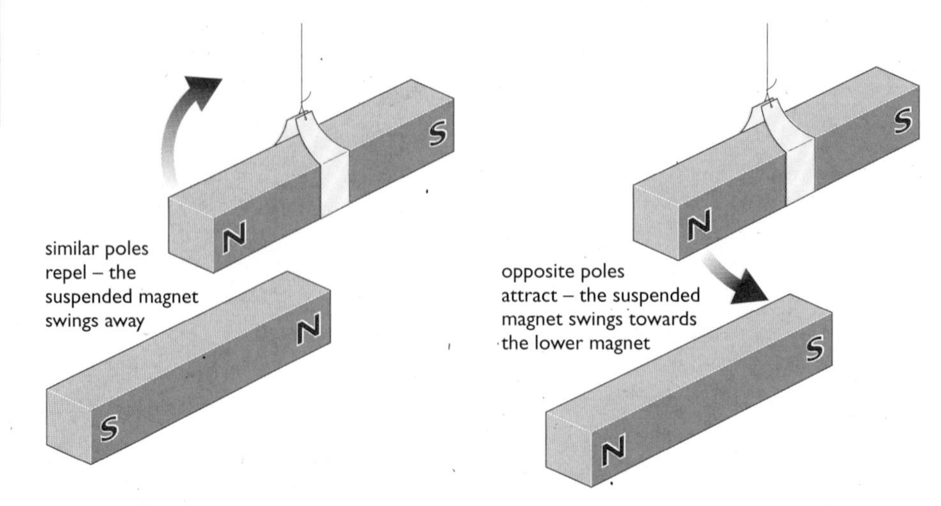

Figure 4.3 *Similar poles repel and opposite poles attract.*

If two similar poles are placed near to each other they **repel**. If two dissimilar (opposite) poles are placed near to each other they **attract**.

Permanent magnets like those shown in Figure 4.3 are made from a **magnetically hard** material such as steel. A magnetically hard material retains its magnetism once it has been **magnetised**. Iron is a **magnetically soft** material and would not be suitable for a permanent magnet. Magnetically soft materials lose their magnetism easily and are therefore useful as temporary magnets.

Magnetic fields

Around every magnet there is a volume of space where magnetism can be detected. This volume of space is called a **magnetic field**. Normally a magnetic field cannot be seen but we can use iron filings or plotting compasses to show its shape and discover something about its strength and direction.

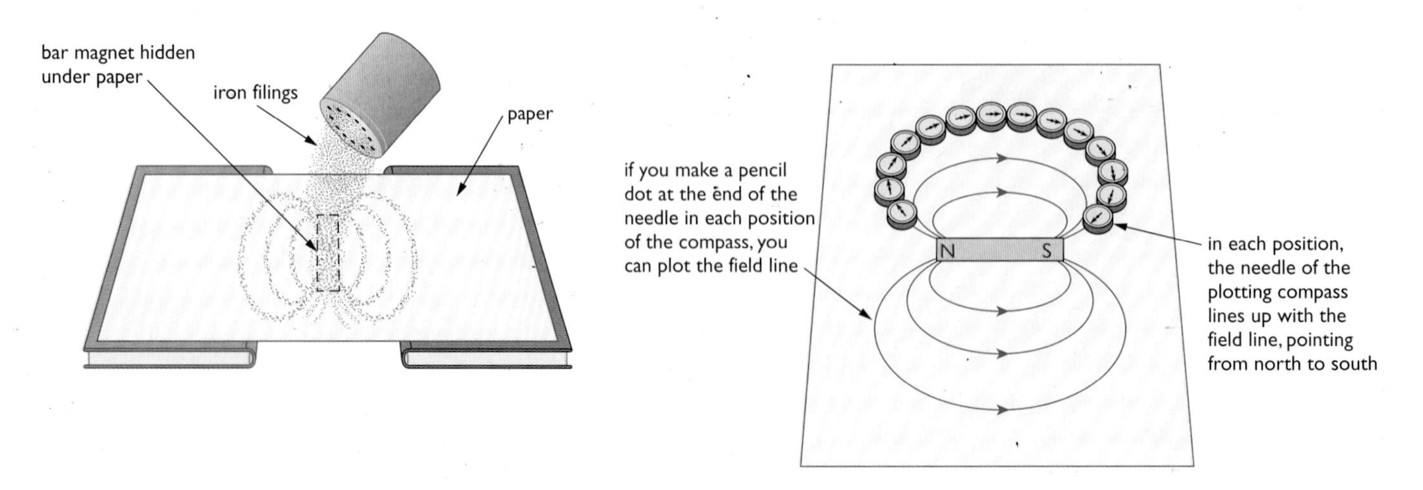

Figure 4.4 *You can see the shape of the magnetic field around a magnet by using iron filings, or a plotting compass.*

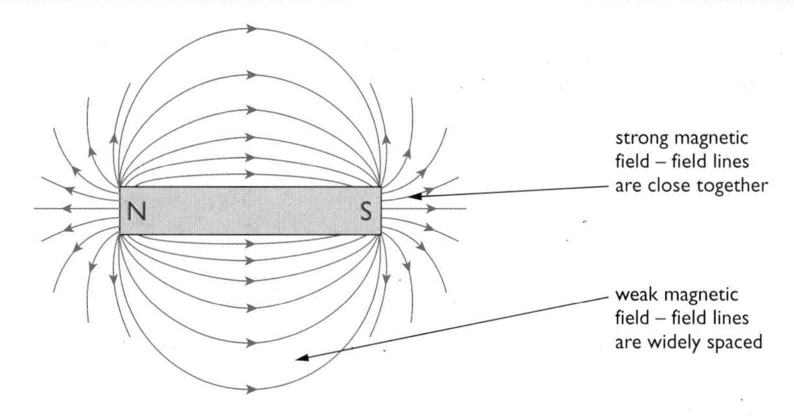

Figure 4.5 *The magnetic field around a bar magnet follows a pattern like this.*

We draw magnetic fields using **lines of force** or **flux lines**. Like contour lines on a map, flux lines don't really exist but they are an aid in helping us visualise the main features of a magnetic field.

The magnetic lines of force:

- show the *shape* of the magnetic field
- show the *direction* of the magnetic field – the field lines "travel" from north to south
- show the *strength* of the magnetic field – the field lines are closest where the magnetic field is strongest.

Electromagnetism

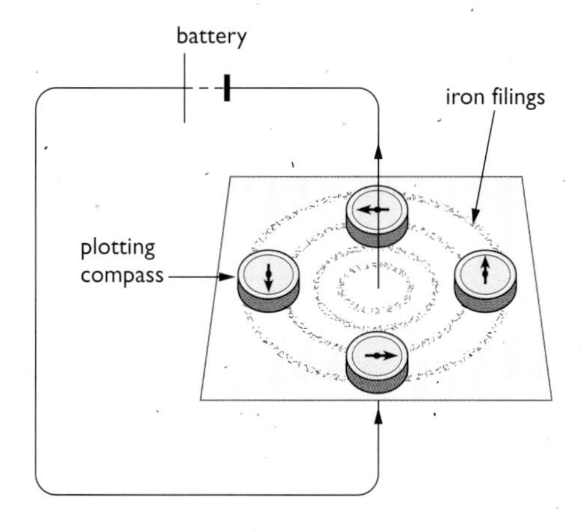

The shape and direction of the magnetic field around a current-carrying wire can be seen using iron filings and plotting compasses. Changing the direction of the current changes the direction of the magnetic field.

Figure 4.6 *A current-carrying wire has a magnetic field around it.*

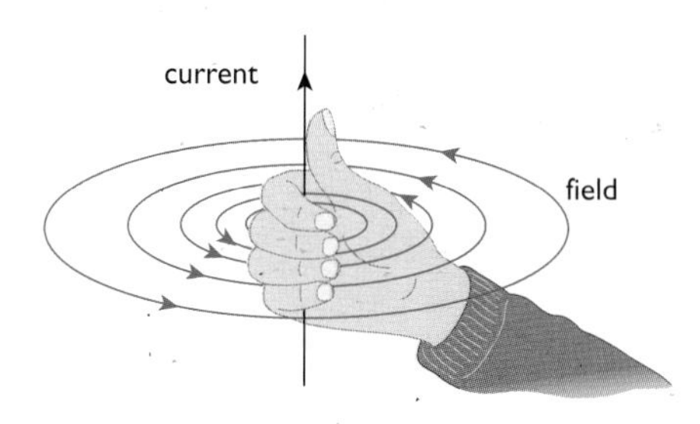

With the thumb of your right hand pointing in the direction of the current, your fingers will curl in the direction of the field.

Figure 4.7 *You can work out the direction of the field using the right hand grip rule.*

When a current flows through a wire a magnetic field is created around the wire. This phenomenon is called **electromagnetism**. The field around the wire is quite weak and circular in shape. The direction of the magnetic field depends upon the direction of the current and can be found using the **right hand grip rule (for fields)**.

The strength of the magnetic field around a current-carrying wire can be increased by:

1 increasing the current in the wire

2 wrapping the wire into a coil or **solenoid** (a solenoid is a long coil).

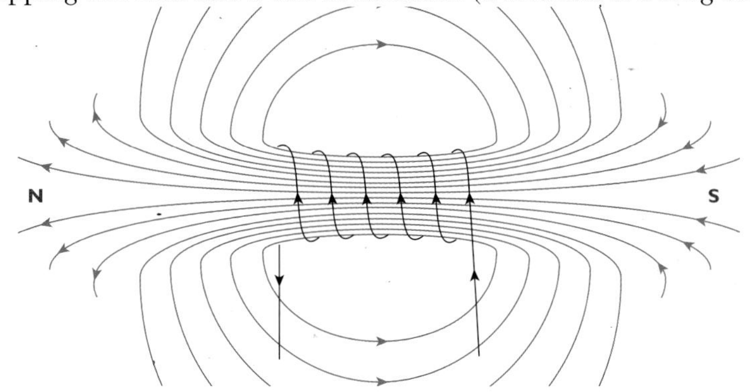

Figure 4.8 *The field around a solenoid looks like this.*

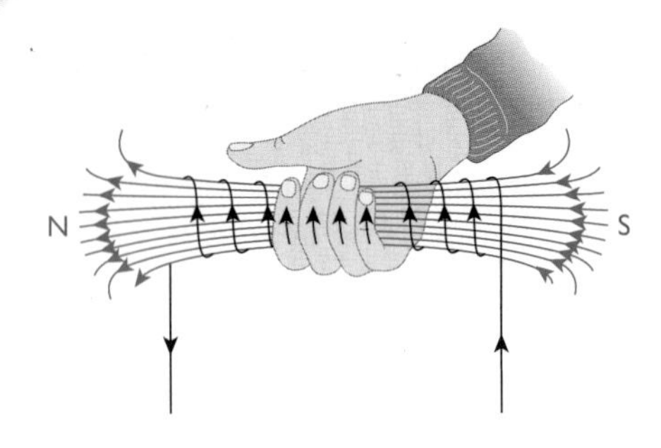

Figure 4.9 *You can work out the polarity of the solenoid by imagining that your right hand is wrapped around it. Your fingers curl in the direction of the current and your thumb points to the north pole of the solenoid.*

The shape of the magnetic field around a solenoid is the same as that around a bar magnet. The positions of the poles can be determined using the **right hand grip rule** (**for poles**).

If the direction of the current flowing through the solenoid is reversed, so too are the positions of the poles.

The strength of the field around a solenoid can be increased by:

1 increasing the current flowing through the solenoid

2 increasing the number of turns on the solenoid

3 wrapping the solenoid around a magnetically soft core such as iron – this combination of soft iron core and solenoid is often referred to as an **electromagnet**.

Using electromagnets

The electric bell

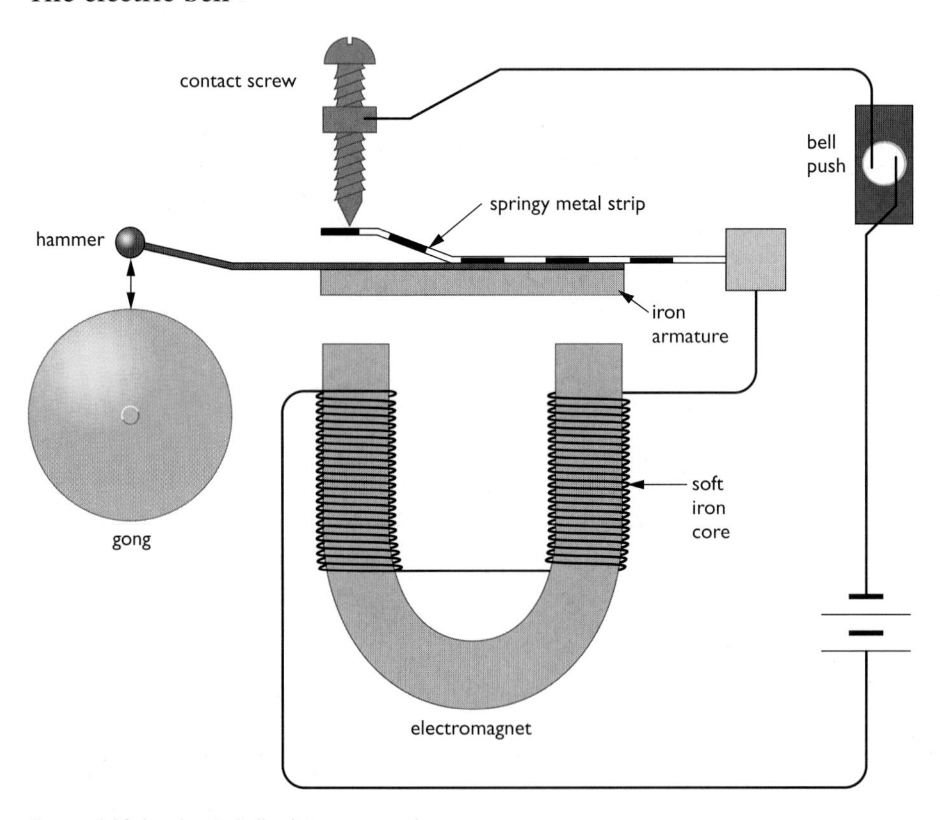

Figure 4.10 *An electric bell relies upon an electromagnet.*

In Figure 4.10, when the bell push is pressed the circuit is complete and current flows. The soft iron core of the electromagnet becomes magnetised and attracts the iron armature. When the armature moves, the hammer strikes the bell and at the same time a gap is created at the contact screw. The circuit is incomplete and current stops flowing. The electromagnet is now turned off so the spring's armature returns to its original position. The circuit is again complete and the whole process begins again.

Circuit breaker

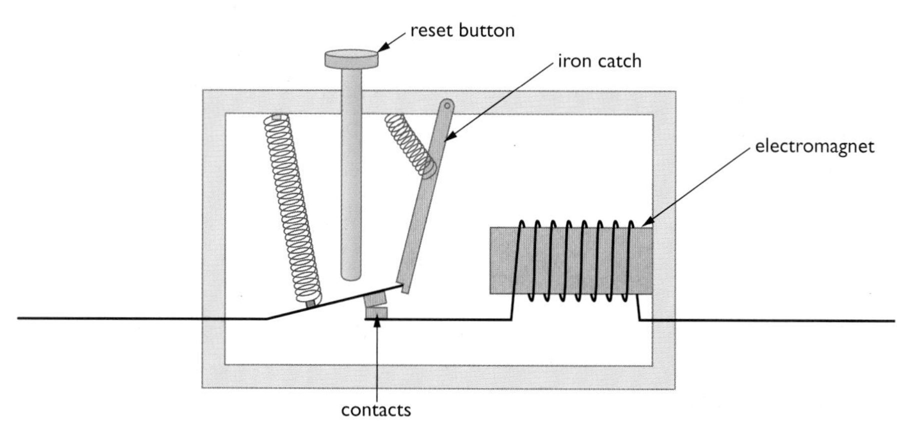

Figure 4.11 *Circuit breakers also use electromagnets.*

The circuit breaker in Figure 4.11 uses an electromagnet to cut off the current should it be larger than a certain value. If the current is too high the electromagnet becomes strong enough to pull the iron catch out of position so that the contacts open and the circuit breaks. Once the problem in the circuit has been corrected the catch is repositioned by pressing the reset button.

The relay switch

When large currents flow in a circuit there is always the danger of the user receiving a severe electric shock. Even turning the circuit on and off can be hazardous. To get around this problem we often use **relay switches**. A relay switch uses a small current in one circuit to turn on a second circuit that may be carrying a much larger current.

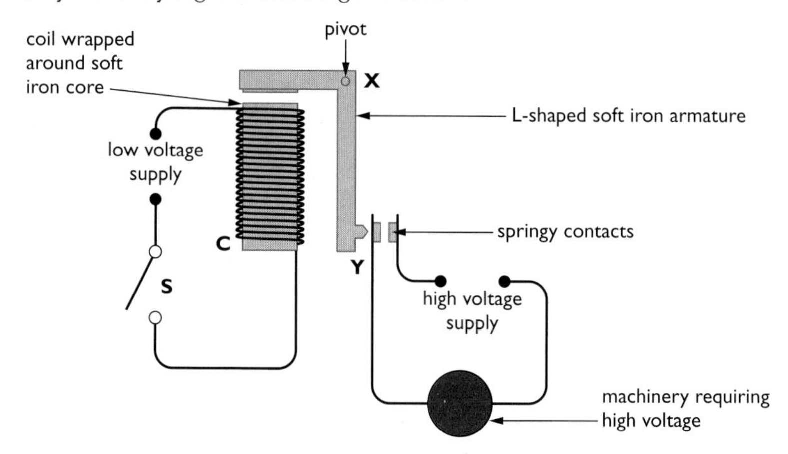

Figure 4.12 *A relay switch uses a small current in one circuit to switch on a larger current in another circuit.*

In Figure 4.12, when switch S is closed a small current flows around the circuit on the left. As current passes through coil C the soft iron core becomes magnetised and attracts the iron armature. Because the armature is pivoted at X, its lower end Y pushes the contacts of the second circuit together. The circuit is complete and current flows without the user coming into contact with the potentially more dangerous circuit. If the switch S is opened the electromagnet is turned off, the iron armature moves back to its original position and the contacts spring apart, turning the right hand circuit off. An arrangement similar to this is used in car ignition circuits.

End of Chapter Checklist

If you haven't got a copy of your specification, read the introduction on page iv.

You will need to be able to do some or all of the following. Check your Awarding Body's specification (syllabus) to find out exactly what you need to know.

- Recall that the strongest parts of a magnet are its poles and most magnets have two poles, a north pole and a south pole. Similar poles repel and opposite poles attract.
- Describe the shape and direction of the magnetic field around a bar magnet.
- Describe the shape and direction of the magnetic field ***a)*** around a current-carrying wire, and ***b)*** around a long coil (solenoid).
- Recall that the strength of an electromagnet can be altered by changing ***a)*** the size of the current flowing, ***b)*** the number of turns on the coil and ***c)*** the material on which the coil is wrapped (core).
- Describe how electromagnets are used in devices such as the electric doorbell, the electromagnetic circuit breaker and the electromagnetic relay switch.

Questions

More questions on electromagnetism can be found at the end of Section A on page 80.

1 ***a)*** Draw a diagram to show the shape of the magnetic field around a solenoid.

b) How does your diagram show where the magnetic field is strong or weak?

c) Explain what happens to the magnetic field if the direction of the current flowing through the solenoid is reversed.

2 In 1819 a scientist named Hans Christian Oersted was using a cell to pass a current through a wire. Close to the wire there was a compass. When current passed through the wire, Oersted noticed – much to his surprise – that the compass needle moved.

a) Why did the compass needle move?

b) When current is passed through a horizontal wire, the needle of a compass placed beneath the wire comes to rest at right angles to the wire, pointing from left to right. In which direction will the compass needle point if it is held above the wire? Explain your answer.

c) Would your answer to part ***b)*** still be correct if the direction of the current in the wire was changed? Explain your answer.

3 The diagram that follows shows a circuit breaker that uses an electromagnet.

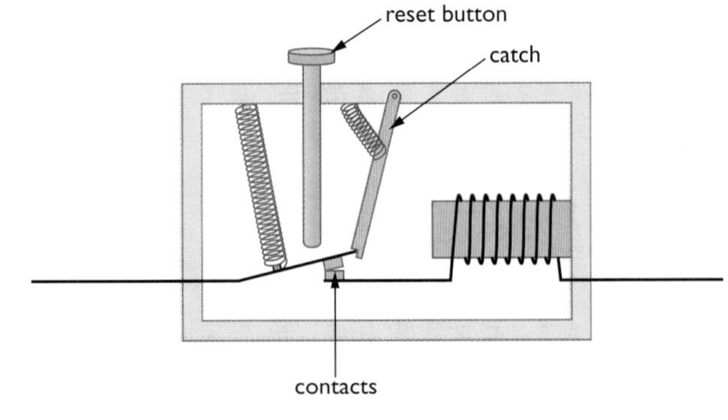

a) Explain why the current is cut off by this circuit breaker if it is larger than a certain value.

b) Give one advantage of this type of circuit breaker compared with a cartridge fuse.

4 The diagram below shows a simple electromagnet.

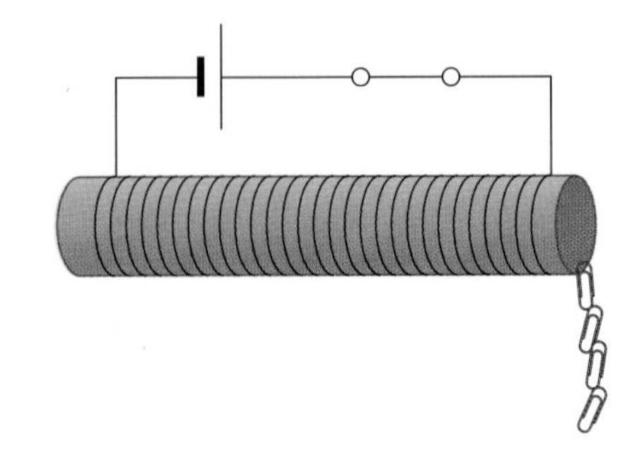

a) Explain why the core of an electromagnet is made of soft iron and not steel.

b) Suggest two ways in which the strength of this electromagnet could be increased.

c) When the switch is closed what type of pole is created at the left end of the electromagnet?

5 The diagram below shows the circuit for an electric bell.

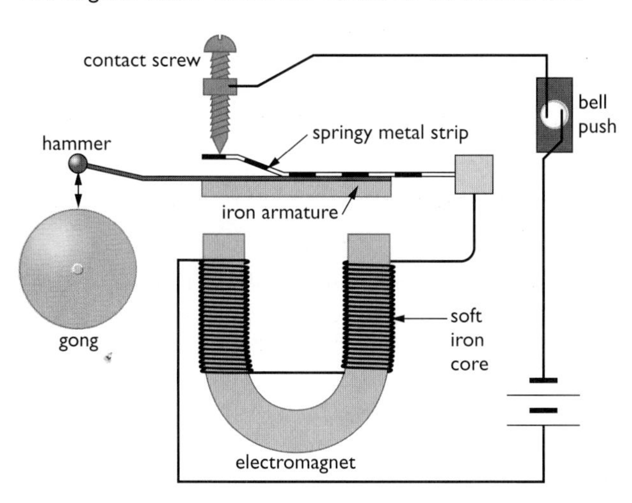

a) Explain in your own words why the bell will not work if the electromagnet is replaced with a permanent magnet.

b) Explain why the core of the electromagnet used in an electric bell must not be made from steel.

6 *a)* What is a relay switch?

b) Where in a car might you find a relay switch?

7 Find out what a reed switch is and how it could be used in a burglar alarm.

Chapter 5: Electric Motors and Electromagnetic Induction

All the trains in the Underground in London, the Metro in Paris and the Subway in New York use electric motors to transport millions of people to and from work each day. In the first part of this chapter, we are going to look at how an electric motor works. An electric motor creates motion from an electric current. Generators, on the other hand, transfer movement into electrical energy, by electromagnetic induction. The second part of this chapter looks at this process.

Figure 5.1 *Electric train in London.*

Movement from electricity

Overlapping magnetic fields

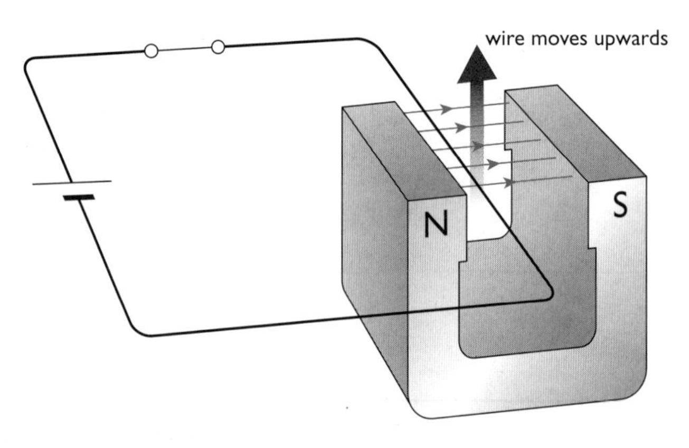

Figure 5.2 *The wire moves as a current flows through it.*

If we pass a current through a piece of wire held at right angles to the magnetic field of a magnet, as shown in Figure 5.2, the wire will move. This motion is the result of a force created by overlapping magnetic fields around the wire and the magnet.

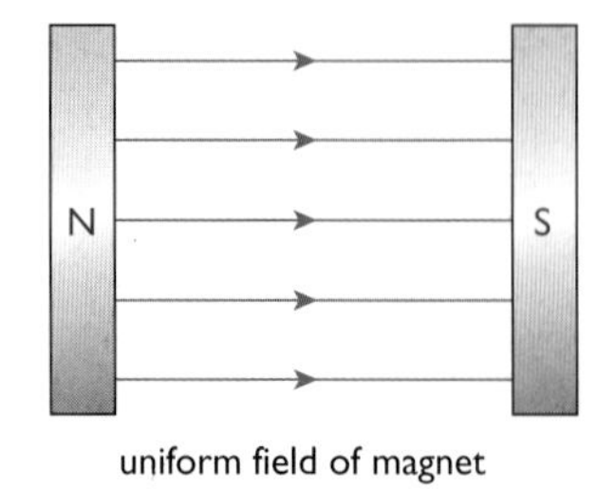

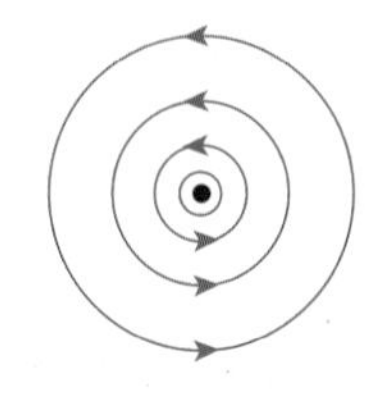

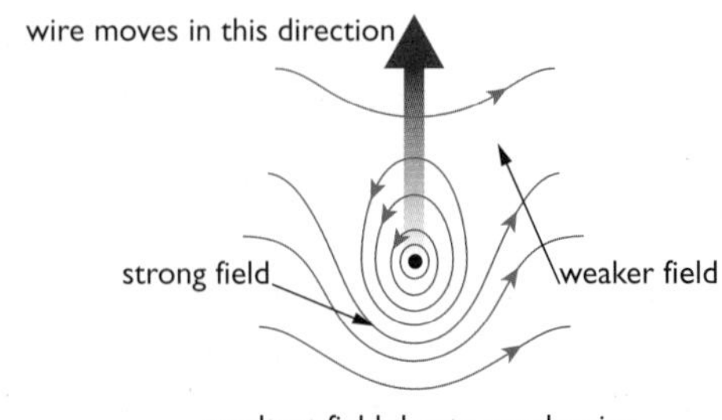

Figure 5.3 *The field around the wire and the field around the magnet overlap.*

As we have seen in Chapter 4, when a current flows along a wire a cylindrical magnetic field is created around the wire. If the wire is placed between the poles of a magnet, the two fields overlap. In certain places, the fields are in the same direction and so reinforce each other producing a strong magnetic field. In other places, the fields are in opposite directions, producing a weaker field. The wire experiences a force, pushing it from the stronger part of the field to the weaker part. This is called the **motor effect**.

If the direction of the current or the direction of the magnetic field is reversed, the wire experiences a force in the opposite direction. We can predict the direction of the motion of the wire using **Fleming's left hand rule**.

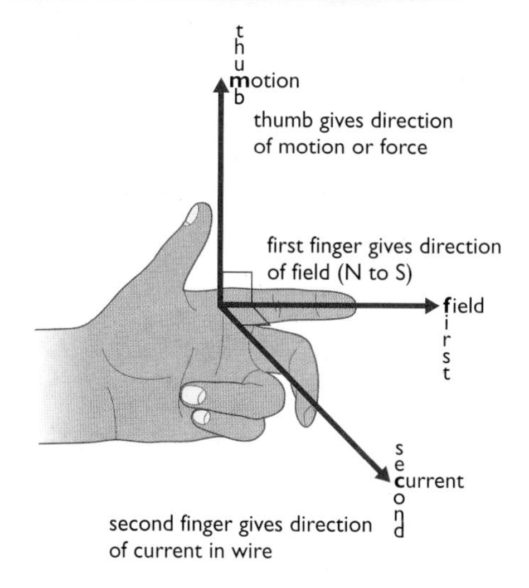

Figure 5.4 *Fleming's left hand rule helps you to work out the direction of the force.*

The moving-coil loudspeaker

The moving-coil loudspeaker uses the motor effect to change electrical energy into sound energy.

Signals from a source, such as an amplifier, are fed into the coil of the speaker as currents that are continually changing in size and direction. The overlapping fields of the coil and the magnet therefore create rapidly varying forces on the wires of the coil, which cause the speaker cone to vibrate. These vibrations create the sound waves we hear.

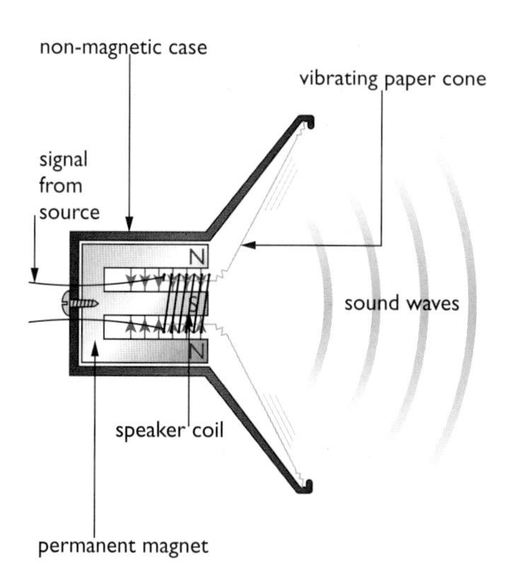

Figure 5.5 *A loudspeaker turns electrical energy into sound energy.*

The electric motor

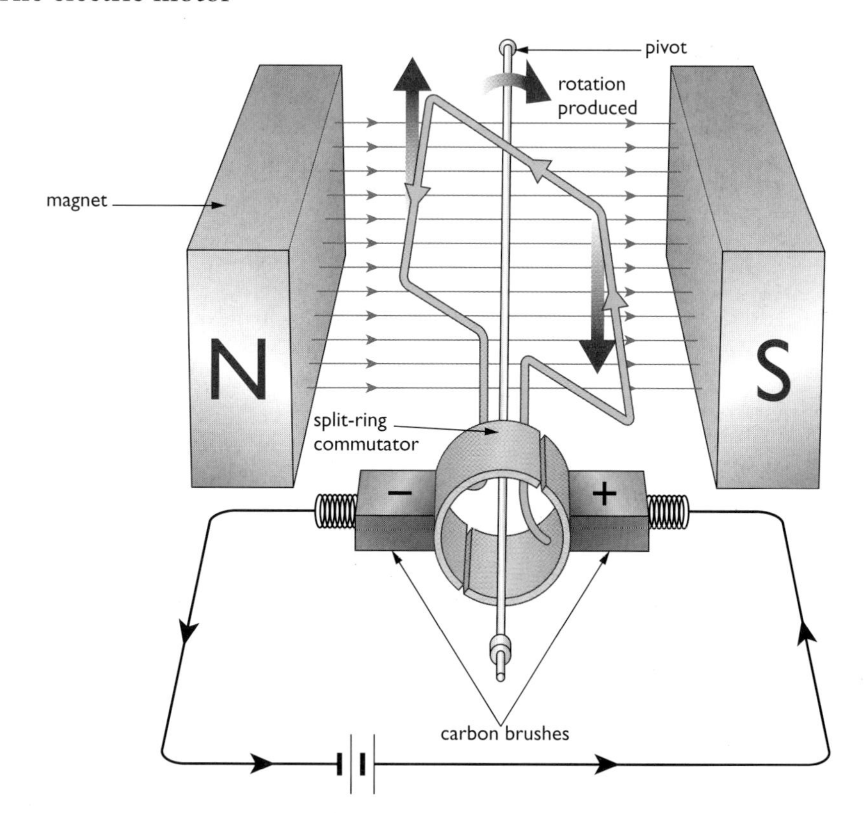

Figure 5.6 *A simple electric motor.*

As current passes around the loop of wire in Figure 5.6, one side of it will experience a force pushing it upwards. The other side will feel a force pushing it downwards, so the loop will rotate. Because of the split ring, when the loop is vertical, the connections to the supply through the brushes swap over, so that the current flowing through each side of the loop

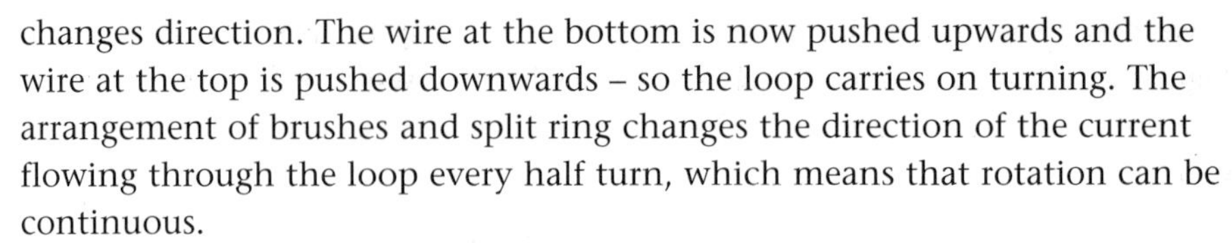

changes direction. The wire at the bottom is now pushed upwards and the wire at the top is pushed downwards – so the loop carries on turning. The arrangement of brushes and split ring changes the direction of the current flowing through the loop every half turn, which means that rotation can be continuous.

Figure 5.7 *A motor.*

To increase the rate at which the motor turns we can:

1 increase the number of turns or loops of wire (to make a coil)

2 increase the strength of the magnetic field

3 increase the current flowing through the loop of wire.

Practical motors differ from that described in Figure 5.6 in several ways.

- The permanent magnets are replaced with curved electromagnets capable of producing very strong magnetic fields.
- The single loop is replaced with several coils of wire wrapped on the same axis. This makes the motor smoother and more powerful.
- The coils are wrapped on a laminated soft iron core. This makes the motor more efficient and more powerful.

Electromagnetic induction

Figure 5.8 *Generators are used to produce electricity for the lights and machinery used on motorway roadworks.*

Motors use electricity to produce movement. Generators make electricity from movement.

The workers shown in Figure 5.8 need electricity for their machines and their lights. Rather than connecting into the mains supply, as we do at home, the workers have their own generator, which produces the electrical energy they need. In fact, the mains supply itself is produced by large generators in power stations. In this next section, you will discover how a generator produces electricity.

The generator

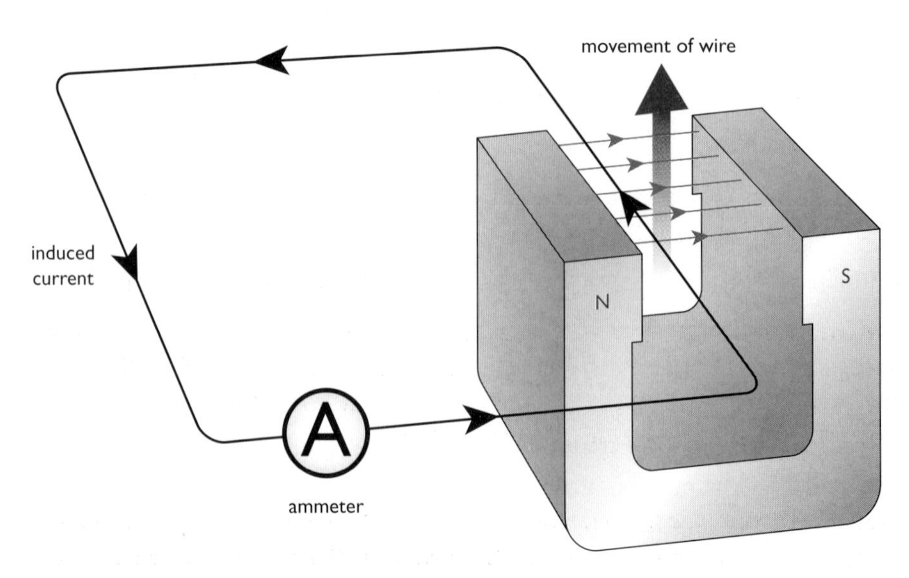

Figure 5.9 *When a wire moves through a magnetic field a voltage is generated in the wire.*

If we move a wire across a magnetic field at right angles, as shown in Figure 5.9, a voltage is **induced** or generated in the wire. If the wire is part of a complete circuit, a current flows. This phenomenon is called **electromagnetic induction**.

The size of the induced voltage (and current) can be increased by:

1 moving the wire more quickly

2 using a stronger magnet

3 wrapping the wire into a coil so that more pieces of wire move through the magnetic field.

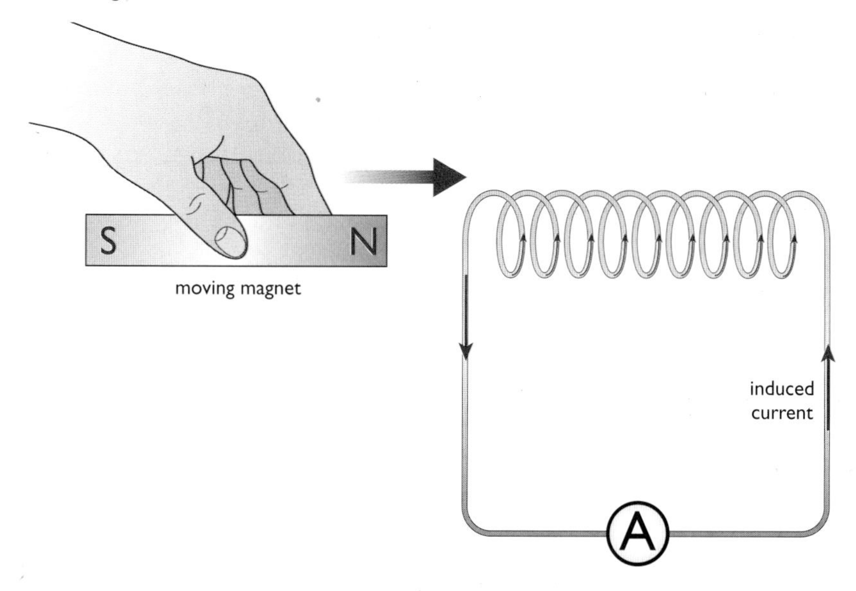

Figure 5.10 *A magnet moving in a coil will generate electricity.*

We can also generate a voltage and current by pushing a magnet into a coil. The size of the induced voltage (and current) can be increased by:

1 moving the magnet more quickly

2 using a stronger magnet

3 using a coil with more turns

4 using a coil with a larger cross-sectional area.

The factors listed above show that:

- a voltage and current are generated when a conductor such as a wire cuts through the magnetic field lines
- the faster the lines are cut the larger the induced voltage and current.

Michael Faraday was the first person to observe how the size of an induced voltage depends upon the rate at which the magnetic lines of flux (field lines) are being cut. He summarised his observations in **Faraday's Law of Electromagnetic Induction**. This states that:

The size of the induced voltage across the ends of a wire (coil) is directly proportional to the rate at which the magnetic lines of flux are being cut.

If we hold the wire or magnet stationary there is no cutting of the field lines so no voltage or current is generated. If we move the wire horizontally between the poles of the magnet, again there are no lines being cut and therefore no voltage or current is generated (Figure 5.11).

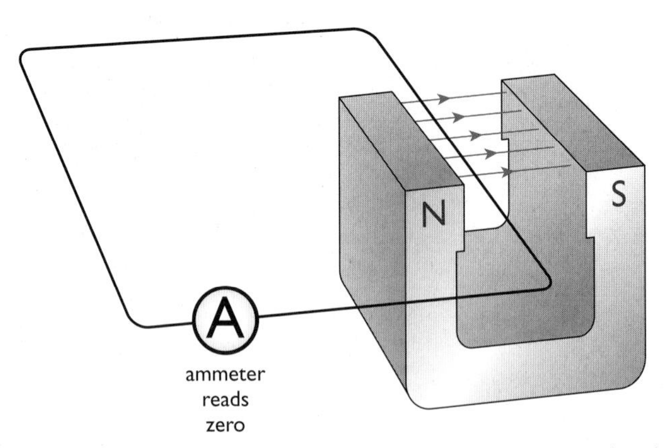

Figure 5.11 *When the field lines are not cut there is no voltage or current produced.*

The direction of the induced current in Figures 5.9 and 5.10 depends upon the direction of the motion. If the direction of movement is reversed so too is the direction of the induced current.

Using generators

Figure 5.12 shows a small generator or dynamo used to generate electricity for a bicycle light.

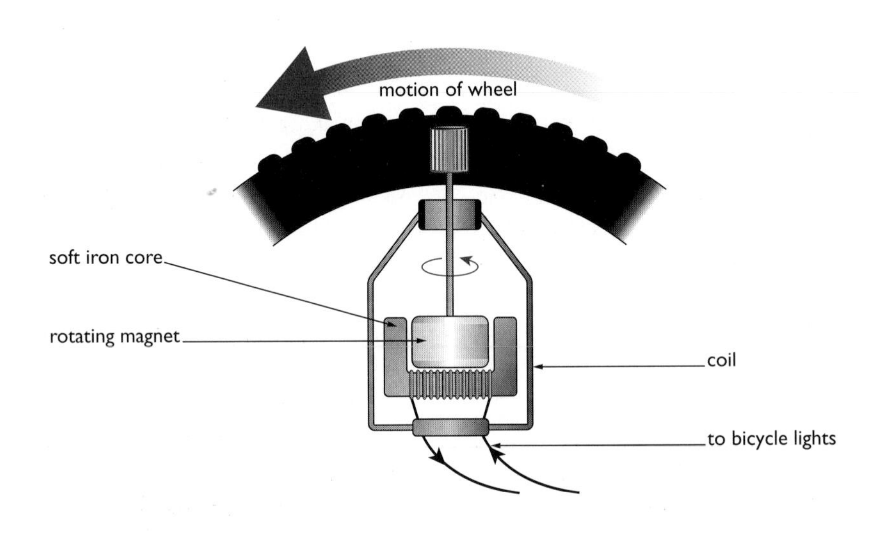

Figure 5.12 *A dynamo is a simple generator.*

As the cyclist pedals, the wheel rotates and a small magnet within the dynamo spins around. As this magnet turns, its magnetic field cuts through the surrounding coil inducing a current in it. This current can be used to work the cyclist's lights.

Figure 5.13 shows a much larger generator used in power stations to generate the mains electricity we use in our homes.

Figure 5.13 *This generator produces electricity on an industrial scale.*

As the coil rotates, its wires cut through magnetic field lines and a current is induced in them. If we watch just one side of the coil, we see that the wire moves up through the field and then down for each turn of the coil. As a result, the current induced in the coil flows first in one direction and then in the opposite direction. This kind of current is called **alternating current**. A generator that produces alternating current is called an **alternator**.

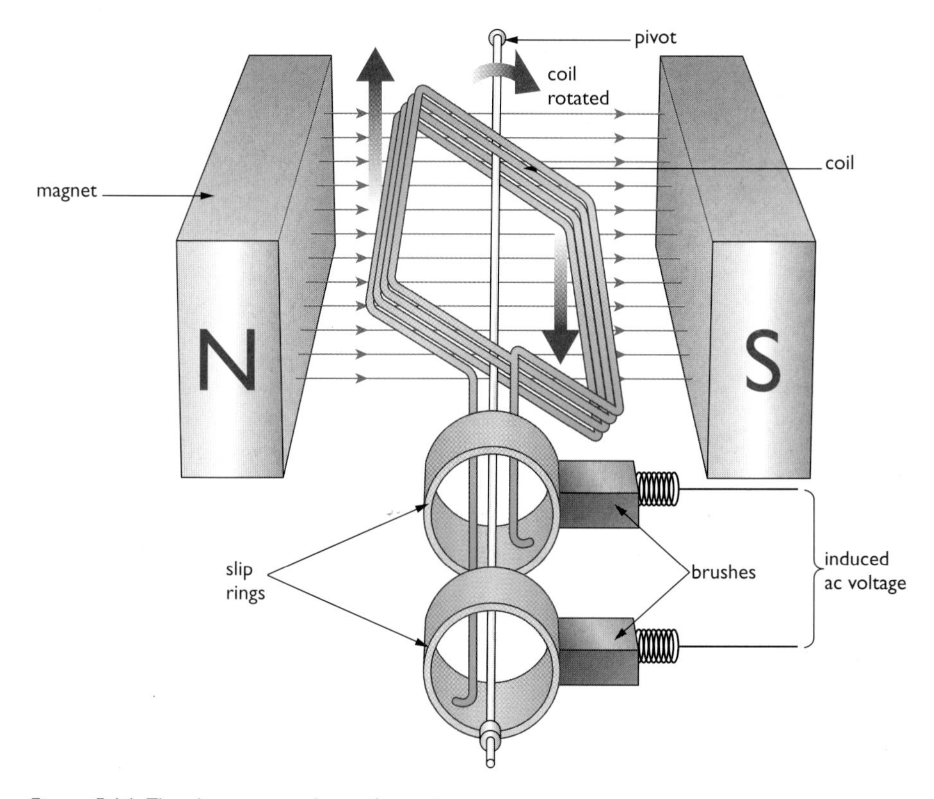

Figure 5.14 *The alternator produces alternating current.*

Most alternators use electromagnets rather than permanent magnets, as these can produce stronger magnetic fields, and it is the electromagnet which is rotated inside a stationary coil.

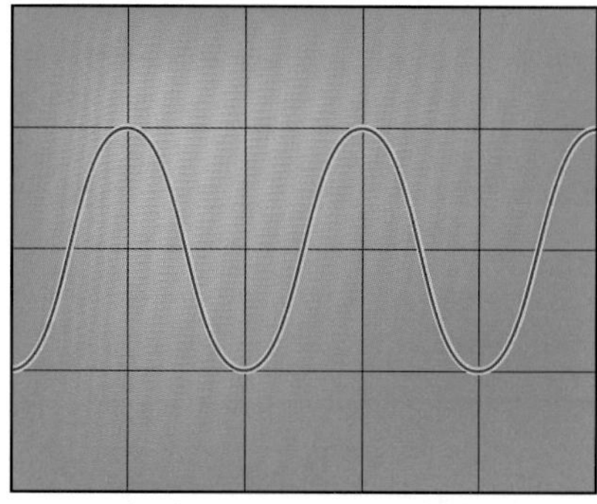

cathode ray oscilloscope (CRO) display for alternating current (ac)

CRO display for direct current (dc)

Figure 5.15 *Direct currents (dc) flow in one direction. Alternating currents (ac) keep changing direction.*

The frequency of an alternating current is the number of complete cycles it makes each second. If an alternator coil rotates twice in a second, the frequency of the alternating current it produces is 2 Hz (2 cycles per second). The frequency of our mains supply is 50 Hz.

The transformer

When alternating current is passed through a coil, the magnetic field around it is continuously changing. As the size of the current in the coil increases the field grows. As the size of the current decreases the field collapses. If a second coil is placed near the first, this changing magnetic field will pass through it. As it cuts through the wires of the second coil, a voltage is induced across that coil. The size and direction of the induced voltage changes as the voltage applied to the first coil changes. (The first coil is more usually called the **primary coil**.) An alternating voltage applied across the primary coil therefore produces an alternating voltage across the **secondary coil**. This combination of two magnetically linked coils is called a **transformer**.

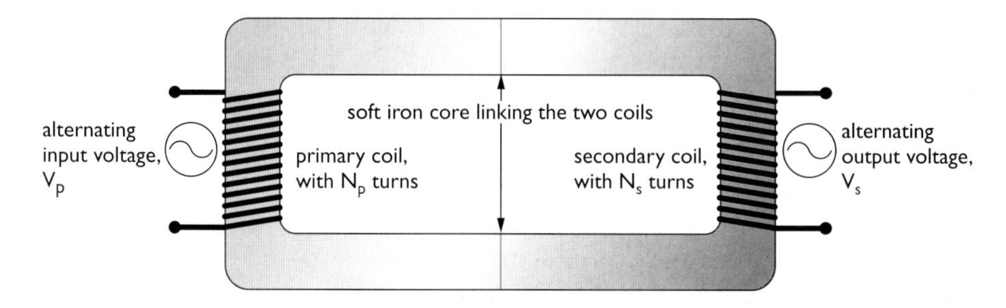

Figure 5.16 *The size of the voltage generated in the secondary coil of a transformer depends on the voltage in the primary coil, and the numbers of turns on each coil.*

Remember that transformers only work if the magnetic field around the primary coil is *changing*. Transformers will therefore only work with ac currents and voltages. They will not work with dc currents and voltages.

The relationship between the voltages across each of the coils is described by the equation:

$$\frac{\text{input voltage}}{\text{output voltage}} = \frac{\text{number of turns on primary coil}}{\text{number of turns on secondary coil}}$$

$$\frac{V_p}{V_s} = \frac{N_p}{N_s}$$

worked example

Example 1

A transformer has 100 turns on its primary coil and 500 turns on its secondary coil. If an alternating voltage of 2 V is applied across the primary, what is the voltage across the secondary coil?

$$\frac{V_p}{V_s} = \frac{N_p}{N_s}$$

$$\frac{2\,V}{V_s} = \frac{100}{500}$$

$$V_s = \frac{500 \times 2\,V}{100}$$

$$V_s = 10\,V$$

A transformer that is used to *increase* voltages, like the one in Example 1, is called a **step-up** transformer. One that is used to *decrease* voltages is called a **step-down** transformer.

If a transformer is 100% efficient then the electrical energy entering the primary coil each second equals the electrical energy leaving the secondary coil each second:

$$P_{in} = P_{out}$$

$$V_p \times I_p = V_s \times I_s$$

worked example

Example 2

When a voltage of 12 V ac is applied across the primary coil of a step-down transformer, a current of 0.4 A flows through the primary coil. Calculate the current flowing through the secondary coil if the voltage induced across it is 2 V ac. Assume that the transformer is 100% efficient.

Using $V_p \times I_p = V_s \times I_s$

$$12\,V \times 0.4\,A = 2\,V \times I_s$$

$$I_s = \frac{12\,V \times 0.4\,A}{2\,V}$$

$$I_s = 2.4\,A$$

Figure 5.17 *Transformers are very efficient.*

Energy losses in a transformer

Transformers are very efficient machines. They are designed to lose very little energy.

1 The wires from which the coils are made are thick and usually copper. As a result they have a low resistance and lose little energy through electrical heating.

2 The primary coil and the secondary coil are linked by an iron core. This concentrates the field lines around the coils so that the maximum number of field lines from the primary cut the secondary coil.

3 The iron core is laminated – it is made of thin sheets insulated from one another so they have a high resistance. This structure reduces any currents induced in the core when the magnetic lines of force from the primary coil pass through it. These swirling, circular induced currents in the core are called **eddy currents**.

Transformers and the National Grid

The National Grid is a network of wires and cables that carries electrical energy from power stations to consumers such as factories and homes. As current passes through a wire, energy is lost in the form of heat. If the current flowing through the wire is kept to a minimum, the heat losses are also reduced. Transformers are used in the National Grid so that the electricity is transmitted as low currents and at high voltages.

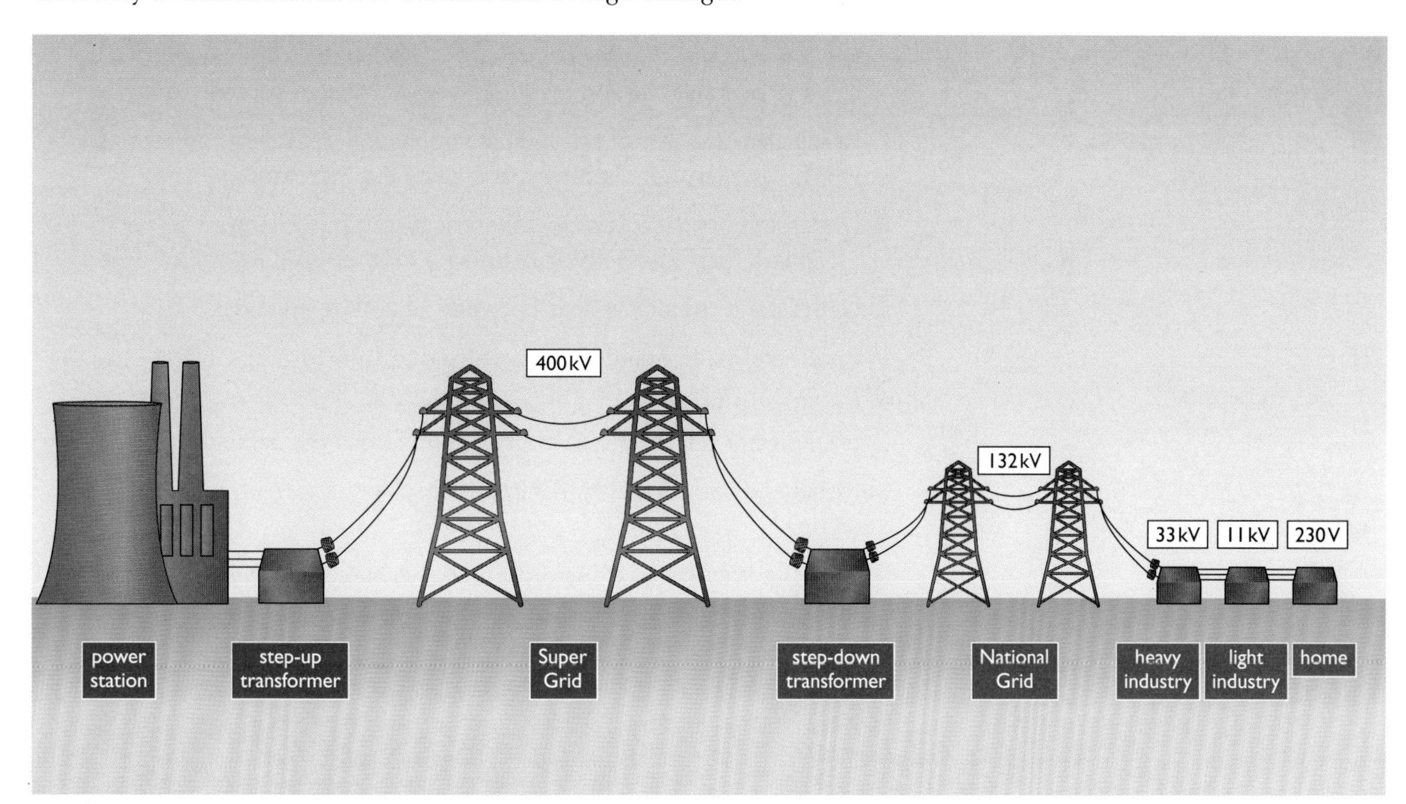

Figure 5.18 *Transformers are used in the National Grid.*

Immediately after generation, electric currents from the alternators are passed through step-up transformers. Here the voltage of the electricity is increased to approximately 400 kV and the size of the electric current is greatly decreased. High voltages like these can be extremely dangerous so the cables are supported high above the ground on pylons. As the cables enter towns and cities they are buried underground. Close to where the electrical energy is needed, the supply is passed through a step-down transformer which decreases the voltage to approximately 230 V, whilst at the same time increasing the current.

End of Chapter Checklist

If you haven't got a copy of your specification, read the introduction on page iv.

You will need to be able to do some or all of the following. Check your Awarding Body's specification (syllabus) to find out exactly what you need to know.

- Understand why a current-carrying conductor placed in a magnetic field experiences a force.
- Recall and use Fleming's left hand rule to predict the direction of the force and consequent movement.
- Recall that the size of the force can be increased by increasing the strength of the magnetic field or the size of the current flowing in the wire.
- Explain how electromagnetic effects are used in the simple dc motor.
- Understand that when a wire cuts through a magnetic field, or vice versa, a voltage is induced across the wire. If the wire is part of a complete circuit this induced voltage will cause a current to flow.
- Recall that the size of the induced voltage increases as the rate at which the magnetic field lines are being cut increases.
- Understand and explain how electricity can be generated by rotating a coil in a magnetic field or rotating a magnet inside a coil of wire.
- Understand the differences between ac and dc currents.
- Recall and understand the structure of a transformer.
- Recall that when an ac voltage is applied to the primary coil of a transformer an ac voltage is produced across the secondary coil.
- Recall and use the relationship $\frac{V_p}{V_s} = \frac{N_p}{N_s}$.
- Describe the use of transformers in the National Grid to reduce energy losses.

Questions

More questions on electric motors and electromagnetic induction can be found at the end of Section A on page 80.

1 The diagram below shows a long wire placed between the poles of a magnet.

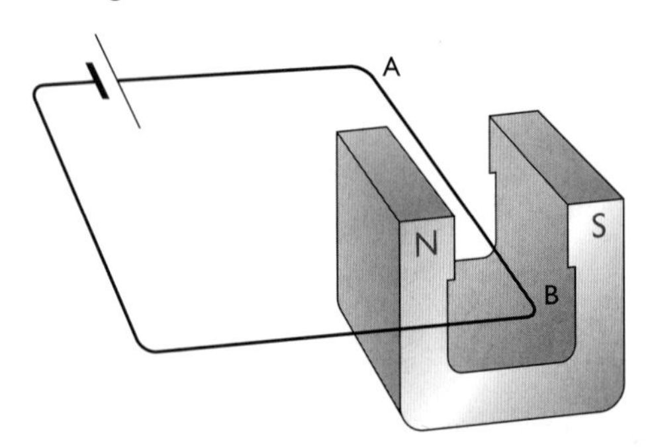

Describe what happens when:

a) current is passed through the wire flowing from A to B

b) the direction of the current is reversed – that is, it flows from B to A

c) with the current flowing from B to A, the poles of the magnet are reversed

d) a larger current is passed through the wire.

2 **a)** Draw a diagram of a simple electric motor.

b) Explain the functions of the split ring and the brushes.

c) Suggest three ways in which the speed of rotation of the motor could be increased.

d) State two ways in which the structure of practical motors is different from that of a simple motor.

3 The diagram below shows a bar magnet being pushed into a long coil. A sensitive meter is connected across the coil.

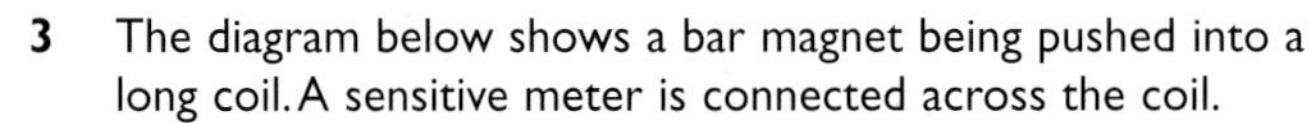

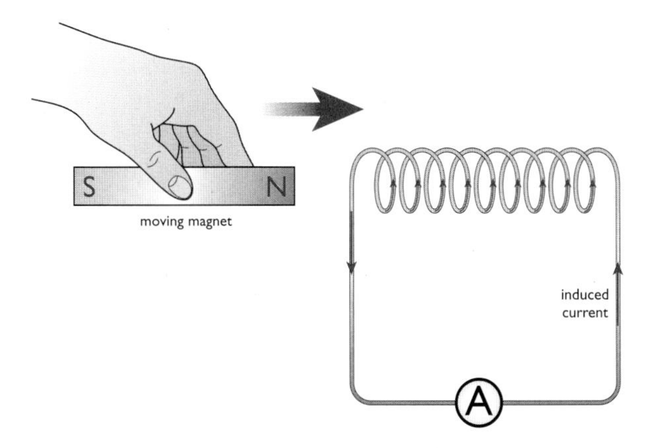

Describe what happens to the meter when:

a) the magnet is pushed into the coil quickly

b) the magnet is held stationary inside the coil

c) the magnet is withdrawn from the coil slowly

d) the magnet is held stationary and the coil is moved towards it.

4 ***a)*** Explain how a bicycle dynamo generates current.

b) Why does the dynamo produce no current when the cyclist has stopped?

c) What is an alternator?

d) Explain using diagrams the difference between ac and dc currents.

e) Explain what is meant by this statement: "The mains supply in the UK has a frequency of 50 Hz."

5 ***a)*** Explain the difference between a step-up transformer and a step-down transformer.

b) Explain where step-up and step-down transformers are used in the National Grid.

c) Explain why transformers are used in the National Grid.

d) Draw a fully labelled diagram of a transformer.

e) Explain those features that reduce energy losses within a transformer.

f) Explain why a transformer will not work if a dc voltage is applied across its primary coil.

g) A transformer has 200 turns on its primary coil and 5000 turns on its secondary coil. Calculate the voltage across the secondary coil when a voltage of 2 V ac is applied across the primary coil.

Section A: Electricity

Chapter 6: Static Electricity

Static electricity is the result of an imbalance of charge.
It can be very useful, but it can also be extremely hazardous.
In this chapter, you will learn about some of the uses and problems associated with static electricity.

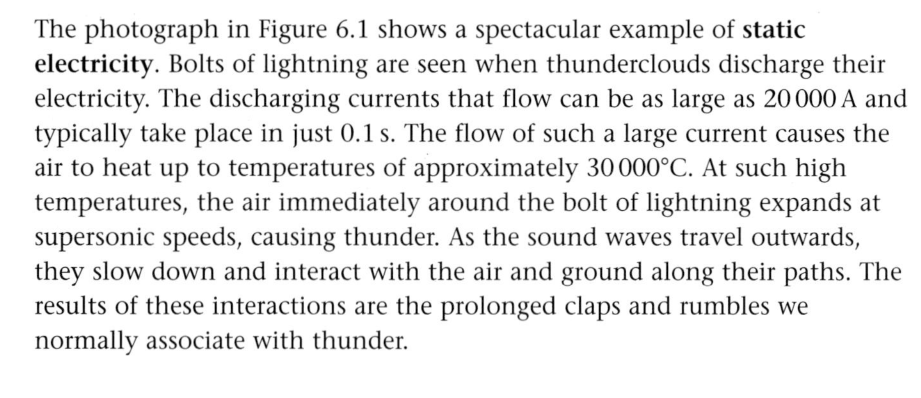

The photograph in Figure 6.1 shows a spectacular example of **static electricity**. Bolts of lightning are seen when thunderclouds discharge their electricity. The discharging currents that flow can be as large as 20 000 A and typically take place in just 0.1 s. The flow of such a large current causes the air to heat up to temperatures of approximately 30 000°C. At such high temperatures, the air immediately around the bolt of lightning expands at supersonic speeds, causing thunder. As the sound waves travel outwards, they slow down and interact with the air and ground along their paths. The results of these interactions are the prolonged claps and rumbles we normally associate with thunder.

Figure 6.1 *Thunderclouds discharge their static electricity as lightning.*

Charges within an atom

All atoms contain small particles called **protons**, **neutrons** and **electrons**. The protons are found in the centre or **nucleus** of the atom and carry a relative charge of +1. The neutrons are also in the nucleus of the atom but carry no charge. The electrons travel around the nucleus in orbits. The electrons carry a relative charge of –1.

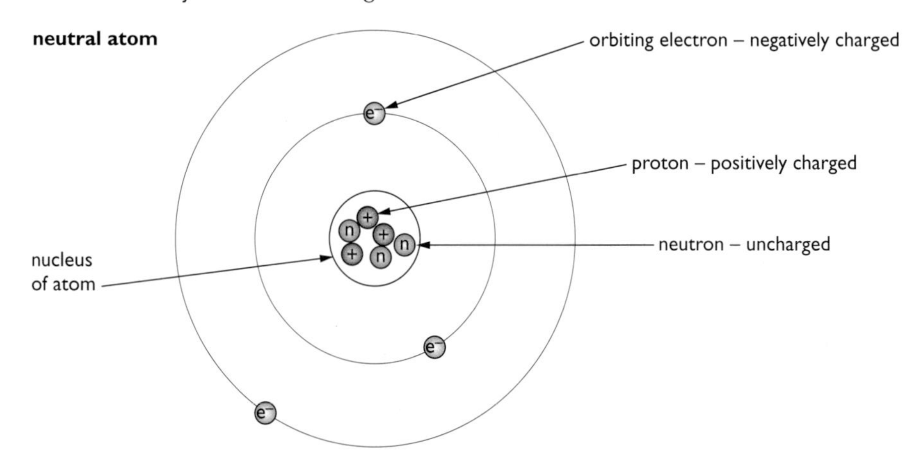

Figure 6.2 *A neutral atom has the same number of negative electrons and positive protons.*

Normally the number of protons in the nucleus is equal to the number of orbiting electrons. The atom therefore has no overall charge. It is **neutral**. If an atom gains extra electrons, it is then negatively charged. If an atom loses electrons, it becomes positively charged. An atom that becomes charged by gaining or losing electrons is called an **ion**.

Charging objects with static electricity

It is important to remember that the rubbing action does not produce or create charge. It simply separates charge – that is, it transfers some electrons from one object to another.

If an uncharged plastic rod is rubbed with an uncharged cloth, it is possible for both of them to become charged. This is sometimes called charging by friction. During the rubbing, electrons from the atoms of the rod may move onto the cloth. There is now an imbalance of charge in both objects. The rod is short of electrons and so is positively charged. The cloth has excess electrons and so is negatively charged.

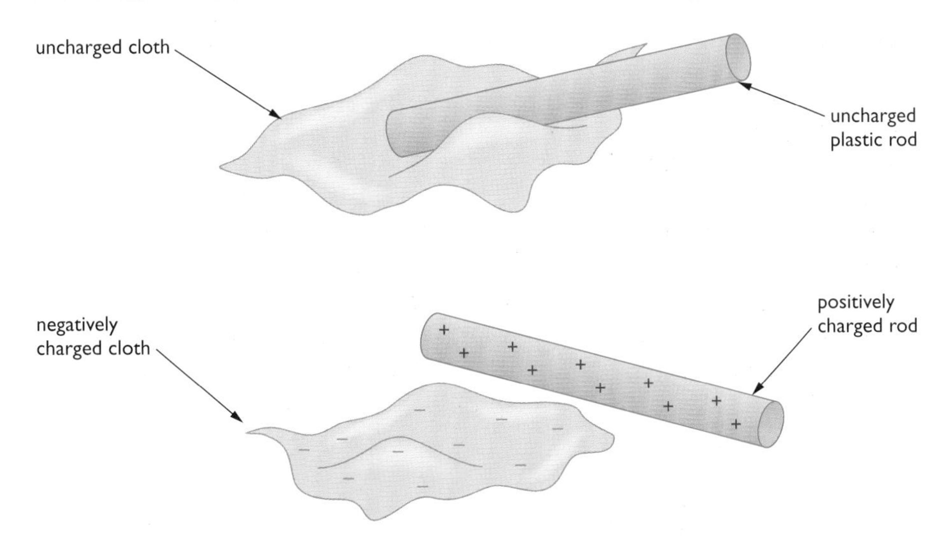

Figure 6.3 *Rubbing a neutral rod with a neutral piece of cloth can result in them both becoming charged.*

Forces between charges

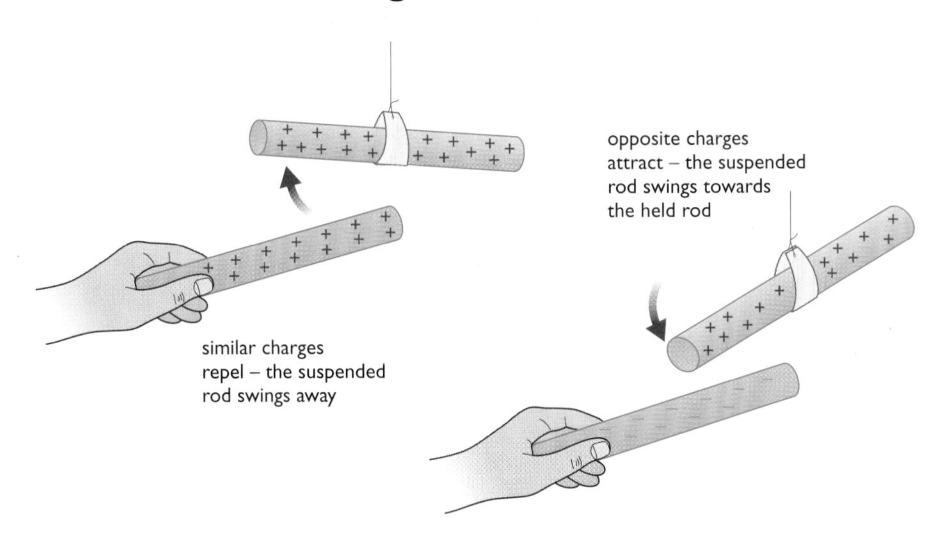

Figure 6.4 *Similar charges repel and opposite charges attract.*

Charged objects can exert forces on other charged objects without being in contact with them. If the charges are similar, the objects repel each other. If the charges are dissimilar (opposite), the objects attract each other.

The photograph in Figure 6.5 shows what can happen if a person is charged with static electricity. The girl has her hands on a Van de Graaff generator. When it is turned on, charges flow onto the large metal dome. Some of the charges flow over her hands and onto all parts of her body including her hair. Each strand of hair has the same type of charge as its neighbour, so there are repulsive forces between the strands. These forces cause her hair to stand on end.

For this demonstration to work, the girl must stand on an insulator to prevent any of the charges she is receiving from the generator from escaping into the floor. At the end of the demonstration the girl steps off the insulator – the charges can now escape and her hair falls. When a path is provided for charges to escape it is called **earthing**.

Figure 6.5 *A build-up of static electricity means that each hair on the girl's head has the same charge, and they repel each other.*

An insulator is a material that does not allow charge to flow through it as there are no free electrons.

Forces between charged and uncharged objects

It is possible for a charged object to attract something that is uncharged.

The balloon experiment

If you charge a balloon by rubbing it against your jumper or your hair and then hold the balloon against a wall you will probably find that the balloon sticks to the wall. There is an attraction between the charged balloon and the uncharged wall.

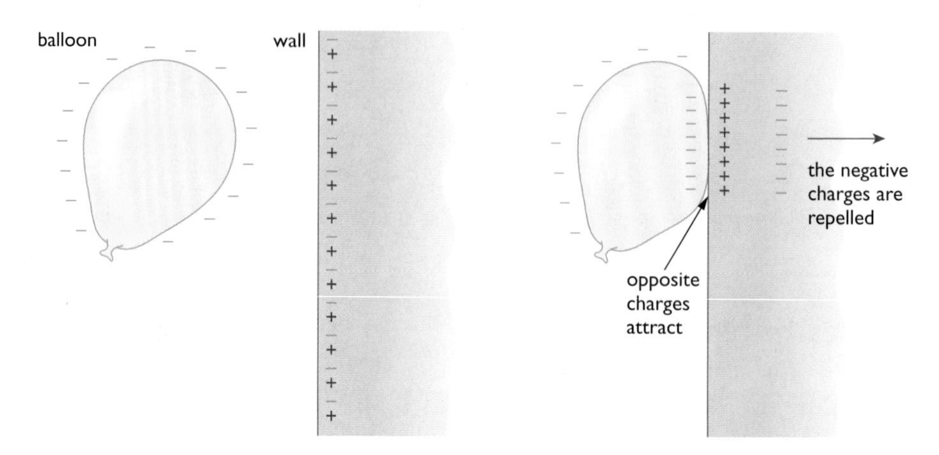

Figure 6.6 *The balloon is negatively charged and induces a positive charge in the wall.*

After the balloon has been charged with static electricity, but before it is brought close to the wall, the charges will be distributed as shown in Figure 6.6. The balloon is charged (we have assumed negatively charged) and the wall is uncharged – that is, it has equal numbers of positive and negative charges.

As the negatively charged balloon is brought closer to the wall some of the negative electrons are repelled from the surface of the wall. This gives the surface of the wall a slight positive charge that attracts the negatively charged balloon.

You can try a similar experiment using a plastic comb and some small pieces of paper. Immediately after combing your hair, the comb is likely to be charged, as shown in Figure 6.7. (In this example, we have assumed that the comb has become positively charged.) As the comb is brought close to the uncharged pieces of paper some of their electrons are attracted to the edges of the paper closest to the comb. There is attraction between these negative parts of the paper and the comb and repulsive forces between the comb and the positive parts of the pieces of paper. The attractive forces are stronger than the repulsive forces because the opposite charges are closer. The paper therefore moves towards the comb.

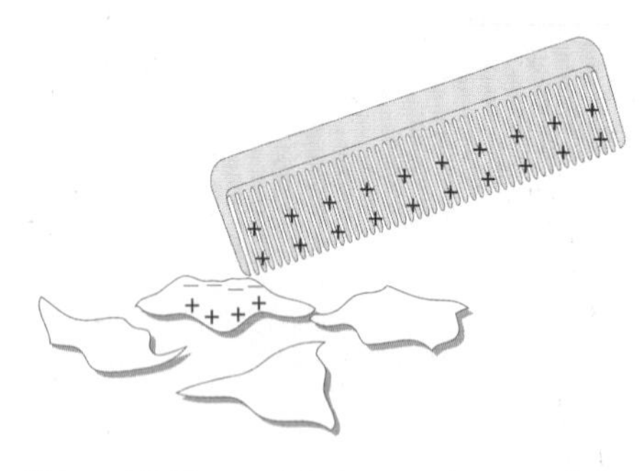

Figure 6.7 *The comb induces a charge in the paper, and the two attract.*

The charges that appear on the pieces of paper are called **induced charges.** When the comb is removed the charges redistribute themselves so that the pieces of paper are once again uncharged.

For you to try

Take a plastic ruler or rod and rub it against a jumper to charge it with static electricity. Turn on a water tap so that the water flows from the tap as slowly as possible but as a continuous flow (not as a series of drops). Hold the charged rod or ruler close to the stream of water but not in it. What happens to the water? Can you explain what is happening?

Uses of static electricity

Electrostatic paint spraying

Painting an awkwardly shaped object such as a bicycle frame with a spray gun can be very time-consuming and very wasteful of paint. Using electrostatic spraying can make the process much more efficient.

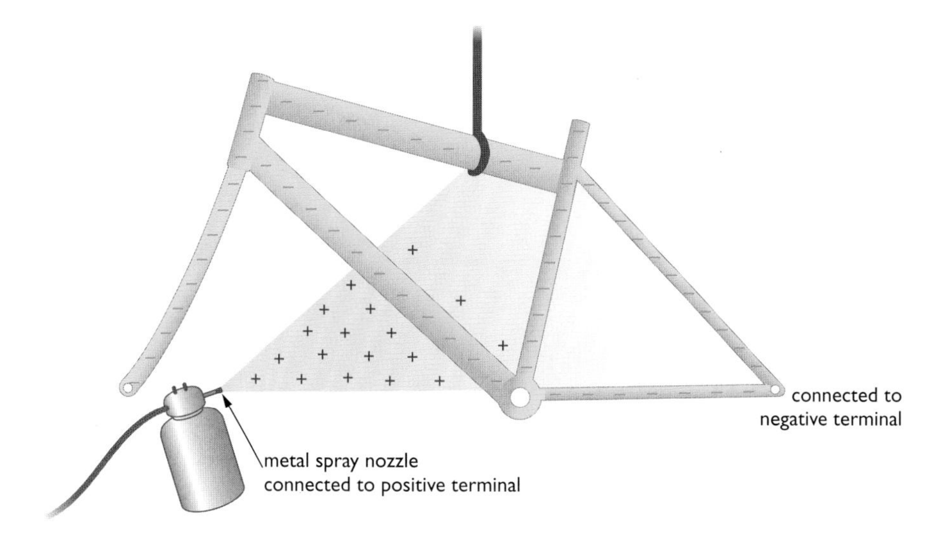

Figure 6.8 *The positive paint is attracted to all parts of the negatively charged object.*

As the droplets of paint emerge from the spray gun, they are charged. As the droplets all carry the same charge they repel and spread out forming a fine spray. The metal bicycle frame has a wire attached to an electrical supply giving the frame the opposite charge. The paint droplets are therefore attracted to the surface of the frame. There is the added benefit that paint is attracted into places, such as tight corners, that might otherwise not receive such a good coating.

Inkjet printers

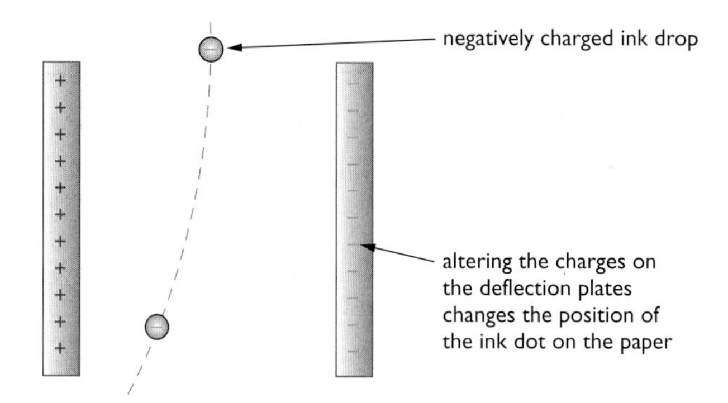

Figure 6.9 *The charged ink droplets are deflected into the correct position on the paper.*

Many modern printers use inkjets to direct a fine jet of ink drops onto paper. Each spot of ink is given a charge so that as it falls between a pair of deflecting plates electrostatic forces direct it to the correct position. The charges on the plates change hundreds of times each second so that each drop falls in a different position, forming pictures and words on the paper as required.

Photocopiers

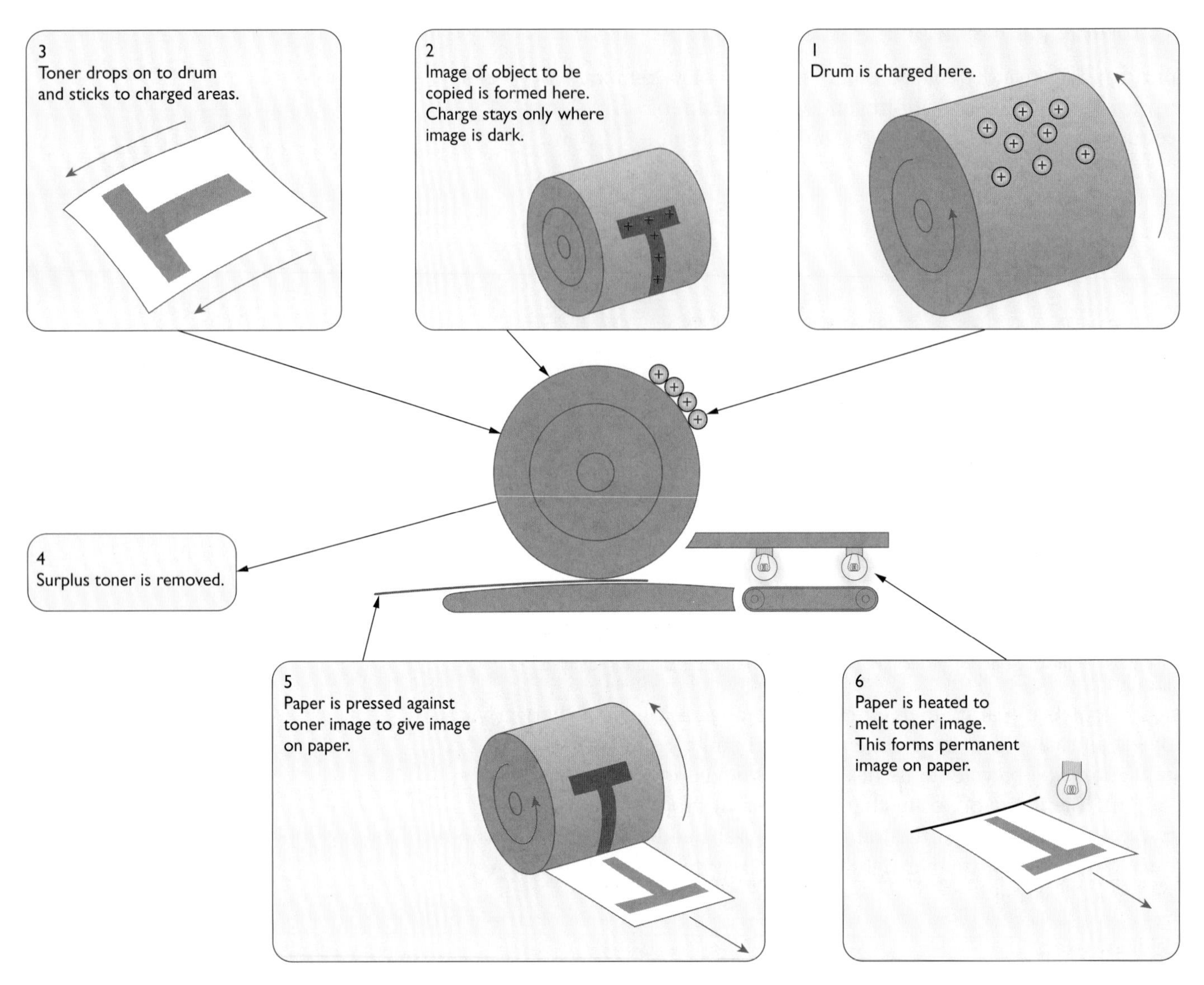

Figure 6.10 *Static electricity is used in photocopiers.*

Positive charges are sprayed onto a rotating drum whose surface is coated with a metal called selenium. A bright light is shone onto the sheet of paper to be copied. The white parts of the paper reflect light onto the drum; the dark or printed parts do not. In those places where light is reflected onto the drum the selenium loses its charge but where no light is reflected onto the drum the charge remains. A fine negatively charged carbon powder called toner is blown across the drum and sticks to just those parts of the drum that are charged. A sheet of paper is now pressed against the drum and picks up the pattern of the carbon powder. The powder is then fixed in place by a heater.

Electrostatic precipitators

Many heavy industrial plants, such as steel-making furnaces and coal-fired power stations, produce large quantities of smoke. This smoke carries small particles of ash and dust into the environment causing health problems and damage to buildings. One way of removing these pollutants from the smoke is to use electrostatic precipitators.

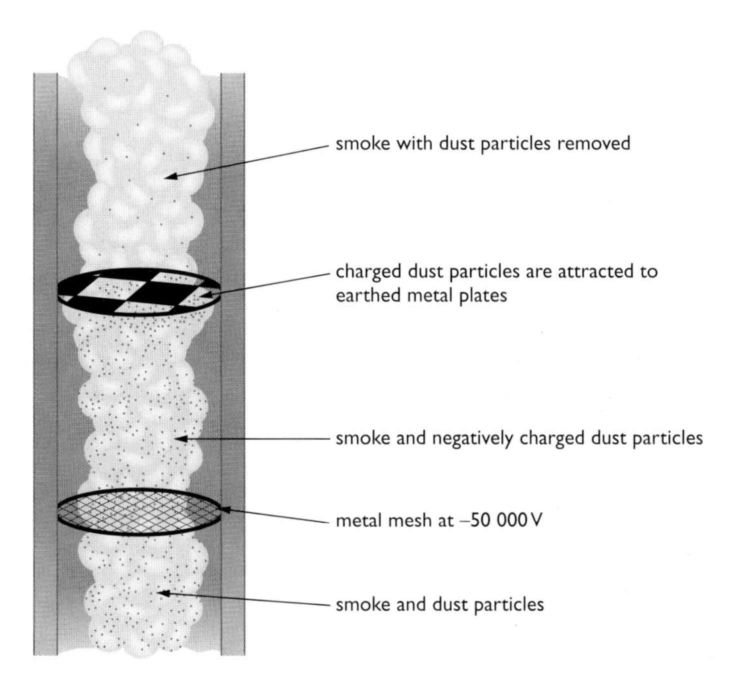

Figure 6.11 *Electrostatic precipitators help to cut down the amount of pollution released into the atmosphere.*

As the smoke initially rises up the chimney it passes through a mesh of wires that are highly charged. (The wires are at a voltage of approximately –50 000 V.) As they pass through the mesh, the ash and dust particles become negatively charged. Higher up the chimney these charged particles are attracted by and stick to large metal earthed plates. The cleaner smoke is then released into the atmosphere. When the earthed plates are completely covered with dust and ash, they are given a sharp rap. The dust and ash fall into collection boxes, which are later emptied.

In a large coal-fired power station, 50–60 tonnes of dust and ash may be removed from smoke each hour!

Problems with static electricity

In some situations the presence of static electricity can be a disadvantage.

- As aircraft fly through the air, they can become charged with static electricity. As the charge on an aircraft increases, so too does the potential difference between it and earth. With high potential differences there is the possibility of charges escaping to earth as a spark during refuelling, which could cause an explosion. The solution to this problem is to earth the plane with a conductor as soon as it lands and before refuelling commences.
- Television screens and computer monitors become charged with static electricity as they are used. These charges attract light uncharged particles – that is, dust.
- Our clothing can, under certain circumstances, become charged with static electricity. When we remove the clothes there is the possibility of receiving a small electric shock as the charges escape to earth.

End of Chapter Checklist

If you haven't got a copy of your specification, read the introduction on page iv.

You will need to be able to do some or all of the following. Check your Awarding Body's specification (syllabus) to find out exactly what you need to know.

- Recall that there are two types of electrical charge – positive and negative.
- Understand that objects that are uncharged contain equal numbers of positive and negative charges.
- Understand how the transfer of electrons between objects can cause them to become charged. Objects that gain electrons become negatively charged. Objects that lose electrons become positively charged.
- Recall that opposite charges attract and similar charges repel.
- Recall some uses of static electricity including electrostatic paint spraying, inkjets, photocopiers and electrostatic smoke precipitators.

Questions

More questions on static electricity can be found at the end of Section A on page 80.

1 ***a)*** What charge is carried by each of these particles?

i) a proton

ii) an electron

iii) a neutron

b) Where inside an atom are each of the three particles mentioned in ***a)*** found?

c) How many protons are there in a neutral atom compared to the number of electrons?

d) What do we call an atom that has become charged by gaining or losing electrons?

e) Describe with diagrams how two objects can be charged by friction (rubbing).

2 Explain the following.

a) A crackling sound can sometimes be heard when removing a shirt or blouse.

b) Sometimes after a journey in a car you can get a mild electric shock when you touch the handle of the door.

c) A plastic comb is able to attract small pieces of paper immediately after it has been used.

d) Dust always collects on the screens of televisions and computer monitors.

3 ***a)*** In a photocopier, why does toner powder stick to some places on the selenium-coated drum but not to others?

b) Explain why ash and dust particles are attracted towards the earthed metal plates of an electrostatic precipitator after they have passed through a highly negatively charged mesh of wires. (**Hint**: Read again about the balloon experiment.)

4 Lightning is caused by clouds discharging their static electricity.

a) Find out:

i) how the clouds become charged

ii) how a lightning conductor works.

b) Suggest two places where it might be *i)* unsafe, and *ii)* safe during a thunderstorm.

5 Find out about Benjamin Franklin and his work on electricity. Why was he a very lucky scientist?

6 Computer chips can be damaged by static electricity. Find out how workers who build and repair computers avoid this problem.

Chapter 7: Electronics and Control

Electronics are used to control all kinds of simple and more complex systems. Although electronic circuits can seem very complicated, they are built up from simple components. Any system has an input stage (such as a sensing circuit), a process stage (involving logic gates, for example), and an output stage (perhaps a switch to turn on some external device, like a heater or fan). In this chapter you will learn about the ways in which a small number of electronic components can be arranged in different ways to create all kinds of control circuits.

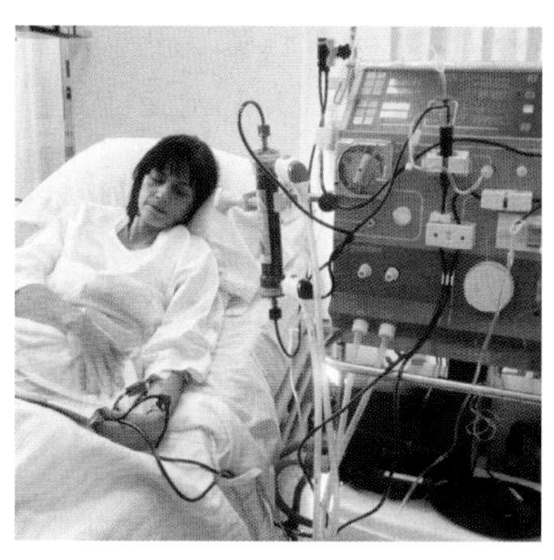

Figure 7.1 *Many areas of our lives now depend on modern electronic control systems.*

Electronic systems are used in nearly every aspect of our lives. Video games and computers are obvious examples of the use of modern electronics. Traffic lights are operated by an electronic control system. Modern cars use electronics to make sure that the engine is working as well as possible. Many homes and buildings have electronic control systems for heating, ventilation and security. Planes rely on electronic control systems not just to keep conditions comfortable for the passengers but also to help the pilot to fly the aircraft. During an operation a patient's condition is monitored by electronic systems and some vital body functions may be controlled by electronics. It is important to know how the systems that affect our lives work.

Digital electronics

Most modern electronic systems are **digital**. They use high and low voltages as signals – "high", in this instance, usually means about 5 V, and "low" is at or close to 0 V. We sometimes use the **binary** number system to represent these two possible signals. Binary uses only the numbers 1 and 0. "High" is represented by the binary number 1 ("logic 1") and "low" by the binary number 0 ("logic 0").

The earliest types of digital computer used switches to process information. Two examples of simple switch circuits are shown in Figures 7.2 and 7.3.

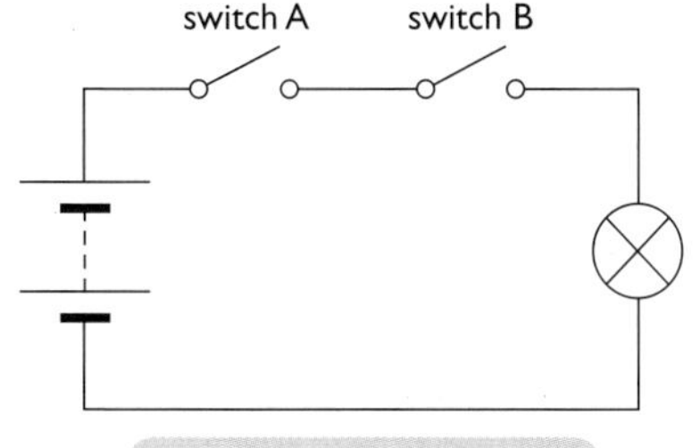

switch A	switch B	bulb
open	open	off
open	closed	off
closed	open	off
closed	closed	on

switch A *and* switch B must be closed to turn the bulb on

Figure 7.2 *Two switches in series. The truth-table to the right shows all the possible combinations of results for this simple system.*

In the rest of this chapter the battery supply will not be drawn out in full. Instead it will be shown by "power rails", as shown in Figure 7.4.

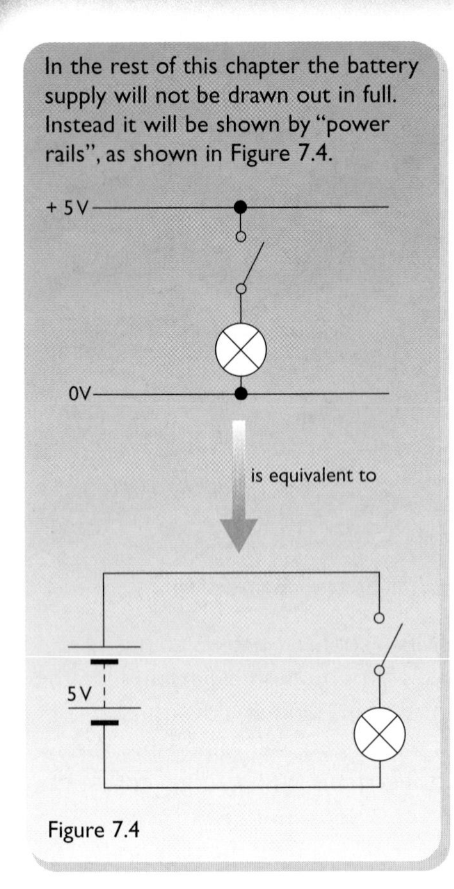

Figure 7.4

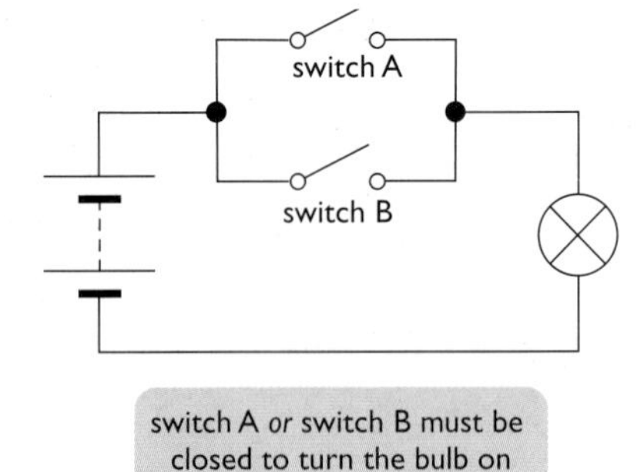

switch A	switch B	bulb
open	open	off
open	closed	on
closed	open	on
closed	closed	on

Figure 7.3 *Two switches in parallel. Again, the truth-table tells us all the possible outcomes for this system.*

The switches have two possible conditions: open (not conducting) and closed (conducting). The bulb also has two conditions in each of these circuits: it is either off or on. The bulb will only be on in the circuit in Figure 7.2 when both switch A *and* switch B are closed. The behaviour of the circuit is shown in the **truth-table** alongside the circuit diagram. The truth-table shows the condition of the lamp for all the possible "open"/"closed" combinations of the two switches. As you can see, with two switches there are four possible combinations. A different arrangement of switches is shown in Figure 7.3, along with its truth-table.

Logic gates

Logic gates are combinations of electronic switches. The switches are operated by voltages applied to the inputs of the logic gates. The inputs can be either "high" (about 5 V) or "low" (about 0 V). The output of a logic gate is also a voltage that can be high or low. Logic gates are usually made in the form of **integrated circuits** (**ICs**). Each IC has many logic gates made on a tiny piece of silicon packaged in plastic.

Figure 7.5 *Some integrated circuits.*

There are three basic logic gates: the **AND** gate, the **OR** gate and the **NOT** gate. The symbols of these gates and their truth-tables are shown in Figure 7.6.

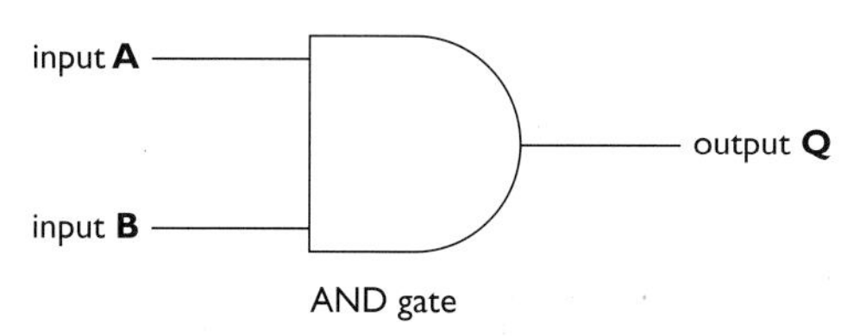

input A	input B	output Q
low	low	low
low	high	low
high	low	low
high	high	high

The output of an AND gate is high (or 1) only when both inputs are high (or 1).
Remember this as output Q is high if both input A *and* input B are high.

Truth-tables usually use the binary numbers 0 and 1 to represent low and high voltages. You should check whether your Awarding Body uses 0 and 1 or "low" and "high" in truth-table questions.

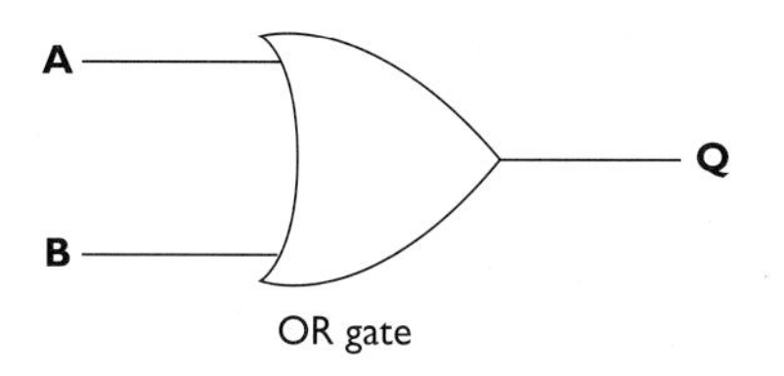

input A	input B	output Q
low	low	low
low	high	high
high	low	high
high	high	high

The output of an OR gate is high (or 1) if either input is high (or 1).
Remember this as output Q is high if either input A *or* input B are high.

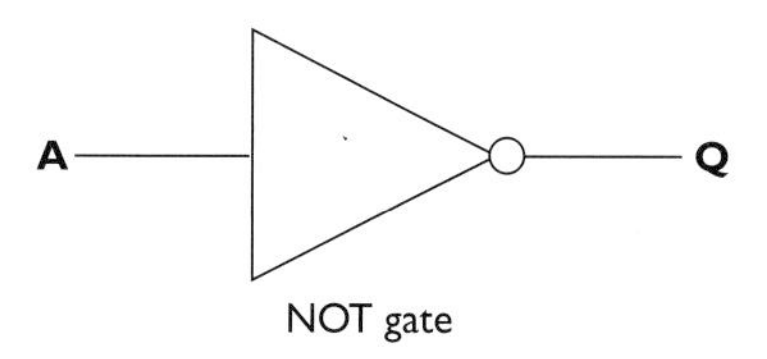

input A	output Q
low	high
high	low

The output of a NOT gate is high (or 1) if its input is low (or 0), and vice versa.
Remember the output of a NOT gate is the inverse of its input.

Figure 7.6 *AND, NOT and OR gates and their truth-tables.*

Control systems using logic gates

Control systems need to have **sensing circuits** that measure the conditions we want to control. The sensing circuits used in digital control systems must produce suitable digital signals – either low or high voltages. We shall see how some sensing circuits work and how they are used to provide input signals for digital control systems. The basic structure of a control system is shown in Figure 7.7.

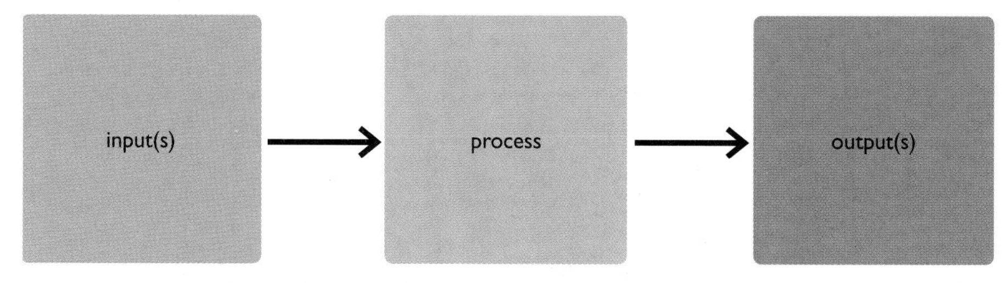

Figure 7.7 *The basic elements of a control system.*

The input section senses changes in some aspect of the environment, like temperature or light level. The output is some device that causes a change in the environment, like a heater or a lamp. The process section determines what change in the inputs will cause the output to change.

The basic sensing circuit

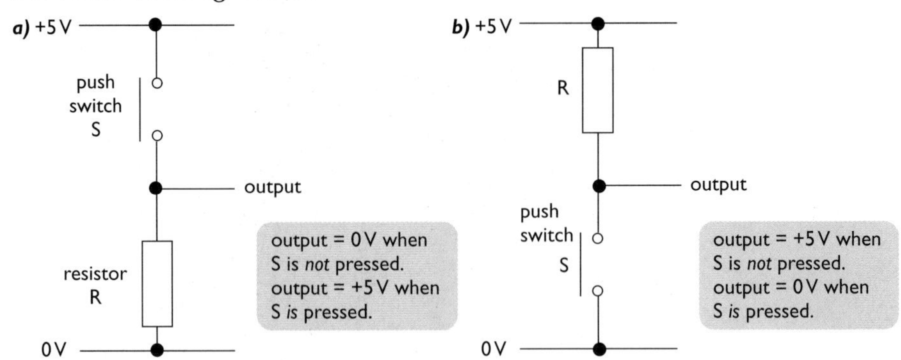

Figure 7.8 *The push switch sensor can operate in two ways.*

The basic sensing circuit has a push switch as its sensor. It is connected in series with a resistor. There are two ways that the sensor can operate. In Figure 7.8a, the resistor holds the output voltage low, at 0 V, when the switch is open (not pressed). When the switch is pressed the output voltage is high at +5 V. In Figure 7.8b, the output is held high by the resistor when the switch is not pressed and is low when the switch is pressed.

Figure 7.9 a) *A push switch.*

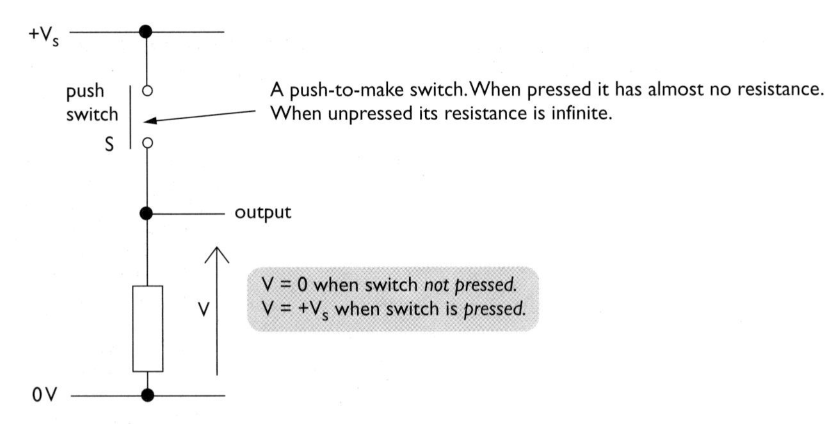

b) *A simple push switch circuit.*

This sensing circuit can use a variety of different types of switch – for example, a reed switch (Figure 7.10), a pressure pad (Figure 7.11) or a tilt switch (Figure 7.12). Pressure pads and tilt switches can be used in burglar alarms. Pressure pads are activated if someone steps on them. Tilt switches are turned on if something is moved or knocked over.

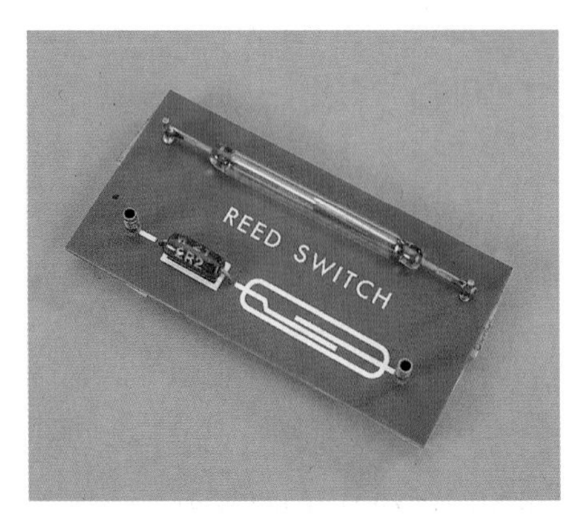

Figure 7.10 a) *A reed switch.*

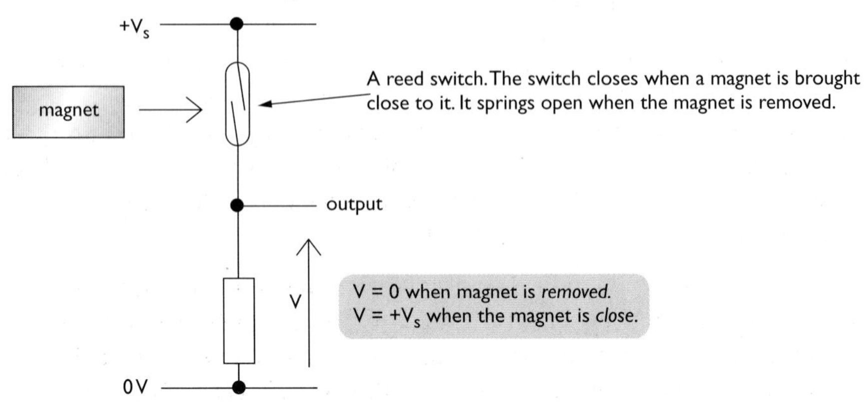

b) *A magnet sensing circuit using a reed switch. This circuit can be used in a burglar alarm to sense if a door or window is open or closed.*

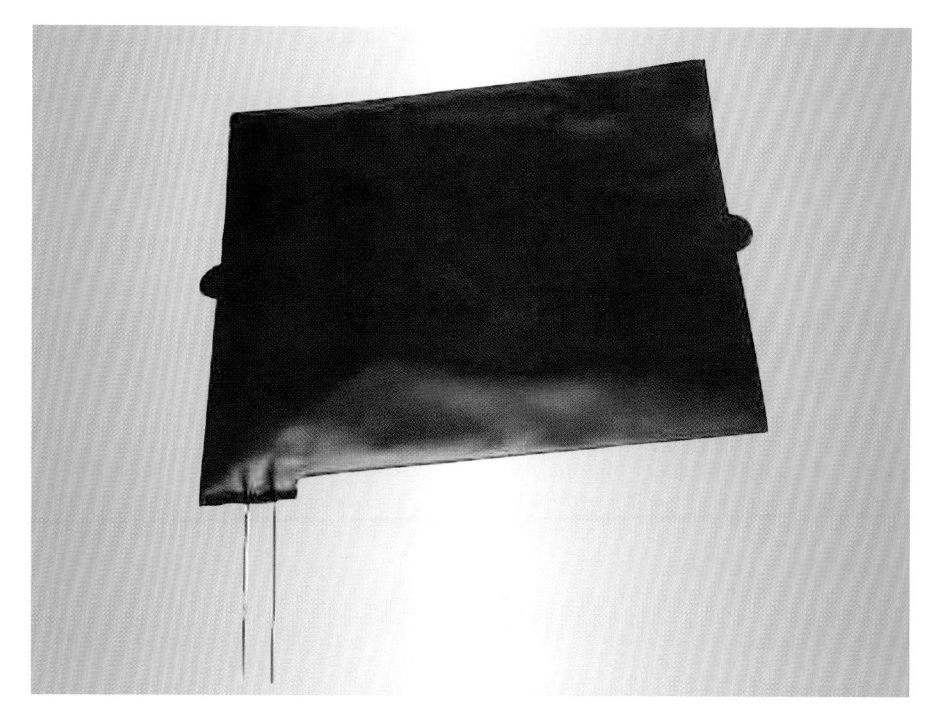

Figure 7.11 *A pressure pad is a large membrane switch that is closed by pressing anywhere on the pad. This kind of switch may be used in burglar alarms, hidden beneath a carpet, to detect if an intruder steps into a room.*

The resistors in these sensing circuits have *no* current flowing through them when the switches are open. The voltage, V, across a resistor is given by V = I × R so if the current I = 0 then V = 0. If the voltage across a resistor is zero then both ends of the resistor are at the *same* voltage. They are used to hold the output voltage of the sensor either high or low. If a simple wire link were to be used instead of a resistor, then pressing the switch in either circuit would cause a short circuit.

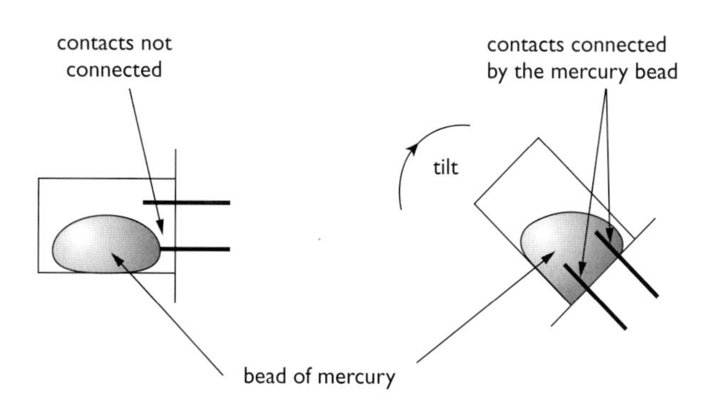

The mercury tilt switch has a bead of mercury inside a sealed capsule. The mercury forms a conducting path between two electrical contacts in the capsule. If the capsule is tilted, the mercury flows off the contacts, thus breaking the connection.

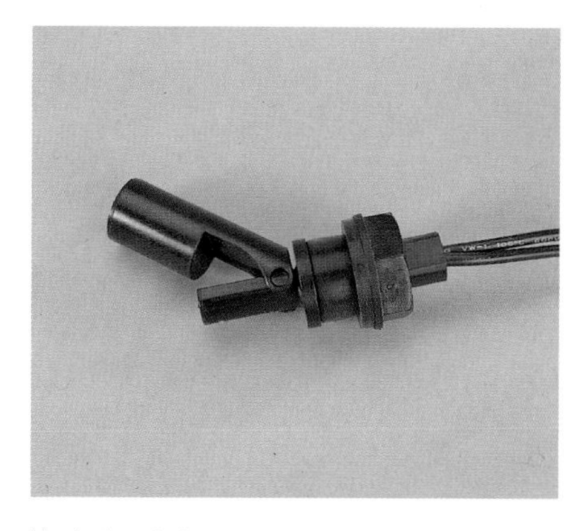

Figure 7.12 a) *The mercury tilt switch is also used in burglar alarms, to detect if things are moved or knocked over.* b) *A tilt switch.*

The potential divider

The sensing circuits we have seen so far have used switches. Switches can be either on (conducting/very low resistance) or off (non-conducting/very high resistance). We shall now look at some sensing circuits that use components whose resistance changes as some aspect of the environment changes. These sensing circuits are basically two resistors connected in series. This arrangement is called a **potential divider**.

The supply voltage V_S is shared between the two resistors in proportion to their resistances – the bigger resistor gets the bigger share of the voltage. We might call this potential divider circuit a "voltage sharer" circuit. The current that flows through each resistor is the same, because they are connected in series. We can calculate the current, I, using Ohm's law (see Chapter 2).

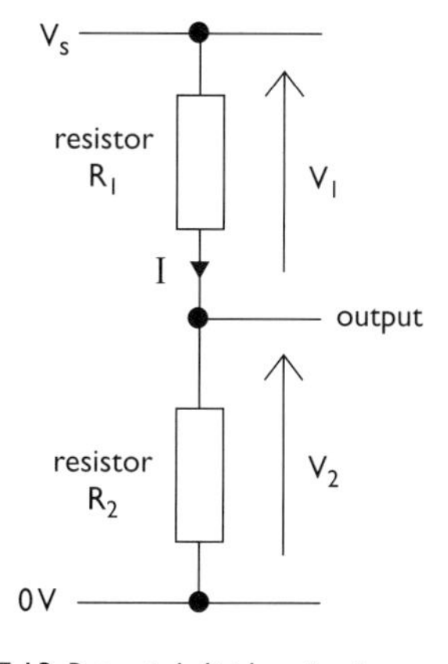

Figure 7.13 *Potential divider circuit.*

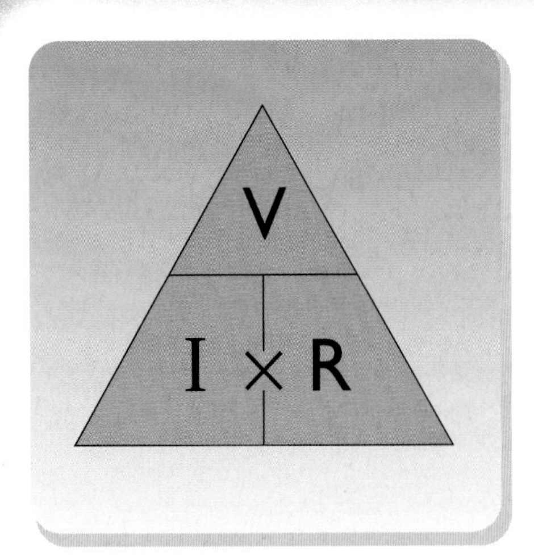

Figure 7.14 *Reminder of Ohm's Law.*

We use the total resistance for the resistors in series, $(R_1 + R_2)$, so:

$$I = \frac{V_s}{(R_1 + R_2)}$$

We could have calculated I using R_2 and the voltage across it, V_2:

$$I = \frac{V_2}{R_2}$$

This shows that:

$$\frac{V_s}{(R_1 + R_2)} = \frac{V_2}{R_2}$$

This can be rearranged to give a formula for the output voltage, V_2, of the potential divider:

$$V_2 = V_s \times \frac{R_2}{(R_1 + R_2)}$$

worked **example**

Example 1

What is the output voltage of the potential divider circuit shown in Figure 7.15?

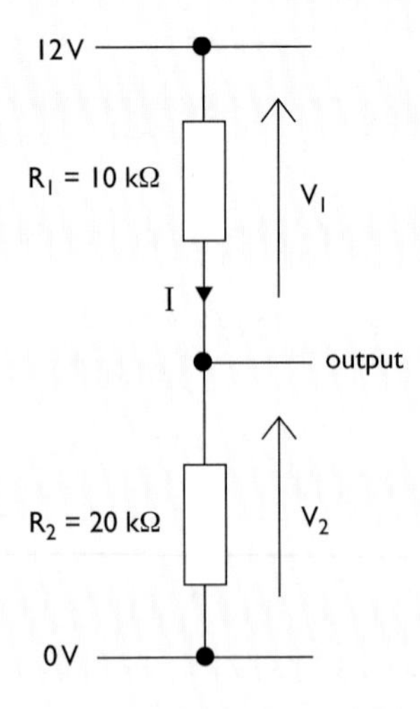

Figure 7.15.

$$V_2 = V_s \times \frac{R_2}{(R_1 + R_2)}$$

$$V_2 = 12\,V \times \frac{20\,k\Omega}{10\,k\Omega + 20\,k\Omega}$$

$$V_2 = 12\,V \times \frac{20\,k\Omega}{30\,k\Omega}$$

So $V_2 = 8\,V$

It is important to remember that the voltage is shared in proportion to the resistances in the circuit – the 20 kΩ resistor has a voltage of 8 V across it and the 10 kΩ has a voltage of 4 V across it. The two voltages add up to the supply voltage of 12 V in this example.

Light sensing circuit

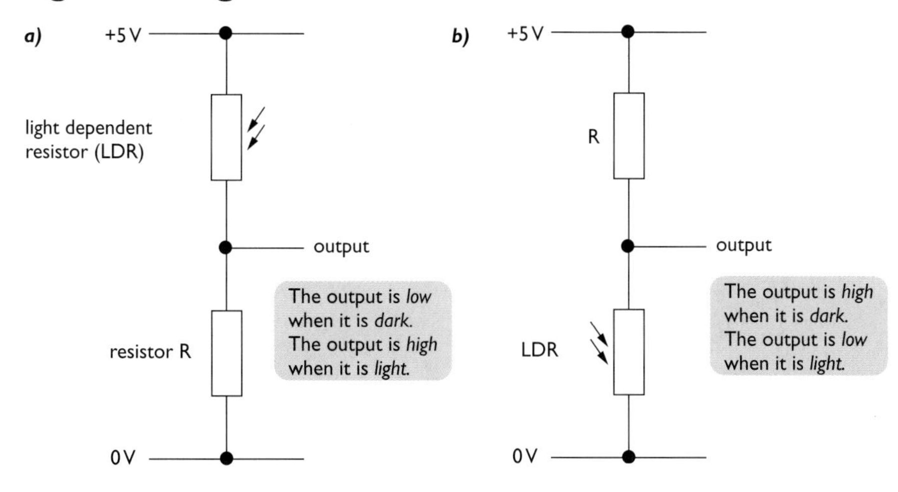

Figure 7.16 *Light sensing circuits.*

The circuits shown in Figure 7.16 use a light-dependent resistor or LDR (see page 17). The resistance of an LDR *decreases* as the light falling on it gets *brighter*; if it gets *darker*, the resistance of the LDR *increases*. The LDR and the resistor R form a potential divider circuit. Remember that the resistors share the supply voltage, with the bigger resistor getting the bigger share of the voltage. As the light falling on the LDR gets brighter, the output voltage of the circuit in Figure 7.16a increases (because the LDR resistance falls and so, therefore, does its share of the voltage). In the circuit in Figure 7.16b, as it gets darker the output voltage increases.

Temperature sensing circuit

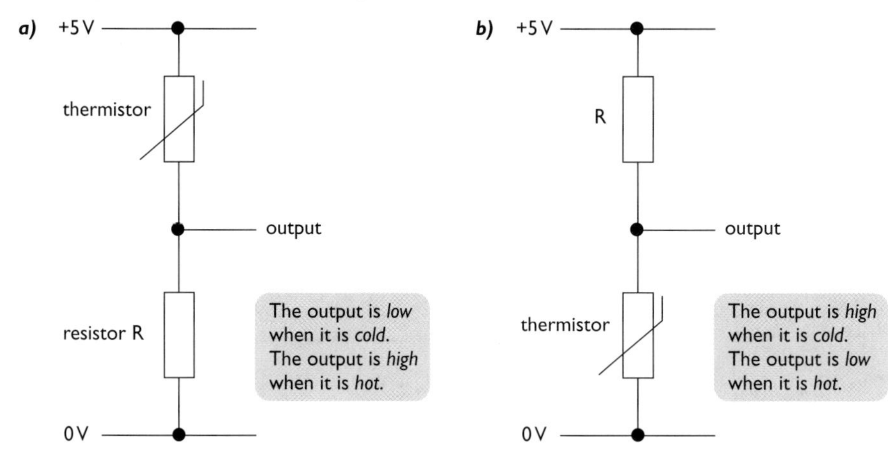

Figure 7.17 *Temperature sensing circuits.*

The circuits shown in Figure 7.17 use a **thermistor**. The resistance of a thermistor *decreases* as its temperature *increases*; if it gets *colder* the resistance of the thermistor *increases*. As in the previous circuits, the thermistor and the resistor R form a potential divider circuit. As the thermistor gets hotter, the output voltage of the circuit in Figure 7.17a increases. In Figure 7.17b, the thermistor and resistor are swapped. This means that as it gets colder the output voltage increases.

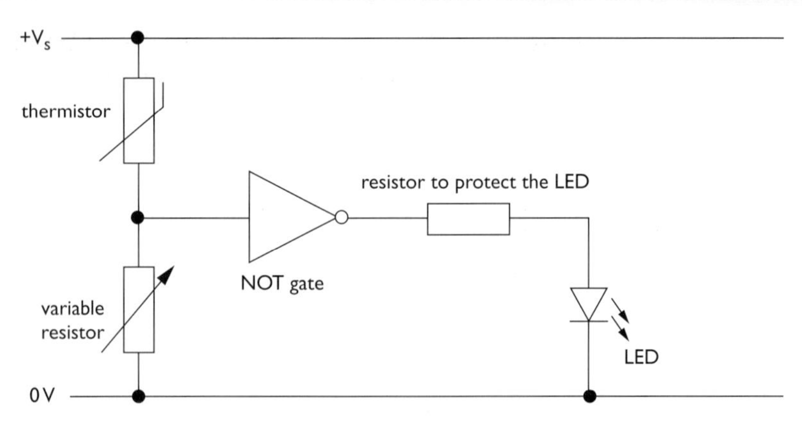

Figure 7.18 *This circuit can be used as an ice detector. The variable resistor can be adjusted so that the LED turns on as a warning when the temperature falls below 0°C.*

In the circuit in Figure 7.18, the sensor input has a variable resistor in series with the thermistor. This makes it possible to adjust the temperature at which the output of the NOT gate switches high and turns on the warning light-emitting diode, or LED (see page 18). So, the circuit could be adjusted to turn the LED on when the temperature falls to freezing point, 0°C.

Figure 7.19 a) *A simple water or moisture sensor is made using two separate interlocking sets of copper tracks on a printed circuit board.*

Moisture sensing circuit

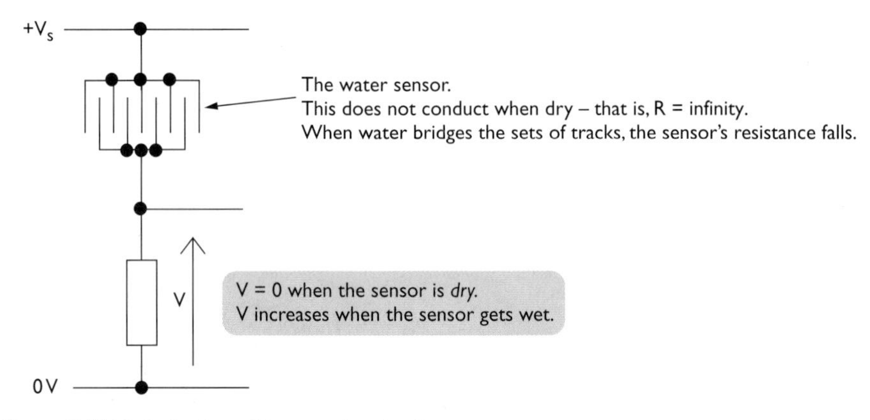

Figure 7.19 b) *A simple moisture sensing circuit.*

Figure 7.19b shows a moisture sensing circuit. The moisture sensor consists of two sets of interlocking copper tracks on an insulator. When the sensor is dry it does not conduct because its resistance is very high. Water usually contains impurities that make it conduct electricity. If water droplets bridge the gaps between the copper tracks the resistance of the moisture sensor is reduced.

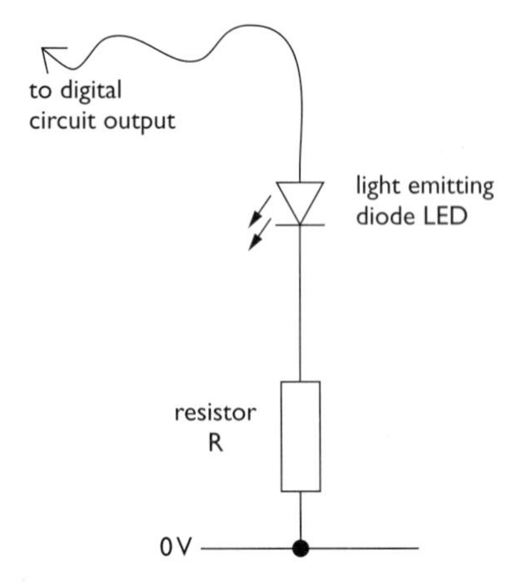

Figure 7.20 *A light-emitting diode (LED) with a protective resistor in series can be used as a simple logic probe.*

A simple output indicator

The circuit shown in Figure 7.20 is a simple way to show the logic level of the output of a digital system. If the output is low (logic level "0"), the light-emitting diode (LED) does not light. If the output is high (at or close to 5 V) then current will flow through the LED, making it light up. The series resistor is used to protect the LED from too much current. LEDs need about 2 V across them to light – if you connect 5 V across them they will burn out.

To work out the correct value of resistor to use in series with an LED you need to know the voltage and current at which the LED is designed to work.

worked example

Example 2

The LED in a logic probe operates at 2 V with a current of 10 mA (10 milliamps or 0.01 A) flowing through it (Figure 7.21). The circuit is run from a 5 V supply. What value of resistor should be connected in series with the LED?

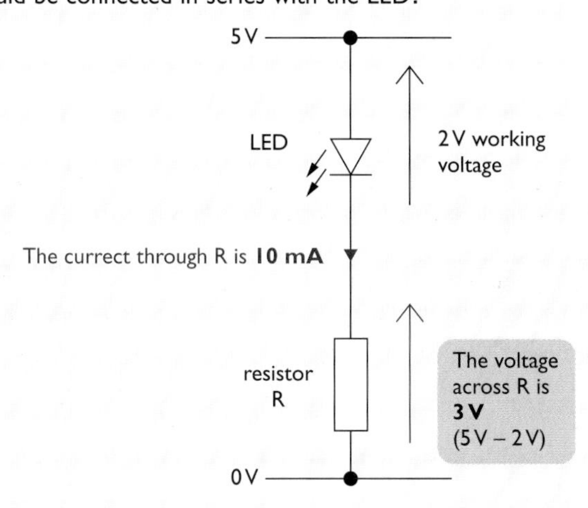

Figure 7.21.

Using Ohm's law:

$$R = \frac{V}{I}$$

The voltage across R is 5 V – 2 V = 3 V.

The current through R is 10 mA.

So:

$$R = \frac{3\,V}{10\,mA}$$

$$R = 300\,\Omega$$

The calculated value for the protective series resistor, R, is 300 Ω. In practice the nearest **preferred value** resistor would be used. This is 330 Ω. Preferred value resistors are explained in Appendix B, with an explanation of the **resistor colour code**. Notice that the nearest *larger* preferred value is chosen for R, as its purpose is to limit the current to a safe *maximum* value.

An example of a simple control circuit

Some control circuits have more than one sensor. The output from such a control circuit depends on the different conditions of its input sensors. This type of control circuit uses logic gates to process the input information. Here is an example of a control system with more than one input.

A gardener needs a system to let her know if the door of a green house has been left open at night. It will light a warning LED if the door is open *and* it is dark. The logic system required is clearly an AND gate. The inputs to the logic system must give a high signal when the door is open, and a high signal when it is dark. The circuit in Figure 7.22 shows how this can be achieved using the sensing circuits we have discussed.

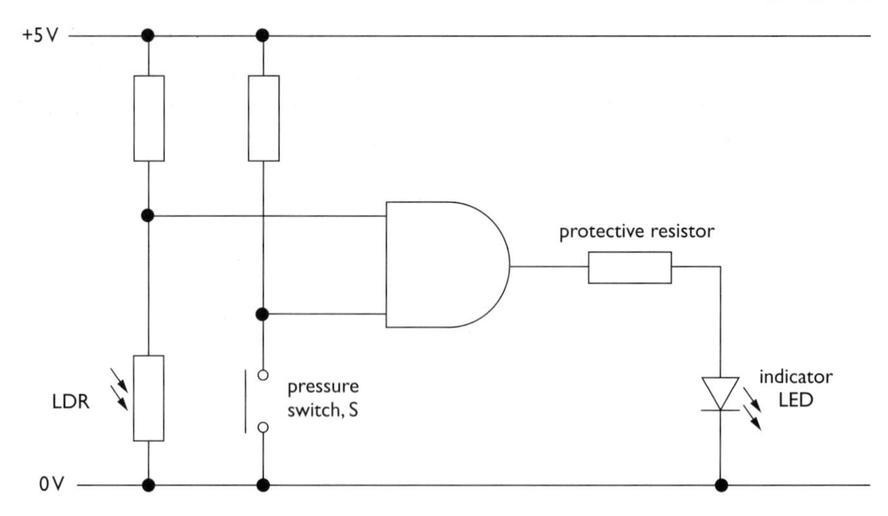

Figure 7.22 *This circuit senses when the door of the greenhouse has been left open in the dark.*

This circuit will light the LED if it is dark (so the LDR has high resistance) *and* the door is left open (so the switch, S, is not pressed), because this means both the inputs to the AND gate will be high causing the output to be high. However, this circuit might be much more useful if it could close the greenhouse door automatically. To do this we would need to make the control system operate a motor.

The relay

The logic gate ICs we use in digital control circuits have a limitation – they cannot supply much power to drive output devices. They output enough current to light an LED but not to operate a heater, a motor or any device that uses a lot of power. Logic gates cannot drive devices that use mains voltage either. The mains voltage would destroy the logic circuits.

To operate devices that need more power we use a device called a relay (Figure 7.23).

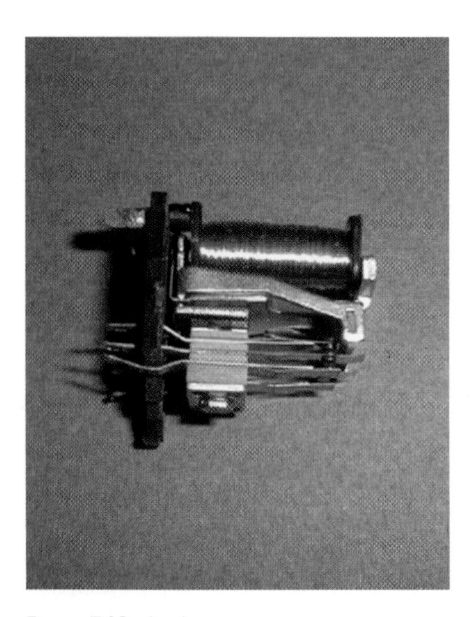

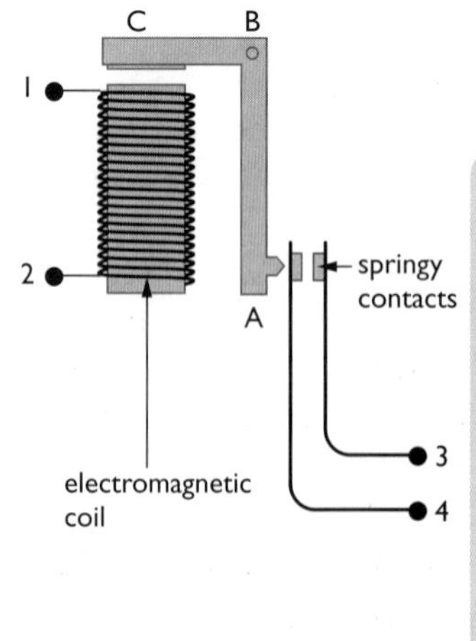

1 and 2 are the coil connections. When a current is passed through the coil it becomes magnetised. C is attracted to the magnetised coil. The lever ABC is pivoted at B. As C moves towards the coil, A moves to the right closing the pair of contacts. When the current through the coil stops flowing the coil demagnetises and the contacts spring apart.

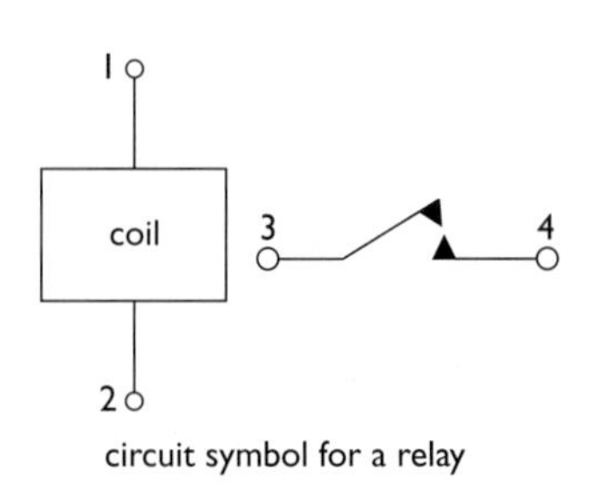

circuit symbol for a relay

Figure 7.23 *A relay.*

The relay has a coil of wire wrapped around an iron core. When a current is passed through the coil it magnetises the core. This attracts a lever that closes a switch. When the current stops flowing through the coil the lever is pulled back to its original position by a spring and the switch opens again.

Relays are operated by small currents but can switch much larger currents on and off with the switch. Relays also *isolate* high voltage circuits like mains heaters from the low voltage circuits used to control them.

Electronically controlled locks use another electromagnetic device called a solenoid. The solenoid is an electromagnet that is used to move something – the bolt in a lock, for example. Solenoids often need quite a large current to operate a bolt or catch, so these would need to be driven by a relay.

The relay shown in Figure 7.23 has a simple on–off switch. Relays with different arrangements of switches are also available.

Using a transistor switch circuit

Relays and solenoids cannot be driven directly from the output of a logic gate, as they need more current than the gate can supply. A **transistor** must be used to boost or amplify the current. Figure 7.24 shows a transistor used as a switching device to control a relay in a logic control circuit.

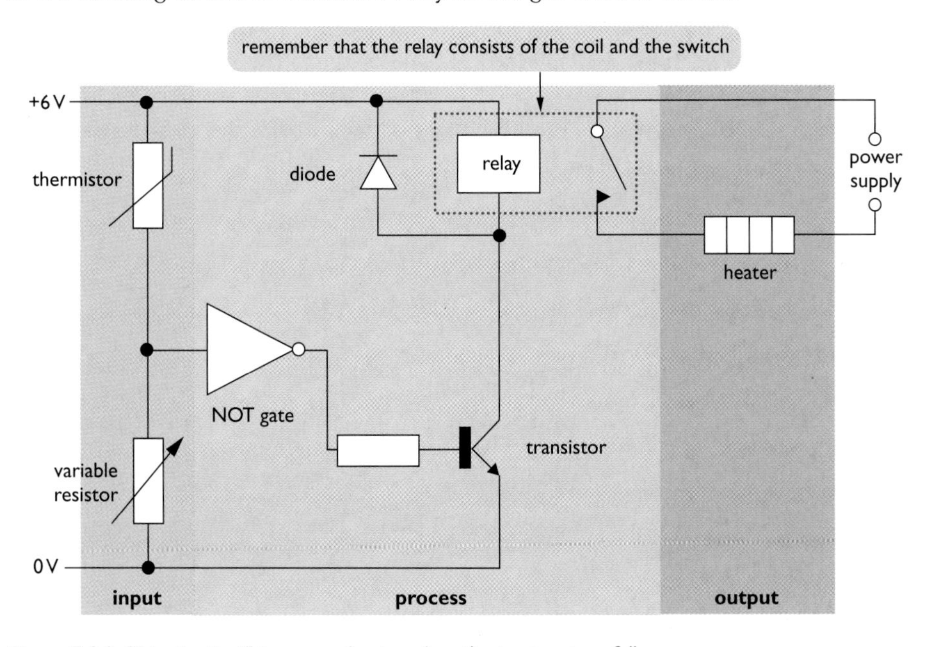

Figure 7.24 *This circuit will turn on a heater when the temperature falls.*

This circuit will turn a heater on when the temperature falls below a set level. The set level of temperature can be adjusted with the variable resistor. The NOT gate cannot supply enough current to work the relay. When the output of the NOT gate is high it switches the transistor on. When the transistor is on current flows through the relay coil and makes the relay switch close. The relay can switch on a high current and voltage device, like a heater, while keeping the low voltage part of the circuit separate from the high voltage part. A diode must be connected in parallel with the relay or solenoid to protect the transistor. Note the way that the diode is connected in parallel with the relay with the arrow pointing towards the positive side of the supply.

A timing circuit

Capacitors can be used in electronic systems to produce simple timers. Some capacitors and the circuit symbols for capacitors are shown in Figure 7.25.

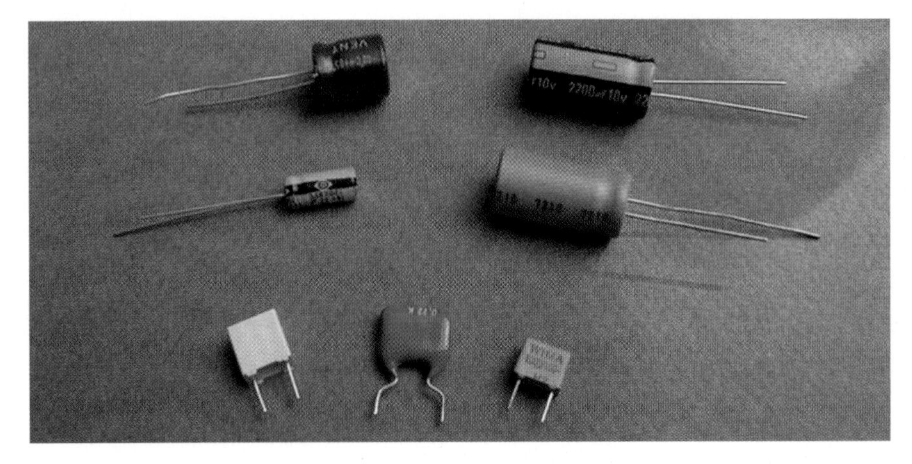

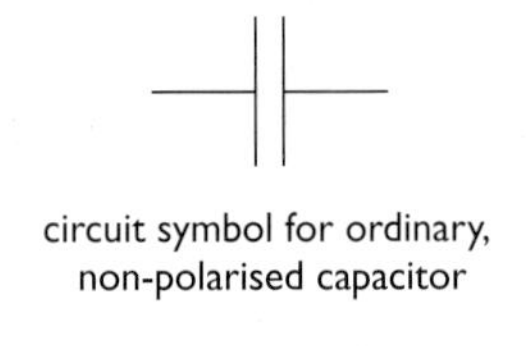

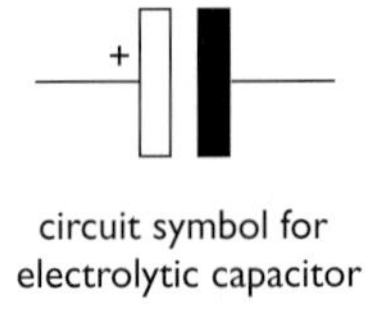

Figure 7.25 *Capacitors can be used in timing circuits.*

Capacitors store electric charge and energy. As they charge up, the voltage across them gets bigger. They stop charging when the voltage has reached the value of the supply voltage that is making them charge up. The time taken for a capacitor to charge up depends on the value of the capacitor (the capacitance of capacitors is measured in farads) and the size of the resistor it charges through. This is put to use in time delay circuits. A simple time delay circuit is shown in Figure 7.26.

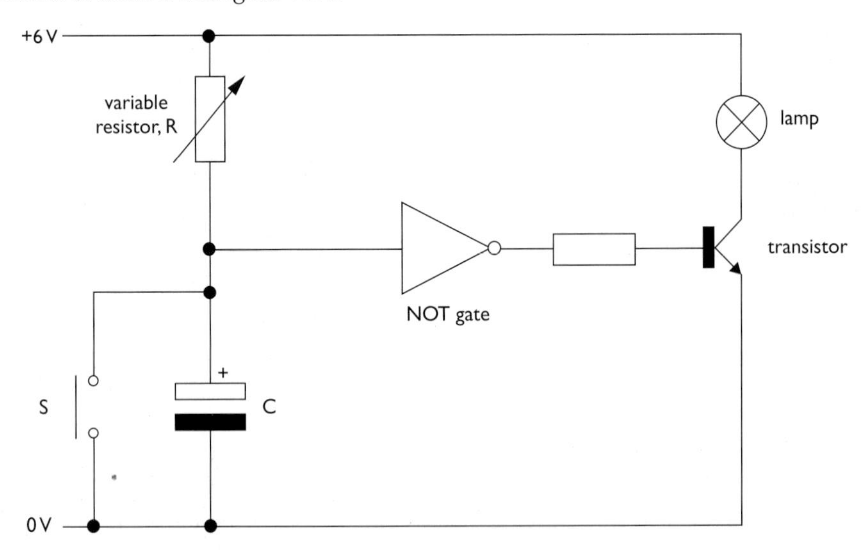

Figure 7.26 *A simple time delay circuit.*

If the circuit has been connected to the power supply for some time the capacitor is fully charged. This means that the input voltage to the NOT gate is high (about 5 V) and its output will be low (0 V). So the transistor will be off or non-conducting, and the lamp will be off. When the switch, S, is pressed the capacitor is discharged and the voltage across it falls to 0 V. The output of the NOT gate will be high, turning the transistor and, therefore, the lamp on. Releasing the switch allows the capacitor to charge up. The lamp will stay on for the time it takes for the voltage across the capacitor to reach the level needed to make the NOT gate output become low once more. Increasing either the capacitance of the capacitor, C, or the resistance of the resistor, R, increases the length of time the lamp stays on after S is pressed and released.

Electronic systems

Electronic systems can be used for very complex tasks, but even the most complex system can be broken up into three basic parts. A reminder of this is shown in Figure 7.27.

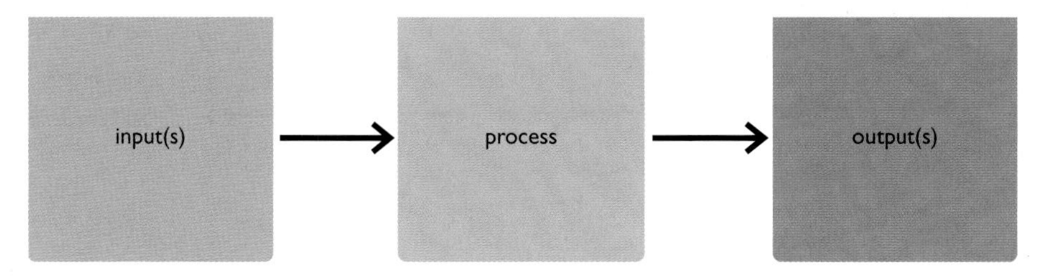

Figure 7.27 *All electronic control systems will have input(s), process and output(s).*

The **process** section uses the information from the **input** to decide what the system output should be. Figure 7.28 shows the circuit we looked at earlier (Figure 7.22) with these sections identified.

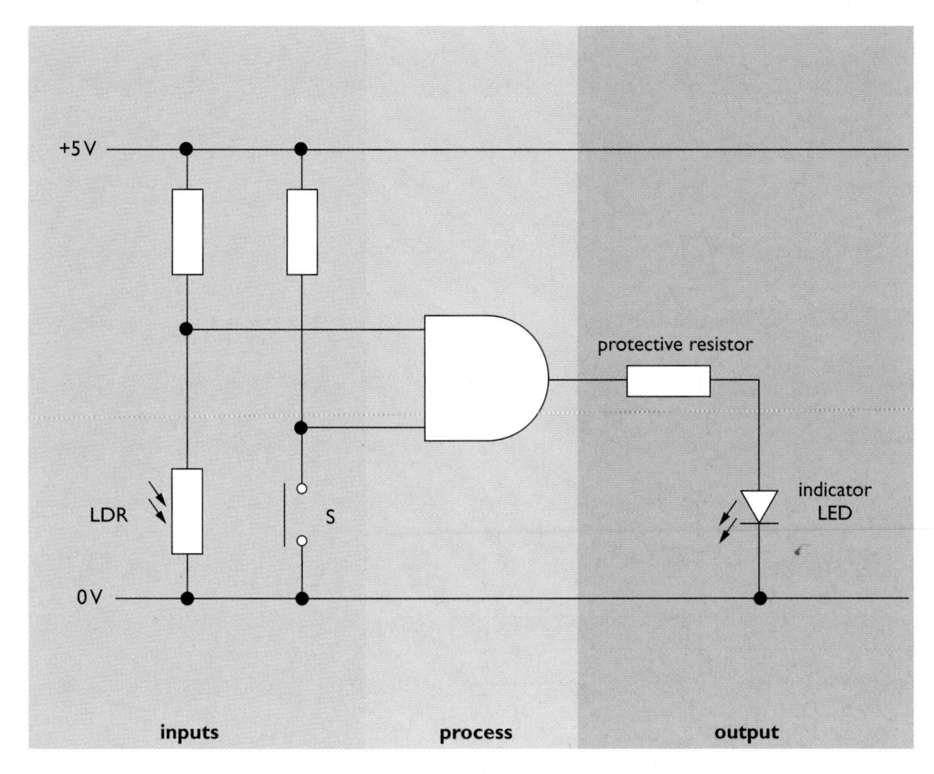

Figure 7.28 *This system can be divided up into input, process and output.*

The system inputs will often include sensors, like the ones we have discussed. The inputs may include continuous **feedback** information about the conditions under control. The information from the feedback is then used to modify the output of the system to ensure that the right conditions are met.

More about truth-tables

Working out the truth-table for a logic circuit

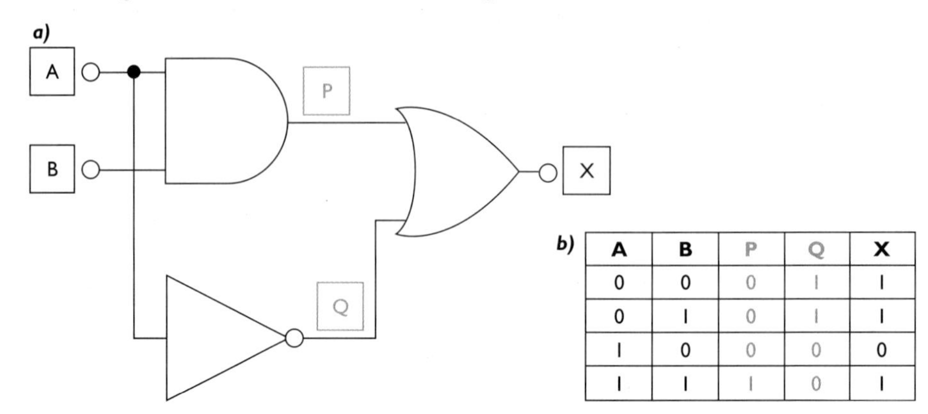

A	B	P	Q	X
0	0	0	1	1
0	1	0	1	1
1	0	0	0	0
1	1	1	0	1

Figure 7.29 *A two-input logic gate circuit, and truth-table.*

Figure 7.29a shows a circuit made up of three logic gates. It has two inputs, labelled A and B, and an output labelled X. The labels P and Q are used to show the intermediate logic levels at two points in the circuit to help make drawing the truth-table easier. The truth-table for the circuit is shown in Figure 7.29b.

The logic level at P is the output of the AND gate with inputs A and B. Therefore P is only 1 when *both* A *and* B are 1. The logic level at Q is the output of a NOT gate with input A, so when A is 1 Q is 0 and vice versa. The output X depends on the inputs P and Q to the final OR gate. If *either* P *or* Q is 1 then X will be 1.

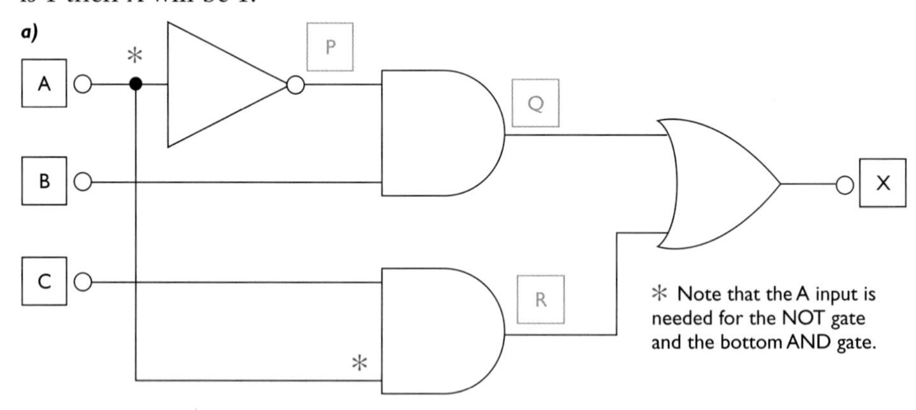

b)

A	B	C	P	Q	R	X
0	0	0	1	0	0	0
0	0	1	1	0	0	0
0	1	0	1	1	0	1
0	1	1	1	1	0	1
1	0	0	0	0	0	0
1	0	1	0	0	1	1
1	1	0	0	0	0	0
1	1	1	0	0	1	1

Explanation:

P (the NOT gate) changes the A input from 0 to 1 and vice versa.

Q is 1 when *both* B *and* P are 1.

R is 1 when *both* B *and* C are 1.

X is 1 when *either* Q *or* R is 1.

Figure 7.30 *A three-input logic system together with its truth-table.*

Logic systems can, and often do, have more than two inputs. Figure 7.30a shows a logic system with three inputs, A, B and C. As there are three inputs there are *eight* possible input combinations, as shown in the truth-table for the system in Figure 7.30b. Intermediate conditions P, Q and R have been included to help you work out the truth-table.

Making logic circuits given the required truth-table

Method 1

Sometimes we have to produce a logic circuit that behaves according to a given truth-table. To see how this is done we shall look at some examples. Figure 7.31 shows the truth-table for a model of the traffic light sequence.

inputs		outputs		
X	Y	R	A	G
0	0	1	0	0
0	1	1	1	0
1	0	0	0	1
1	1	0	1	0

The outputs are the *red*, *amber* and *green* lights of an ordinary set of traffic lights. The inputs can come from two simple switches. As the inputs are changed in the sequence shown, the traffic lights go through the order *red*, *red* and *amber*, *green*, then *amber*.

Figure 7.31 *Traffic light sequence model truth-table.*

Each of the traffic lights – red, amber and green – must turn on and off as set out in the truth-table. If you study the truth-table, looking at just one output at a time, you can often identify the required logic gate. This is shown in Figure 7.32a. The logic circuit for the green output is a little more complicated, requiring two logic gates. This is explained in Figure 7.32b. The complete solution for the traffic light system is given in Figure 7.32c.

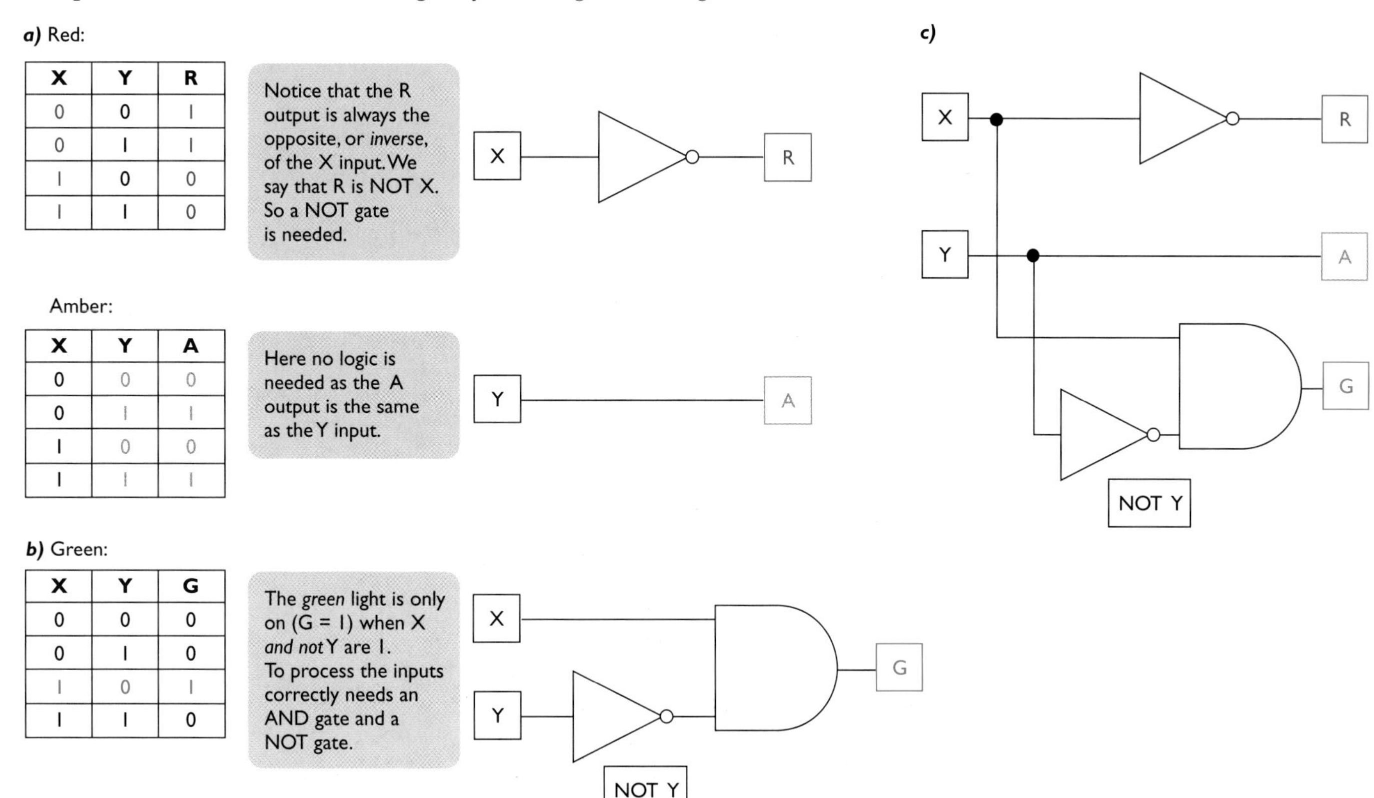

X	Y	R
0	0	1
0	1	1
1	0	0
1	1	0

X	Y	A
0	0	0
0	1	1
1	0	0
1	1	1

X	Y	G
0	0	0
0	1	0
1	0	1
1	1	0

Figure 7.32 a) *The logic circuits for the red and amber lights,* b) *the logic circuit for the green light,* c) *the complete solution.*

Now let us consider an example of a system with three inputs. Remember that this means that the truth-table has eight lines. We shall show how to produce a logic gate circuit for the truth-table shown in Figure 7.33a.

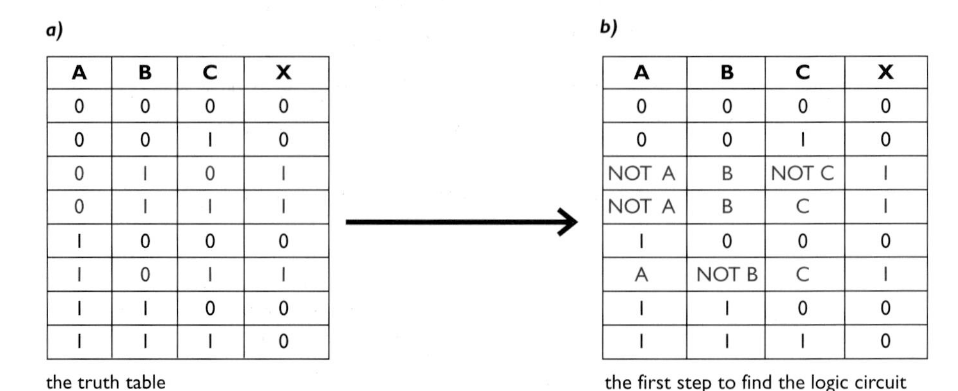

a)

A	B	C	X
0	0	0	0
0	0	1	0
0	1	0	1
0	1	1	1
1	0	0	0
1	0	1	1
1	1	0	0
1	1	1	0

the truth table

b)

A	B	C	X
0	0	0	0
0	0	1	0
NOT A	B	NOT C	1
NOT A	B	C	1
1	0	0	0
A	NOT B	C	1
1	1	0	0
1	1	1	0

the first step to find the logic circuit

Figure 7.33 a) *Truth-table for a three-input logic system,* b) *first stage of the solution.*

Figure 7.33b shows the first step we must take to produce a logic circuit for this truth-table. The lines where X = 1 have been picked out in red. If an input in one of these lines is 0, (for example, A in the first of the X = 1 lines) write NOT A. This has been done for each term in the three X = 1 lines. So, for the first of these lines X will be 1 *only when* NOT A AND B AND NOT C are 1. This step has been repeated for the other two lines where the output X = 1.

We can now write a complete word equation to describe the input conditions that will make the output X = 1:

X is 1 only when
(NOT A AND B AND NOT C)
OR (NOT A AND B AND C)
OR (A AND NOT B AND C)
are 1

Figure 7.34 shows you how to convert the first term (NOT A AND B AND NOT C) into a logic gate circuit.

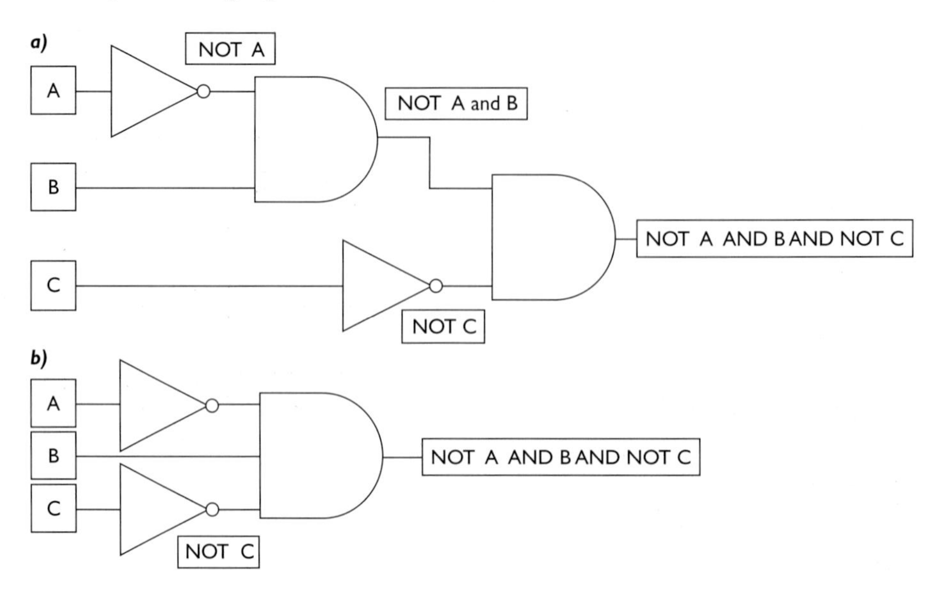

Figure 7.34 a) *Using two-input AND gates,* b) *using a three-input AND gate.*

Each of the parts of the word equation in brackets needs NOT gates and AND gates. The solution for the first part using AND gates with two inputs is shown in Figure 7.34a. Logic gates can have more than two inputs. Figure 7.34b shows the solution using a three-input AND gate. A three-input AND gate only has an output of 1 when *all three* of its inputs are 1.

The complete solution is shown in Figure 7.35, using three-input AND gates for simplicity. The other two bracketed parts of the word equation are produced in the same way as the first, and all three parts are then combined using a three-input OR gate. (You could use two two-input OR gates in the same way as shown with AND gates in Figure 7.34a.)

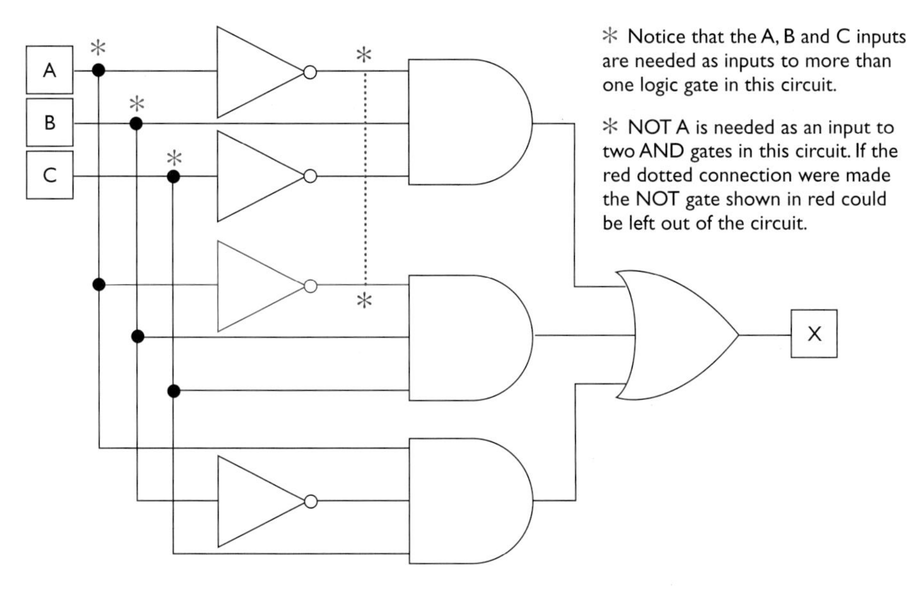

Figure 7.35 *The complete solution – the logic circuit that will produce the truth-table in Figure 7.33.*

It is possible to produce a simpler logic gate solution than that shown in Figure 7.35. However, this method will always enable you to get a correct answer, if not the simplest answer, to any problem like this.

Method 2

Our final example shows a different approach that involves spotting patterns. It often gives simpler solutions than the method described above. You can use the above method if you find it difficult to spot the patterns in the truth-table. The truth-table for the system in this example is shown in Figure 7.36.

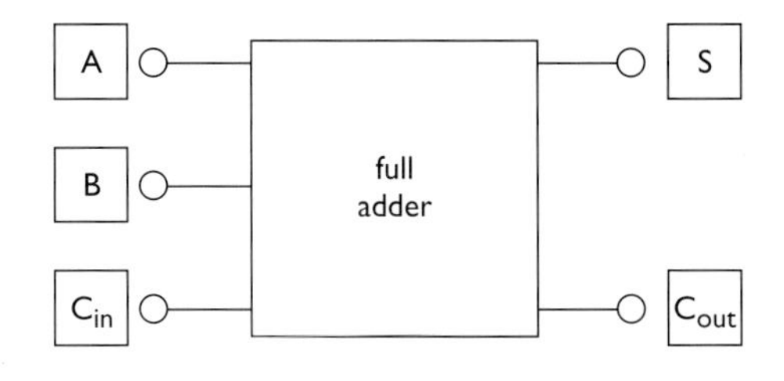

inputs			outputs	
A	B	C_{in}	S	C_{out}
0	0	0	0	0
0	0	1	1	0
0	1	0	1	0
0	1	1	0	1
1	0	0	1	0
1	0	1	0	1
1	1	0	0	1
1	1	1	1	1

C_{in} is the digit that is carried in from the previous column of the addition. S is the sum for that column and C_{out} is the carry digit that must be added in to the next column in the addition. The use of 'carry in' and 'carry out' is the same as in the addition of decimal numbers.

Figure 7.36 *Full-adder system and its truth-table.*

The outputs are the sum, S, and the carry, C_{out}, that result from the addition of three binary digits, or "bits", A, B and C_{in}.

The system is called a full-adder circuit. Full-adder circuits are used in microprocessors to add binary numbers together. You do not need to know about full-adders and microprocessors, but you may need to produce a logic circuit for an eight line truth-table.

Producing a logic gate circuit for this truth-table is more complicated. However, if you follow some simple rules you will be able to do it quite easily. The solution for the C_{out} output and the circuit is shown in Figure 7.37. You may notice that a three-input AND gate could be used to simplify the circuit.

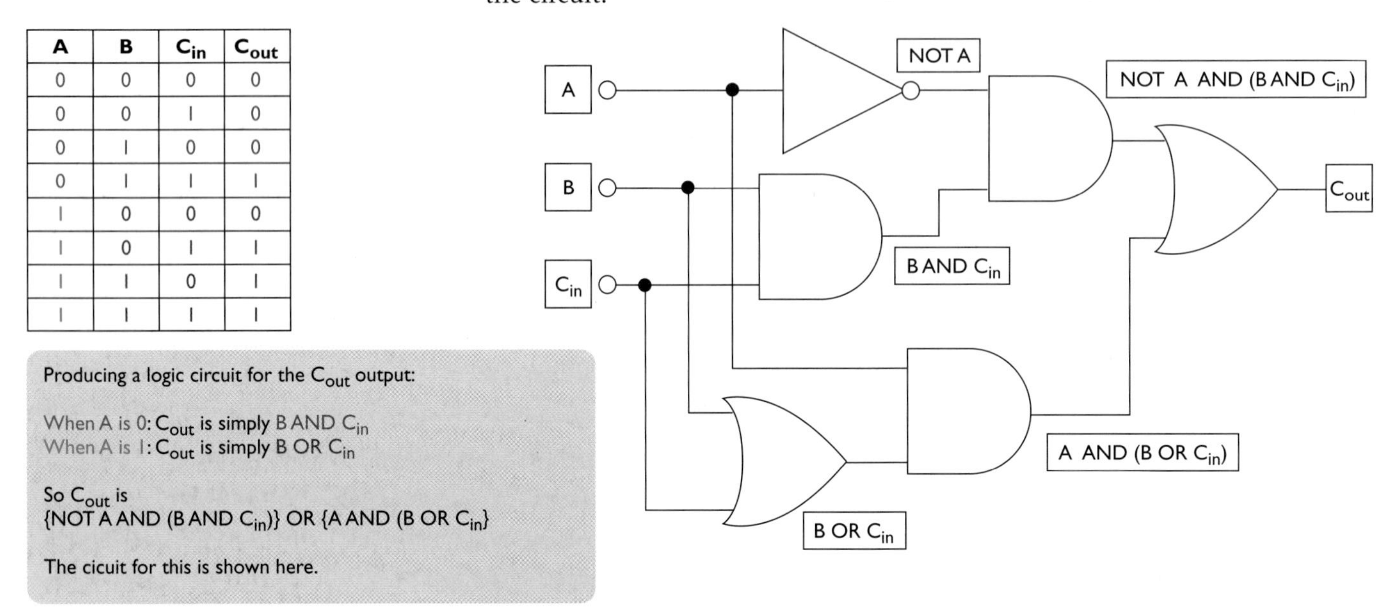

A	B	C_{in}	C_{out}
0	0	0	0
0	0	1	0
0	1	0	0
0	1	1	1
1	0	0	0
1	0	1	1
1	1	0	1
1	1	1	1

Producing a logic circuit for the C_{out} output:

When A is 0: C_{out} is simply B AND C_{in}
When A is 1: C_{out} is simply B OR C_{in}

So C_{out} is
{NOT A AND (B AND C_{in})} OR {A AND (B OR C_{in}}

The cicuit for this is shown here.

Figure 7.37 *Solution and logic circuit for the C_{out} output of the full-adder.*

The solution for the S output is left for you to do as an exercise.

Two more logic gates

Two more logic gates are widely used in digital electronic circuits. These are the **NAND** and the **NOR** gate. We shall learn the symbols and truth-tables for each and then look at a useful circuit that is made using them.

The NAND gate

The symbol and truth-table for a NAND gate are shown in Figure 7.38.

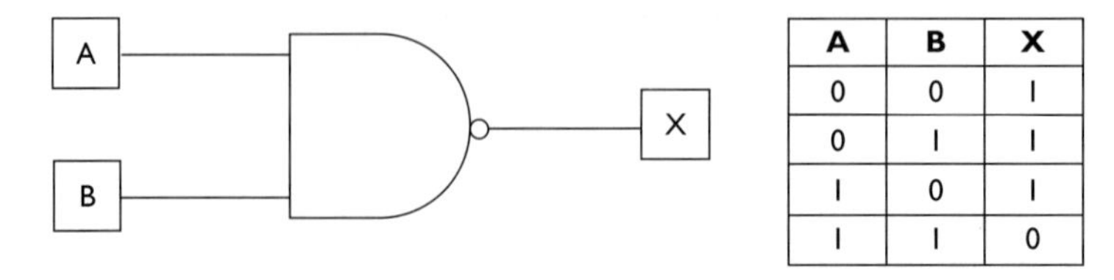

A	B	X
0	0	1
0	1	1
1	0	1
1	1	0

Figure 7.38 *The symbol and truth-table for a NAND gate.*

The NAND gate behaves in the opposite way to an AND gate. The AND gate output is only 1 when *both* A *and* B are 1 – the NAND gate output is only 0 (NOT 1) when *both* A *and* B are 1. You may find it helpful to think of a NAND gate as an AND gate followed by a NOT gate, as shown in Figure 7.39.

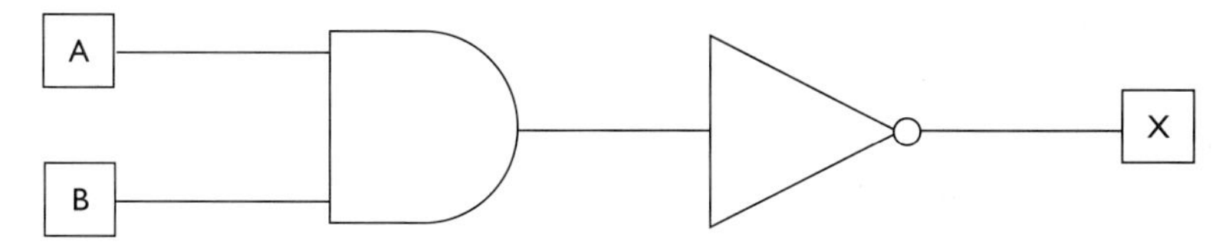

Figure 7.39 *A NAND gate behaves like an AND gate followed by a NOT gate.*

The NOR gate

The symbol and truth-table for a NOR gate are shown in Figure 7.40.

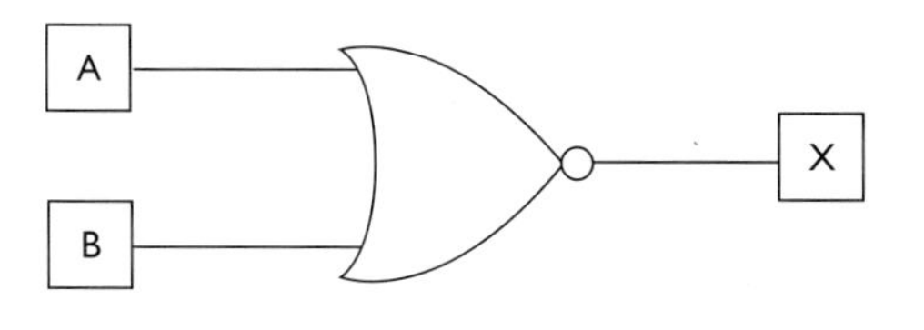

A	B	X
0	0	1
0	1	0
1	0	0
1	1	0

Figure 7.40 *The symbol and truth-table for a NOR gate.*

The NOR gate behaves in the opposite way to an OR gate. The OR gate output is 1 when *either* A *or* B are 1 – the **NOR** gate output is 0 (NOT 1) when *either* A *or* B are 1. You may find it helpful to think of a NOR gate as an OR gate followed by a NOT gate, as shown in Figure 7.41.

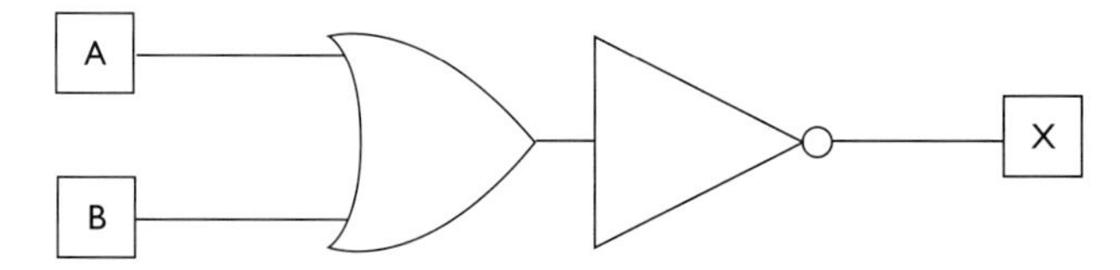

Figure 7.41 *A NOR gate behaves like an OR gate followed by a NOT gate.*

The bistable latch

NAND gates (or NOR gates) can be connected to make a circuit called the **bistable latch**. This circuit has two stable output conditions. We can set the output in one of these conditions and it will stay in this condition until we change, or reset, the circuit. The circuit "remembers" its output is either 1 or 0 until we want to change it. The bistable latch is the basic unit of memory in digital computers. The NAND gate version of the bistable latch is shown in Figure 7.42.

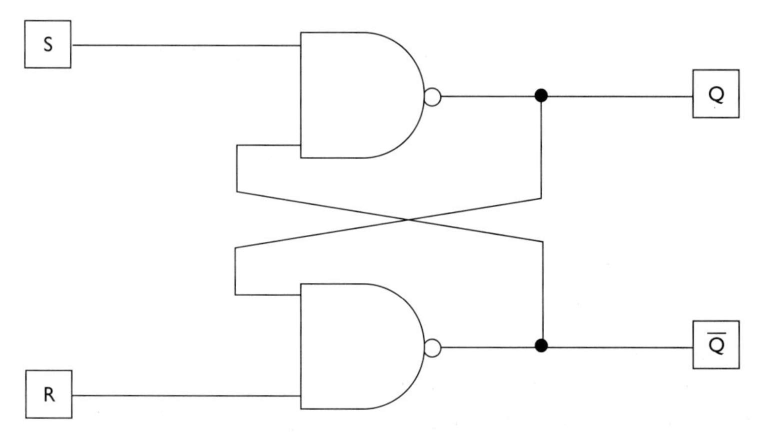

Figure 7.42 *A bistable latch circuit made using NAND gates.*

The bistable circuit has two inputs, S and R, and two outputs, $\overline{Q}$ and Q. The inputs are SET and RESET; you will see why they are called these in a moment. The bar over the second output means that it is always the *opposite* of the first output – when $\overline{Q}$ is 1, Q is 0 and vice versa.

In the NAND version of the bistable latch the two inputs, S and R, are held at 1 (logic 1). The SET and RESET operations involve making *either* S *or* R logic 0 for a brief instant, but *never both at the same time*. As we explain how the circuit works it is important to remember the truth-table for a NAND gate – if either input to a NAND gate is 0 then its output must be 1. Figure 7.43 shows setting the Q output to 1.

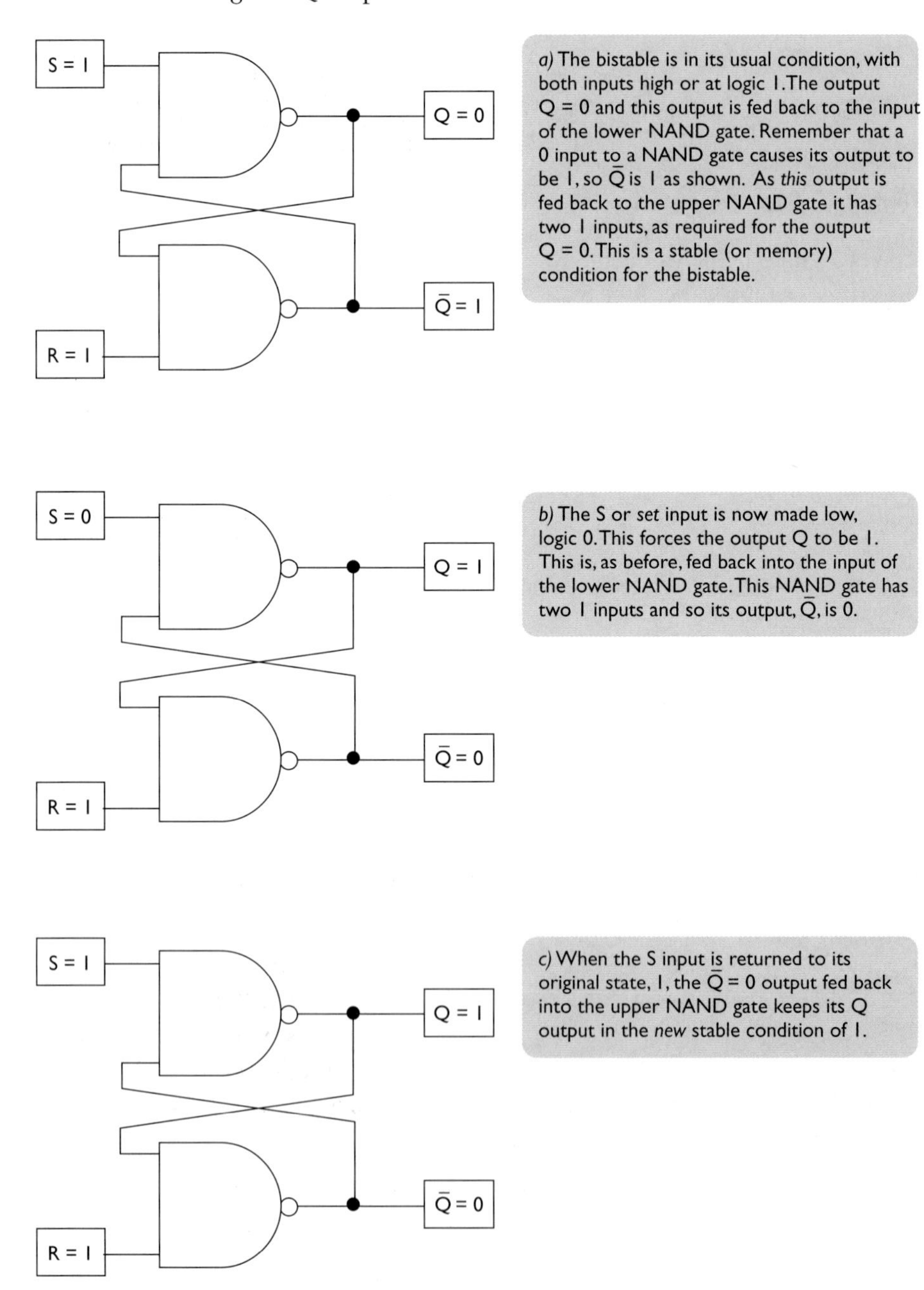

Figure 7.43 *Setting Q to 1*

The sequence shown in Figure 7.44 shows how bringing the R or RESET input low (logic 0) resets the Q output to 0.

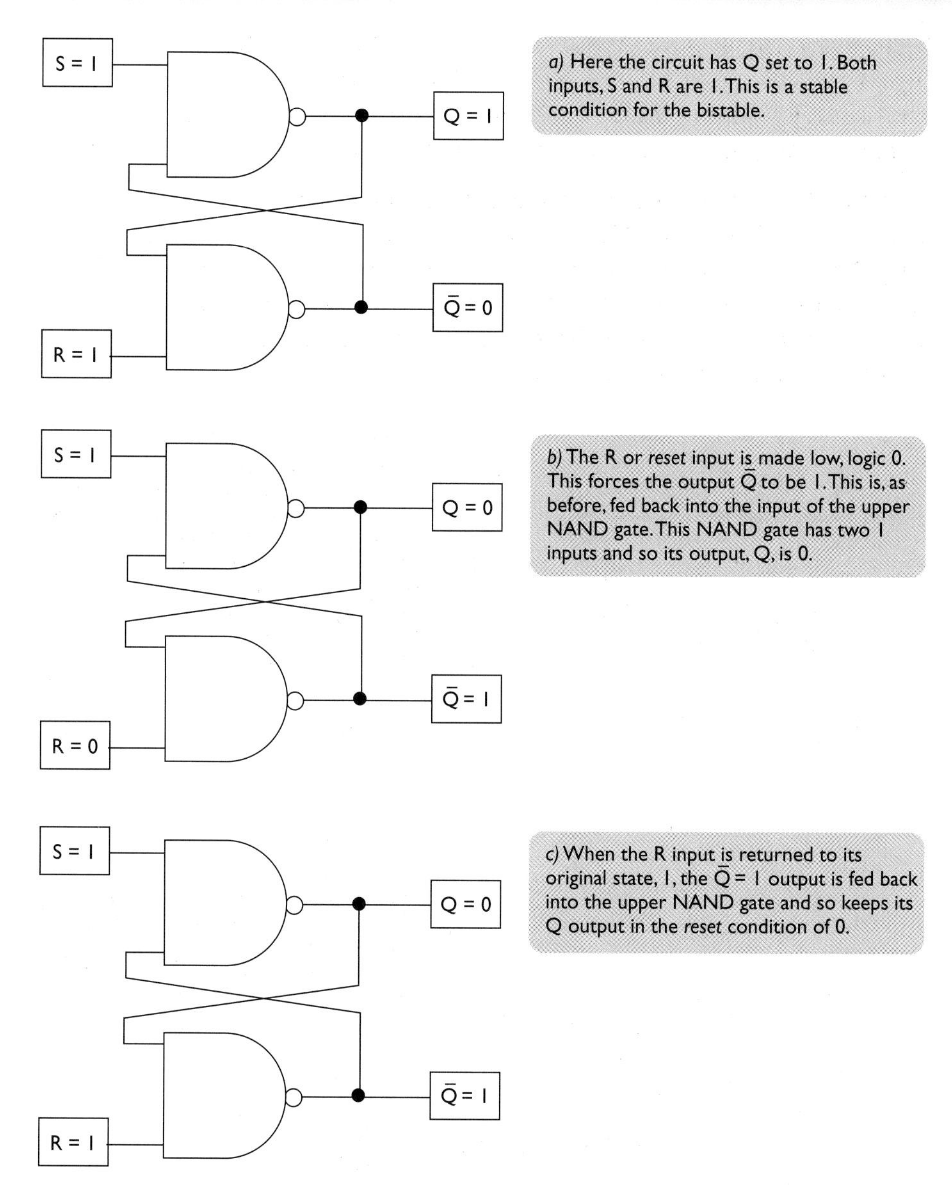

Figure 7.44 *Resetting Q to 0.*

The bistable latch can be made in a similar way from two NOR gates. The circuit is shown in Figure 7.45.

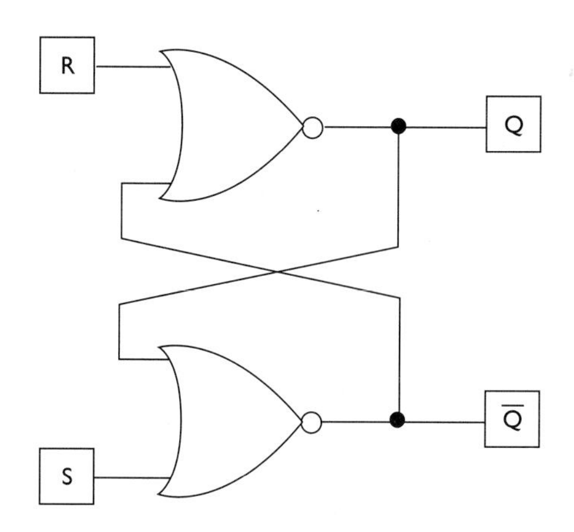

Figure 7.45 *A bistable latch made using NOR gates.*

The key point to remember with the NOR gate version of the bistable latch is that if either input of a NOR gate is 1 then its output will be 0. The normal input condition for a NOR gate bistable latch is with both S and R low, at logic 0. Making the S or SET input briefly high (logic 1) causes Q = 1 and making the R or RESET input briefly high (logic 1) causes Q = 0. Both output conditions are stable when S = R = 0.

Note the differences between the NOR and NAND gate versions of the bistable latch:

- the S and R inputs are the other way round in the NOR version
- the S and R inputs are normally held at 0 in the NOR version
- the set and reset operations are achieved by making the required input momentarily high (logic level 1) in the NOR version.

ideas
evidence

Electronics in society

Electronics influences nearly every part of life in modern society. The number of applications of electronics has multiplied at an amazing rate since the invention of the transistor in 1947. As with all managed developments in science, we must be aware of the disadvantages as well as the advantages of electronics. Here are some of the issues of concern about the increase in use of electronics in our lives.

CCTV

Whenever you go into a bank or a building society you will be on TV! All such places use CCTV (closed circuit television) to film and record people. In the event of a robbery the police can use the information collected in this way to track and catch the criminals. None of us would consider this to be a bad thing (apart from the criminals!). CCTV is now used in many other public places for similar reasons. Some people are concerned that this represents an invasion of privacy. Information is gathered about our movements and we are not aware that it is happening most of the time. In some countries this information could be used to track individuals who disagree with the government of the country and used to suppress their rights as individuals. We should be aware of the issue and make sure that people are able to protect their privacy as individuals without losing the security that CCTV can bring to society.

Mobile phones

Another major advance in electronics has allowed millions of people the convenience of a mobile phone. Again the benefits are obvious. Apart from the simple convenience of being able to talk to our friends whenever we want to, mobile phones can provide life-saving links to the emergency services. The possible drawbacks of this new technology include the health risk to mobile phone users and the increase in muggings and thefts that mobile phones have caused. The health risk to users is uncertain but should not be ignored. Further advances in electronics will, undoubtedly, make a mobile phone useless in the hands of anyone but its rightful owner.

Internet

The Internet has meant that people all over the world can have access to all kinds of information at any time – it provides us with an enormous database that can be rapidly accessed. A problem with the Internet is that the information available is not regulated in anyway. This means that the information may be wrong or offensive. It is possible for people to deliberately mislead others or to post information onto the world wide web that you would not want to see or allow young children to see. Another potential problem with the Internet is the ease with which viruses can be spread from computer to computer in a very short time.

As with any new advance in science and technology we should welcome the benefits of electronics, but be aware of the possible disadvantages. With planning and foresight the advantages can outweigh the disadvantages.

End of Chapter Checklist

If you haven't got a copy of your specification, read the introduction on page iv.

You will need to be able to do some or all of the following. Check your Awarding Body's specification (syllabus) to find out exactly what you need to know.

- Know that a digital signal can have one of two possible voltages: high (logic 1) or low (logic 0).
- Understand that the output of a logic gate will be high or low depending on the condition of its input signals.
- Recall the truth-tables for AND, OR, NOT, NAND and NOR logic gates.
- Know how to use switches (including magnetic, tilt and pressure switches), LDRs, thermistors and moisture sensors in series with resistors to provide input signals for logic gates and logic gate systems.
- Calculate the output voltage of potential dividers made from resistors, LDRs and thermistors.
- Know how to use an LED with a protective series resistor as an output indicator for logic gates.
- Appreciate that electronic control systems consist of input, processor and output stages and be able to identify these stages in a control circuit diagram.
- Know why relays are necessary to switch high power or high voltage loads on and off.
- Understand the need for transistor switches as output drivers for logic gate circuits.
- Appreciate the use of capacitors in simple timing circuits.
- Draw truth-tables for logic gate systems with up to three inputs.
- Design logic gate circuits to perform according to a given truth-table with up to eight rows (three inputs).
- Know how to use NOR and NAND gates to make bistable latch circuits and understand how they operate.
- Know how to use a variable resistor in a potential divider circuit to adjust the operating level of the output.

Questions

More questions on electronics and control can be found at the end of Section A on page 80.

1 Copy and complete the following sentences.

a) The resistance of a thermistor _________ as its temperature falls.

b) LDR stands for _________ _________ _________.

c) The resistance of an LDR decreases as the brightness of the light falling on it _________.

2 Explain how the output voltage in the following circuits changes:

a) as the LDR is covered up

b) as the thermistor is warmed

c) as the resistance of the variable resistor is increased.

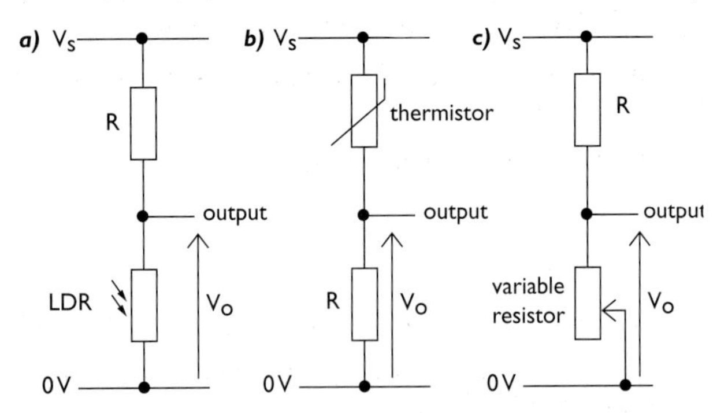

3 **a)** Copy and complete the following logic circuit to demonstrate how an LED can be used to show the output condition of the logic gate.

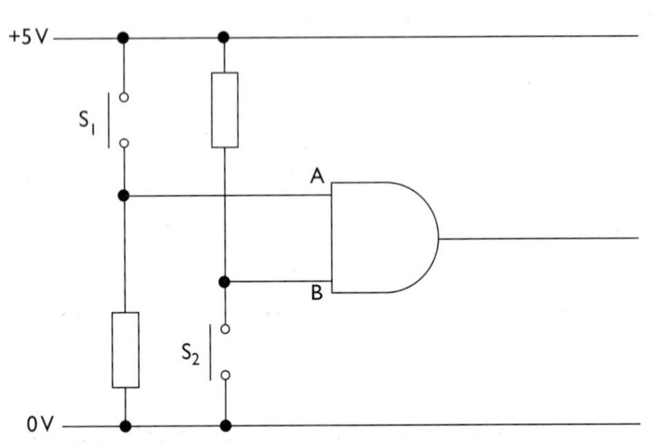

b) Name the logic gate shown in the circuit and draw its truth-table.

c) State the output condition of the logic gate if both the switches in the circuit are closed.

4 The circuit below is part of a control circuit for a mains powered electric fan.

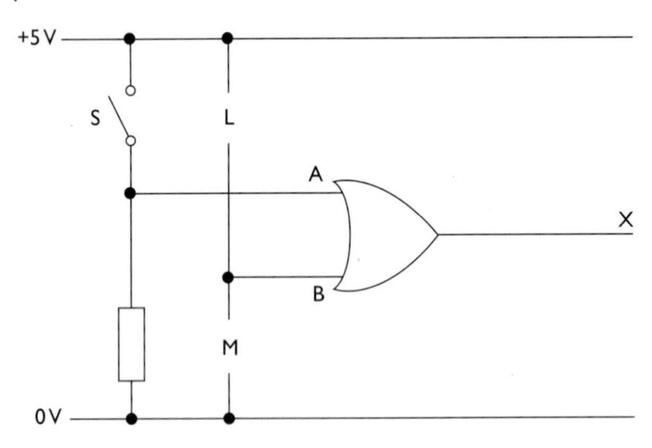

a) Draw a truth-table for the logic gate in the circuit.

b) Copy the circuit showing:

i) a suitable device to connect to the output, X, of the logic gate to operate the electric fan

ii) components in positions L and M to complete the circuit so that the fan switches on when the temperature reaches a certain level.

c) How could you make it possible to adjust the temperature at which the fan switches on?

d) What is the effect of closing the switch, S, in the circuit?

5 Draw the symbols and truth-tables for:

a) a NAND gate

b) a NOR gate.

6 Copy and complete the truth-table for the logic gate system shown below.

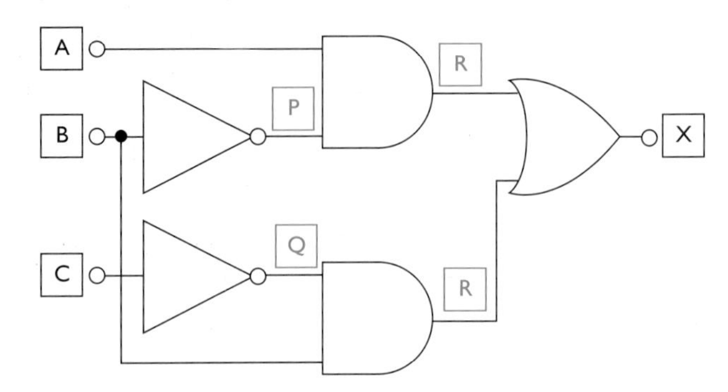

A	B	C	P	Q	R	S	X
0	0	0					
0	0	1					
0	1	0					
0	1	1					
1	0	0					
1	0	1					
1	1	0					
1	1	1					

Reminder: In this truth-table we are using "0" to represent low voltage inputs and "1" to represent high voltage inputs.

7 Design logic gate circuits for the following truth-tables (each requires the use of one NOT gate and one AND gate).

a)

A	B	Q
0	0	0
0	1	0
1	0	1
1	1	0

b)

A	B	Q
0	0	0
0	1	1
1	0	0
1	1	0

8 The truth-tables below have eight lines and require three inputs. Design logic gate circuits that behave according to each table. You will need to use NOT gates, two three-input AND gates and a two-input OR gate for each circuit.

a)

A	B	C	Q
0	0	0	0
0	0	1	0
0	1	0	1
0	1	1	0
1	0	0	0
1	0	1	1
1	1	0	0
1	1	1	0

b)

A	B	C	Q
0	0	0	1
0	0	1	1
0	1	0	0
0	1	1	0
1	0	0	0
1	0	1	0
1	1	0	0
1	1	1	0

9 Simplify the logic gate circuit for question **8 *b)*** to just two NOT gates and one two-input AND gate.

10 The diagram below shows a bistable latch circuit.

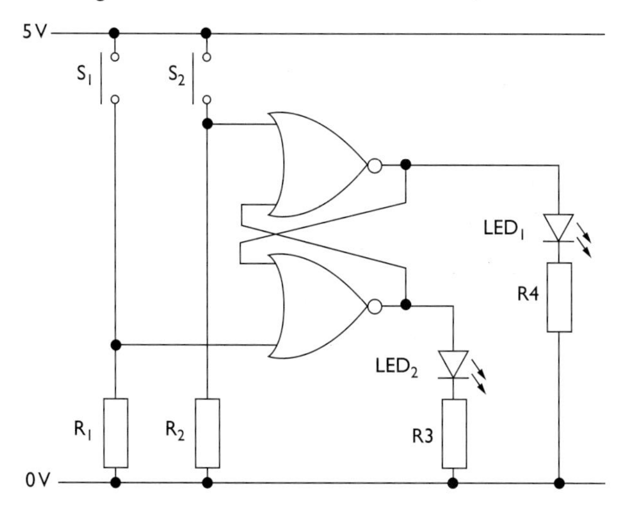

a) Draw the truth-table for a NOR gate.

b) With switches S_1 and S_2 open, as shown, and LED_1 off, what must the condition of LED_2 be?

c) What is the purpose of the resistors R_3 and R_4 in this circuit?

d) What happens when S_1 is

i) pressed

ii) then released?

e) What effect does pressing S_2 have on the outputs?

11 Work out the output voltages of the potential dividers shown below.

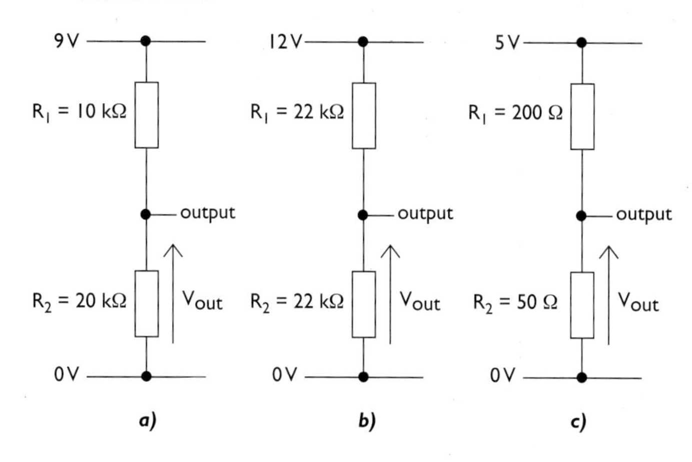

12 The Internet is a widely used tool in schools. Give two advantages and two disadvantages of the use of the Internet in education.

End of Section Questions

1 Bill set up the circuit shown below to investigate how the resistance of a bulb changes as the current flowing through it changes.

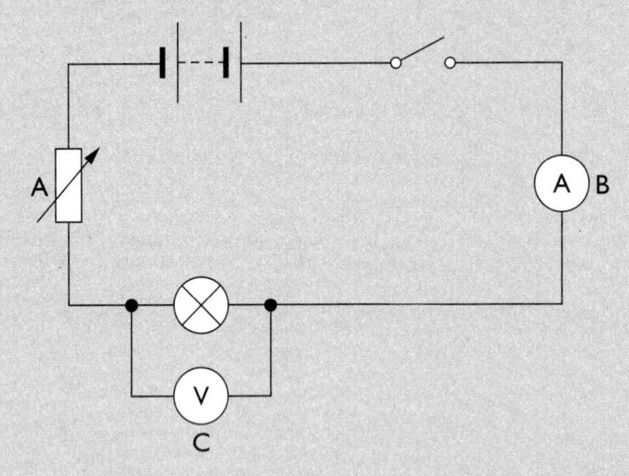

a) What are the names of the instruments labelled B and C? *(2 marks)*

b) What is the name of the component labelled A? *(1 mark)*

c) What is the purpose of A in this circuit? *(1 mark)*

d) What energy transfer takes place as current flows through the bulb? *(1 mark)*

Bill takes a series of readings. He measures the voltage across the bulb and the current passing through it. He then plots the graph shown below.

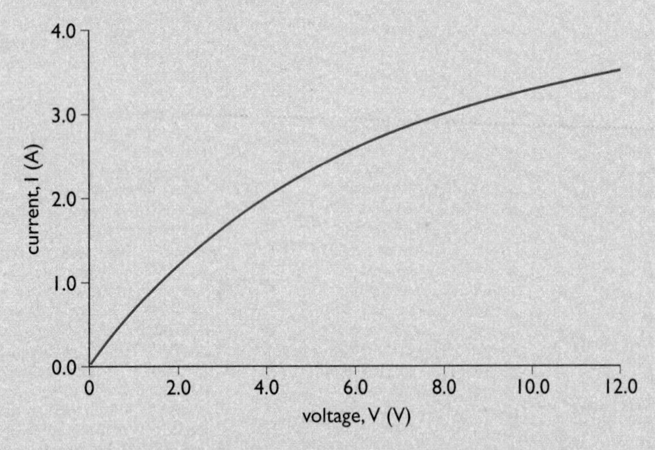

e) What current passes through the bulb when a voltage of 6 V is applied across it? *(1 mark)*

f) What voltage is applied across the bulb when a current of 2.0 A passes through it? *(1 mark)*

g) Calculate the resistance of the bulb when a current of 2.0 A passes through it. *(2 marks)*

h) What happens to the resistance of the bulb as the current passing through it increases? *(1 mark)*

Total 10 marks

2 A simple series circuit containing a 12 V battery, a 4 Ω resistor and a 6 Ω resistor was constructed as shown below.

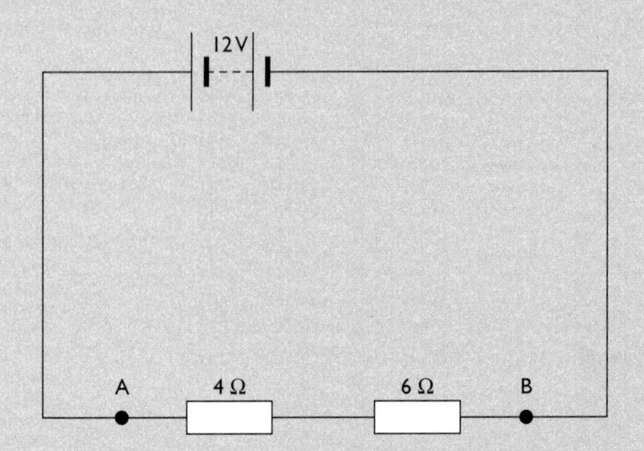

a) Calculate the resistance between points A and B. *(2 marks)*

b) Calculate the current that flows between points A and B. *(2 marks)*

c) Calculate the total charge that flows between A and B in 5 s. *(2 marks)*

d) Calculate the voltage across the 4 Ω resistor. *(2 marks)*

e) Calculate the rate at which energy is dissipated by the 4 Ω resistor. *(2 marks)*

Total 10 marks

3 The diagram below shows how the electrical energy is transmitted from a power station to our homes.

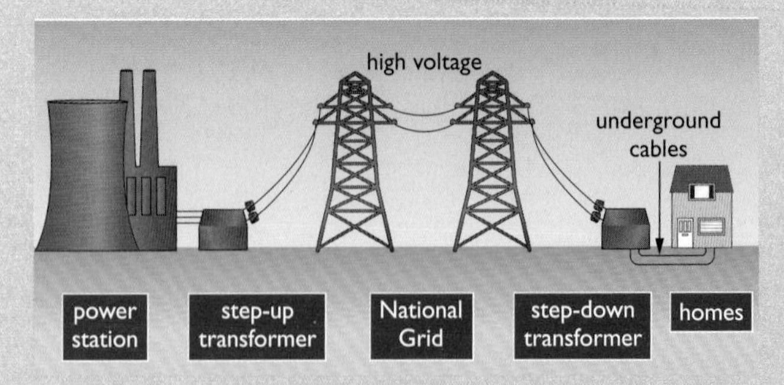

a) Why is the voltage of the supply increased before transmission? *(1 mark)*

b) Why is the voltage of the supply decreased before entering our homes? *(1 mark)*

c) A step up transformer has 100 turns on its primary coil and 20 000 turns on is secondary coil. Calculate the output voltage of the transformer if the input voltage is 12 V ac. *(3 marks)*

d) Assuming that the transformer is 100% efficient, calculate the current flowing through the secondary coil if the current flowing through the primary coil is 10 A. *(3 marks)*

e) Suggest two ways in which the design of a transformer helps reduce energy losses. *(2 marks)*

Total 10 marks

4 The diagram below shows a relay circuit. These circuits can be used to turn other circuits carrying high currents on and off safely.

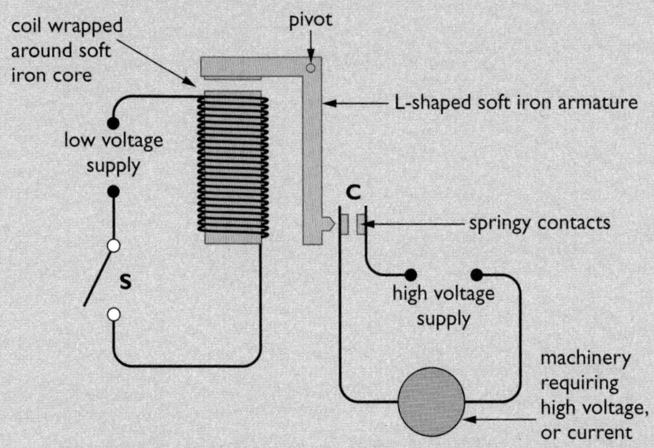

a) What is created around the coil when switch S is closed? *(1 mark)*

b) What happens to the soft iron core around which the coil is wrapped? *(1 mark)*

c) Explain in your own words why, when the contacts at C are closed, the second high voltage circuit is turned on. *(3 marks)*

d) Explain in detail what happens if switch S is then opened. *(3 marks)*

e) Why would the relay circuit not work if the soft iron core was replaced with one made of steel? *(1 mark)*

Total 9 marks

5 An electric kettle is rated at 2 kW when connected to a 240 V electrical supply.

a) Calculate the current that will flow when the kettle is turned on. *(3 marks)*

b) What value fuse should be included in the circuit of the kettle? Assume that the fuses available are 3 A, 5 A, and 13 A. *(1 mark)*

c) Modern kettles often have double insulation. Explain what this means and how it provides extra safety for the user. *(2 marks)*

d) Calculate the resistance of the heating element of the kettle. *(3 marks)*

e) Calculate the heat produced by the element each minute. *(3 marks)*

f) Calculate the number of units of electrical energy used if the above kettle is turned on for a total of 5 h each week. *(2 marks)*

g) Calculate the cost per year of using this kettle if the cost of each unit is 7 p. *(2 marks)*

Total 16 marks

6 **a)** Explain in detail how insulating materials can be charged by friction. *(4 marks)*

b) When an aircraft lands it is important that it is earthed before it is refuelled.

i) Explain why the aircraft should be earthed. *(3 marks)*

ii) Suggest one way in which the aircraft could be earthed. *(1 mark)*

c) Explain why electrostatic painting of objects such as bicycle frames makes good economic sense. *(3 marks)*

d) Describe briefly how an inkjet makes use of some of the properties of static electricity. *(3 marks)*

Total 14 marks

7 **a)** The circuit shown below is a sensor input sub-system to a control system.

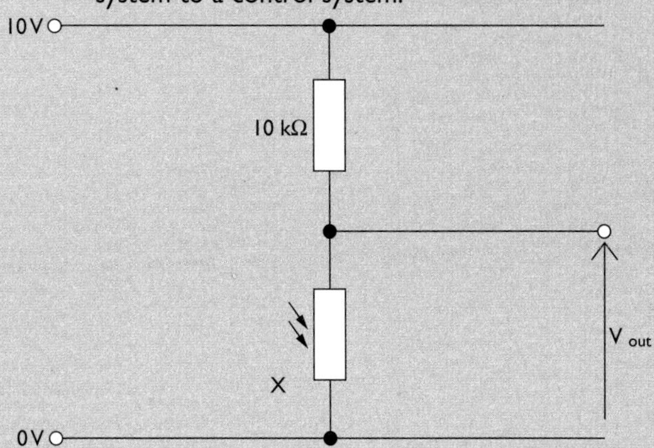

i) Name component X and explain how it senses changes in its environment. *(3 marks)*

ii) If the output voltage, V_{out}, is 8 V, calculate the value of the resistance of component X. *(2 marks)*

The 10 kΩ resistor is replaced with one colour coded ORANGE WHITE ORANGE GOLD.

(The order of the colours in the colour code is BLACK BROWN RED ORANGE YELLOW GREEN BLUE VIOLET GREY WHITE. **Hint**: See Appendix B) *iii)* State the value of this resistor. *(2 marks)*

iv) Use your answer to *iii)* to calculate the new value of V_{out}. *(2 marks)*

b) The circuit shown below uses the sensor circuit from part ***a)*** above to control a security light that switches on at dusk and turns off at dawn.

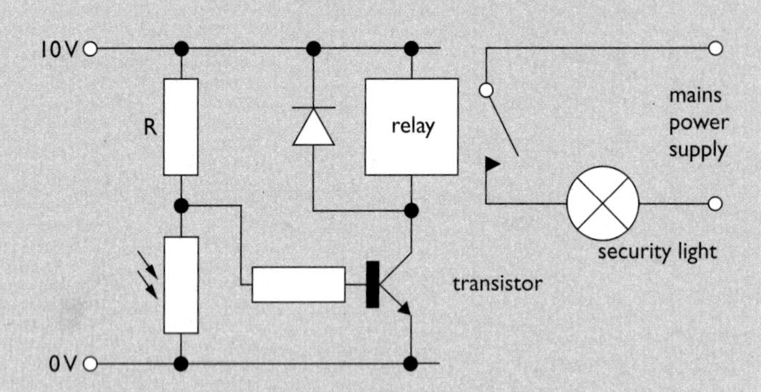

i) What is the purpose of the transistor in this circuit? *(1 mark)*

ii) Why is it necessary to use a relay in this circuit? *(2 marks)*

Total 12 marks

8 ***a)*** *i)* Copy and complete the truth-table for the logic circuit shown below. The logic circuit has inputs A and B and output X. P, Q and R are the outputs of the gates that make up the system. *(4 marks)*

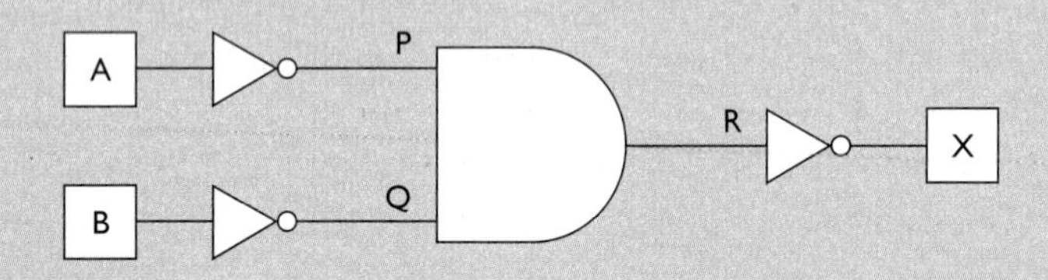

A	B	P	Q	R	S	X
0	0					
0	1					
1	0					
1	1					

ii) Name the single logic gate that can replace this logic circuit. *(1 mark)*

iii) Draw the symbol for the logic gate you have named in *ii)*. *(1 mark)*

b) The diagram shows a seven segment LED display. This consists of seven LEDs that can be lit in combinations to display the numbers 0–9 and some letters. The segments are identified by the letters "a" to "g", as shown.

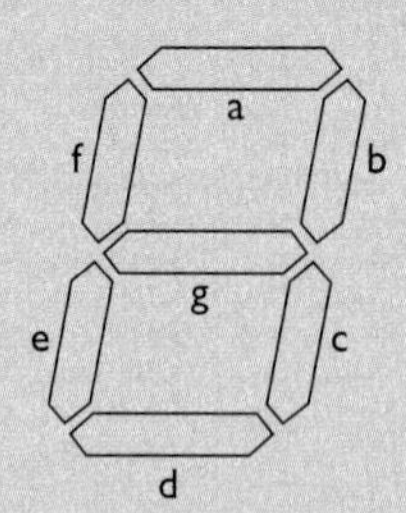

A student wants to display the numbers 1, 2, 3 or 4. The number displayed is selected using two inputs, A and B. The truth-table showing how the inputs affect the outputs is shown below, together with details of the segments that must be lit to form each number.

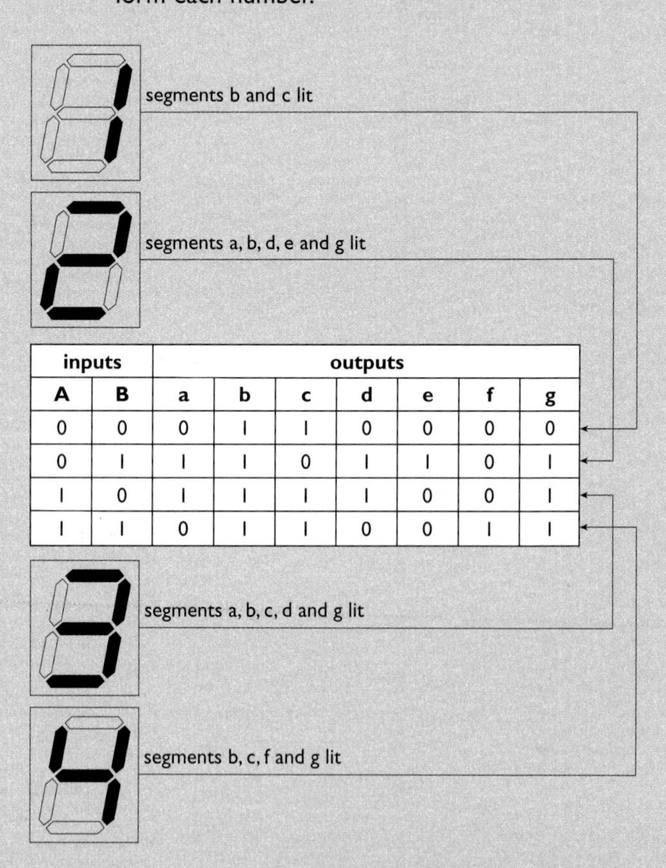

inputs		outputs						
A	B	a	b	c	d	e	f	g
0	0	0	1	1	0	0	0	0
0	1	1	1	0	1	1	0	1
1	0	1	1	1	1	0	0	1
1	1	0	1	1	0	0	1	1

i) Look at the truth-table for the output g (controlling the "g" segment of the display). Name the logic gate that produces this output. *(1 mark)*

ii) Draw a logic circuit that would provide the output for segment "e". *(3 marks)*

iii) Use your answer for *ii)* to produce a simple logic circuit to provide the output for segment "c". *(1 mark)*

Total 11 marks

Chapter 8: Speed and Acceleration

It is very useful to be able to make predictions about the way moving objects behave. In this chapter you will learn about some equations of motion that can be used to calculate the speed and acceleration of objects, and the distances they travel in a certain time.

Figure 8.1 *The world is full of speeding objects.*

Speed is a term that is used a great deal in everyday life. Action films often feature high-speed chases. Speed is a cause of fatal accidents on the road. Sprinters strive for greater speed in competition with other athletes. Rockets must reach a high enough speed to put communications satellites in orbit around the Earth. This chapter will explain how speed is defined and measured and how distance–time graphs are used to show the movement of an object as time passes. We shall then look at changing speed – acceleration and deceleration. We shall use velocity–time graphs to find the acceleration of an object. We shall also see how to find how far an object has travelled using its velocity–time graph.

Speed

If you were told that a car travelled 100 kilometres in 2 hours you would probably have no difficulty in working out that the speed (or strictly speaking the *average* speed – see page 84) of the car was 50 km/h. You would have done a simple calculation using the following definition of speed:

$$\text{speed} = \frac{\text{distance travelled}}{\text{time taken}}$$

This is usually written using the symbol v for speed (or velocity), s for distance travelled and t for time:

$$v = \frac{s}{t}$$

Units of speed

Typically the distance travelled might be measured in metres and time taken in seconds, so the speed would be in metres per second (m/s). Other units can be used for speed, such as kilometres per hour (km/h), or centimetres per second (cm/s). In physics the units we use are metric, but you will be used to measuring speed in miles per hour (mph). All cars show speed in both mph and kph (km/h). Exam questions should be in metric units, so remember that m is the abbreviation for *metres* (and *not miles*).

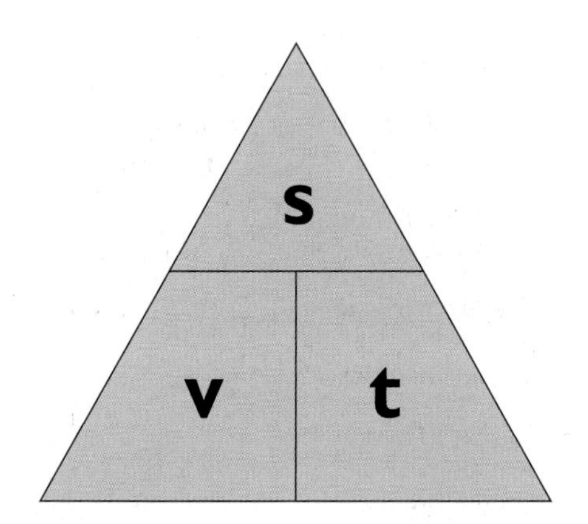

Figure 8.2 *You can use the triangle method for rearranging equations like s = v × t.*

Reminder: To use the triangle method to rearrange an equation, cover up the thing you want to find. For example, in Figure 8.2, if you wanted to work out how long (t) it took to travel a distance (s) at a given velocity (v), covering t in Figure 8.2 leaves d/v, or distance divided by speed.

Rearranging the speed equation

The speed equation can be rearranged to give two other useful equations:

$$\text{distance travelled, } s = \text{speed, } v \times \text{time taken, } t$$

and

$$\text{time taken, } t = \frac{\text{distance travelled, } s}{\text{speed, } v}$$

Average speed

The equation you used to work out the speed of the car, on page 83, gives you the **average speed** of the car during the journey. It is the total distance travelled, divided by the time taken for the journey. If you look at the speedometer in a car you will see that the speed of the car changes from instant to instant as the accelerator or brake is used. The speedometer therefore shows the **instantaneous speed** of the car.

Speed trap!

Suppose you want to find the speed of cars driving down your road. You may have seen the police using speed guns to check that drivers are keeping to the speed limit. Speed guns use microprocessors (computers on a "chip") to produce an instant reading of the speed of a moving vehicle, but you can conduct a very simple experiment to measure car speed.

Measure the distance between two points along a straight section of road with a tape measure or "click" wheel. Use a stopwatch to measure the time taken for a car to travel the measured distance. Figure 8.3 shows you how to operate your "speed trap".

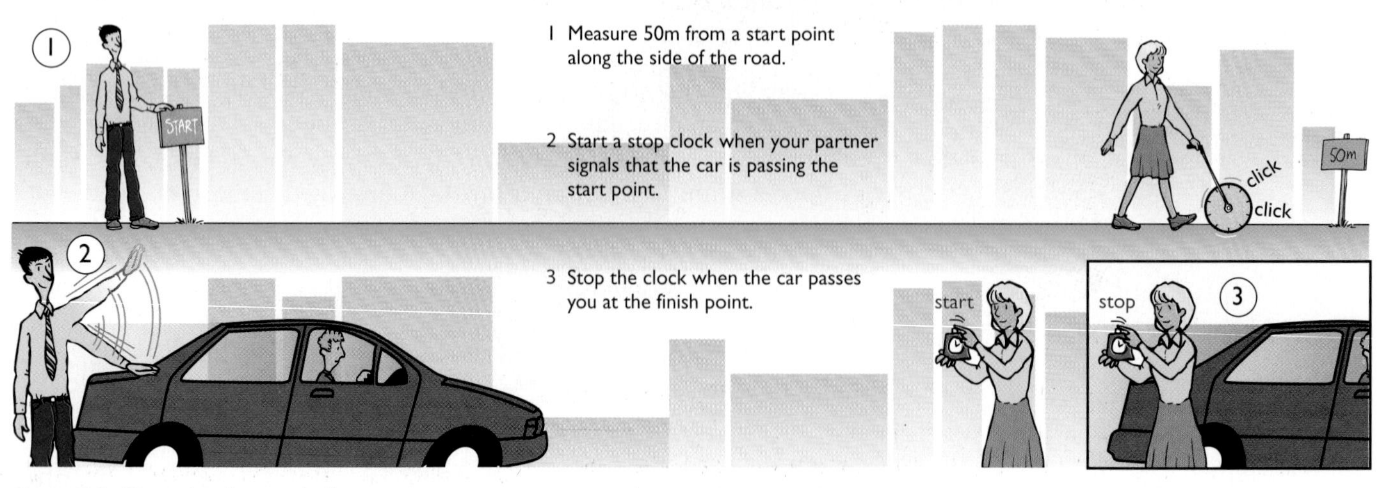

Figure 8.3 *Measuring the speed of a car.*

Using the measurements made with your speed trap, you can work out the speed of the car. Use the equation:

$$\text{speed} = \frac{\text{distance travelled}}{\text{time taken}}$$

So the speed of the car in this experiment is:

$$\text{speed} = \frac{50\,\text{m}}{3.9\,\text{s}} = 12.8\,\text{m/s}$$

You can convert a speed in m/s into a speed in km/h.

If the car travels 12.8 metres in one second it will travel

12.8 × 60 metres in 60 seconds (that is, one minute) and

12.8 × 60 × 60 metres in 60 minutes (that is, 1 hour), which is

46 080 metres in an hour or 46.1 km/h (to one decimal place).

We have multiplied by 3600 (60 × 60) to convert from m/s to m/h, then divided by 1000 to convert from m/h to km/h (as there are 1000 m in 1 km).

Rule: to convert m/s to km/h simply multiply by 3.6.

Distance–time graphs

Figure 8.4 *A car travelling at constant speed.*

Figure 8.4 shows pictures of a car travelling along a road. They show the car at 0.5 second intervals. The distances that the car has travelled from the start position after each 0.5 s time interval are marked on the picture. The picture provides a record of how far the car has travelled as time has passed. We can use the information in this sequence of pictures to plot a graph showing the distance travelled against time (Figure 8.5).

Time from start (s)	0.0	0.5	1.0	1.5	2.0	2.5
Distance travelled from start (m)	0.0	6.0	12.0	18.0	24.0	30.0

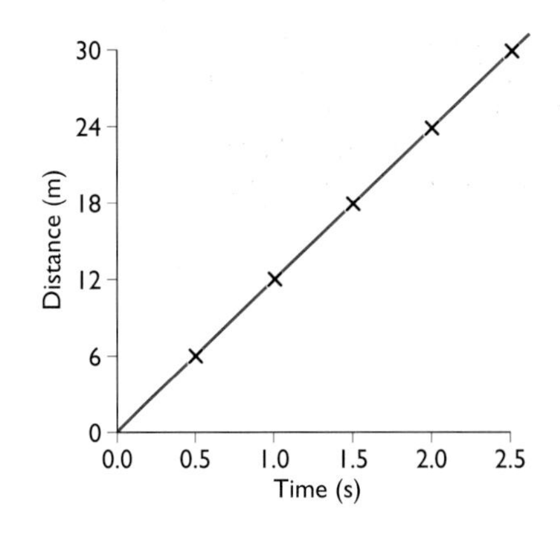

Figure 8.5 *Distance–time graph for the travelling car in Figure 8.4.*

The **distance–time graph** tells us about how the car is travelling in a much more convenient form than the sequence of drawings in Figure 8.4. We can see that the car is travelling equal distances in equal time intervals – it is moving at a steady or **constant speed**. This fact is shown immediately by the fact that the graph is a *straight line*. The slope or **gradient** of the line tells us the speed of the car – the steeper the line the greater the speed of the car. So, in this example:

$$\text{speed} = \text{gradient} = \frac{AB}{BC} = \frac{40\,\text{m}}{1.6\,\text{s}} = 25\,\text{m/s}$$

Speed and velocity

Some distance–time graphs look like the one shown in Figure 8.6. It is a straight line, showing that the object is moving with constant speed, but the line is sloping down to the right rather than up to the right. The gradient of such a line is negative because the distance that the object is

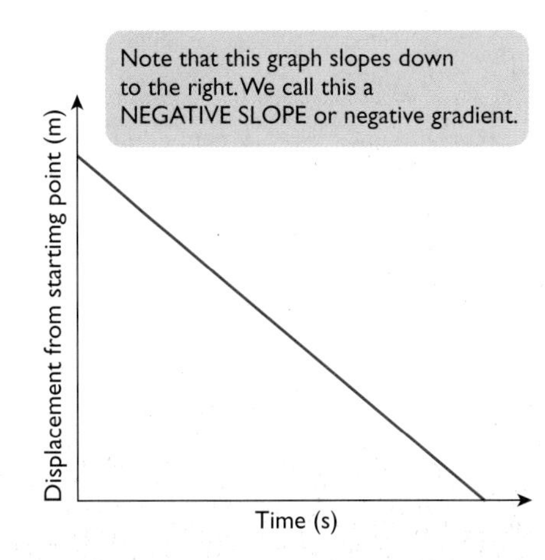

Figure 8.6 *In this graph distance is decreasing with time.*

A vector is a quantity that has both size *and* direction. Displacement is distance travelled in a *particular direction*.

Force is another example of a vector. The size of a force *and* the direction in which it acts are both important.

from the starting point is now *decreasing* – the object is retracing its path back towards the start. Physicists use the word **displacement** to mean "distance travelled in a particular direction" from a specified starting point. (The symbol **s** is used to represent displacement in equations.) So if the object was originally travelling in a northerly direction, the negative gradient of the graph means that it is now travelling south. Displacement is an example of a **vector**.

Velocity is also a vector. Velocity is speed in a particular direction. If a car travels at 50 km/h around a bend its speed is constant but its velocity will be changing for as long as the direction that the car is travelling in is changing.

$$\textbf{velocity} = \frac{\textbf{increase in displacement}}{\textbf{time taken}}$$

worked example

Example 1

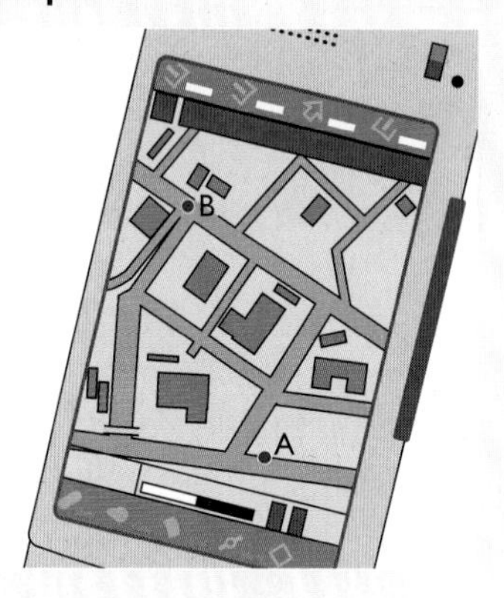

Figure 8.7 *The screen of a global positioning system (GPS). A GPS is an aid to navigation that uses orbiting satellites to locate its position on the Earth's surface.*

The GPS in Figure 8.7 shows two points on a journey. The second point is 3 km north west of the first. If a walker takes 45 minutes to travel from the first point to the second, what is the average velocity of the walker?

Write down what you know:
increase in displacement is 3 km north west
time taken is 45 min (45 min = 0.75 h).
Use:

$$\text{velocity} = \frac{\text{increase in displacement}}{\text{time taken}}$$

$$\text{average velocity} = \frac{3\text{ km}}{0.75\text{ h}}$$

$$= 4.0\text{ km/h north west}$$

Acceleration

Figure 8.8 *Acceleration …*

… constant speed …

… and deceleration.

Figure 8.8 shows some objects whose speed is changing. The plane must accelerate to reach take-off speed. The curling stone decelerates only very slowly when it slides across the ice. When the egg hits the ground it is forced to decelerate (decrease its speed) very rapidly. Rapid deceleration can have destructive results.

Acceleration is the rate at which objects increase their speed. It is defined as follows:

$$\textbf{acceleration} = \frac{\textbf{increase in velocity}}{\textbf{time taken}}$$

This is written as an equation:

$$a = \frac{v - u}{t}$$

where a = acceleration, v = the velocity after accelerating for a time, t, from a starting velocity of u. (Why u? Simply because it comes before v!) v is usually called the "final velocity" and u the "initial velocity".

Acceleration, like velocity, is a vector because the *direction* in which the acceleration occurs is important as well as the size of the acceleration.

Units of acceleration

Velocity is measured in m/s, so increase in velocity is also measured in m/s. Acceleration, the **rate** of increase in velocity with time, is therefore measured in m/s/s (read as "metres per second per second"). We normally write this as m/s^2 (read as "metres per second squared"). Other units may be used – for example, cm/s^2.

It is good practice to include units in equations – this will help you to supply the answer with the correct unit.

worked example

Example 2

A car is travelling at 20 m/s. It accelerates steadily for 5 s, after which time it is travelling at 30 m/s. What is its acceleration?

Write down what you know:

initial or starting velocity, u = 20m/s

final velocity, v = 30m/s

time taken, t = 5s

Use: $a = \frac{v - u}{t}$

$$a = \frac{30\ m/s - 20\ m/s}{5\ s}$$

$$a = \frac{10\ m/s}{5\ s}$$

$$= 2\ m/s^2$$

The car is accelerating at $2\ m/s^2$.

Deceleration

Deceleration means slowing down. This means that a decelerating object will have a smaller final velocity than its starting velocity. If you use the equation for finding the acceleration of an object that is slowing down the answer will have a negative sign. A negative acceleration simply means deceleration.

worked example

Example 3

An object strikes the ground travelling at 40 m/s. It is brought to rest in 0.02 s. What is its acceleration?

Write down what you know:

initial velocity, u = 40 m/s

final velocity, v = 0 m/s

time taken, t = 0.02 s

As before, use:

$$a = \frac{v - u}{t}$$

$$a = \frac{0 \text{ m/s} - 40 \text{ m/s}}{0.02 \text{ s}}$$

$$a = \frac{-40 \text{ m/s}}{0.02 \text{ s}}$$

$$= -2000 \text{ m/s}^2$$

So the acceleration is -2000 m/s^2.

In Example 3, we would say that the object is decelerating at 2000 m/s^2. This is a very large deceleration. Later, in Chapter 10, we shall discuss the consequences of such a rapid deceleration!

Measuring acceleration

When a ball is rolled down a slope it is clear that its speed increases as it rolls – that is, it accelerates. Galileo was interested in how and why objects like the ball rolling down a slope speeded up, and devised an interesting experiment to learn more about acceleration. A version of his experiment is shown in Figure 8.9.

Galileo was an Italian scientist who was born in 1564. He developed a telescope, which he used to study the motion of the planets and other celestial bodies.

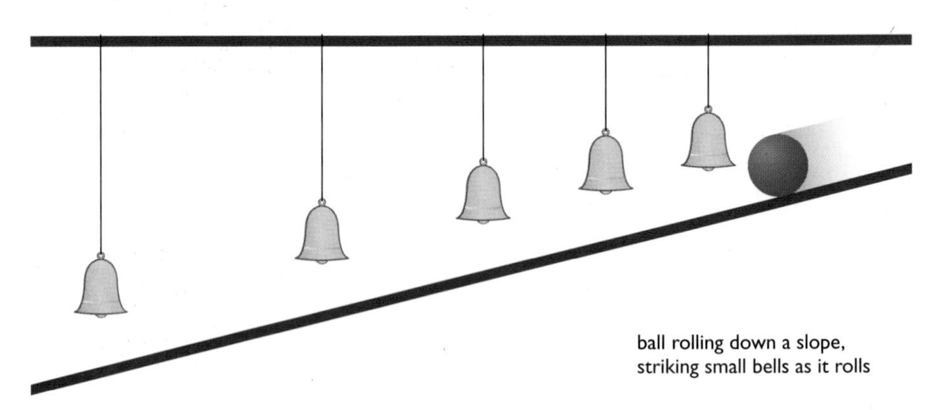

Figure 8.9 *Galileo's experiment.*

Galileo wanted to discover how the distance travelled by a ball depends on the time it has been rolling. In this version of the experiment, a ball rolling down a slope strikes a series of small bells as it rolls. By adjusting the positions of the bells carefully it is possible to make the bells ring at equal intervals of time as the ball passes. Galileo noticed that the distances travelled in equal time intervals increased, showing that the ball was

Though Galileo did not have a clockwork timepiece (let alone an electronic timer), he used his pulse and a type of water clock to achieve timings that were accurate enough for his experiments.

travelling faster as time passed. Galileo did not possess an accurate way of measuring time (there were no digital stopwatches in seventeenth century Italy!) but it was possible to judge *equal* time intervals accurately simply by listening.

Galileo also noticed that the distance travelled by the ball increased in a predictable way. He showed that the rate of increase of speed was steady or uniform. We call this **uniform acceleration**. Most acceleration is non-uniform – that is, it changes from instant to instant – but we shall only deal with uniformly accelerated objects in this chapter.

Velocity–time graphs

The table below shows the distances between the bells in an experiment such as Galileo's. If the time interval between each bell ringing is 0.5 s, the average speed (or velocity, since the direction is always down the slope) can easily be calculated using $v = \frac{s}{t}$ (see page 84). This is done in the second row of the table.

Distance travelled (cm)	5	10	15	20	25
Average velocity (cm/s)	10	20	30	40	50

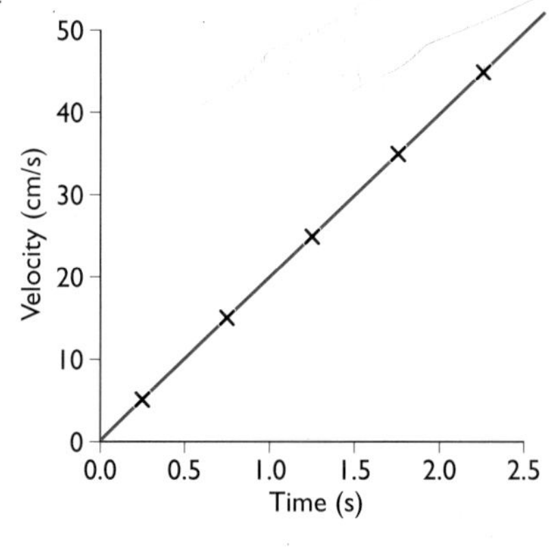

Figure 8.10 *Velocity–time graph for an experiment in which a ball is rolled down a slope. (Note that as we are plotting* average *velocity, the points are plotted in the middle of each successive 0.5 s time interval.)*

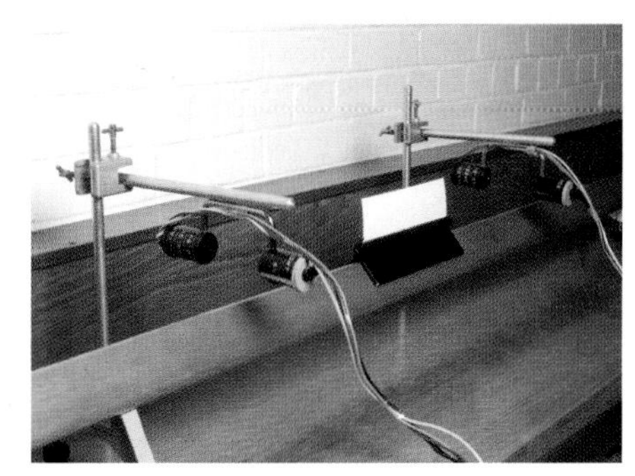

Using the information in the table we can draw a graph showing how the velocity of the ball is changing with time. The graph, shown in Figure 8.10, is called a **velocity–time graph**.

The graph in Figure 8.10 is a straight line. This tells us that the velocity of the rolling ball is increasing by equal amounts in equal time periods. We say that the acceleration is uniform in this case.

A modern version of Galileo's experiment

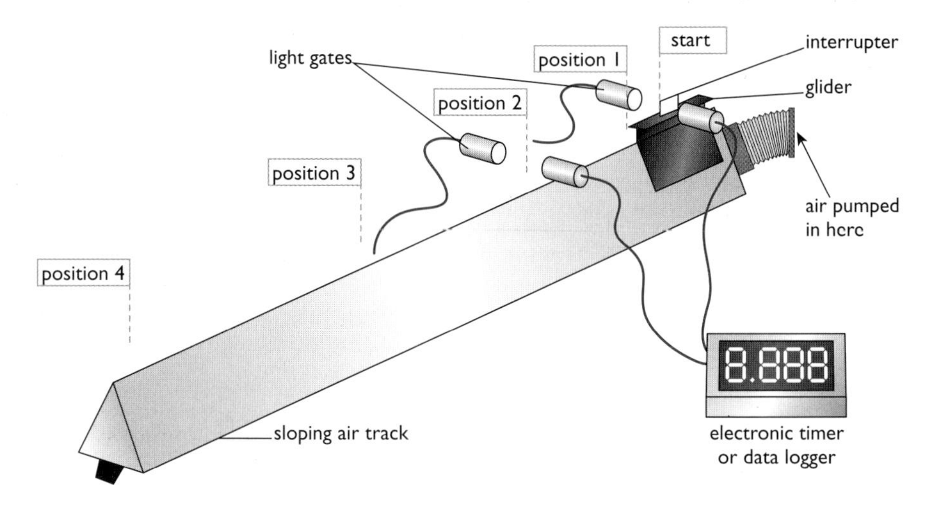

Figure 8.11 *Measuring acceleration.*

Today we can use data loggers to make accurate direct measurements that are collected and manipulated by a computer. A spreadsheet programme can be used to produce a velocity–time graph. The diagram shows a glider on a slightly sloping air-track. The air-track reduces friction because the glider rides on a cushion of air that is pumped continuously through holes along the air-track. As the glider accelerates down the sloping track the white card mounted on it breaks a light beam, and the time that the glider takes to pass is measured electronically. If the length of the card is measured, and this is entered into the spreadsheet, the velocity of the glider can be calculated by the spreadsheet programme using $v = \frac{s}{t}$.

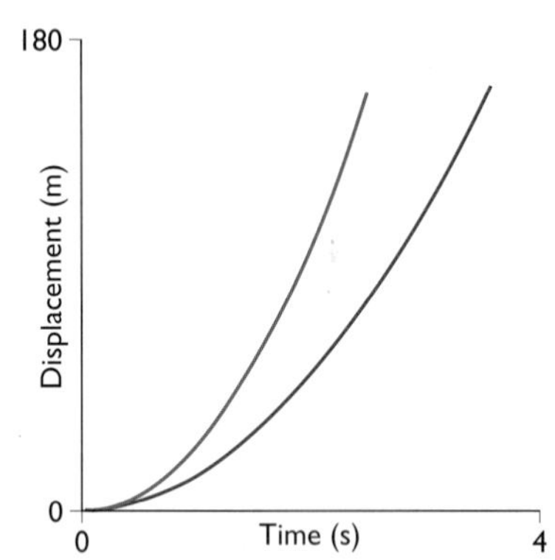

Airtrack at 1.5°		Airtrack at 3.0°	
Time (s)	Disp (cm)	Time (s)	Disp (cm)
0.0	0	0.00	0
0.9	10	0.63	10
1.8	40	1.26	40
2.7	90	1.90	90
3.6	160	2.50	160

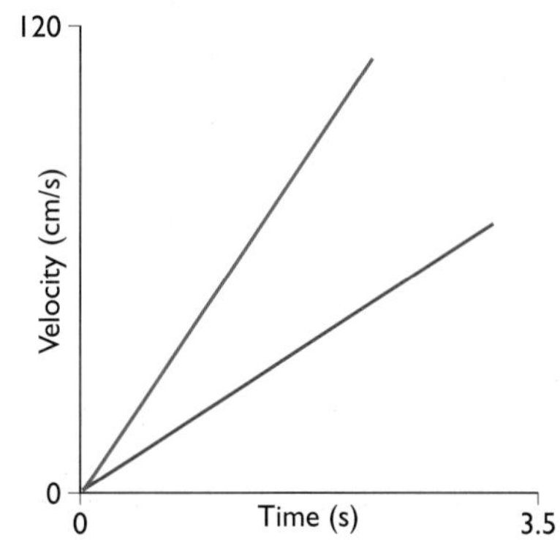

Airtrack at 1.5°		Airtrack at 3.0°	
Time (s)	Av Vel. (cm/s)	Time (s)	Av Vel. (cm/s)
0.00	0.0	0.00	0.0
0.45	11.1	0.32	15.9
1.35	33.3	0.95	47.6
2.25	55.6	1.56	79.4
3.15	77.8	2.21	111.1

Figure 8.12 *Results of two air-track experiments. (Note, once again, that because we are plotting average velocity in the velocity–time graphs, the points are plotted in the middle of each successive time interval.)*

Figure 8.12 shows some velocity–time graphs for two experiments done using the air-track apparatus. In each experiment the track was given a different slope. The steeper the slope of the air-track the faster the glider accelerates. This is clear from the graphs: the greater the acceleration the steeper the gradient of the graph.

More about velocity–time graphs

Gradient

The results of the air-track experiments in Figure 8.12 show that the slope of the velocity–time graph depends on the acceleration of the glider. The slope or gradient of a velocity–time graph is found by dividing the increase in the velocity by the time taken for the increase, as shown in Figure 8.13. Increase in velocity divided by time is, you will recall, the definition of acceleration (see page 87), so we can measure the acceleration of an object by finding the slope of its velocity–time graph. The meaning of the slope or gradient of a velocity–time graph is summarised in Figure 8.14.

Tips

1 When finding the gradient of a graph, draw a **big** triangle.
2 Choose a convenient number of units for the length of the base of the triangle to make the division easier

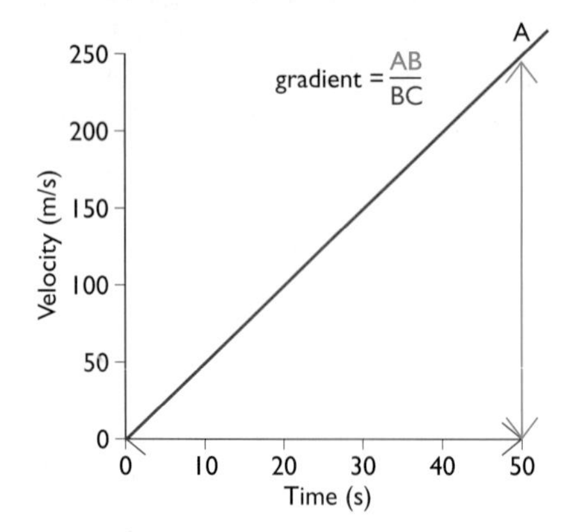

Figure 8.13 *Finding the gradient of a velocity–time graph.*

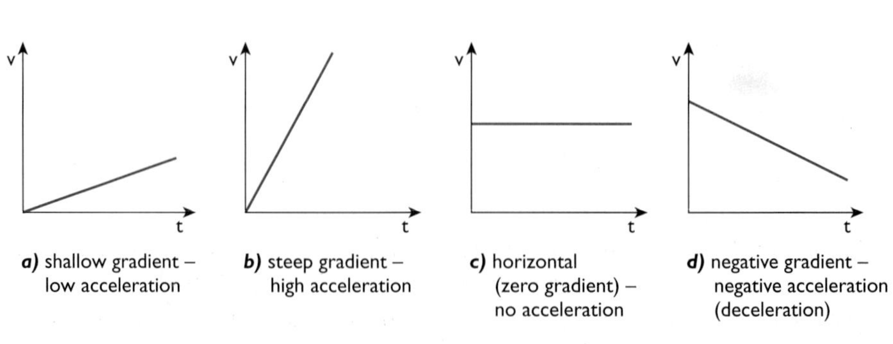

Figure 8.14 *The gradient of a velocity–time graph gives you information about the motion of an object at a glance.*

Area under a velocity–time graph

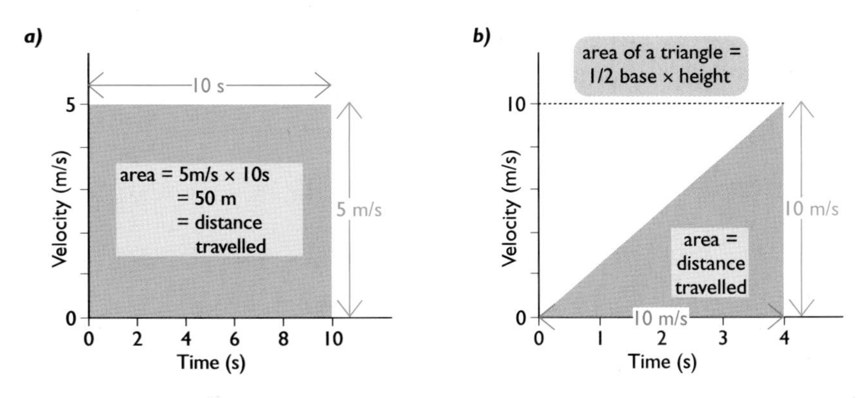

Figure 8.15 a) *An object travelling at constant velocity,* b) *An object accelerating at a constant rate.*

Figure 8.15a shows a velocity–time graph for an object that travels with a constant velocity of 5 m/s for 10 s. A simple calculation shows that in this time the object has travelled 50 m. This is equal to the shaded area under the graph. Figure 8.15b shows a velocity–time graph for an object that has accelerated at a constant rate. Its average velocity during this time is given by:

$$\text{average velocity} = \frac{\text{initial velocity} + \text{final velocity}}{2} \text{ or } \frac{u + v}{2}$$

In this example the average velocity is, therefore:

$$\text{average velocity} = \frac{0 \text{ m/s} + 10 \text{ m/s}}{2}$$

which works out to be 5 m/s. If the object travels, on average, 5 metres in each second it will have travelled 20 metres in 4 seconds. Notice that this, too, is equal to the shaded area under the graph (given by the area formula for a triangle: area = $\frac{1}{2}$ base × height).

The area under a velocity–time graph is equal to the distance travelled by (displacement of) the object in a particular time interval.

A further equation for distance travelled

We have already seen that:

$$a = \frac{v - u}{t}$$

Rearranging this equation, we can see that:

$$\mathbf{v = u + at}$$

The final velocity, v, after accelerating at a m/s^2 for t s is equal to the initial velocity, u, plus the increase in velocity (a × t). If we look at the velocity–time graphs in Figure 8.16, we can see that the distance travelled (displacement) can be worked out from the area under the graphs using another equation.

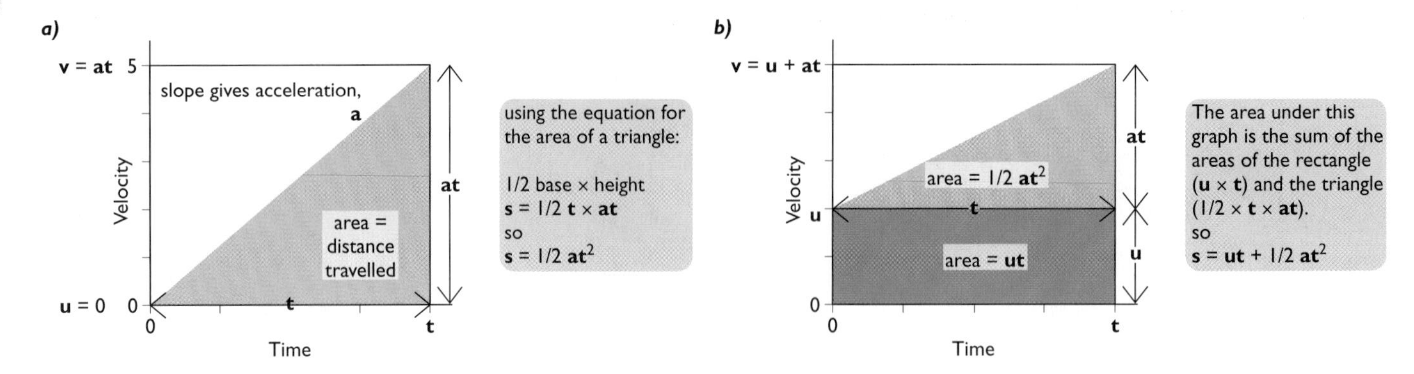

Figure 8.16 a) *Velocity–time graph for an object accelerating from rest (u = 0),* b) *Velocity–time graph for an object accelerating from an initial velocity, u.*

Figure 8.16a shows the velocity–time graph for an object that has accelerated at a constant rate, a, from rest for t seconds. The slope of the graph is the acceleration, a. The velocity increase is acceleration multiplied by time (a × t). As the area under a velocity–time graph gives us the distance travelled – or displacement, s – we can see that:

$$s = \frac{1}{2}at^2$$

Figure 8.16b shows a more general case. Here the object has accelerated uniformly from an initial velocity of u to a final velocity of v in a time, t. You should notice that the increase in velocity is a × t as above, so the final velocity, v, is the initial velocity, u, plus the increase in velocity, a × t. The area under the graph can be divided up into a rectangle and a triangle, so the total area can be calculated easily. This area tells us the distance travelled, or displacement, s:

$$\mathbf{s = ut + \frac{1}{2}at^2}$$

Remember that the distance (displacement) will be in metres if you have used the correct units for the other quantities in the equation: m/s for u, m/s^2 for a and s for t.

A final equation of motion

There are *five* different quantities that we use in equations of uniformly accelerated motion:

initial velocity, u
final velocity, v
displacement, s
acceleration, a
time taken, t

We have seen the following equations using these quantities:

$$\text{displacement} = \text{average velocity} \times \text{time, or } s = \frac{u + v}{2} \times t$$

$$v = u + at$$

$$s = ut + \frac{1}{2}at^2$$

You will notice that all of these equations involve time, t. We can combine and rearrange the first two equations to get a fourth equation of motion that does not have t in it.

Rearrange the second equation, $v = u + at$ to give:

$$t = \frac{v - u}{a}$$

Then substitute for **t** in the first equation, so:

$$s = \frac{u + v}{2} \times \frac{v - u}{a}$$

$$2as = (u + v)(v - u)$$

$$2as = v^2 - u^2$$

The result is usually written:

$$\mathbf{v^2 = u^2 + 2as}$$

End of Chapter Checklist

If you haven't got a copy of your specification, read the introduction on page iv.

You will need to be able to do some or all of the following. Check your Awarding Body's specification (syllabus) to find out exactly what you need to know.

- speed, $v = \frac{\text{distance travelled, s}}{\text{time taken, t}}$
- The units of speed are metres per second, m/s (also km/h and cm/s).
- Distance–time graphs for objects moving at constant speed are straight lines.
- The gradient of a distance–time graph gives the speed.
- Distance travelled in a specified direction is called displacement. Displacement is a vector quantity.
- Velocity is speed in a specified direction. It is also a vector quantity.
- acceleration $= \frac{\text{increase in velocity}}{\text{time taken}}$ or $a = \frac{v - u}{t}$
- The units of acceleration are metres per second squared, m/s^2 (also centimetres per second squared, cm/s^2, for smaller rates of acceleration).
- Acceleration is a vector.
- Velocity–time graphs of objects moving with constant velocity are horizontal straight lines.
- The gradient of a velocity–time graph gives acceleration; negative gradient (graph line sloping down to the right) indicates deceleration.
- The area under a velocity–time graph is equal to the distance travelled (displacement).
- average velocity $= \frac{\text{initial velocity + final velocity}}{2}$ or $\frac{u + v}{2}$
- The equations $s = ut + \frac{1}{2}at^2$ and $v^2 = u^2 + 2as$.

Questions

More questions on speed and acceleration can be found at the end of Section B on page 139.

1 A sprinter runs 100 metres in 12.5 seconds. Work out her speed:

a) in m/s ***b)*** in km/h.

2 A jet can travel at 350 m/s. How far will it travel at this speed in:

a) 30 seconds

b) 5 minutes

c) half an hour?

3 A snail crawls at a speed of 0.0004 m/s. How long will it take to climb a garden cane 1.6 m high?

4 Look at the following sketches of distance–time graphs of moving objects.

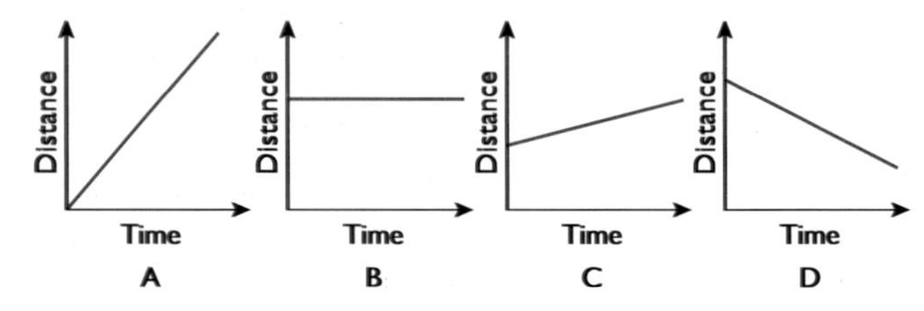

In which graph is the object

a) moving backwards

b) moving at the slowest speed

c) moving at the fastest speed

d) not moving at all?

5 Sketch a distance–time graph to show the motion of a person walking quickly, stopping for a moment, then continuing to walk slowly in the same direction.

6 Plot a distance–time graph using the data in the following table. Draw a line of best fit and use your graph to find the speed of the object concerned.

Distance (m)	0.00	1.60	3.25	4.80	6.35	8.00	9.60
Time (s)	0.00	0.05	0.10	0.15	0.20	0.25	0.30

7 The diagram below shows a trail of oil drips made by a car as it travels along a road. The oil is dripping from the car at a steady rate of one drip every 2.5 seconds.

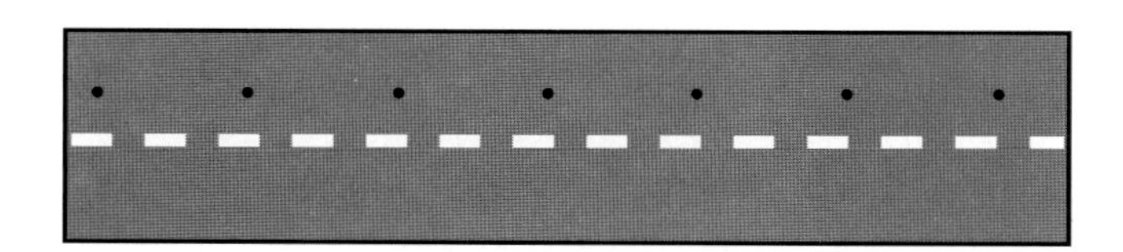

a) What can you tell about the the way the car is moving?

b) The distance between the first and the seventh drip is 135 metres. What is the average speed of the car?

8 A car is travelling at 20 m/s. It accelerates uniformly at 3 m/s^2 for 5 s.

a) Draw a velocity–time graph for the car during the period that it is accelerating. Include numerical detail on the axes of your graph.

b) Calculate the distance the car travels while it is accelerating.

9 Explain the difference between the following terms:

a) *average speed* and *instantaneous speed*

b) *speed* and *velocity*.

10 A sports car accelerates uniformly from rest to 24 m/s in 6 s. What is the acceleration of the car?

11 Sketch velocity–time graphs for:

a) an object moving with a constant velocity of 6 m/s

b) an object accelerating uniformly at 2 m/s^2 for 10 s

c) an object decelerating at 4 m/s^2 for 5 s.

12 A plane starting from rest accelerates at 3 m/s^2 for 25 s. By how much has the velocity increased after:

a) 1 s *b)* 5 s *c)* 25 s?

13 Look at the following sketches of velocity–time graphs of moving objects.

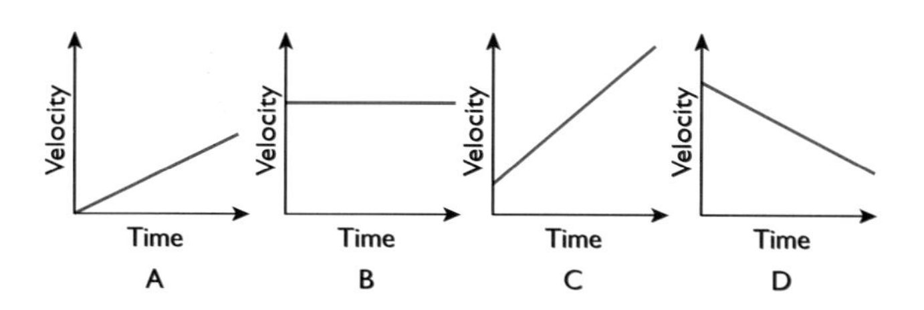

In which graph is the object:

a) not accelerating

b) accelerating from rest

c) decelerating

d) accelerating at the greatest rate?

14 Sketch a velocity–time graph to show how the speed of a car travelling along a straight road changes if it accelerates uniformly from rest for 5 s, travels at a constant speed for 10 s, then brakes hard to come to rest in 2 s.

15 Plot a velocity–time graph using the data in the following table.

Velocity (m/s)	0.0	2.5	5.0	7.5	10.0	10.0	10.0	10.0	10.0	10.0
Time (s)	0.0	1.0	2.0	3.0	4.0	5.0	6.0	7.0	8.0	9.0

Draw a line of best fit and use your graph to find:

a) the acceleration during the first 4 s

b) the distance travelled in

i) the first 4 s

ii) the last 5 s

of the motion shown

c) the average speed during the 9 seconds of motion shown.

16 The leaky car from question 7 is still on the road! It is still dripping oil but now at a rate of one drop per second. The trail of drips is shown on the diagram below.

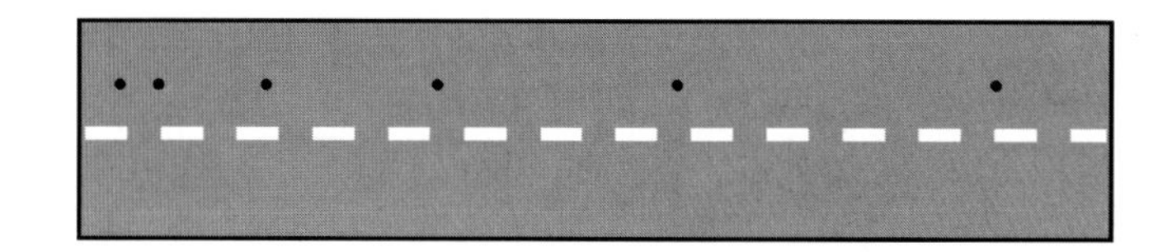

The distance between the first and second oil drip is 0.5 m. Does the spacing of the oil drips show that the car is accelerating at a steady rate? Explain how you would make and use measurements from the oil drip trail to determine this. Work out the rate of acceleration of the car.

Section B: Forces and Motion

Chapter 9: Forces

Forces are acting on us, and on objects all around us, all the time. In this chapter you will learn about different kinds of forces and how they affect the way objects move.

Figure 9.1 *Forces include pushing, pulling, falling and squashing.*

Forces are simply pushes and pulls. Sometimes we can see their effects quite clearly. In Figure 9.1, the bulldozer is pushing the earth; the tug is pulling the tanker; the bungee jumper is being pulled to Earth by gravity, and then (hopefully before meeting the ground) being pulled back up by the stretched elastic rope; the force applied by one robot would permanently change the shape of the other. In this chapter we shall discuss different types of forces and look at their effects on the way objects move. We shall also see how Galileo tested the widely accepted ideas of Aristotle by doing experiments, and showed that they were wrong. Even the greatest of thinkers sometimes make mistakes!

All sorts of forces

If you are to study forces, first you need to spot them! As we have already said, sometimes they are easy to see and their effect is obvious. Look at Figure 9.2 and try to identify any forces that you think are involved.

Figure 9.2 *What forces do you think are working here?*

You will immediately see that the man is applying a force to the car – he is pushing it. This is an example of a **contact** force – a force acting between two objects in direct contact. But there are quite a few more forces in the picture. To make the task a little easier we shall confine our search to just those forces *acting on the car*. We shall also ignore forces that are very small and therefore have little effect.

The man is clearly struggling to make the car move. This is because there is a force acting on the car trying to stop it moving. This is the force of friction between the moving parts in the car engine, gearbox, wheel axles and so on. This unhelpful force opposes the motion that the man is trying to achieve. However, when the car engine is doing the work to make the car go, the friction between the tyres and the road surface is vital. On an icy road even powerful cars may not move forward because there is not enough friction between the tyres and the ice.

Another force that acts on the car is the pull of the Earth. We call this a **gravitational** force or simply **weight**. If the car were to be pushed over the edge of a cliff, the effect of the gravitational force would be very clear as the car plunged towards the sea. This leads us to realise that yet another force is acting on the car in Figure 9.2 – the road must be stopping the car from being pulled into the Earth. This force, which acts in an upward direction on the car, is called the reaction force. (A more complete name is **normal reaction** force. Here the word "normal" means acting at 90° to the road surface.) All four forces that act on the car are shown in Figure 9.3.

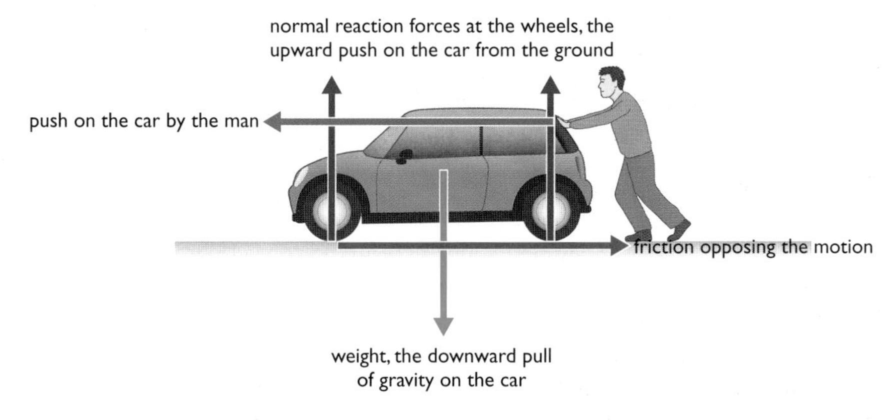

Figure 9.3 *There are four forces at work.*

You will have realised by now that it is not just the size of the force that is important – the direction in which the force is acting is important, too.

Force is another example of a **vector**.

Units of force

The unit used to measure force is the newton (N), named after Sir Isaac Newton who made a study of forces that forms the foundation of our understanding today.

A force of one newton will make a mass of one kilogram accelerate at one metre per second squared.

This is explained more fully later (see Chapter 10). To give you an idea of the size of the newton, the force of gravity on a kilogram bag of sugar (its weight) is about 10 N; an average sized apple weighs 1 N.

Some other examples of forces

Figure 9.4 *More forces!*

It is not always easy to spot forces acting on objects. In Figure 9.4, the parachute is causing the parachutist to descend more slowly because an upward force acts on the parachute called **air resistance** or **drag**. Air resistance is like friction – it tries to oppose movement of objects through the air. Designers of cars, high-speed trains and other fast-moving objects try to reduce the effects of this force. Objects moving through liquids also experience a drag force – fast-moving animals that live in water have streamlined shapes to reduce this force.

The hot air balloons in Figure 9.4 are carried upwards in spite of the pull of gravity on it because of a force called **upthrust**. This is the upward push of the surrounding air on the balloon. An upthrust force also acts on objects immersed in liquids.

The compass needle, which is a magnet, is affected by the **magnetic** force between it and the other magnet. Magnetic forces are used to make electric motors rotate, to hold fridge doors shut, and in many other situations.

More types of force, such as electric and nuclear forces, are mentioned in other chapters of this book. The rest of this chapter will look at the *effects* of forces.

Figure 9.5 *Aristotle (top) and Galileo.*

Aristotle was a Greek philosopher who lived in the fourth century BC (384–322 BC). His theories on motion were shown to be wrong by Galileo Galilei, an Italian astronomer, mathematician and scientist who lived from 1564 to 1642.

Why do things move?

The Greek philosopher Aristotle wrote about forces and motion in the fourth century BC. His ideas were accepted and largely unchallenged for over 1900 years until the Italian scientist Galileo started to conduct experiments to test them.

Aristotle *wrongly* classified motion into two types:

1 natural motion – for example, the movement of the stars, the upward movement of smoke and the movement of falling objects

2 violent motion – the movement of an object that is made to move by the visible application of a force. Aristotle said that a force must act on an object to keep it moving.

Aristotle also stated that the speed of falling objects was proportional to their weight. This would mean that heavy objects fall faster than light objects. We now know that this is *not* the case.

It should be pointed out that Aristotle and many Greek philosophers did not conduct experiments. Galileo, however, *was* an experimental scientist and one of his famous experiments involved dropping cannonballs of different weights from a tall tower (some say it was the Leaning Tower of Pisa). He found that the balls, when released at the same time, struck the ground at more or less the same time, regardless of mass. This was quite different from predictions based on Aristotle's ideas, and marked the beginning of a much more accurate understanding of the relationship between force and motion.

Figure 9.6 *Galileo was said to have used the Leaning Tower of Pisa as the site of his experiments.*

More than one force

As we saw earlier, in most situations there will be more than just one force acting on an object. If we push a book across a table, a number of forces act on it, as shown in Figure 9.7.

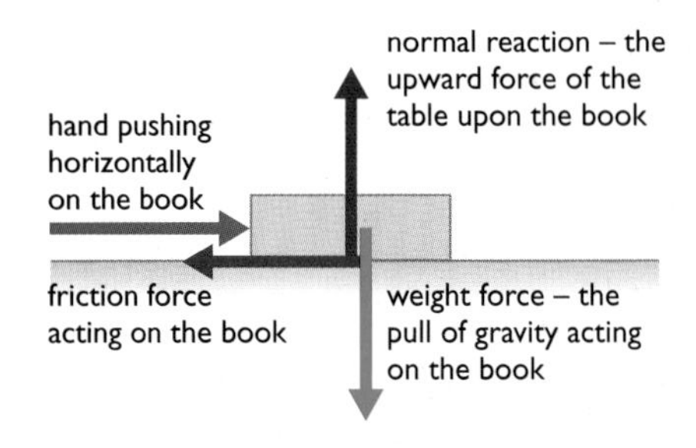

Figure 9.7 *The forces acting on a book resting on a table are in equilibrium.*

Aristotle argued that the book would remain "naturally" at rest unless a "violent" force was applied to it, causing it to move "unnaturally". He also said that the book would only move while the "violent" force was acting on the book – this seemed like common sense as, when you stop pushing the book, it stops moving. We now understand that the book stops because of the friction force acting on it while it is moving.

It is possible to do experiments in the science laboratory in which the friction force on a moving object is reduced to a very low value. Such an object can be set in motion with a small push and it will continue to move at a constant speed even when the force is no longer acting on it. An experiment like this is shown in Figure 9.8.

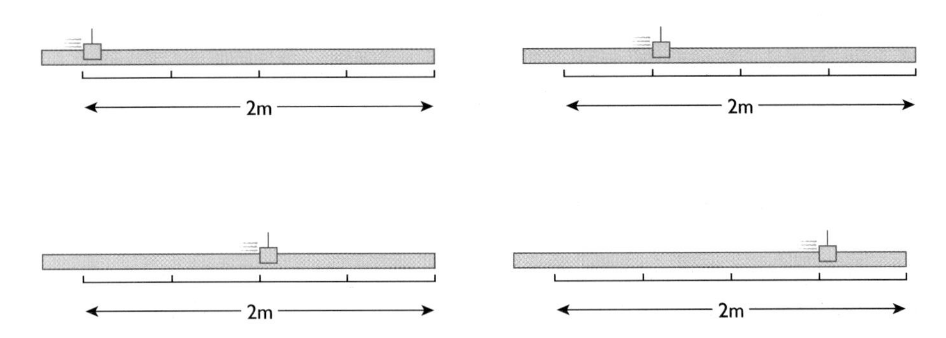

Figure 9.8 *A linear air-track reduces friction dramatically. The glider moves equal distances in equal time intervals. Its velocity is constant.*

You may also have seen scientists working on space stations demonstrating that objects keep moving in a straight line at constant speed, once set in motion. They do this in space because the objects are weightless and there is no friction acting on them.

Balanced and unbalanced forces

Figure 9.9 *Balanced forces … and unbalanced forces.*

Figure 9.9 shows two situations in which forces are acting on an object. In the tug of war contest the two teams are both pulling on the rope in opposite directions. For much of the time the rope doesn't move because the two forces are **balanced**. This means that the forces are the same size but act in opposite directions along the line of the rope. There is no **resultant** force in one direction or the other. When the forces acting on something are balanced, the object *does not change the way it is moving*. In this case if the rope is stationary, it *remains* stationary. Eventually, one of the teams will tire and its pull will be smaller than that of the other team. When the forces acting on the rope are **unbalanced** the rope will start to move in the direction of the greater force. There will be a resultant force in that direction. Unbalanced forces acting on an object cause it to *change the way it is moving*. The rope was stationary and the unbalanced forces acting on it caused it to *accelerate*.

The rocket car in Figure 9.9 is designed to have an enormous acceleration from rest. As soon as it starts to move the forces that oppose motion – friction and drag – must be overcome. The thrust of the rocket engine is, to start with, much greater than the friction and drag forces. This means that the forces acting on the car in the horizontal direction are unbalanced and the result is a change in the way that the car is moving – it accelerates! Once the friction forces balance the thrust the car no longer accelerates – it moves at a steady speed.

Friction

Figure 9.10 *The ice skater can glide because friction is low. The cars need friction to grip the road.*

Friction is the force that causes moving objects to slow down and finally stop. The kinetic energy of the moving object is converted to heat as work is done by the friction force. For the ice skater in Figure 9.10 the force of friction is very small so she is able to glide for long distances without having to do any work. It is also the force that allows a car's wheels to grip the road and make it accelerate – very quickly in the case of the drag racing car in Figure 9.10.

Scientists have worked hard for many years to develop some materials that reduce friction and others that increase friction. Reducing friction means that machines work more efficiently (wasting less energy) and do not wear out so quickly. Increasing friction can help to make tyres that grip the road better and to make more effective brakes.

Friction occurs when solid objects rub against other solid objects and also when objects move through fluids (liquids and gases). Sprint cyclists and Olympic swimmers now wear special fabrics to reduce the effects of fluid friction so they can achieve faster times in their races. Sometimes fluid friction is very desirable – for example, when someone uses a parachute after jumping from a plane!

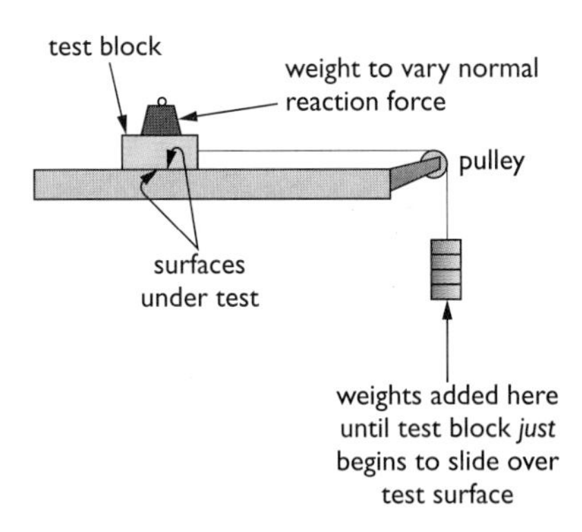

Figure 9.11 *This apparatus can be used to investigate friction.*

Investigating friction

The simple apparatus shown in Figure 9.11 can be used to discover some basic facts about friction. The weight on the nylon line running over the pulley pulls the block horizontally along the track and friction opposes this force. The weight is increased until the block *just* starts to move; this happens when the pull of the weight *just* overcomes the friction force. The rig can be used to test different factors that may affect the size of the friction, such as the surfaces in contact – the bottom of the block and the surface of the track. If the track surface is replaced with a rough surface, like a sheet of sandpaper, the force required to overcome friction will be greater.

Figure 9.12 *These parachutists are in free fall.*

If you repeat this experiment with a model car, you will discover that changing the track surface for a rougher one will have little or no effect on the pull needed to make the model car move. This is because the tyre on the wheel does *not* slip over the surface – the tyre *grips* the surface. The car rolls because the friction force between the axles and the axle-bearings is small and it is *here* that the friction force is overcome. The distinction between the friction force between the tyres and the track surface and the friction in the bearings is a key point in understanding how cars brake.

Falling objects

Galileo was interested in falling objects and so was Newton. If you drop a feather and a hammer at the same time, they do not normally hit the ground at the same time. This seems to disagree with the results of Galileo's legendary Leaning Tower of Pisa experiment. In this experiment cannon balls with different masses fell at the same rate and struck the ground at the same time. The feather and hammer do not usually arrive at the ground together because of the effects of air resistance. The astronaut Dave Scott was able to repeat the feather and hammer experiment on the Moon, where there is no air and therefore no air resistance. The feather and the hammer landed at the same time, which confirmed Galileo's findings.

Figure 9.13 *Astronaut Dave Scott demonstrated that objects with different mass fall with the same acceleration on the Moon. There is no atmosphere to cause air resistance on the Moon.*

The acceleration due to gravity

When a small object – like a steel ball-bearing or a table tennis ball – is dropped through a small distance near the Earth's surface it accelerates. Figure 9.14 shows a sequence of images of a falling table tennis ball taken with a time interval of 0.1 s between each image.

We can calculate the acceleration of the falling ball, using the measurements shown on Figure 9.14.

Measure the distance between two of the images of the table tennis ball – say, the second and the third. This is the distance that the ball travelled during the second interval of one tenth of a second. The average velocity during this time is found by dividing the distance travelled, 14.7 cm, by the time taken, 0.1 s. This gives an average velocity of 147 cm/s or 1.47 m/s over the interval. If we repeat the calculation for the next tenth of a second, between images 3 and 4, we find that the average velocity has increased to 2.45 m/s. We can then use the equation for acceleration:

$$\text{acceleration, a} = \frac{\text{final velocity, v} - \text{initial velocity, u}}{\text{time taken, t}}$$

We find that:

$$a = \frac{2.45 \text{ m/s} - 1.47 \text{ m/s}}{0.1 \text{ s}} = 9.8 \text{ m/s}^2$$

We often use an approximate value of 10 m/s^2 for the acceleration due to gravity in questions. (It makes calculations easier!)

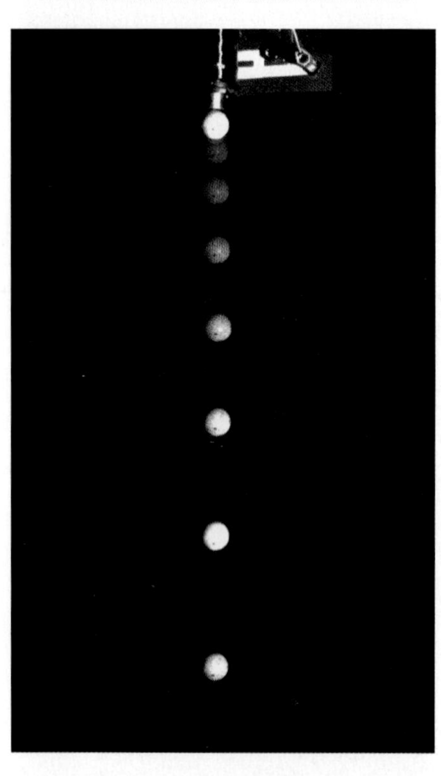
Figure 9.14 *The falling ball accelerates towards the ground.*

The result of this experiment gives us a value for the acceleration caused by the force of gravity. We use the symbol g to represent the acceleration due to gravity.

If there were no air to resist the motion of objects falling through it, *all* objects falling near the Earth's surface would accelerate at 9.8 m/s^2.

Weight

The **weight** of an object is force that acts on it because of gravity. We can work out the weight of an object by using Newton's rule:

force, F (in N) = mass, m (in kg) × acceleration, a (in m/s^2)
F = ma

If we drop an object it accelerates at approximately 10 m/s^2. If the object has a mass of 0.1 kg we can work out the force that must be acting on it to cause the acceleration (its weight):

$$F = ma$$
$$F = 0.1 \text{ kg} \times 10 \text{ m/s}^2$$
$$F = 1 \text{ N}$$

The weight of the 0.1 kg mass is 1 N.

Generally, we use the equation:

weight, W (in N) = mass of object, m (in kg) × acceleration due to gravity, g (in m/s²)

$$W = mg$$

The value of **g** depends on how strong a planet's gravity is. For example, objects on the Moon accelerate at about 1.6 m/s². The numerical value of g is, therefore, a measure of how strong the gravitational pull on an object is. It is sometimes called gravitational field strength and its value is measured in N/kg, the gravitational force per unit mass.

Air resistance and terminal velocity

An object moving through air experiences a friction force which opposes the movement. This force is called **air resistance** or **drag** force. The size of the drag force acting on an object depends on its shape and its speed. Cars are designed to have a low "drag coefficient". The drag coefficient is a measure of how easily an object moves through the air. High-speed trains have a streamlined shape so that air flows more smoothly around them. Streamlined, smooth surfaces produce less drag.

It is particularly important to make fast-moving objects streamlined because the drag force increases with the speed of the object. The fact that drag increases with speed affects the way that dropped objects accelerate, because the faster they get the greater the force opposing their motion becomes.

Objects falling through the air experience two significant forces: the weight force (that is, the pull of gravity on the object) and the opposing drag force.

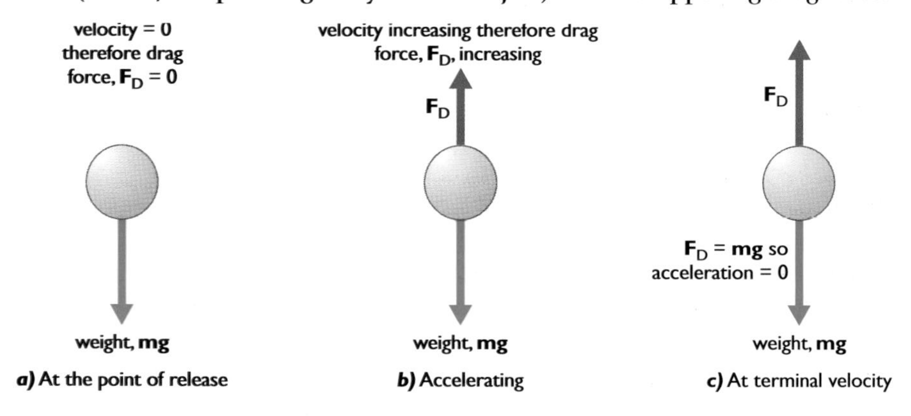

Figure 9.15 *How the forces acting on a body change as its velocity changes.*

In Figure 9.15a the object has just been released and has a starting velocity of 0 m/s. This means that there is no drag. (Remember that the drag force acts on *moving* objects.) The resultant downward-acting force is just the weight force. This force makes the object accelerate towards the Earth.

Figure 9.15b shows the object now moving. Because it is moving it has a drag force, F_D, acting on it. The drag force acts *upwards* against the movement. This means that the resultant downward force on the object is $(mg - F_D)$. You can see that the drag force has made the resultant downward force smaller, so the acceleration is smaller. All the time that the object is

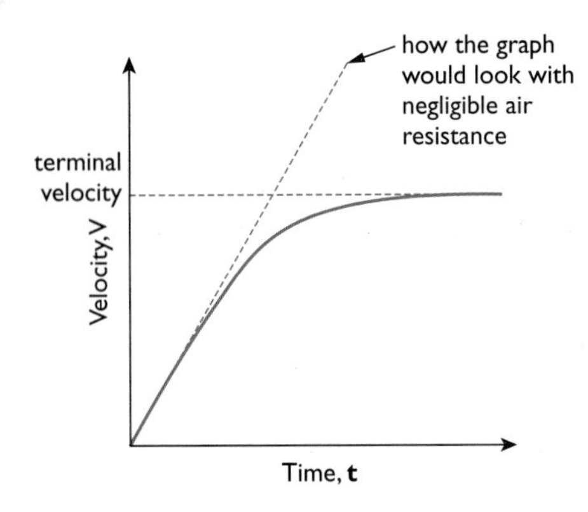

Figure 9.16 *The velocity–time graph for an object accelerating until it reaches terminal velocity.*

accelerating it is getting faster. The faster the object moves the bigger the drag force is.

In Figure 9.15c the drag force has increased to the point where it exactly balances the weight force – since the resultant force on the object is now zero its acceleration is also zero. The object has reached its **terminal velocity** and although it is still falling it will not get any faster. Figure 9.16 shows a velocity–time graph for an object falling through air and reaching terminal velocity.

Parachutes

When a skydiver jumps from a plane at high altitude she will accelerate for a time and eventually reach terminal velocity. Typically this will be between 150 and 200 kph. When she opens her parachute this will cause a sudden increase in the drag force. At this velocity (around 200 kph) the drag force of the parachute is greater than the weight of the skydiver. This means that the resultant force acting on the parachutist acts *upwards* and, for a while, she will decelerate. As she slows down the size of the drag force decreases and, eventually, a new terminal velocity is reached. Obviously the new terminal velocity depends on the design of the parachute, but it must be slow enough to allow the parachutist to land safely. Figure 9.17 shows a velocity–time graph for a skydiver.

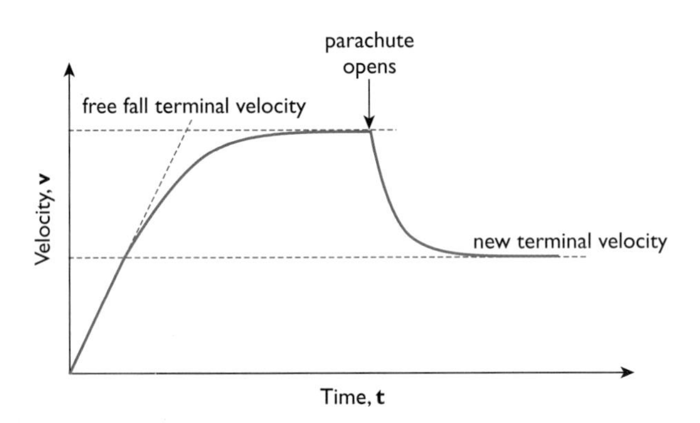

Figure 9.17 *Velocity–time graph for a free-fall parachutist reaching terminal velocity, then opening the parachute.*

Modelling terminal velocity

Objects have to accelerate to quite high speeds in air to reach terminal velocity. This makes demonstrating the effect in a laboratory difficult. However, objects falling through liquids also experience a drag force that increases with speed. The sizes of drag forces in liquids are much higher than in gases. This means that objects falling through liquids have a much lower terminal velocity than objects falling through air, and can be used to model terminal velocity. You can use a tall measuring cylinder filled with water and drop small diameter (1–2 mm) glass beads into it. Alternatively, use a much thicker liquid like glycerine and use small diameter ball-bearings. As well as demonstrating terminal velocity this presents plenty of scope for investigations.

End of Chapter Checklist

If you haven't got a copy of your specification, read the introduction on page iv.

You will need to be able to do some or all of the following. Check your Awarding Body's specification (syllabus) to find out exactly what you need to know.

- Know that forces are pushes and pulls and that they can change the way an object is moving or change the shape of an object.
- Recognise the following types of force: contact, reaction, weight, upthrust, friction, drag, magnetic.
- Know that force is a vector and is measured in newtons.
- Recall Aristotle's theories about motion.
- Understand how Galileo demonstrated, by experiment, the errors in Aristotle's ideas.
- Know that more than one force may act on an object.
- Know that if the forces on an object are balanced, or if no force is acting on the object, the way it is moving *will not change*; if the forces acting on an object are unbalanced, the way it is moving *will change*.
- Recognise that friction is a force that acts to oppose the motion of an object.
- Know that friction acts when an object moves through a liquid or gas.
- Know that friction is essential in certain situations but in others it wastes energy and causes wear.
- Know that weight = mass × acceleration due to gravity.
- Know that all objects fall with the same acceleration *in the absence of air resistance*.
- Know that objects falling through liquids and gases experience a drag force that increases with their speed.
- Understand that a falling object reaches terminal velocity because the drag force on the object balances its weight.
- Understand that the drag force also depends on the shape of the falling object. That is, objects must be smooth or streamlined to reduce drag.

Questions

More questions on forces can be found at the end of Section B on page 139.

1 Name the force that:

a) causes objects to fall towards the Earth

b) makes a marble rolled across level ground eventually come to rest

c) stops a car sinking into the road surface.

2 Name two types of force that oppose motion.

3 Aristotle thought that heavy objects fall faster than light ones.

a) What will happen if you test this idea by releasing a feather and a marble at the same time from the same height above the ground?

b) Do you think that this confirms Aristotle's understanding of the effect of force on motion?

4 The diagram below shows a block of wood on a sloping surface. Copy the diagram and label all the forces that act on the block of wood.

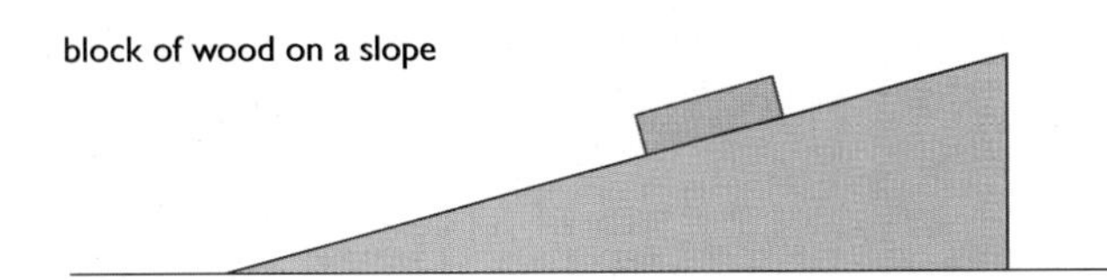

5 A car is travelling along a level road at constant velocity (that is, its speed and direction are not changing). Draw a labelled diagram to show the forces that act on the car.

6 *a)* Why is it vital that there is a friction force that opposes motion when two surfaces try to slide across one another?

b) Give two examples of things that it would be impossible to do without friction.

7 Copy the diagrams below and label the direction in which friction is acting on the objects.

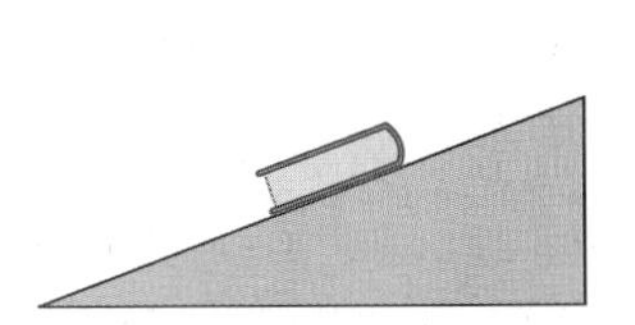

a) a book on a sloping surface

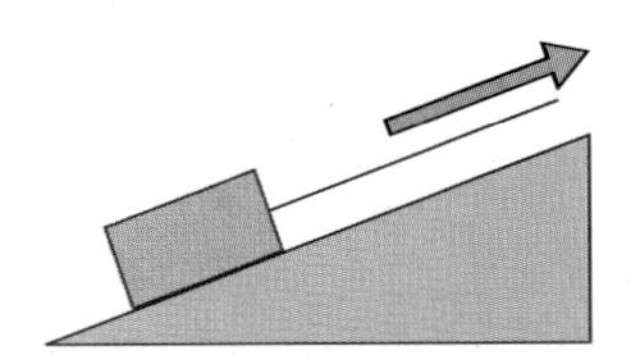

b) a block of wood being pulled up a slope

8 The drawing below shows a car pulling a caravan. They are travelling at constant velocity.

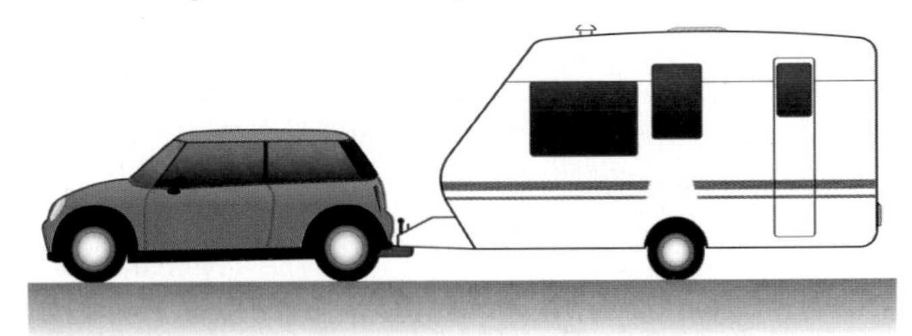

a) Copy the drawing. Label the forces that are acting on the caravan with arrows showing the direction that they act in.

b) Label the forces that act on the car.

9 Calculate the weight of an apple of mass 100 grams:

a) on the Earth

b) on the Moon.

10 What factors affect the drag force that acts on a high-speed train?

11 Describe an experiment to demonstrate terminal velocity. Say what measurements you need to take in your experiment to show that a falling object has reached terminal speed.

12 Look at the velocity–time graph for a free-fall parachutist shown below.

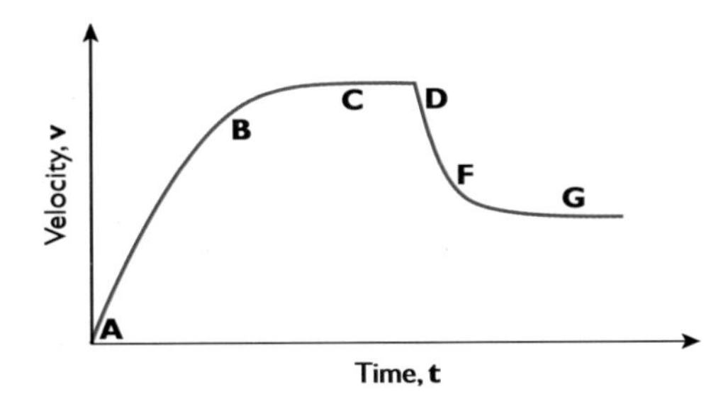

State and explain the direction and relative size of the resultant force acting on the parachutist at each of the points A–G labelled on the graph.

Chapter 10: Force and Acceleration

The way an object moves depends upon its mass and the resultant force acting upon it. In this chapter you will find out how Newton's Laws of Motion help us to predict the way an object will move, particularly in the context of car safety.

Figure 10.1 *This aircraft has only a short distance to travel before taking off.*

The aircraft in Figure 10.1 must accelerate to a very high speed in a very short time. The resultant force on the plane causes the acceleration. The forces that act horizontally on the aircraft are the friction force between the wheels and the deck, and air resistance, when the aircraft starts to move. At the start, the forward thrust of the aircraft engines is much greater than air resistance and friction, so there is a large resultant force to cause the acceleration. When the aircraft lands on the carrier deck it must decelerate to a halt in a short distance. Parachutes and drag wires are used to provide a big resultant force acting in the opposite direction to the aircraft's movement. In this chapter we look at how acceleration is related to the force acting on an object.

Newton's Laws of Motion – a first look

Scientists use the word "law" very cautiously. Only when a hypothesis (idea) has been tested many times independently by careful experiment is it raised to the status of a "law". Einstein showed that in special situations Newton's Laws break down, but they are still accurate enough to predict the way objects respond to forces with a high degree of accuracy.

Sir Isaac Newton lived from 1642 to 1727. He made many famous discoveries and some important observations about how forces affect the way objects move. The first observation, called Newton's 1st Law, was:

Things don't speed up, slow down or change direction unless you push (or pull) them.

Newton didn't put it quite like that, of course! He said that a body would continue to move in a straight line at a steady speed unless a resultant force was acting on the body. We have already discussed the idea of resultant force when considering a tug of war contest in Chapter 9. If the forces acting on an object are balanced then a stationary object stays in one place, and a moving object continues to move in just the same way as it did before the forces were applied.

This idea sounds obvious but it is not our everyday experience of how things move. Some form of friction usually acts on moving objects, causing them to slow down and eventually stop. Newton's first law is only obvious when we see objects moving with no resultant force acting on them. This is not a common situation! It is more noticeable when friction forces are very small. For example, if you throw a stone across the surface of a frozen pond it keeps moving in a straight line for a long time before coming to a halt.

Newton then asked another obvious question: how does the acceleration of an object depend on the force that you apply to it? Again Newton's formal statement of the answer (Newton's 2nd Law of Motion) sounds (and is) complicated, but the basic findings are quite simple:

The bigger the force acting on an object, the faster the object will speed up.

Objects with greater mass require bigger forces than those with smaller mass to make them speed up (accelerate) at the same rate.

The force referred to is the resultant force acting on the object.

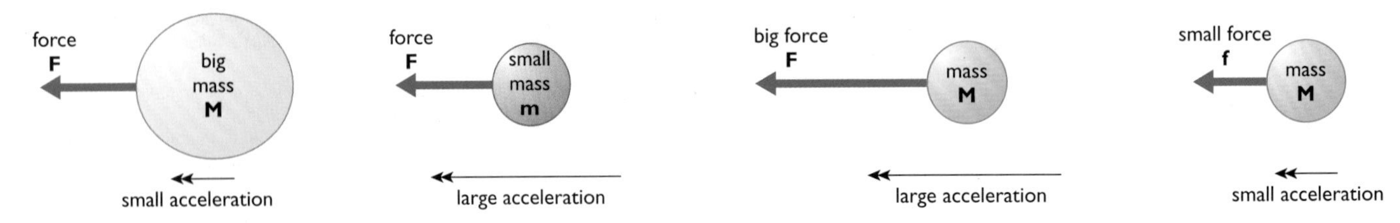

a) When the same force is applied to objects with different mass, the smaller mass will experience a greater acceleration.

b) Different sized forces are applied to objects with the same mass. The small force produces a smaller acceleration than the large force.

Figure 10.2 *The acceleration of an object is affected by both its mass and the force applied to it.*

Another of Newton's important discoveries about forces is this:

When you push something it pushes back just as hard, but in the opposite direction.

This is called Newton's 3rd Law and is usually stated as "For every action there is an equal and opposite reaction". When you sit down, your weight pushes down on the seat. The seat pushes back on you with an equal, but upward, force. An experiment to give you the idea of what this law means is shown in Figure 10.3.

In Figure 10.3, person X is clearly pushing person Y but it is not obvious that Y is pushing X back. When *both* X and Y move it is clear that X has been affected by a force pushing him to the left. The force felt by X is the reaction force.

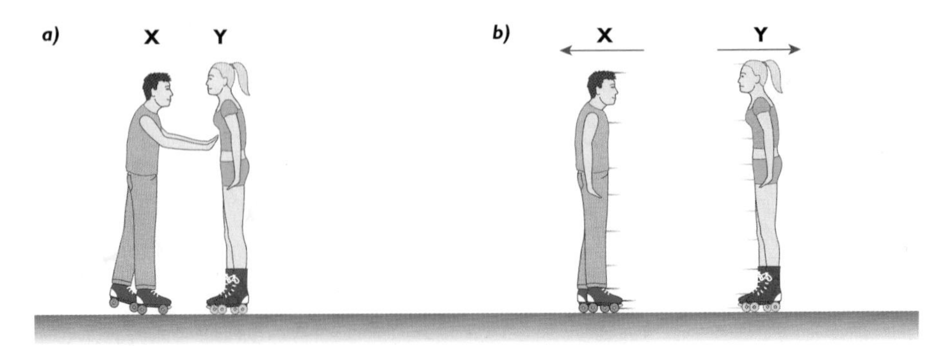

Figure 10.3 *For every action there is an equal and opposite reaction.*

Investigating force, mass and acceleration

The experiment described in Figure 10.4 shows how the relationship between force, mass and acceleration can be investigated. It uses an air-track so that friction is reduced as much as possible. This means that the only effective force acting on the glider is the force provided by the pull of the weight on the nylon line. As the force is constant, the acceleration of the glider will be constant.

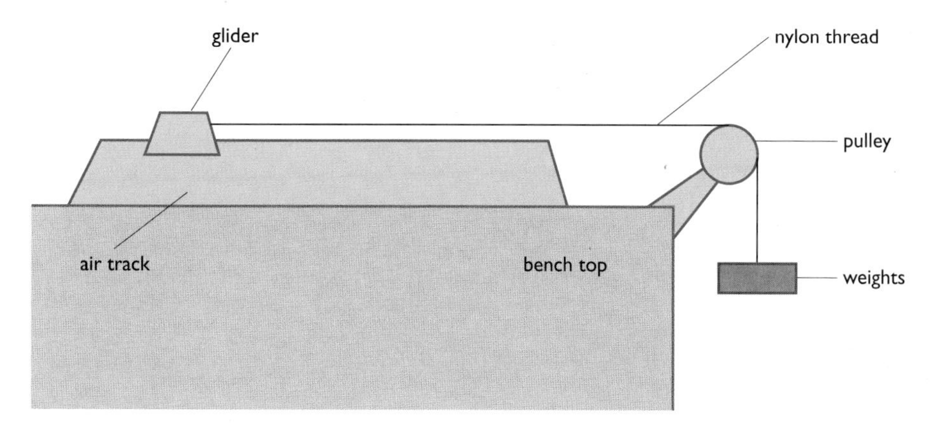

Figure 10.4 *You can use the air-track to find the acceleration caused by a particular force.*

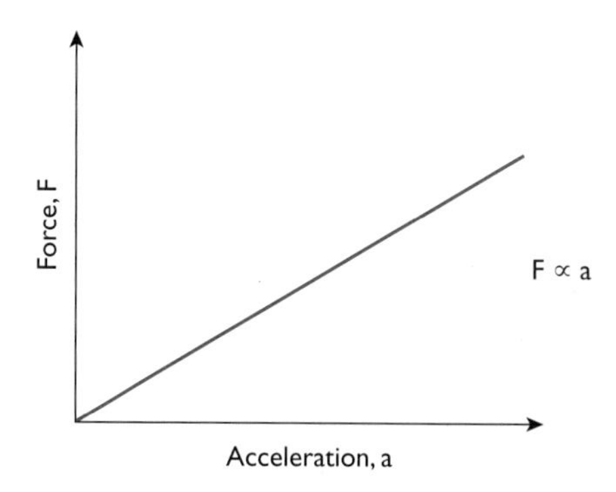

Figure 10.5 *Force is proportional to acceleration.*

The acceleration of the glider can be measured by taking a series of pictures using a digital video camera. Alternatively, since the acceleration is constant, we can use the equation $s = \frac{1}{2}at^2$ from Chapter 8 ($u = 0$ as the glider starts from rest). Measure the time, t, to travel a distance, s, and substitute into the rearranged formula $a = 2s/t^2$.

The acceleration can be found by measuring the distance travelled from the start for each image. Since the time between each image is known, a graph of displacement against time can be drawn. The gradient of the displacement–time graph gives the velocity at a particular instant, so data for a velocity–time graph can be obtained. The gradient of the velocity–time graph produced is the acceleration of the glider.

Figure 10.5 shows a graph of force against acceleration when the mass of the glider is constant and the accelerating force is varied.

The graph is a straight line passing through the origin, which shows that:

force is proportional to acceleration

$$F \propto a$$

So doubling the force acting on an object doubles its acceleration.

In a second experiment, the accelerating force is kept constant and the mass of the glider is varied.

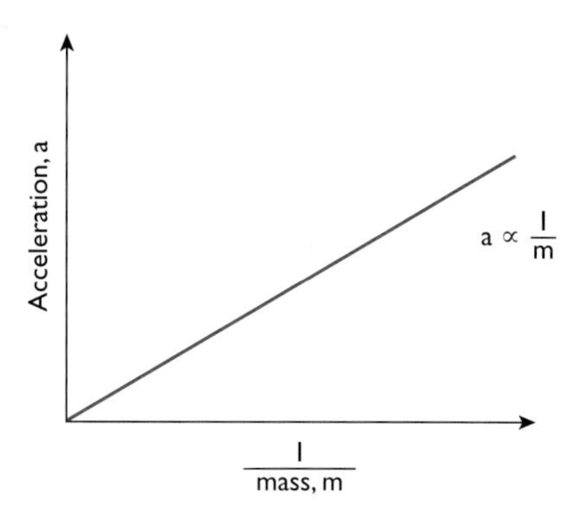

Figure 10.6 *Acceleration is inversely proportional to mass.*

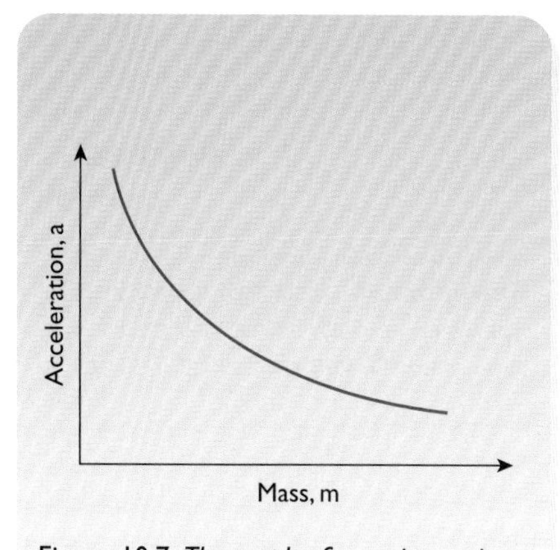

Figure 10.7 *The graph of a against m is a curve. Plotting a against 1/m gives a straight line. This makes it easier to spot the way that a is affected by m.*

Figure 10.6 shows the acceleration of the glider plotted against 1/m. This is also a straight line passing through the origin, showing that:

acceleration is inversely proportional to mass

$$a \propto \frac{1}{m}$$

This means that for a given resultant force acting on a body, doubling the mass of the body will halve the acceleration.

Combining these results gives us:

force, F (in N) = mass, m (in kg) × acceleration, a (in m/s²)

$$F = ma$$

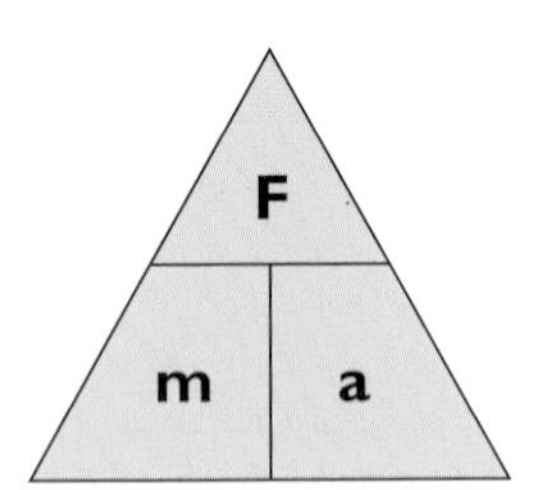

Figure 10.8 *The equation from Newton's 2nd Law of Motion can be rearranged using the triangle method.*

Force is measured in newtons (N), mass is measured in kilograms (kg), and acceleration is measured in metres per second squared (m/s^2). From this we see that:

One newton is the force needed to make a mass of one kilogram accelerate at one metre per second squared.

Car safety and Newton's Laws of Motion

Deceleration in a collision

If you are designing a car for high acceleration, the equation F = ma tells you that the car should have low mass and the engine must provide a high accelerating force. You must also consider the force needed to stop the car.

When a moving object is stopped, it decelerates.

A negative acceleration is a deceleration.

If a large deceleration is needed then the force causing the deceleration must be large, too. Usually a car is stopped by using the brakes in a controlled way so that the deceleration is not excessive. In an accident the car may collide with another vehicle or obstacle, causing a very rapid deceleration.

***worked* example**

Reminder: **v** is the final velocity, **u** is the initial velocity and **t** is the time for the change in velocity to take place.

Example 1

A car travelling at 20 m/s (45 mph) collides with a stationary lorry and is brought to rest in just 0.02 s. Calculate the deceleration of the car.

$$\text{acceleration} = \frac{\text{increase in velocity}}{\text{time taken}}$$ (see page 87)

$$a = \frac{v - u}{t}$$

$$a = \frac{0\ m/s - 20\ m/s}{0.02\ s}$$

$$a = \frac{-20\ m/s}{0.02\ s}$$

$$a = -1000\ m/s^2$$

The minus sign in Example 1 for velocity indicates that the velocity has decreased.

A person of mass 50 kg in the car would experience the same deceleration when she came into contact with a hard surface in the car. This could be the dashboard or the windscreen. Calculate the force that the person experiences.

$F = ma$

$F = 50\ \text{kg} \times 1000\ \text{m/s}^2$

$F = 50\,000\ \text{N}$

This huge force would undoubtedly have unfortunate effects on the person!

Figure 10.9 *Cars are designed to crumple.*

Cars are now designed to have a rigid passenger cell or compartment with crumple zones in front and behind. The crumple zones, as the name suggests, collapse during a collision and increase the time during which the car is decelerating. For instance, if the deceleration time in Example 1 above is increased from 0.02 s to 1 s, then the impact causes a much smaller force of just 1000 N to act on the passenger, greatly increasing her chances of survival.

Crumple zones are just one of the safety features now used in modern cars to protect the passengers in an accident. They only work if the passengers are wearing seat belts so that the reduced deceleration applies to their bodies too. Without seat belts, the passengers will continue moving forward (obeying Newton's 1st Law, see page 107) until they come into contact with some part of the car or with a passenger in front. If they hit something that does not crumple they will be brought to rest in a very short time, which – as we have seen in Example 1 – means a large deceleration and, therefore, a large force acting on them.

Cars are now fitted with air-bags to reduce the forces acting on passengers during collisions, again by extending the time of deceleration. Air-bags are triggered by devices called accelerometers that detect the rapid deceleration that occurs during a collision.

Friction and braking

Brakes on cars and bicycles work by increasing the friction between the rotating wheels and the body of the vehicle, as shown in Figure 10.10.

Figure 10.10 *These bicycle brakes and motorcycle disc brakes each work using friction. Friction is necessary if we want things to stop.*

The "tread" of a tyre is the grooved pattern moulded into the rubber surface. It is designed to keep the rubber surface in contact with the road by throwing water away from the tyre surface.

The friction force between the tyres and the road will depend on the condition of the tyres and the surface of the road. (It also depends on the weight of the vehicle.) If the tyres have a good tread, are properly inflated and the road is dry, the friction force between the road and the tyres will be at its maximum.

Unfortunately, we do not always travel in ideal conditions. If the road is wet or the tyres are in bad condition the friction force will be smaller. If the brakes are applied too hard, the tyres will not grip the road surface and the car will skid. Once the car is skidding the driver no longer has control and it will take longer to stop. Skidding can be avoided by applying the brakes appropriately, so that the wheels do not lock. Most modern cars are fitted with ABS (anti-lock braking system) to reduce the chance of a skid occurring. ABS is a computer-controlled system that senses when the car is about to skid and momentarily releases the brakes.

Safe stopping distance

The Highway Code gives stopping distances for cars travelling at various speeds. The **stopping distance** is the sum of the **thinking distance** and the **braking distance**. The faster the car is travelling the greater the stopping distance will be.

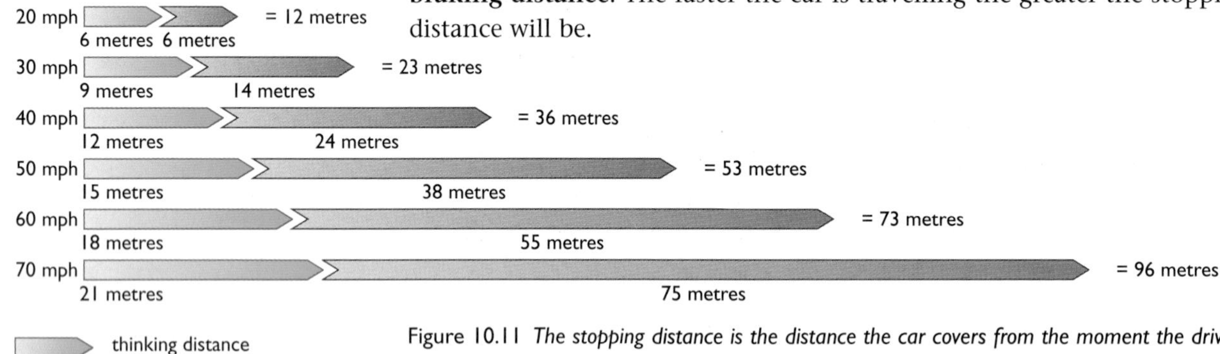

Figure 10.11 *The stopping distance is the distance the car covers from the moment the driver is aware of the need to stop to the point at which the vehicle comes to a complete halt.*

1 Thinking distance

When a driver suddenly sees an obstacle ahead, it takes time for him or her to respond to the new situation before taking any action, such as braking. This time is called **reaction time** and will depend on the person driving the car. It will also depend on a number of other factors including whether the driver is tired or under the influence of alcohol or other drugs that slow reaction times. Poor visibility may also make it difficult for a driver to identify a hazard and so cause him or her to take longer to respond. Clearly, the longer the driver takes to react, the further the car will travel before braking even starts – that is, the longer the thinking distance will be. Equally clear is the fact that the higher the car's speed, the further the car will travel during this "thinking time". If the distance between two cars is not at least the thinking distance then, in the event of an emergency stop by the vehicle in front, a violent collision is inevitable.

2 Braking distance

With ABS braking, in an emergency you brake as hard as you can. This means that the braking force will be a maximum and we can work out the deceleration using the equation below.

> **Reminder:** The equation F = ma can be rearranged using the triangle method (see page 110).

F = ma, rearranged to give:

$a = \frac{F}{m}$

It is worth pointing out here that vehicles with large masses, like lorries, will have smaller rates of deceleration for a given braking force – they will, therefore, travel further while braking.

Chapter 8 shows that the distance travelled by a moving object can be found from its velocity–time graph. The area under the graph gives the distance travelled. Look at the velocity–time graphs in Figures 10.12 and 10.13.

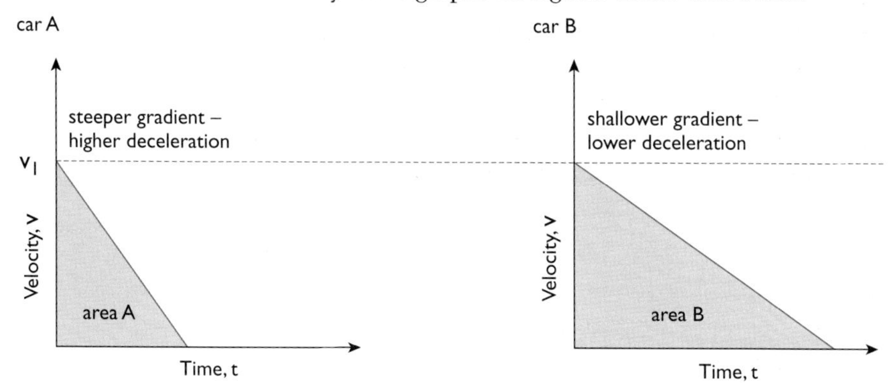

Figure 10.12 *Velocity–time graphs for two cars braking at different rates from the same speed, v_1, to rest.*

Figure 10.12 shows two cars, A and B, braking from the same velocity. Car A is braking harder than car B and comes to rest in a shorter time. Car B travels further before stopping, as you can see from the larger area under the graph. Remember that the maximum rate of deceleration depends on how hard you can brake without skidding – in poor conditions it will be lower.

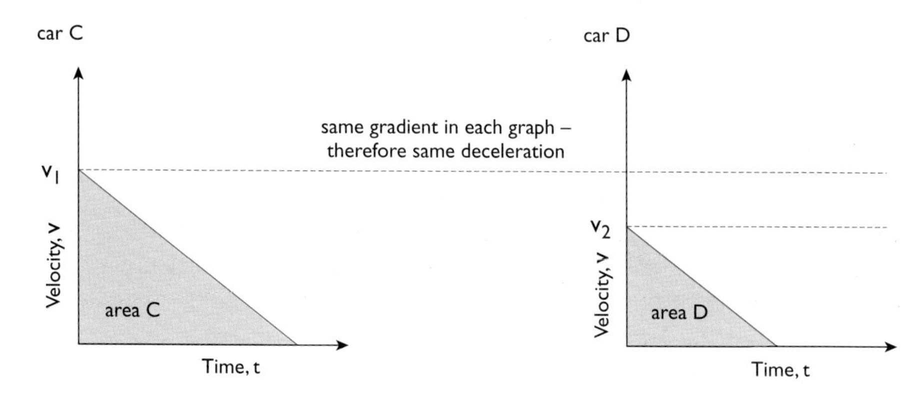

Figure 10.13 *Velocity–time graphs for two cars braking at the same rate to rest, from different speeds, v_1 and v_2.*

Figure 10.13 shows two cars, C and D, braking at the same rate, as you can see from the gradients. Car C is braking from a higher velocity and so takes longer to stop. Again, the greater area under the graph for car C shows that it travels further whilst stopping than car D.

Vehicles *cannot* stop instantly! Remember also that the chart in Figure 10.11 shows stopping distances in *ideal* conditions. If the car tyres or brakes are in poor condition, or if the road surface is wet, icy or slippery because of oil spillage, then the car will travel further before stopping.

End of Chapter Checklist

If you haven't got a copy of your specification, read the introduction on page iv.

You will need to be able to do some or all of the following. Check your Awarding Body's specification (syllabus) to find out exactly what you need to know.

- Know that objects keep moving in a straight line at constant speed unless a resultant force acts on them.
- Know that when a resultant force acts on an object along the line in which the object is moving, then the object will accelerate (or decelerate if the force acts in the opposite direction to that of the motion of the object).
- Know that for a particular object the bigger the resultant force acting on it the bigger the acceleration.
- Know that if the *same* force is applied to objects of different mass then the more massive the object the smaller its rate of acceleration will be.
- Recall and use the equation force = mass × acceleration, F = ma, rearranging it as necessary.
- Understand that stopping distance for a moving vehicle is the sum of the thinking distance and the braking distance.
- Know that the thinking distance is the time before the brakes are applied and that it is affected by tiredness, drugs and poor visibility.
- Know that the distance travelled while thinking depends on the speed of the vehicle.
- Know that the braking distance is the distance travelled by the vehicle *after* the brakes have been applied.
- Know that the braking distance (and, therefore, the stopping distance) is affected by brake and tyre condition and the condition of the road surface.
- Appreciate that the mass of a vehicle will also affect its braking distance.
- Be able to use Newton's Laws and graphs of motion to calculate stopping distances.

Questions

More questions on force and acceleration can be found at the end of Section B on page 139.

1 What is meant by a resultant force? Illustrate your answer with an example.

2 What three different effects can a resultant force have on an object?

3 Rockets burn fuel to give them the thrust needed to accelerate. As the fuel burns the mass of the rocket gets smaller. Assuming that the rocket motors provide a constant thrust force, what will happen to the acceleration of the rocket as it burns its fuel?

4 ***a)*** What force is required to make an object of mass 500 g accelerate at 4 m/s^2? (Take care with the units!)

b) An object accelerates at 0.8 m/s^2 when a resultant force of 200 N acts upon it. What is the mass of the object?

c) What acceleration is produced by a force of 250 N acting on a mass of 25 kg?

5 Crumple zones in cars are designed to reduce the forces during a collision. Explain how they do this. Refer to Newton's Laws of Motion in your answer.

6 Parachutists bend their legs and roll with the fall on landing. They do this to reduce the chance of injury. Explain why this technique means that they are less likely to suffer broken bones on hitting the ground.

7 Explain the meaning of the following terms used in *The Highway Code* in the section about stopping vehicles in an emergency:

a) thinking distance

b) braking distance

c) overall stopping distance.

8 What factors affect the braking distance of a vehicle?

9 The diagram below shows the velocity–time graph for a car travelling from the instant that the driver sees an obstacle in the road ahead.

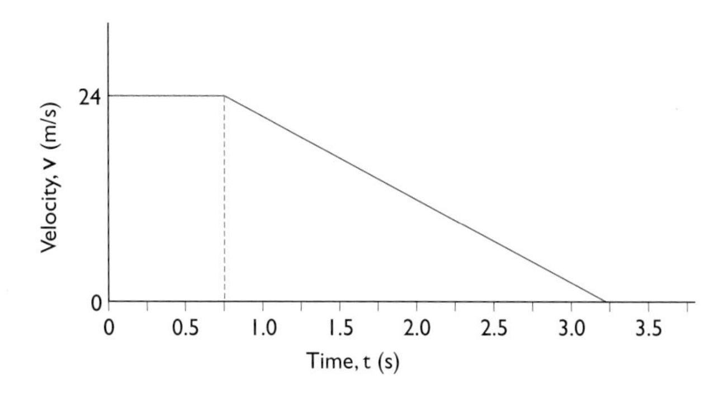

Use the graph to find out:

a) how long the driver takes to react to seeing the obstacle (reaction time)

b) how far the car travels in this reaction time

c) how long it takes to bring the car to a halt once the driver starts braking

d) the total distance the car travels before stopping.

Chapter 11: The Turning Effect of Forces

A force can have a turning effect – it can make an object turn around a fixed pivot point. When the anticlockwise turning effects of forces are balanced by turning forces in the clockwise direction, the object will not turn – it is in balance.

Figure 11.1 *Turning effects are used in many places.*

Forces applied to objects can make them accelerate or decelerate. In the examples in Figure 11.1, the forces acting are having a **turning effect**. They are tending to make the objects, like the see-saw, turn around a fixed point called a **pivot** or **fulcrum**. We use this turning effect of forces all the time. In our bodies the forces of our muscles make parts of our bodies turn around joints like our elbows or knees. When you turn a door handle, open a door or lever the lid off a tin of paint with a screwdriver you are using the turning effect of forces. Understanding the turning effect of forces is important. Sometimes we want things to turn or rotate – the see-saw wouldn't be much fun if it didn't. However, sometimes we want the turning effects to balance so that things don't turn – it would be disastrous if the crane did not balance!

Opening a door

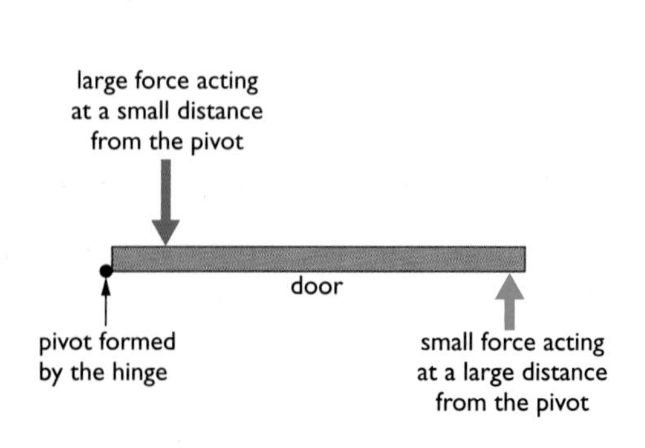

Figure 11.2 *The distance of the force from the pivot is crucial.*

Challenge a partner (perhaps one who thinks that he or she is strong!) to try to hold a door closed while you try to open it – then explain the rules! They can apply a pushing force but *no further than 20 cm* from the hinge, while you try to pull the door open by pulling on the handle. You will be able to open the door quite easily.

You will realise that you have an advantage because the turning effect of your force doesn't just depend on the *size* of the force you apply but also on the *distance* from the hinge or pivot at which you apply it. You have the advantage of greater leverage.

The moment of a force

The turning effect of a force about a hinge or pivot is called its **moment**. The moment of a force is defined like this:

moment of a force (in Nm) = force, F (in N) × distance from pivot, d (in m)
moment = Fd

The moment of a force is measured in newton metres (Nm) because force is measured in newtons and the distance to the pivot is in metres. We need to be precise about what distance we measure when calculating the moment of a force. Look at the diagrams in Figure 11.3.

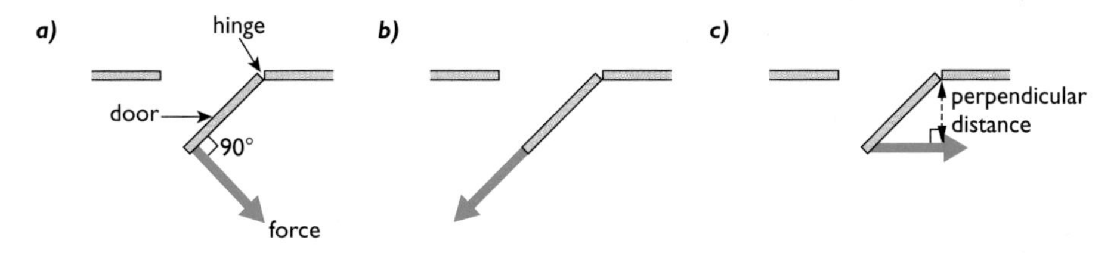

Figure 11.3 *For a force to have its biggest effect it should be at 90° to the lever.*

If you think about the simple door opening "competition" we discussed earlier you will realise that, for a force to have the biggest turning effect, it should be applied as in Figure 11.3a – that is, its line of action should be perpendicular (at 90°) to the lever.

In Figure 11.3b the force has *no turning effect at all* because the line along which the force is acting passes through the pivot.

Figure 11.3c shows how the distance to the pivot must be measured to get the correct value for the moment. The distance is the *perpendicular distance from the line of action of the force to the pivot.*

In balance

An object will be in balance (that is, it will not try to turn about a pivot point) if:

sum of anticlockwise moments = sum of clockwise moments

For example, Figure 11.4 shows two children sitting on a see-saw.

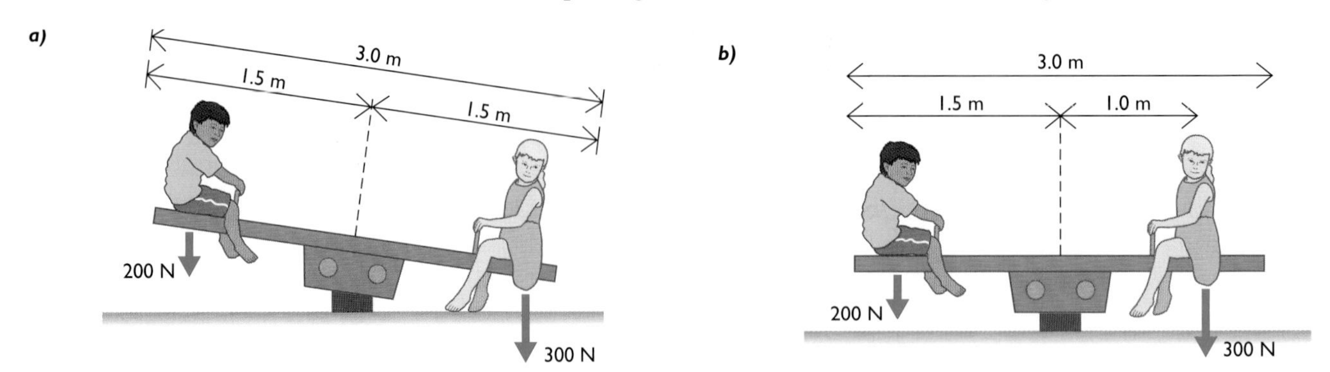

Figure 11.4 *The moment of the heavier child is reduced when he sits closer to the pivot, so that it balances the moment of the lighter child.*

In Figure 11.4a:

the anticlockwise moment = 200 N × 1.5 m = 300 Nm

the clockwise moment = 300 N × 1.5 m = 450 Nm

So the see-saw is not balanced and tips down to the right as it rotates clockwise about the pivot.

In Figure 11.4b:

the anticlockwise moment = 200 N × 1.5 m = 300 Nm

the clockwise moment = 300 N × 1.0 m = 300 Nm

So the see-saw is now balanced.

Look back at Figure 11.1. The load arm is long so that the crane can reach across a construction site and move loads backwards and forwards along the length of the arm. The weight of the long load arm and the load must be counterbalanced by the large concrete blocks at the end of the short arm that projects out behind the crane controller's cabin. The **counterbalance** weights must be large because they are positioned close to the pivot point, where the crane tower supports the crosspiece of the crane. Without careful balance the turning forces on the support tower could cause it to bend and collapse.

Centre of mass

Try balancing a ruler on your finger, as shown in Figure 11.5.

When the ruler is balanced, the anticlockwise moment is equal to the clockwise moment, but there are no other forces acting in this situation than the weight of the ruler itself. We know that the weight of the ruler is due to the pull of the Earth's gravity on the mass of the ruler. The mass of the ruler is evenly spread throughout its length. It is not, therefore, surprising to find that the ruler balances at its centre point.

We say that the **centre of mass** of the ruler is at this point – it is the point where the whole of the mass of the ruler appears to act. This means that the weight force acts at this point, so if we support the ruler at this point there is no turning moment in any direction about the point, and it balances.

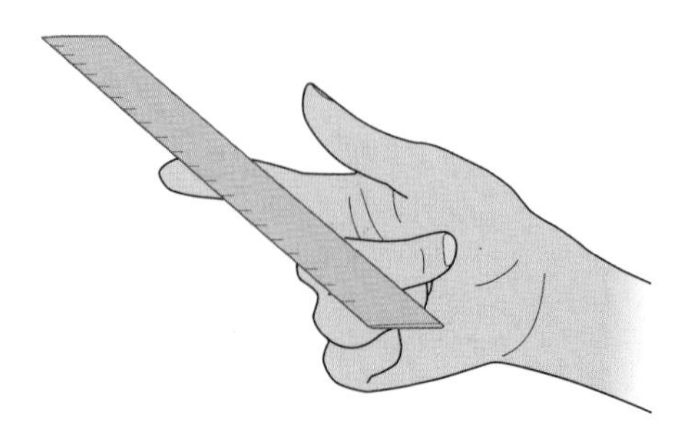

Figure 11.5 *Can you find the point at which the ruler balances?*

Finding the centre of mass of an irregular sheet of card

If you have a symmetrical sheet of card or any other uniform material then finding its centre of mass is quite straight forward – it will be located where the axes of symmetry cross, as shown in Figure 11.6.

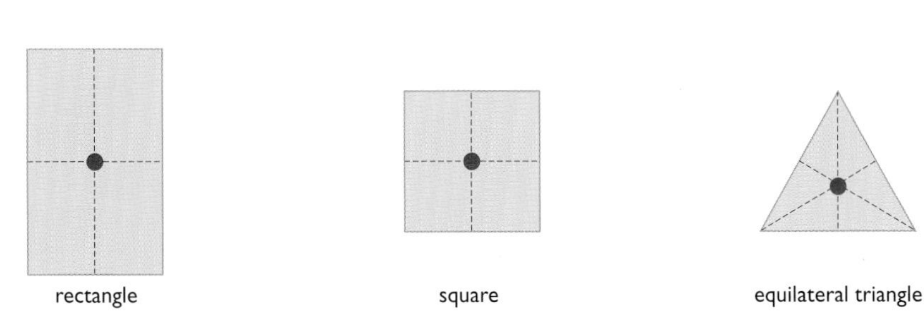

Figure 11.6 *The centre of mass for these three regular shapes is where the axes of symmetry cross. An axis of symmetry can be found using a plane mirror. If you place a plane mirror along one of the dotted lines in any of the above shapes, the reflection in the mirror looks exactly like the original.*

If the sheet of card has an irregular shape then this method clearly cannot be applied. To find the centre of mass of an irregularly shaped sheet simply suspend it freely by a point on its edge, as shown in Figure 11.7a.

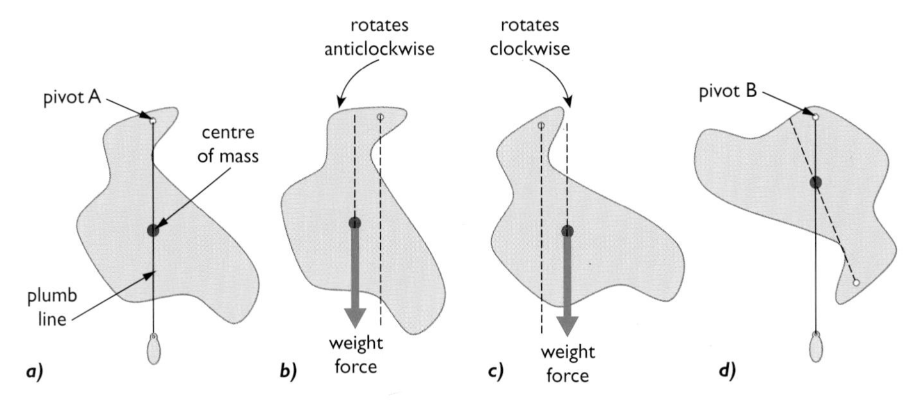

Figure 11.7 *Finding the centre of mass for an irregularly shaped object.*

The point of suspension is a pivot point so the shape will come to rest in the balance position. The shape is in balance when there is no turning effect about the pivot. This will be when the centre of mass of the shape is *vertically below the pivot.* If it were not, then the line of action of the weight force, passing through the centre of mass, would not pass through the pivot and there would be a turning effect. Figures 11.7b and c show that if the centre of mass is not directly below the pivot, there is a turning effect acting on the shape – so it rotates, until the centre of mass is below the pivot and the system is in balance.

If a plumb line (a small mass on the end of a thread) is hung from the pivot point, a line can be drawn perpendicularly down from the pivot – the centre of mass must lie somewhere along this line. If this procedure is repeated, suspending the shape from a different pivot point as shown in Figure 11.7d, a second perpendicular line can be drawn on the sheet. The centre of mass will lie at the point where the two lines cross.

Objects not pivoted at the centre of mass

A simple see-saw is a uniform beam pivoted in the middle. The centre of mass of a uniform beam is in the middle, so the see-saw is pivoted through its centre of mass. When an object is not pivoted through its centre of mass the weight of the object will produce a turning effect. This is shown in Figure 11.8.

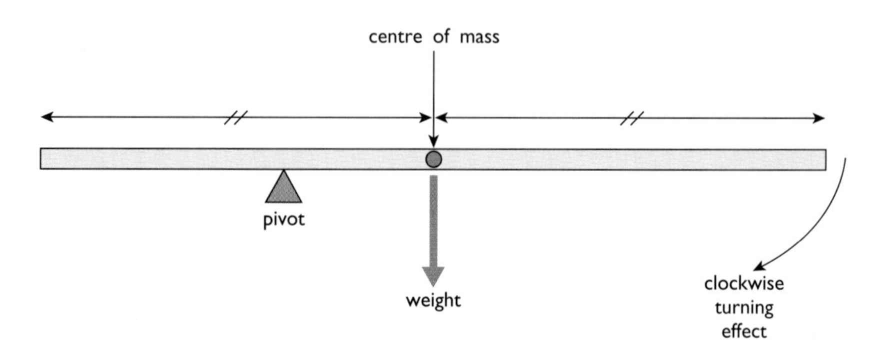

Figure 11.8 *The weight of the beam causes a clockwise turning effect about the pivot.*

Centre of mass and stability

The position of the centre of mass of an object will affect its stability. A **stable** object is one that is difficult to push over; when pushed and then released it will tend to return to its original position.

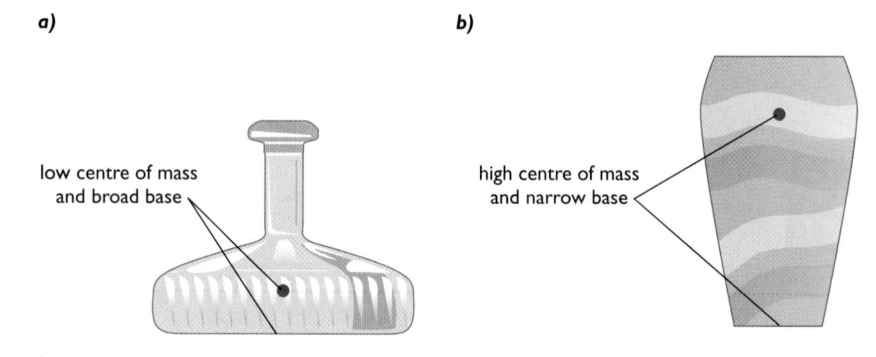

Figure 11.9 *Stable and unstable objects.*

The shape in Figure 11.9a is typical for a ship's decanter (bottle). Bottles used on ships need to be difficult to knock over for obvious reasons! It is stable because, when tipped, its low centre of mass and wide base result in a turning moment that tries to pull it back to its original position. The object in Figure 11.9b has a higher centre of mass and smaller base, so it is much less stable. Only a small displacement is needed to make it topple over. Designers of robots for "Robot Wars" should aim for a low centre of mass and a wide base, or design robots that can work either way up!

End of Chapter Checklist

If you haven't got a copy of your specification, read the introduction on page iv.

You will need to be able to do some or all of the following. Check your Awarding Body's specification (syllabus) to find out exactly what you need to know.

- Understand that forces can have a turning effect or moment on a body.
- Know that the moment of a force is given by force × perpendicular distance from force to pivot.
- Understand and apply the condition for balance to an object: anticlockwise acting moments = clockwise acting moments.
- Know what is meant by centre of mass.
- Find the centre of mass of regularly and irregularly shaped sheets of material by simple experiment.
- Appreciate that stable objects have low centres of mass and wide bases.

Questions

More questions on the turning effect of forces can be found at the end of Section B on page 139.

1 Look at the diagram below. It shows various forces acting on objects about pivots. Which one has the largest moment? Put the diagrams in order starting with the largest moment.

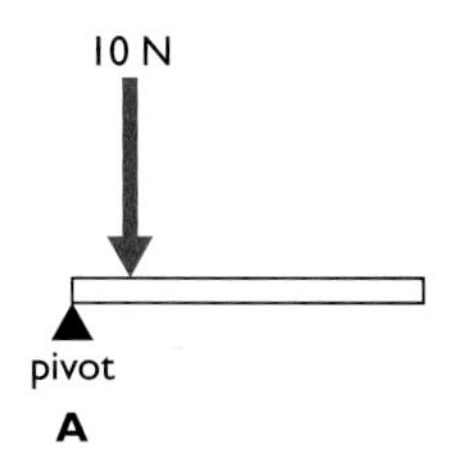

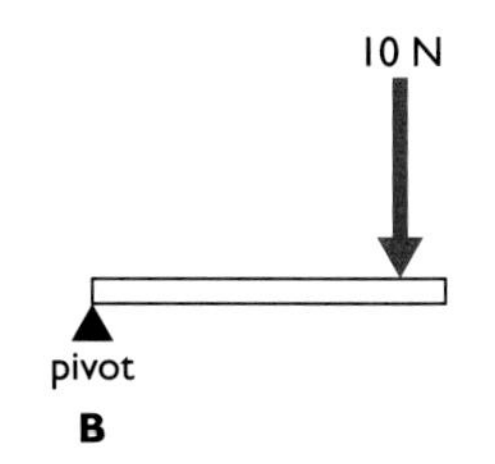

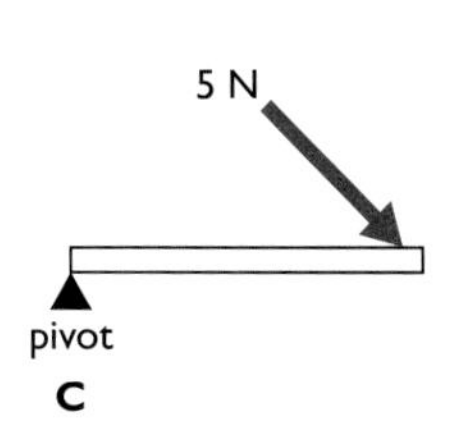

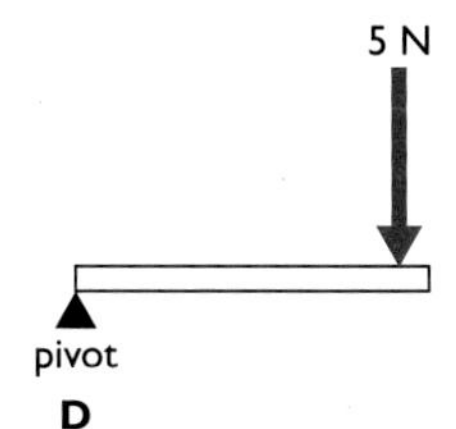

2 ***a)*** Which see-saw in the diagram below is balanced?

b) In which direction will the unbalanced see-saws tip?

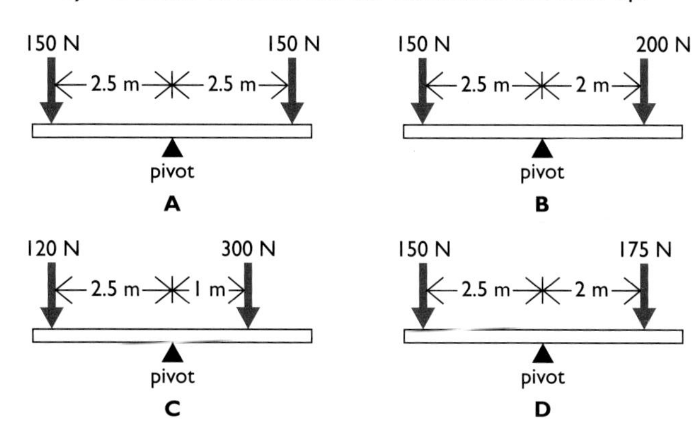

3 Stability is a very important safety feature in public vehicles like buses. To ensure that buses are stable they are tested as follows.

The bus is tilted until it topples and the angle at which it becomes unstable is noted. The test is done with the lower deck empty and with the upper deck carrying a full load of passengers (not real ones, of course).

a) Explain the effect of loading the bus in this way on its stability.

b) Why do you think that buses are tested under these loading conditions?

Chapter 12: Momentum

Momentum is possessed by masses in motion – it is calculated by multiplying the mass of an object by its velocity. In this chapter you will learn that, when objects speed up or slow down, the rate of change of momentum is proportional to the force causing the change. You will also see that momentum is conserved in collisions and explosions.

Do not confuse momentum, a property of moving masses, with moment, the turning effect of a force.

Figure 12.1 *Newton's cradle.*

Newton's cradle is an entertaining toy but it also demonstrates a physics conservation law. When one of the balls is drawn back a short way and released it swings and collides with the remaining group of balls. After the impact the ball at the opposite end springs away and swings out as far as the first ball was drawn back to start with. If two balls are drawn back and released then two balls will move away at the opposite end as the collision occurs. The moving ball has momentum, and momentum is conserved in collisions. This chapter is about **momentum** and what is meant by conservation of momentum.

Momentum

We talk about objects "gaining momentum" in everyday speech. When we say this we are usually trying to get across an idea of something becoming more difficult to stop. Sometimes we use the word in a way that is very close to the way it is defined in physics – for example, if a car starts rolling down a hill we might say that the car is "gaining momentum" as it speeds up.

In physics, momentum is a quantity possessed by *masses in motion*. Momentum is a measure of how difficult it is to stop something that is moving. We calculate the momentum of a moving object using the formula:

momentum (in kg m/s) = mass, m (in kg) × velocity, v (in m/s)
momentum = mv

Momentum is a **vector** quantity and is measured in kilogram metres per second (kg m/s) – provided that mass in the above equation is measured in kg and velocity in m/s.

You can see from the equation that the more mass an object has, the more momentum it will have when moving. The momentum of a moving object also increases with its speed.

Newton's Second Law of Motion and momentum

We have already discussed the relationship F = ma discovered by Newton (see page 102). A more precise statement of Newton's discovery would be that when a resultant force acts on an object it causes a change in the momentum of the object in the direction of the resultant force. Newton discovered that the rate of change of momentum of an object is proportional to the force applied to that object. This means that if you double the force acting on an object its momentum will change twice as quickly. Figure 12.2 shows the effect of the thrust force on the velocity of a space shuttle, and, therefore, on its momentum.

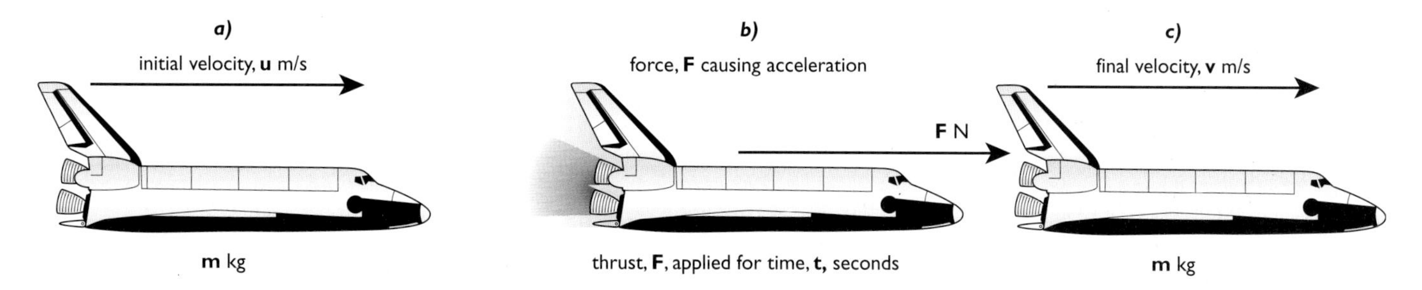

Figure 12.2 *The thrust of the rocket motor makes the velocity and, therefore, the momentum of the shuttle increase.*

initial momentum of object	$= mu$
and final momentum of object	$= mv$
therefore increase in momentum	$= mv - mu$
so, rate of increase of momentum	$= \dfrac{(mv - mu)}{t}$

As stated above, Newton identified a *proportional* relationship between the rate of increase of momentum and the force applied, but with the system of units we use the relationship appears as shown:

$$\mathbf{F = \frac{(mv - mu)}{t}}$$

If you look at this equation you will notice it can be rearranged to give the more familiar equation $F = ma$:

$F = (mv - mu)/t$
$F = m(v - u)/t$
since $(v - u)/t = a$,
then $F = ma$

The rearrangement is possible because we have assumed that the mass of the object involved is constant. This will not always be the case in real situations – for example, when a space shuttle is launched the mass of the rocket changes continuously as fuel is burned and rocket stages are jettisoned. However, you will not meet problems like this at GCSE level.

Momentum and collisions

We can express Newton's Second Law in terms of momentum change, as follows:

force × time = increase in momentum

This simply says that a bigger force applied to an object for a longer time will result in a greater change in the momentum of the object.

The term **impulse** is used for the product (force × time).

Consider what happens when two balls collide, as shown in Figure 12.3.

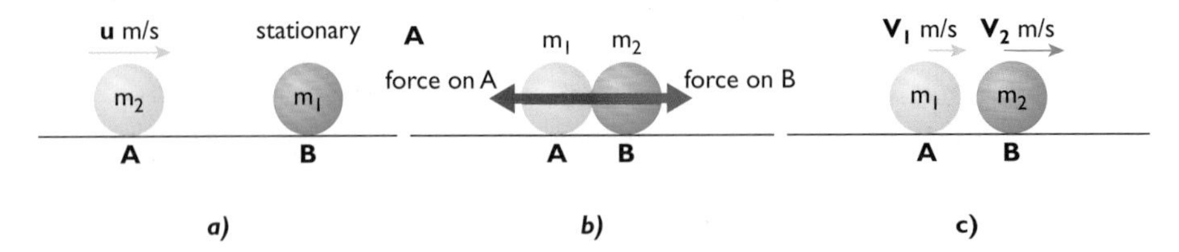

Figure 12.3 a) *Moving ball A, mass* $\mathbf{m}_1$, *rolls towards stationary ball B, mass* $\mathbf{m}_2$, b) *during the impact each ball exerts a force on the other – equal in size and opposite in direction,* c) *the balls after the collision.*

During the time the two balls are in contact each exerts a force on the other (Newton's Third Law about action and reaction, see page 108). The forces act in *opposite* directions and obviously act for the *same* amount of time. This means that $F \times t$ for each is the same size, but opposite in direction. The increase in momentum of ball B is *exactly* balanced by the decrease in momentum of ball A, so the total momentum of the two balls is unchanged before and after the collision – *momentum is conserved.*

In any system, momentum is always conserved provided no external forces act on the system. This means that when two snooker balls collide the momentum of the balls is conserved if no friction forces act on them. The presence of friction means that the balls will eventually slow down and stop, thus ending up with no momentum. Although the balls have "lost" momentum something else will, inevitably, have gained an equal amount of momentum! As the balls are slowed by the friction of the snooker table they, in turn, cause a friction force to act on the table. The table gains some momentum. However, the large mass of the table means that the effect is unnoticeable.

momentum before the collision = momentum after the collision

Elastic and inelastic collisions

When objects collide, some of the movement energy they possess is converted into other forms, typically heat and sound. Collisions in which no kinetic (movement) energy is "lost" by being transformed into other types of energy are called **elastic collisions**. Some practical collisions are close to elastic in that a very small proportion of the kinetic energy is "lost". Collisions between molecules in a gas are usually taken as perfectly elastic. (This is why the molecules in a container of air keep on moving – they don't end up in a pile at the bottom of the container.)

When a ball bounces off the ground, the collision is **partially elastic**: the ball rebounds, regaining its original shape, but loses some of its kinetic energy in the collision.

When two objects collide and stick together, the collision is said to be **inelastic**. Inelastic collisions are usually used for exam questions because they are simpler to do!

worked example

Example 1

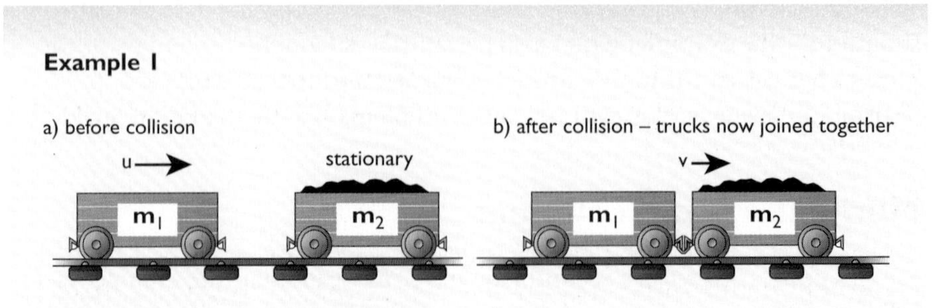

Figure 12.4 *Railway trucks in collision.*

A railway truck with a mass of 5000 kg rolling at 3 m/s collides with a stationary truck of mass 10 000 kg. The trucks join together. At what speed do they move after the collision?

We shall assume that friction forces are small enough to ignore, so we can apply the law of conservation of momentum:

momentum before the collision = momentum after the collision

so, momentum of A before collision	+	momentum of B before collision	=	momentum of A and B moving together after collision

$$m_1 \times u + m_2 \times 0 = (m_1 + m_2) \times v$$

where m_1 is the mass of truck A, u is its velocity *before* the collision, m_2 is the mass of truck B (at rest before the collision so its velocity is 0), and v is the velocity of the two trucks *after* the collision.

Substituting these values gives:

$$5000\,\text{kg} \times 3\,\text{m/s} + 0 = (5000\,\text{kg} + 10\,000\,\text{kg}) \times v$$

$$\text{so } v = \frac{15\,000\,\text{kg m/s}}{15\,000\,\text{kg}} = 1\,\text{m/s}$$

After the collision the trucks move with a velocity of 1 m/s in the same direction that the original truck was travelling.

Explosions

The conservation of momentum principle can be applied to explosions. An explosion involves a release of energy causing things to fly apart. The momentum before and after the explosion is unchanged, though there will be a huge increase in movement energy. A simple demonstration of a safe "explosion" is shown in Figure 12.5.

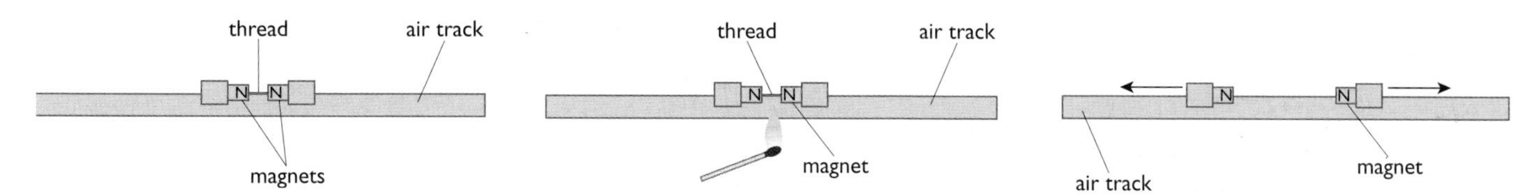

Figure 12.5 *The momentum before and after the "explosion" is unchanged although there is an increase in movement energy.*

In this demonstration, the two gliders are fitted with magnets trying to push them apart, but they are held together by a thread. When the thread is burned through, the gliders spring apart. If the gliders have the same mass, you will notice that they move off with the same speed but in opposite directions – the gliders have gained equal amounts of momentum but in opposite directions, so there is no overall change in the momentum before and after the "explosion".

Rockets

Rocket motors use the principle of conservation of momentum to propel spacecraft through space. They produce a continuous, controlled explosion that forces large amounts of fast-moving gases (produced by the fuel burning) out of the back of the rocket. The spacecraft gains an equal amount of momentum in the opposite direction to that of the moving exhaust gases. You can see the same effect if you blow up a balloon and release it without tying up the end!

End of Chapter Checklist

If you haven't got a copy of your specification, read the introduction on page iv.

You will need to be able to do some or all of the following. Check your Awarding Body's specification (syllabus) to find out exactly what you need to know.

- Know that the momentum of a moving object is found by multiplying its mass by its velocity.
- Know that momentum is a vector quantity – it has direction as well as size.
- Know that when objects collide they exert equal forces on each other but in opposite directions and that this causes the momentum change for each body to be the same but opposite in direction.
- Calculate the momentum of bodies before and after collisions in a straight line.
- Understand that in collisions or explosions the total momentum of the moving objects is the same before and after the collision or explosion, provided no external forces are acting.
- Know that kinetic energy is conserved in *perfectly elastic* collisions but normally the total kinetic energy of bodies after a collision will be less than before (because collisions are usually not perfectly elastic so kinetic energy is converted into sound, heat, and so on).
- Know that when colliding objects stick together the collision is described as inelastic.
- Understand that momentum conservation is used to drive spacecraft using rockets.

Questions

More questions on momentum can be found at the end of Section B on page 139.

1 Work out, giving your answers in kg m/s, the momentum of the following moving objects:

a) a bowling ball of mass 6 kg travelling at 8 m/s

b) a ship of mass 50 000 kg travelling at 3 m/s

c) a tennis ball of mass 60 g travelling at 180 km/h.

2 *a)* How does an elastic collision differ from partially elastic and inelastic collisions?

b) Give an example of:

i) an elastic collision

ii) an inelastic collision.

3 An air rifle pellet of mass 2 g is fired into a block of plasticine mounted on a model railway truck. The truck and plasticine have a mass of 0.1 kg. The truck moves off after the pellet hits the plasticine with an initial velocity of 0.8 m/s. Calculate the momentum of the plasticine and truck after the collision. Hence work out the velocity of the pellet just before it hits the plasticine.

4 A rocket of mass 1200 kg is travelling at 2000 m/s. It fires its engine for 1 minute. If the forward thrust provided by the rocket engines is 10 kN (10 000 N), what is the increase in momentum of the rocket? From this, work out the increase in velocity of the rocket and its new velocity after firing the engines.

Chapter 13: Force and Energy

Forces can cause objects to change shape. When forces act on a resilient or elastic object, such as a spring or an elastic band, the object changes shape temporarily. The work done by the force in changing the shape is stored as elastic potential energy in the object. In this chapter you will learn how to calculate the energy stored in a stretched spring.

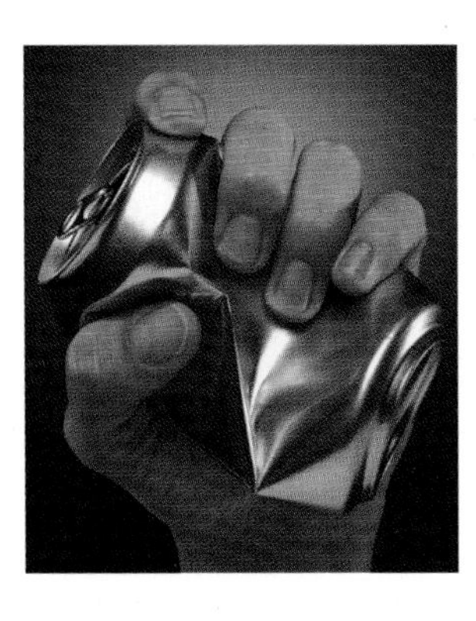

Figure 13.1 *Forces can cause changes in shape.*

We have seen that forces can make things start to move, accelerate or decelerate. The examples in Figure 13.1 show another effect that forces can have – they can change the shape of an object.

Sometimes the change of shape is temporary, as in the suspension spring in the mountain bike. Sometimes the shape of the object is permanently changed, like the crushed can or a car in collision with another object. A temporary change of shape may provide a useful way of absorbing and storing energy, as in the spring in the clockwork torch. A permanent change may mean the failure of a structure like a bridge to support a load. In this chapter we look at temporary changes in the lengths of springs and elastic bands.

Temporary changes of shape

If you apply a force to an elastic band, its shape changes – the band stretches and gets longer. All materials will stretch a little when you put them under **tension** (that is, pull them) or shorten when you **compress** or squash them. You can stretch a rubber band quite easily, but a huge force is needed to cause a noticeable extension in a piece of steel of the same length.

Some materials, like glass, do not change shape easily and are **brittle**, breaking rather than stretching noticeably. **Resilient** or **elastic** materials do not break easily and tend to return to their original shape when the forces acting on them are removed. Other materials, like putty and plasticine, are not resilient but **plastic**, and they change shape permanently when even quite small forces are applied to them.

We shall look at resilient materials, like rubber and metals formed into springs, in the first part of this chapter.

Springs

Springs are coiled lengths of certain types of metal, which can be stretched or compressed by applying a force to them. They are used in many different situations. Sometimes they are used to absorb bumps in the road as suspension springs in a car or cycle. In beds and furniture they are used to make sleeping and sitting more comfortable. They are also used in door locks to hold bolts and catches closed, and to make doors close automatically. Sometimes they are used in measuring devices like spring balances or bathroom scales.

To choose the right spring for a particular application, we must understand some important features of springs. A simple experiment with springs shows us that:

Springs change length when a force acts on them, and return to their original length when the force is removed.

This is true provided you do not over-stretch them! If springs are stretched beyond a certain point they do not spring back to their original length.

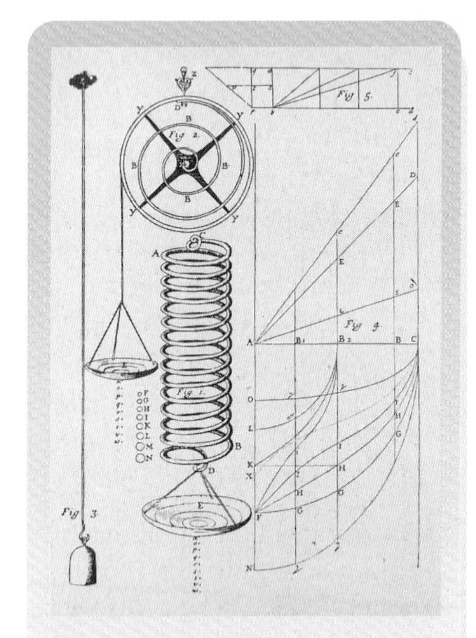

Figure 13.2 *Robert Hooke (1635–1703) was a contemporary of Sir Isaac Newton. This is a drawing of the apparatus Hooke used in his experimental work on the extension of a spring.*

Hooke's Law

Hooke discovered another important property of springs. He used simple apparatus like that shown in Figure 13.3a.

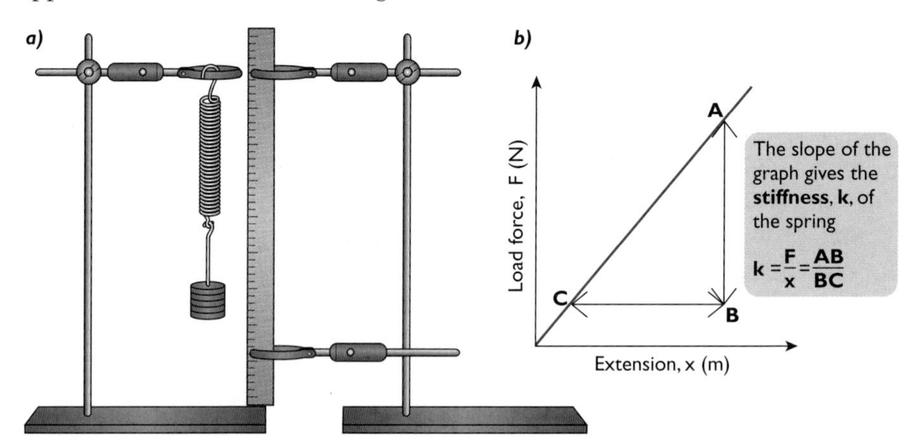

Figure 13.3 *The relationship between load and extension.*

Hooke measured the increase in length (extension) produced by different load forces on springs. The graph he obtained by plotting force against extension looked like that in Figure 13.3b. This straight line passing through the origin shows that the extension of the spring (x) is proportional to the force. This relationship is known as *Hooke's Law*.

The slope of the graph obtained from an experiment like that shown in Figure 13.3b depends on the particular spring under investigation and is called the **stiffness** of the spring. The stiffness of a spring is usually given the symbol k.

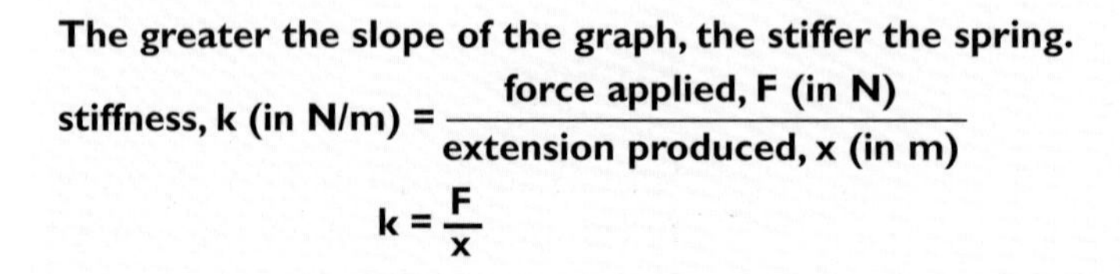

The greater the slope of the graph, the stiffer the spring.

$$\textbf{stiffness, k (in N/m)} = \frac{\textbf{force applied, F (in N)}}{\textbf{extension produced, x (in m)}}$$

$$k = \frac{F}{x}$$

The stiffness of a spring is measured in newtons per metre (N/m).

Example 1

A spring has a stiffness, k, of 100 N/m. What force will make it extend by 1 cm?

We can rearrange the formula to give:

$F = kx$

Substituting k = 100 N/m and x = 0.01 m (you must convert to metres), gives:

$F = 100\,N/m \times 0.01\,m$

$F = 1\,N$

So a force of 1 N will cause the spring, of stiffness 100 N/m, to extend by 0.01 m or 1 cm.

worked example

Example 2

A spring has a stiffness of 20 N/m. Its natural (unloaded) length is 4 cm. What load force will cause it to extend to a length of 8 cm?

The extension, x, is (8 – 4) cm = 0.04 m

Using F = kx gives:

$F = 20\,N/m \times 0.04\,m$

$= 0.8\,N$

So a force of 0.8 N will cause the spring, of stiffness 20 N/m, to extend to 8 cm.

worked example

We can use this simple relationship between the extension of a spring and the force causing it (F = kx) in force-measuring instruments like newton meters (spring balances) or bathroom scales. Although these instruments are measuring the force acting on the mass of an object in the pull of Earth's gravity, they are normally calibrated in units of mass.

Remember that force is measured in newtons, while mass is measured in kilograms.

Elastic bands

Elastic bands are usually made of rubber. If you stretch a rubber band with increasing load forces, you get a graph like that shown in Figure 13.4. The graph is not a straight line, showing that rubber bands do not obey Hooke's law.

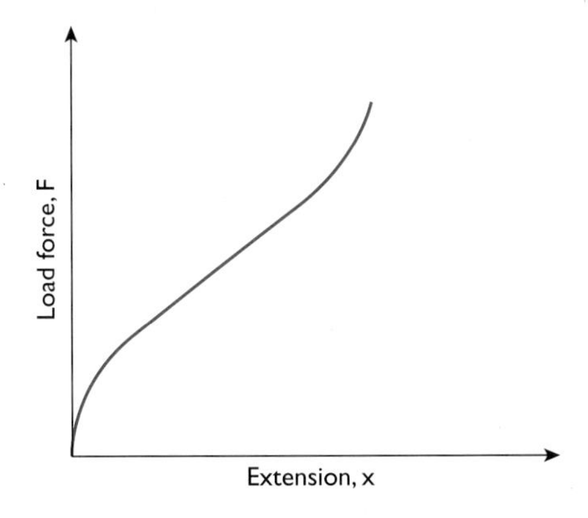

Figure 13.4 *Rubber bands do not obey Hooke's Law – the extension is not directly proportional to the force causing it.*

Reminder: work done = force × distance

Elastic potential energy

To stretch a spring or an elastic band, you must apply a force through a distance. This means that you are doing work.

The work done in stretching is stored as elastic potential energy in the spring or elastic band. We can calculate the work done in stretching a spring from its force–extension graph, as shown in Figure 13.5.

The work done in stretching the spring by a distance of **x** metres is stored as eleastic potential energy in the spring. This energy is given by the area under the force–extension graph for the spring, which is calculated using the area formula for a triangle ($\frac{1}{2}$ base × height):

$$\text{elastic potential energy stored} = \frac{1}{2}\ \text{force} \times \text{extension}$$

$$EPE = \frac{1}{2} Fx$$

Since F = kx, we can substitute for F giving:

$$EPE = \frac{1}{2} (kx) \times x$$

So:

$$EPE = \frac{1}{2} kx^2$$

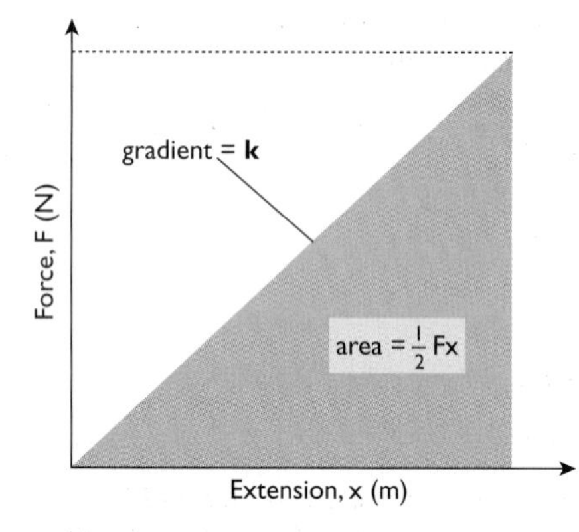

Figure 13.5 *Finding the energy stored in a stretched spring.*

where **x** is the extension produced when a force acts on a spring of stiffness k.

The work done – and, therefore, the energy stored – is measured in joules (J), provided the extension is in metres (m) and the spring stiffness is measured in N/m. The area under the force–extension graph of an elastic band is also equal to the work done in stretching the band (and therefore the energy stored in it). As the graph is not a straight line (see Figure 13.4), the area will not be given with complete accuracy by using the formula for a triangle.

The "spring" of the air

If you have ever used a bicycle pump, trapping the air inside by putting your finger over the end, you will have discovered that air is springy! If you squash the air by pushing the piston in, and then let go, the piston springs back. This is a property of all gases. Robert Boyle was one of a number of scientists interested in the properties of gases. The properties of gases, including their "springiness", are dealt with in Chapter 29.

End of Chapter Checklist

If you haven't got a copy of your specification, read the introduction on page iv.

You will need to be able to do some or all of the following. Check your Awarding Body's specification (syllabus) to find out exactly what you need to know.

- Know that some materials are resilient (or elastic).
- Know and be able to use the equation F = kx.
- Be able to find the stiffness, k, of a spring from its force–extension graph.
- Know that the work done in stretching or compressing a spring is equal to the area under its force–extension graph.
- Recall and use the formula for elastic potential energy, EPE = $\frac{1}{2}$ kx^2.

Questions

More questions on force and energy can be found at the end of Section B on page 139.

1 What force will make a spring with a stiffness, k, of 50N/m stretch by 3cm?

2 A spring has a stiffness of 40 N/m. What extension results from hanging a load of 0.5 N from it?

3 If a spring is stretched by 5 cm when a force of 10 N is applied to it, what is the stiffness of the spring?

4 The force–extension graphs for three different springs, A, B and C, are shown below.

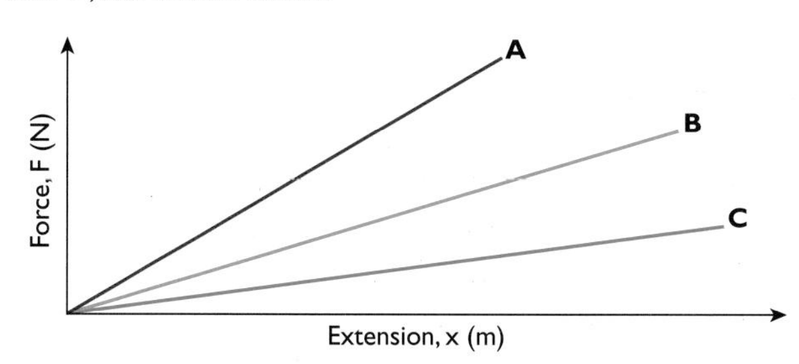

a) Which spring is the stiffest?

b) How do you measure the stiffness of a spring from its force–extension graph?

5 The three force–extension graphs shown below are all drawn to the same scale.

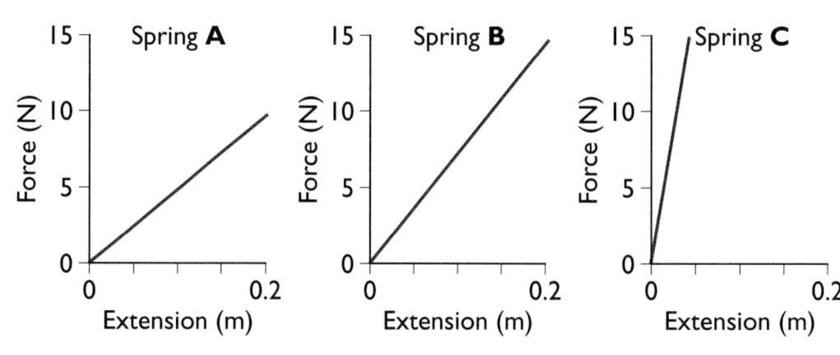

a) Which spring is the stiffest?

b) Which spring is storing the most elastic potential energy?

6 The information in the following table was obtained from an experiment with a spring of original (unstretched) length 5 cm.

Load force on spring (newtons)	Length of spring (cm)	Extension of spring (cm)
0	5.0	
2	5.8	
4	6.5	
6	7.4	
8	8.3	
10	9.0	

a) Copy and complete the table by calculating the extensions produced by each load.

b) Use your table to plot a force–extension graph for the spring.

c) Use your graph to calculate the stiffness of the spring.

d) How much energy is stored in the spring when it is stretched by a load of 10 N?

e) In a separate experiment, an object is hung from the spring and it stretches to a new length of 7.6 cm. Use your graph to find how much the object weighs, in newtons.

Section B: Forces and Motion

Chapter 14: Projectiles and Circular Motion

In everyday life, moving objects don't always follow a straight path. An object travels in a curved path when a resultant force acts on it to make it change its direction. In this chapter you will look at how forces make objects move in this way.

Figure 14.1 *Both the roller coaster and the football follow curved paths.*

This chapter is about objects following curved rather than straight paths. You will remember that objects travel in straight lines at constant speed unless a resultant force acts on them. An object will only travel in a curved path if a force acts on it in such a way as to make it change direction. This happens when the force does not act along the line of motion of the object. A football will follow a curved path when it has been kicked because of the pull of gravity. As the football moves through the air, both the direction and speed of its movement will change. This kind of motion is called **projectile** motion. Motion around a circular path can take place at constant speed with the direction of travel changing continuously. This chapter looks at how projectiles, like balls and rockets, move. We shall also consider the force that must act on a body to make it follow a circular path at constant speed – the force that makes "white-knuckle" rides like one shown in Figure 14.1 possible. The same type of force keeps satellites in orbit around the Earth.

a)

direction of motion

horizontal component of the ball's velocity

ball

verticle component of the ball's velocity

gravitational force acting on the ball

F

v

velocity of the ball at this moment in time

b)

horizontal component of the ball's velocity is *unchanged*

verticle component of the ball's velocity has *increased*

new velocity of the ball at a later time

v'

Figure 14.2 *The velocity of the moving ball can be resolved (broken down) into a vertical component and a horizontal component.*

Projectiles and how they move

If you throw a ball, it will follow a curved path towards the ground, unless you throw it vertically upwards. Spray water from a hosepipe and it, too, will follow a curved path. The curved path involves movement in two dimensions: the horizontal and the vertical. Typically, when an object is in flight, it will be travelling at an angle to the horizontal. It will have a component, or part of its movement, taking place in the vertical direction and a component in the horizontal direction.

For example, Figure 14.2a includes a **vector diagram** for a moving ball. The instantaneous velocity of the ball is v. This velocity can be resolved (broken down) into a vertical component and a horizontal component. A gravitational force, F, acts on a ball as it moves. Since this pull of gravity acts at right angles to the horizontal component of the ball's velocity, it has no effect on this component. The horizontal component remains constant and unchanged because no force acts on the ball in the horizontal direction. (This assumes that air resistance is too small for any noticeable effect – an assumption that is not correct for fast-moving projectiles.) The vertical velocity will increase as the force is causing the ball to accelerate in the vertical direction.

Figure 14.2b shows a vector diagram for the ball's motion a short time later. The horizontal velocity component is the same but the vertical component of velocity has increased. The result is that both the size and the direction of the velocity of the ball have changed.

Figure 14.3 shows the path of a ball as it is thrown into the air, rises to its maximum height, and then falls back to the ground.

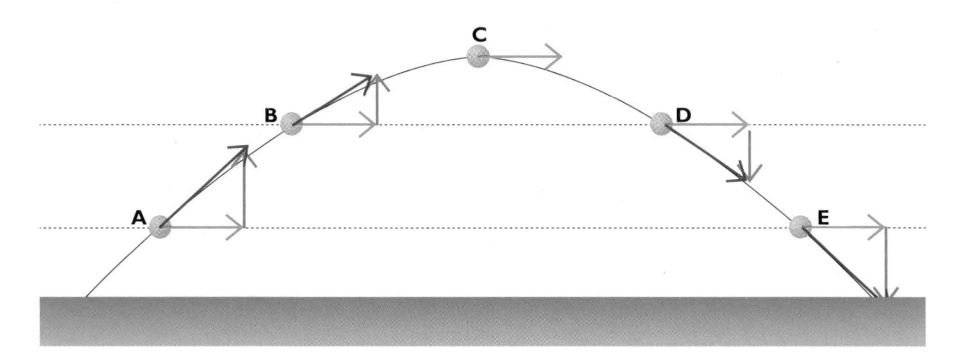

Figure 14.3 *As a ball flies through the air its horizontal velocity does not change but its vertical velocity does.*

The ball is shown at various points along the path of flight. The ball's horizontal velocity component, shown by the blue vector, remains constant throughout. The vertical component of its velocity is shown by the green vector. Note that the force of gravity causes the vertical component to change throughout the flight. It has a decreasing size in the upward direction until it reaches C, when it has been brought to rest in its vertical motion. It then has an increasing size in the downward direction.

Reminder: A vector is a physical quantity with both size and direction. We use arrows to represent vectors. The lengths of the arrows show the relative sizes of the vectors and the directions are shown by the directions of the arrows.

The actual velocity of the ball is shown by the dark red vectors. This is always the vector sum of the vertical and horizontal components of the velocity. It is changing in size and direction, as we discussed earlier.

Projectiles and the equations of motion

We can calculate how a projectile will move using the equations of motion (see Chapter 8). We do this by considering the horizontal part of the motion and the vertical part of the motion separately.

The horizontal motion is easy to calculate as we assume that the horizontal velocity is constant. (Remember that, to do this, we assume that no horizontal force acts on the projectile.) If we know the horizontal component of velocity, we can work out how far the projectile travels in this direction by multiplying horizontal velocity × time of flight. The time of flight will depend on the vertical motion.

To calculate how long the projectile is in flight, we use the equations of motion in the vertical direction. The downward acceleration of objects moving in the Earth's gravity is constant at 10 m/s^2 (approximately).

worked example

Example 1

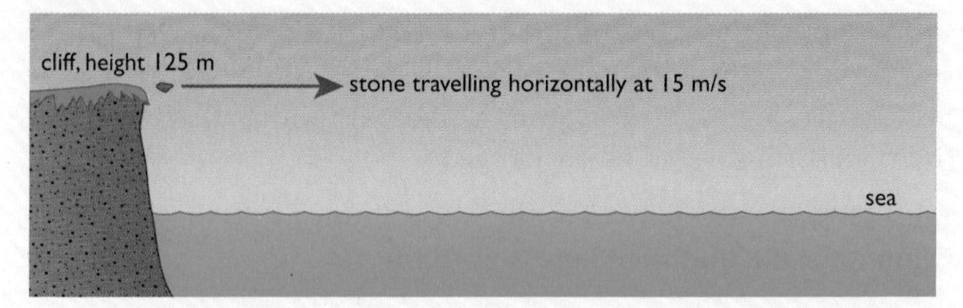

Figure 14.4 *Can you work out how far from the cliff the stone will be when it hits the water?*

A boy kicks a stone off a cliff 125 metres high. The stone leaves the boy's foot travelling horizontally at 15 m/s. How long does it take to hit the sea below the cliff? How far from the cliff does the stone splash into the sea?

The vertical velocity of the stone is initially zero, so u = 0 m/s. (Remember that the horizontal motion has no effect on the vertical speed.)

The vertical distance travelled (cliff height) at the end of the flight is s = 100 m.

The vertical acceleration, due to gravity, is $a = 10\ \text{m/s}^2$.

We want to find the time, t, that it takes for the stone to reach the sea. Since we know u, a, and s, and want to find t, we use the formula:

$$s = ut + \frac{1}{2}at^2$$ (see page 92)

Substituting values gives:

$$125\ \text{m} = 0 \times t + \frac{1}{2} \times 10\ \text{m/s}^2 \times t^2$$

$$125\ \text{m} = 5\ \text{m/s}^2\ t^2$$

$$25\ \text{s}^2 = t^2$$

$$\therefore t = 5\ \text{s}$$

The stone takes 5 s to hit the sea. Its time of flight is therefore 5 s, so it travels horizontally at 15 m/s for 5 s. To find how far the stone travels from the cliff before plunging into the sea we use:

horizontal distance = horizontal velocity × time

= 15 m/s × 5 s

= 75 m

In Example 1, the stone was projected horizontally. This made the calculation easier because the vertical component of its velocity was zero. It is more usual for projectiles to be projected at an angle to the horizontal, as shown in Figure 14.3. This is the path a javelin would follow when thrown by a field sports athlete, or the path of a shell fired from a cannon. We can work out how these objects move in exactly the same way.

1 Assume that the horizontal component of their velocity is constant throughout their flight.

2 Use the equations of motion to work out the effect of gravity on the vertical component of their motion.

Motion in a circle – acceleration at constant speed!

When an object moves around a circular path at constant speed it is continuously accelerating because its direction is changing all the time.

Remember that acceleration is the rate of change of velocity and that, since velocity is a vector, its direction is important, too.

For a force to act on an object without causing its speed to change, the force must act at right angles to the direction of travel. An example of circular motion with constant speed is spinning a conker around on a piece of string, Figure 14.5a.

When you spin an object like this you will feel the tension in the string. This force pulls on you along the line of the string outwards from your fingers. Therefore, it pulls on the conker along the line of the string towards your fingers. The other force acting on the conker is the pull of gravity. The two forces acting on the conker are shown in the vector diagram in Figure 14.5b. The conker spins around a horizontal circle. This means that it is not moving up or down so the resultant force acting on it in the vertical direction must be zero. The direction of the combined effect of these forces – the resultant force on the conker – acts horizontally *towards the centre of the circular path.*

The resultant force on the conker acts towards the centre of the circular path that the conker follows. This force is, therefore, always at right angles to the instantaneous direction of travel of the conker. Because the force is perpendicular to the line of travel it has no effect on the size of the conker's velocity, but it causes the direction to change continuously.

For a body to move along a circular path at constant speed the resultant force on the body must act towards the centre of the circular path. This force is called the centripetal force.

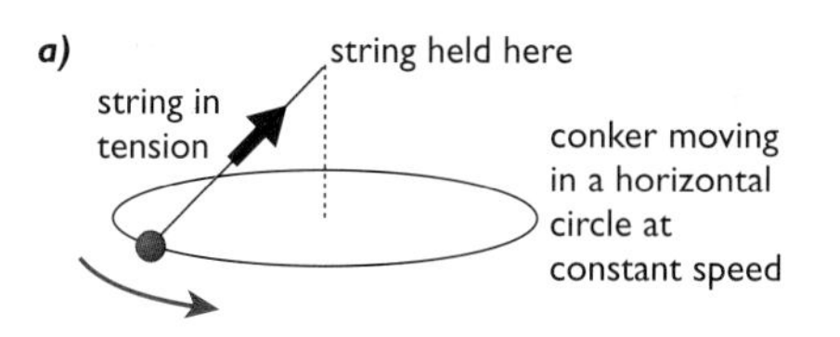

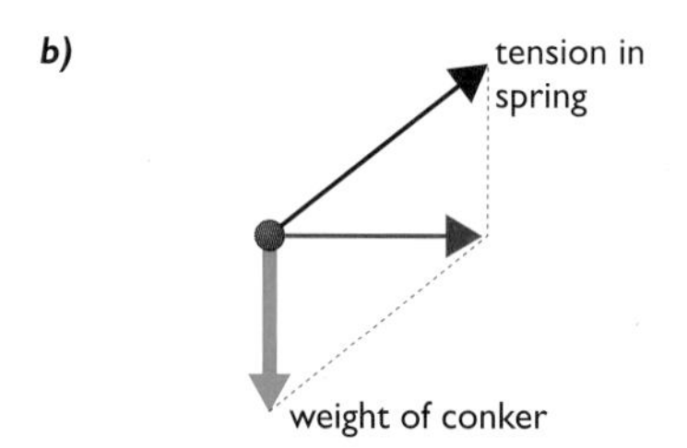

The **resultant** of these two forces acts towards the centre of the circle providing the **centripetal force**

Figure 14.5 *The conker is moving with constant speed, but is continuously accelerating.*

Adding vectors: The resultant or sum of two vectors, like the forces in the example in Figure 14.5, can be found by completing the parallelogram formed by the two forces. The resultant will then be the diagonal, as shown in Figure 14.5b. The vectors must be drawn to scale with the correct angle between them.

More examples of circular motion

Whenever something travels a circular path there must be a force towards the centre of the circular path acting on it – the **centripetal** force. This force is provided in different ways – for example, by gravity on an orbiting satellite, by friction on a car's tyres as it rounds a bend, by the electrostatic force between a proton in a hydrogen atom and its orbiting electron, and by the lift force on the wings of an aircraft when it banks to turn.

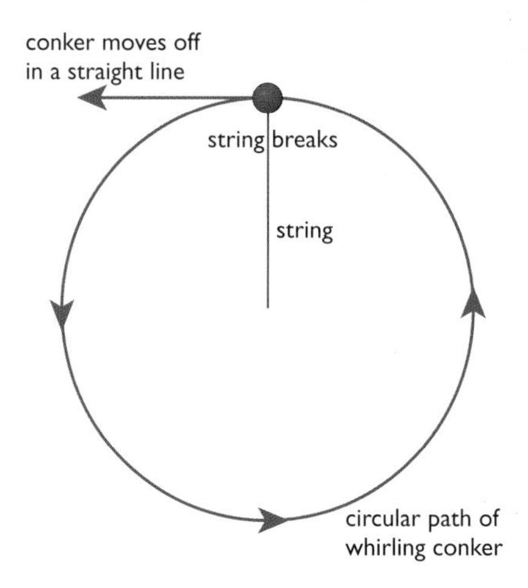

Figure 14.6 *If the centripetal force stops, the object moves off in a straight line.*

Figure 14.7 *As the road was icy when this car went round the bend, there was no friction between the tyres and road.*

Figure 14.8 *Friction between the tyres and the road provides the centripetal force as this motorcycle turns a corner.*

If the centripetal force suddenly stopped acting on the object then it would continue to move in the way it was moving at the instant the centripetal force ceased. This sometimes happens! If the string breaks when you whirl a conker around in a circle the conker will fly off in a straight line at a tangent to the circle. If a car hits ice on the road as it turns a bend, the friction between the tyres and the road will be suddenly reduced, and the car will skid off the road.

Factors affecting the size of the centripetal force

The size of the centripetal force required to keep an object following a circular path depends on the following factors:

- the *mass* of the object – the bigger the mass, the bigger the centripetal force needed
- the *speed* of the object – the faster it moves, the greater the centripetal force required
- the *radius* of the curved path – a smaller radius means a larger centripetal force must act on the object.

Whenever an object is spinning in a circular path, these factors must be taken into account. Next time you see a windmill spinning think about the force required to keep the sails following a circular path – what would happen if the wind made the sails spin round too fast? The same problem faces the designers of the turbine blades that have to spin at very high speed in the jet engines on aircraft.

Figure 14.9 *Communications satellites are in geostationary orbits.*

Satellites and circular motion

The planets in the Solar System orbit the Sun in nearly circular paths. Some of the planets have natural satellites or moons. The Earth is circled by our own Moon, as well as by artificial satellites. Some of these satellites are used for communication around the world and are in special orbits. These are called **geostationary orbits** – the satellite stays in the same position above the Earth's surface as it orbits and the Earth turns on its axis. Communications satellites are discussed in Section D.

The orbital speed of a satellite will depend on the circumference ($2\pi r$) of its path – it is calculated using the following equation:

$$\text{orbital speed} = \frac{2\pi \times \text{orbital radius}}{\text{time period}}$$

$$v = \frac{2\pi r}{T}$$

For a geostationary orbit, T must be 24 hours (86 400 seconds), since it must complete one orbit each time the Earth completes one rotation on its axis.

We shall now work out the size of the gravitational force needed to keep a satellite in orbit. We have already said that the required centripetal force is dependent on the mass, speed and radius of the circular path of the object. The exact relationship is:

$$\text{centripetal force} = \frac{\text{mass, } F_C \times (\text{orbital speed, u})^2}{\text{radius, r}}$$

$$F_C = \frac{mv^2}{r}$$

This centripetal force, F_C, is provided by the gravitational force on the satellite. The size of the gravitational pull on a satellite is proportional to its mass. It also depends on the distance between the centres of mass of the planet and the satellite. The gravitational force on a satellite decreases rapidly as the height above the Earth increases. However, the higher the orbit of a satellite, the faster it must move in order to complete one orbit in 24 hours. The equation shows that the faster the satellite moves, the bigger the force needed to keep it in a circular orbit around the Earth.

The result is that if the orbit is too high the force of gravity will not be enough to keep the satellite circling the Earth once a day.

Similarly, if the orbit is too low the pull of gravity will be more than that needed to allow the satellite to keep in its once-daily orbit. If the orbit is too low the geostationary satellite will be pulled back to Earth.

The radius required for a stable geostationary orbit is about 42 000 km. This means that geostationary satellites are about 35 000 km above the Earth's surface.

We can calculate the speed of a satellite in a geostationary orbit by substituting the value for the radius in metres into the equation:

$$v = \frac{2\pi r}{T}$$

$$v = \frac{2\pi \times 42\,000\,000\,\text{m}}{86\,400\ \text{s}}$$

So, v = 3054 m/s

We can use this value for the speed to calculate the centripetal force provided by gravity on a 500 kg satellite:

$$F_C = \frac{mv^2}{r}$$

$$F_C = \frac{500\ \text{kg} \times (3054\ \text{m/s})^2}{42\,000\,000\,\text{m}}$$

$F_C = 111\,\text{N}$

Satellites don't have to be geostationary or geosynchronous (orbiting in time with the Earth's rotation). The first artificial satellite to orbit the Earth was Sputnik 1, launched by Russia in 1957. Sputnik orbited the Earth once every 98 minutes at a height of about 500 km. Lower orbits mean that the satellite must travel faster to be in a stable orbit. Although the atmosphere at this height is very thin, it does slow satellites down.

At a lower speed, the force needed to keep the satellite following a circular orbit gets smaller. Gravity, of course, does not change, so the actual force on the satellite is bigger than the centripetal force needed, so the satellite is pulled towards the Earth.

End of Chapter Checklist

If you haven't got a copy of your specification, read the introduction on page iv.

You will need to be able to do some or all of the following. Check your Awarding Body's specification (syllabus) to find out exactly what you need to know.

- Know that a resultant force on an object can cause it to change its direction of motion.
- Describe the curved path followed by objects thrown (other than vertically) from the Earth's surface – the path of a projectile.
- Calculate the time of flight and range of objects projected horizontally in the Earth's gravitational field.
- Know that objects travelling in a circular path at constant speed are accelerating because their direction of travel changes continuously.
- Know that for circular motion at constant speed to occur there must be a resultant force on a body acting towards the centre of the circular path – the centripetal force.
- Know how the size of the centripetal force is related to the mass and speed of the body, and to the radius of its circular path.
- Identify the type of force that provides the centripetal force in a variety of different examples of circular motion.
- Use the equations for the orbital speed of, and centripetal force acting on, an object performing circular motion.

Questions

More questions on projectiles and circular motion can be found at the end of Section B on page 139.

1 How can a force make a body accelerate at constant speed?

2 What force or forces act on a javelin as it flies through the air?

3 If a force acts on an object at right angles to the direction it is travelling:

a) what feature of its motion is unchanged?

b) what feature of its motion changes?

4 Describe the differences between the motion of a projectile (an object thrown at an angle to the horizontal) and an object performing circular motion. You should refer to the force acting in each case and the way it affects the velocity of the objects.

5 Work out the speed of a conker spinning in a circle of radius 0.3 m at a rate of 10 complete revolutions per second. (Take the value of π as 3.14.)

6 Use the formula:

$$F_C = \frac{mv^2}{r}$$

to calculate the centripetal force required to make a satellite of mass 200 kg follow a circular orbit of radius 7000 km if it completes one orbit every 100 minutes.

End of Section Questions

1 The diagram shows a displacement–time graph for a person taking a walk.

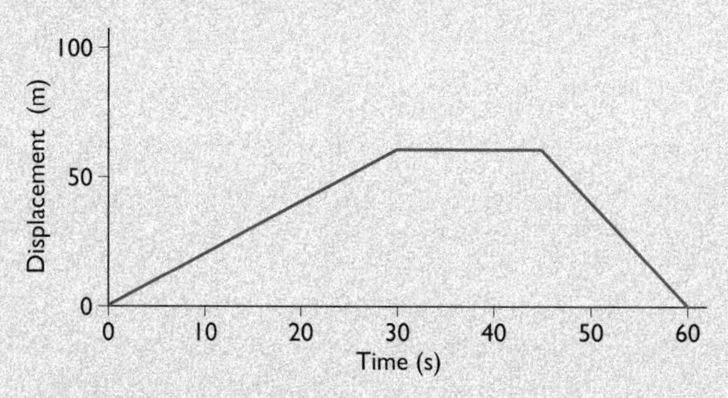

a) How fast is the person walking during the first 30 seconds of the walk? Show how you obtained your answer. *(3 marks)*

b) Describe what the person is doing during the next 15 seconds. *(1 mark)*

c) What is the person doing during the last 15 seconds of the walk? *(3 marks)*

d) What is the average velocity of the person during the first 45 seconds of the walk? *(2 marks)*

Total 9 marks

2 The diagram shows a velocity–time graph for a space shuttle during the early part of a flight.

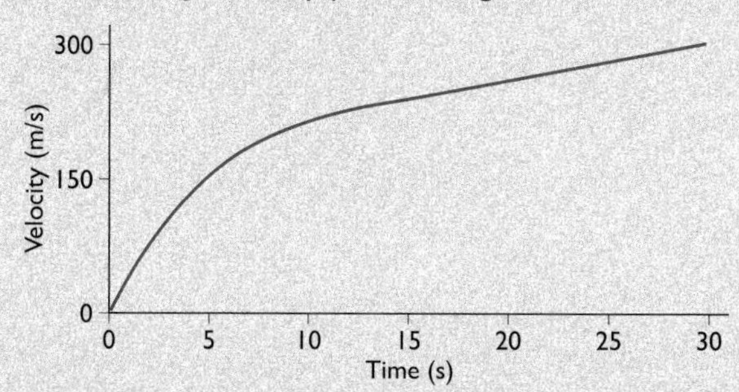

a) Calculate the acceleration during the first 5 seconds of the flight. Show all steps in your working. *(3 marks)*

b) Estimate the distance travelled during the first 30 seconds of the flight. Explain your method. *(4 marks)*

c) What is the average velocity of the shuttle during the 30-second period shown on the graph? *(3 marks)*

Total 10 marks

3 A parachutist jumps from a plane flying at a height of 5000 metres. She falls for 30 seconds before opening her parachute. Take the acceleration due to gravity as 9.8 m/s^2.

a) If we ignore the effect of air resistance, how fast will the parachutist be falling after 30 seconds? *(2 marks)*

b) How far will she have fallen in the 30-second interval, assuming we may ignore air resistance? *(3 marks)*

c) In practice, we cannot ignore air resistance. Use a sketch graph to show how the velocity of the parachutist will *actually* change as she falls. Explain the shape of your sketch graph. *(4 marks)*

Total 9 marks

4 The diagram shows the position of a ball rolling down a slope at intervals of 0.1 s. The slope is marked with a scale dividing the slope into 3 cm distances.

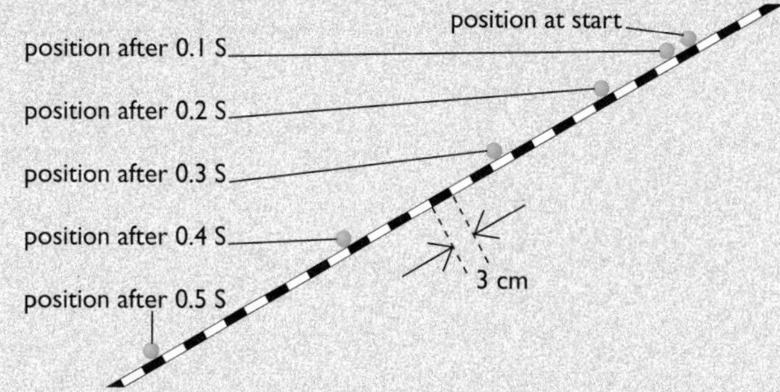

a) What is the average velocity during the interval 0.1 s to 0.2 s? *(3 marks)*

b) Calculate the average velocity of the ball during the interval 0.4 s to 0.5 s. *(2 marks)*

c) Use your answers to ***a)*** and ***b)*** to calculate the acceleration of the ball down the slope. *(4 marks)*

d) Is the acceleration of the ball uniform? Explain your answer. *(4 marks)*

Total 13 marks

5 A car travelling at 40 m/s brakes and decelerates uniformly to rest. The car has a mass of 700 kg. The car takes 10 s to come to rest.

a) Work out the deceleration of the car. *(3 marks)*

b) What is the average braking force acting on the car? *(2 marks)*

c) How much energy is "lost" during the braking process? *(2 marks)*

d) What becomes of this "lost" energy? *(2 marks)*

Total 9 marks

6 Copy this diagram of a wooden float tethered to the seabed so that it is completely submerged.

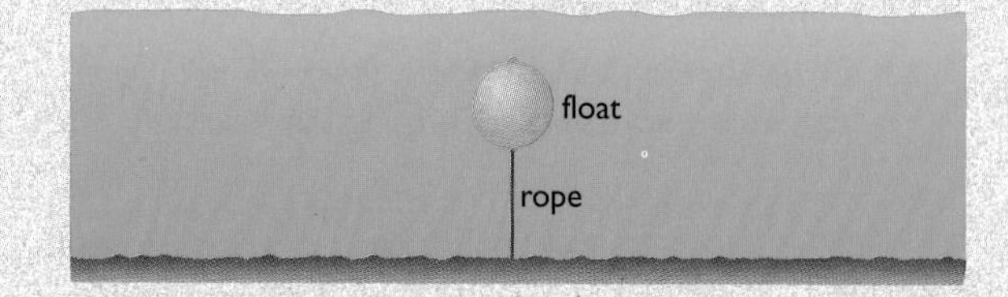

a) Label all the forces that act *on the float.* Indicate the direction in which each force acts. *(4 marks)*

b) The float is not moving. What does this tell us about the forces acting on it? *(2 marks)*

c) The rope securing the float to the seabed breaks. Describe what happens to the float. *(2 marks)*

Total 8 marks

7 The diagram below shows the hand on a swimming pool clock. The hand is balanced so that the turning effect of the weight of each side of the hand is the same. This makes it easier for the low-power electric motor to keep the hand moving round at a steady rate. The hand measures 80 cm from tip to tip.

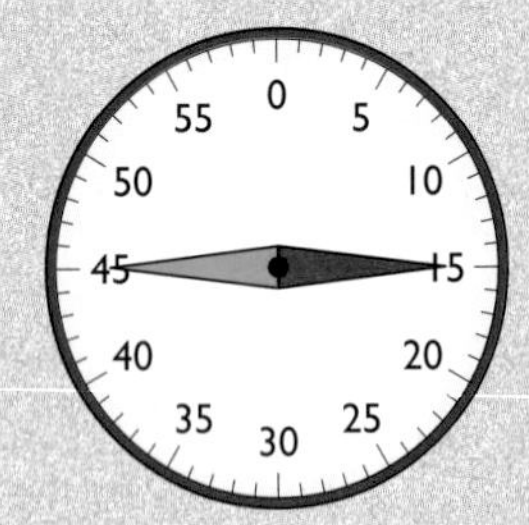

a) Define the *moment* or turning effect of a force. *(2 marks)*

b) A clock of this type is checked and the hands are found to be unbalanced. The result is that there is a clockwise moment of 0.05 Nm. You are provided with a number small self-adhesive weights, each weighing 0.1 N. Show how you would use these to re-balance the hand, so that the turning moment about the centre spindle is zero. You should use a diagram to illustrate your answer. *(5 marks)*

Total 7 marks

8 The following diagram shows a simple experiment using a spring. The spring is loaded with weight and the length of the spring is measured for each different load force applied. The results of the experiment are given in the table.

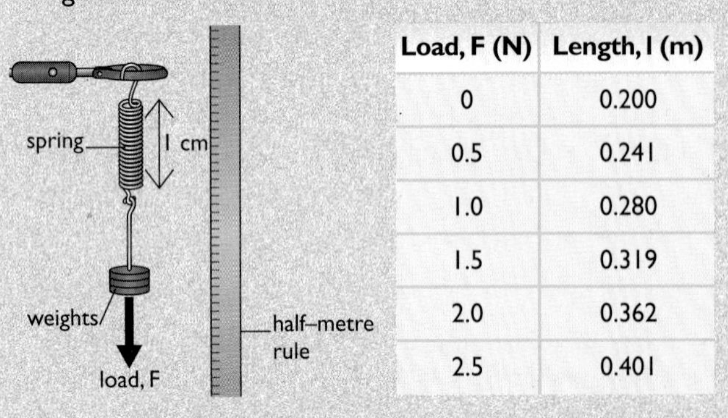

Load, F (N)	Length, l (m)
0	0.200
0.5	0.241
1.0	0.280
1.5	0.319
2.0	0.362
2.5	0.401

a) Draw a graph of load force (vertical axis) against extension (horizontal axis) using the results. *(4 marks)*

b) Use your graph to calculate:

i) the *stiffness* (or spring constant) of the spring *(3 marks)*

ii) the energy stored in the spring when its extension is 0.4 m. *(3 marks)*

Total 10 marks

9 a) Define *momentum*. *(2 marks)*

b) A cannon, of mass 200 kg, fires a cannon ball with a mass of 10 kg. The cannon ball leaves the barrel of the cannon with a velocity of 60 m/s. Using the law of conservation of momentum, explain what happens to the cannon when the cannon ball is fired. Your answer should include a calculation. *(6 marks)*

Total 8 marks

10 The diagram shows three examples of circular motion.

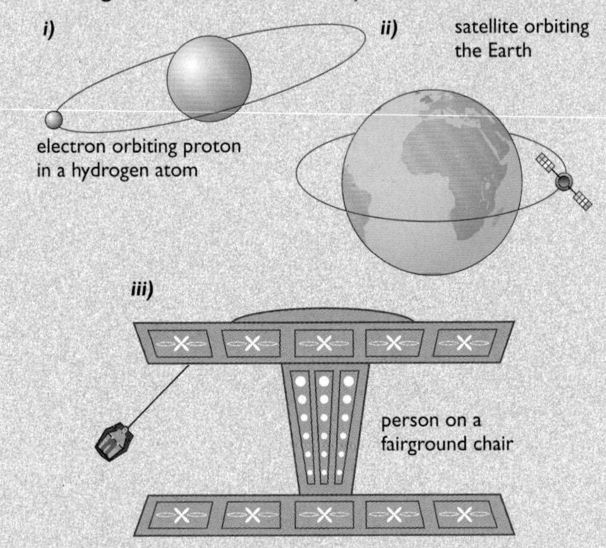

a) What provides the centripetal force:

i) on the electron orbiting the proton in a hydrogen atom *(1 mark)*

ii) on the satellite orbiting the planet *(1 mark)*

iii) on the child being whirled round in a chair on the end of a chain in the fairground ride? *(1 mark)*

The child in the fairground ride has a mass of 30 kg. The time for one complete circuit of the ride is 20 seconds. The radius of the circular path is 10 metres. Take π as 3.14

b) What is the speed of the child? *(3 marks)*

c) What is the centripetal force on the child? *(3 marks)*

d) What would happen if the speed of rotation of the ride were to be increased? *(3 marks)*

Total 12 marks

Chapter 15: Introduction to Waves

Waves, such as sound waves and light waves, affect all aspects of our lives. We also use many types of waves to our advantage, particularly in the field of communication. In this chapter, you will learn about some basic features and properties of waves.

Figure 15.1 *Mobile phones rely on waves.*

Speaking to someone on the other side of the Earth using a mobile phone may only be a matter of being able to dial the correct number, but the technology that had to be developed for this to happen was based on a thorough understanding of the properties of waves. In this section we will be looking at the basic properties of waves and seeing how these properties can be used to our advantage.

What are waves?

Waves are a means of moving energy from place to place. This movement of energy takes place with no matter being transferred.

Figure 15.2 *Waves are produced if you drop a large stone into a pond. The waves spread out from the point of impact, carrying energy to all parts of the pond. But the water in the pond does not move from the centre to the edges.*

Transverse waves

Waves can be produced in ropes and springs. If you waggle one end of a slinky spring from side to side you will see waves travelling through it. The energy carried by these waves moves along the slinky from one end to the other, but if you look closely you can see that the coils of the slinky are vibrating *across* the direction in which the energy is moving. This is an example of a **transverse wave**.

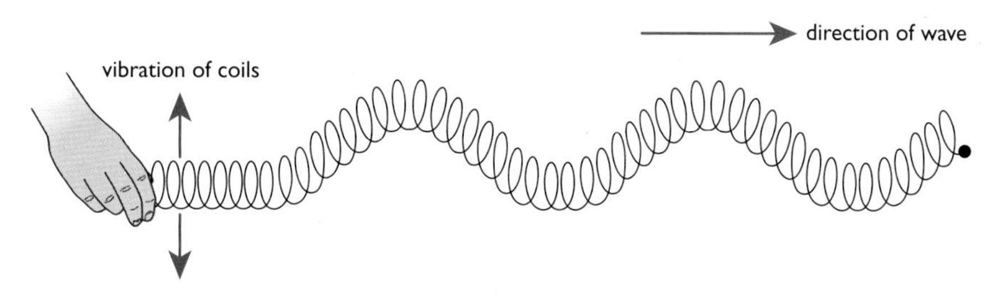

Figure 15.3 *A transverse wave vibrates at right angles to the direction in which the wave is moving.*

A transverse wave is one that vibrates, or **oscillates**, at right angles to the direction in which the energy or wave is moving. Examples of transverse waves include light waves and waves travelling on the surface of water.

Longitudinal waves

If you push and pull the end of a slinky in a direction parallel to its axis, you can again see energy travelling along it. This time however the coils of the slinky are vibrating in directions that are *along its length*. This is an example of a **longitudinal wave**.

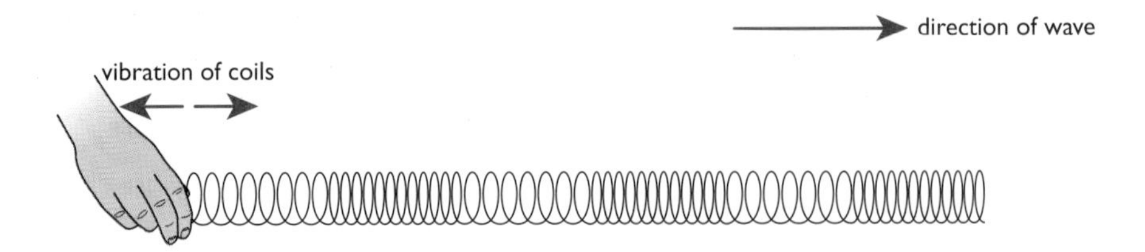

Figure 15.4 *A longitudinal wave vibrates along the direction in which the wave is travelling.*

A longitudinal wave is one in which the vibrations, or oscillations, are along the direction in which the energy or wave is moving. Examples of longitudinal waves include sound waves and seismic P-waves in earthquakes.

Describing waves

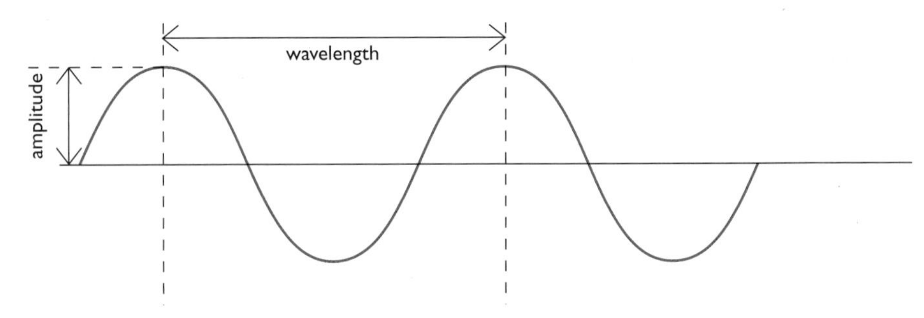

Figure 15.5 *A wave has amplitude and wavelength.*

When a wave moves through a substance, it causes the particles in the substance to move from their equilibrium or resting position. The maximum movement of particles *from their resting position* caused by a wave is called its **amplitude** (A).

The distance between a particular point on a wave and the same point on the next wave (for example, from crest to crest) is called the **wavelength** (λ).

If the source that is creating a wave vibrates quickly it will produce a large number of waves each second. If it vibrates more slowly it will produce fewer waves each second. The number of waves produced each second by a source, or the number passing a particular point each second, is called the **frequency** of the wave (f). Frequency is measured in hertz (Hz). A wave source that produces five complete waves each second has a frequency of 5 Hz.

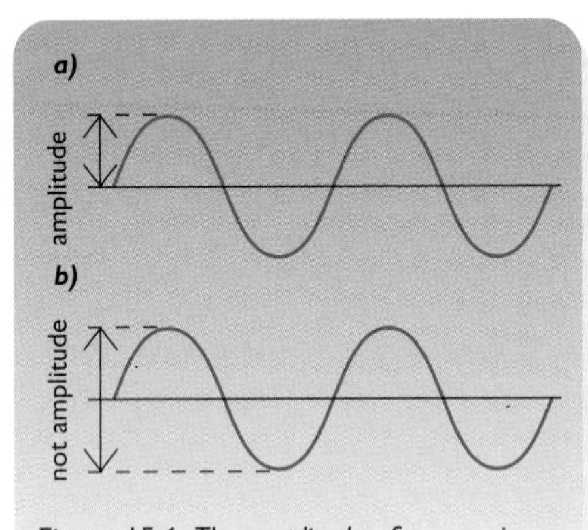

Figure 15.6 *The amplitude of a wave is as shown in a), and not as in b).*

λ is the Greek letter lambda and is the usual symbol for wavelength.

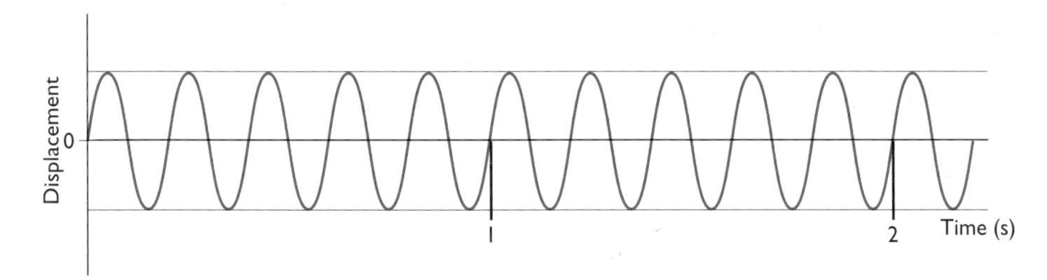

Figure 15.7 *This graph shows a wave with a frequency of 5 Hz.*

The time it takes for a source to produce one wave is called the period of the wave (T). It is related to the frequency (f) of a wave by the equation:

$T = 1/f$

worked example

Example 1

Calculate the period of a wave with a frequency of 200 Hz.

$T = 1/f$

$T = 1/200$

$= 0.005$ or 5ms (1000ms = 1s)

The wave equation

There is a relationship between the wavelength (λ), the frequency (f) and the wave speed (v) that is true for all waves:

$$v = f\lambda$$

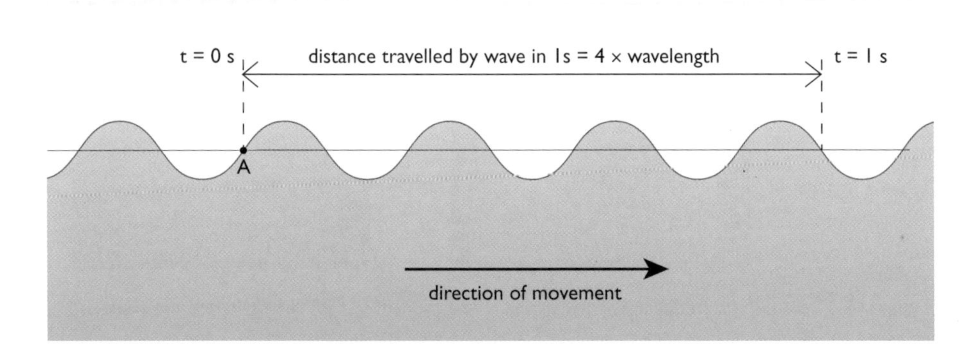

Figure 15.9 *A wave with a frequency of 4 Hz.*

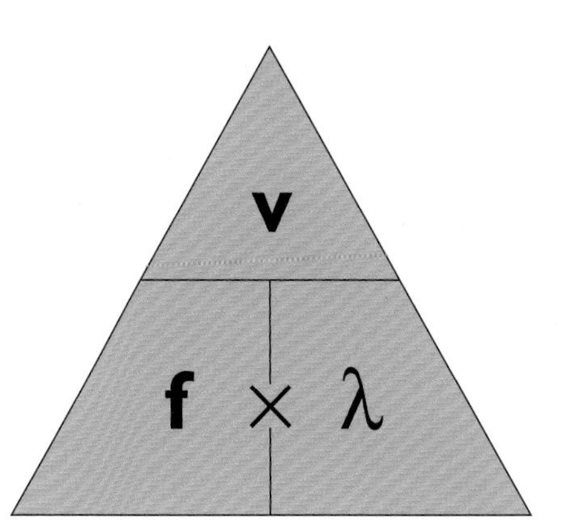

Figure 15.8 *You can use the triangle method for rearranging equations like v = fλ.*

Imagine that you have created water waves with a frequency of 4 Hz. This means that four waves will pass a particular point each second. If the wavelength of the waves is 3 m, then the waves travel 12 m each second. The speed of the waves is therefore 12 m/s:

$v = f\lambda$

$v = 4\,\text{Hz} \times 3\,\text{m}$

$= 12\,\text{m/s}$

worked example

Example 2

A tuning fork creates sound waves with a frequency of 170 Hz. If the speed of sound in air is 340 m/s, calculate the wavelength of the sound waves.

$v = f\lambda$

So $\lambda = v/f$

$\lambda = 340\text{m/s} / 170\text{Hz}$

$\lambda = 2\text{m}$

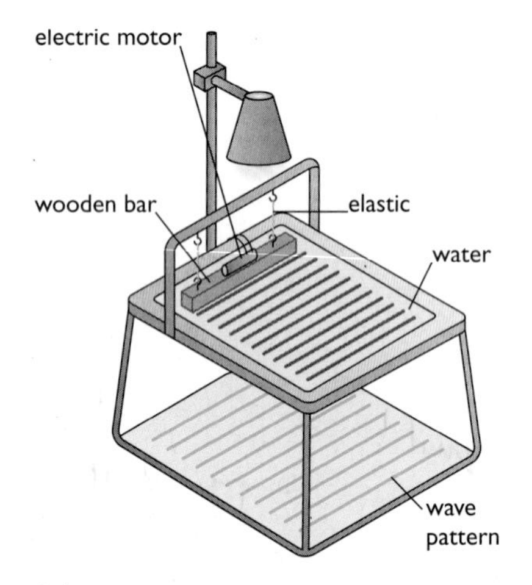

Figure 15.10 *The light shines through the water and we can see the patterns of the waves.*

The ripple tank

We can study the behaviour of water waves using a ripple tank.

When the motor is turned on the wooden bar vibrates creating a series of ripples on the surface of the water. A light placed above the tank creates patterns of the water waves on the floor. By observing the patterns we can see how the water waves are behaving.

Wavelength and frequency

At low speeds, the motor on the ripple tank produces a small number of waves each second. The frequency of the waves is small and the pattern shows that the waves have a long wavelength.

At higher frequencies, the water waves have shorter wavelengths.

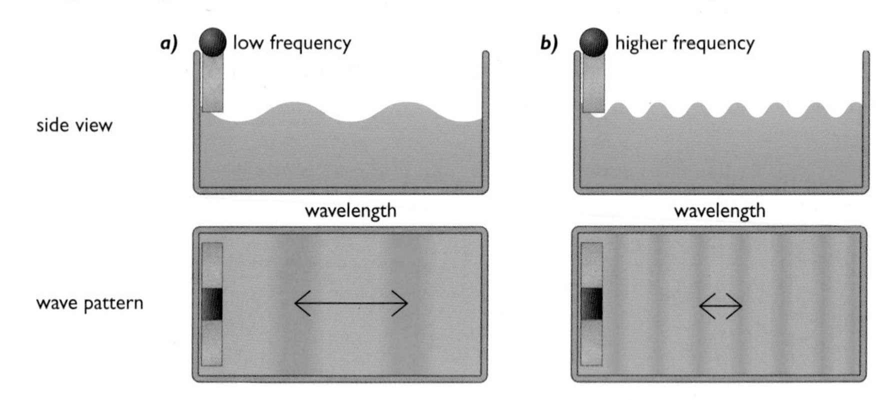

Figure 15.11 *When the frequency of the waves is low, the wavelength is long. When the frequency is higher, the wavelength is shorter.*

Reflection

When waves strike a straight or flat barrier, the angle at which they leave the barrier surface is equal to the angle at which they meet the surface. That is, the waves are reflected from the barrier at the same angle as they struck it.

The angle of incidence is equal to the angle of reflection.

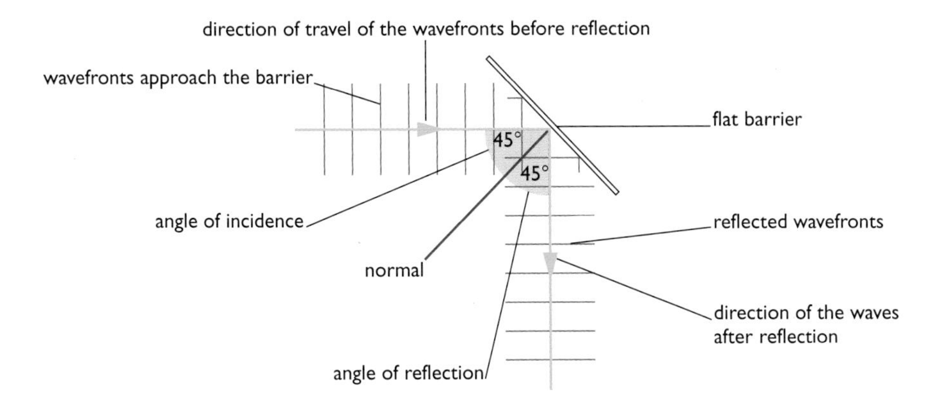

Figure 15.12 *Waves striking a flat barrier are reflected. The angle at which they strike the barrier is the same as the angle at which they are reflected.*

A normal is a line drawn at right angles to a surface.

When the waves strike a concave barrier, they become curved and are made to converge.

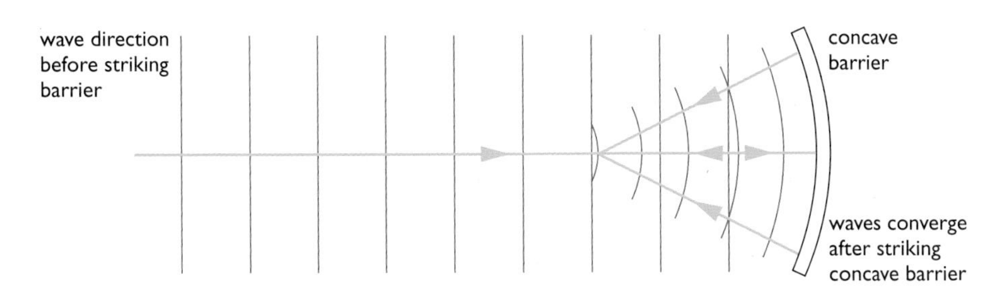

Figure 15.13 *Waves striking a concave barrier are reflected backwards and converge.*

Figure 15.14 *Radio telescope dishes have a concave shape so that the signals they receive are made to converge onto a detector.*

When the waves strike a convex barrier, they are made to diverge (spread out).

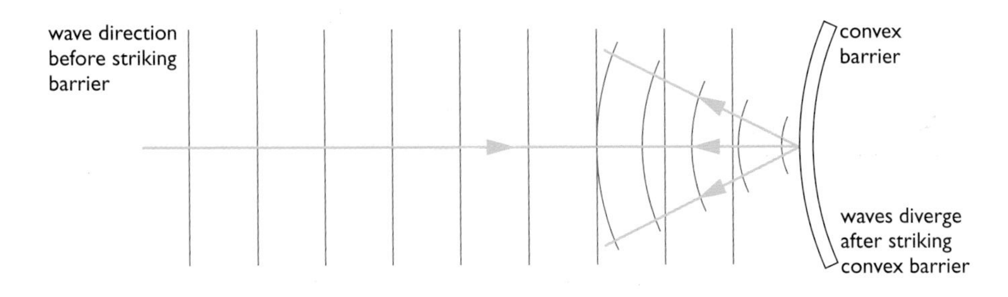

Figure 15.15 *Waves striking a convex barrier are reflected backwards and spread out.*

Refraction

If a small glass plate is placed in the centre of the ripple tank, the depth of the water here is reduced. As waves enter this region we can see that their wavelength becomes shorter. The frequency of the waves is unaltered. It follows therefore, from the formula $v = f\lambda$, that the waves are travelling more slowly in the shallower water. As the waves enter the deeper water again their wavelength increases, indicating that their speed has increased.

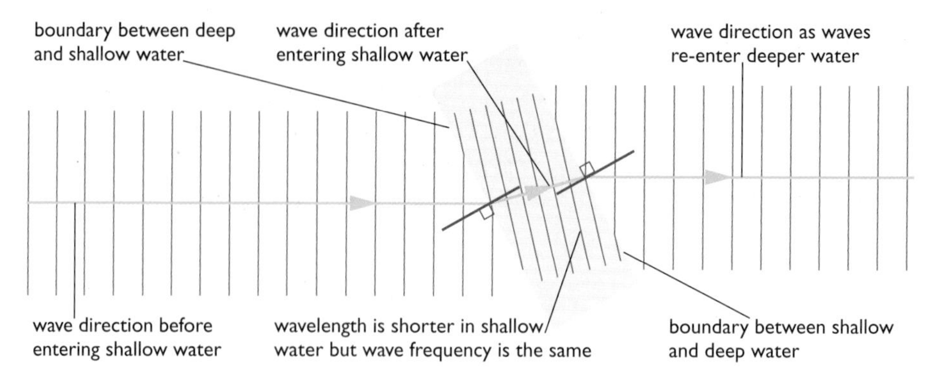

Figure 15.16 *As waves enter shallow water their wavelength becomes shorter. If the boundary between the deep and shallow water is at an angle, the waves are refracted.*

If the boundary between the shallow water and the deep water is at an angle to the direction in which the waves are moving, the direction of the waves changes. We say the waves have been **refracted** or have undergone **refraction**. The waves bend towards the normal as they enter the shallow water and are slowed down. The waves bend away from the normal as they leave the shallow water and enter the deeper water.

Diffraction

If a barrier with a large gap is placed in the path of the waves, the majority of the waves passing through the gap continue through in a straight line. There are regions to the left and right of the gap where there are no waves.

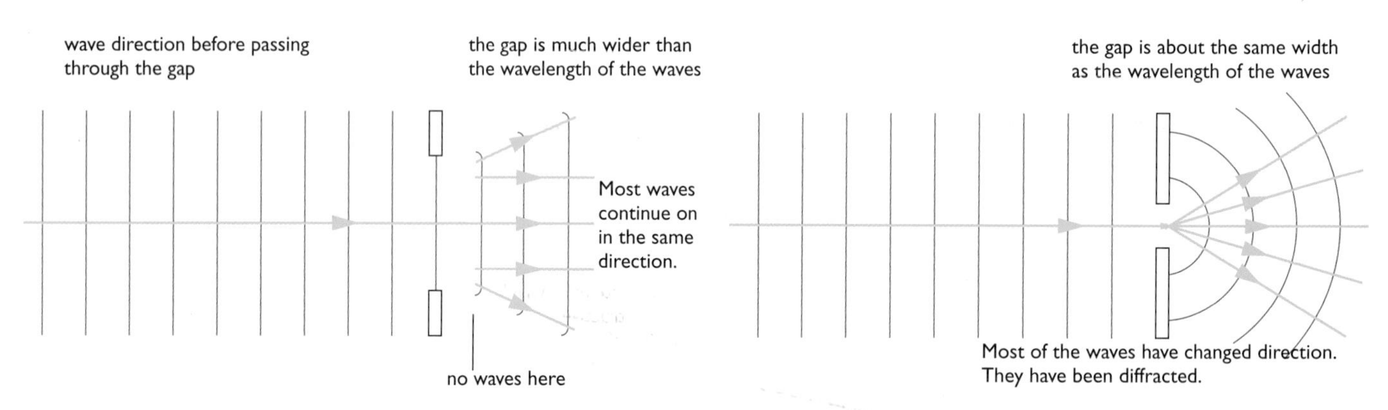

Figure 15.17 *Most waves passing through a large gap in a barrier continue in a straight line.*

Figure 15.18 *If the gap in the barrier is the same size as the wavelength, then the waves spread out.*

If the size of the gap is adjusted so that it is equal to the wavelength of the water waves, the wave pattern shows that there are waves to the left and the right of the gap where previously they had been absent. The waves have spread out from the "straight-on" direction.

This spreading out is called **diffraction**. Diffraction is a property that is demonstrated by all waves. The effect is most noticeable when the wavelength of the waves is approximately equal to the size of the aperture (hole) through which they are moving. Examples of diffraction include radio waves that are diffracted around hills and sound waves that diffract as they pass through doorways.

Interference

If two or more waves overlap they may combine to create a new wave. This phenomenon is called **interference**. You can demonstrate interference in a ripple tank by connecting two dippers to the vibrating bar. The dippers each produce a circular wave pattern, which overlap to produce an **interference pattern**.

In those places where the waves arrive in step (that is, their crests arrive together), their overlapping produces a new wave with a larger amplitude. This is known as **constructive** interference.

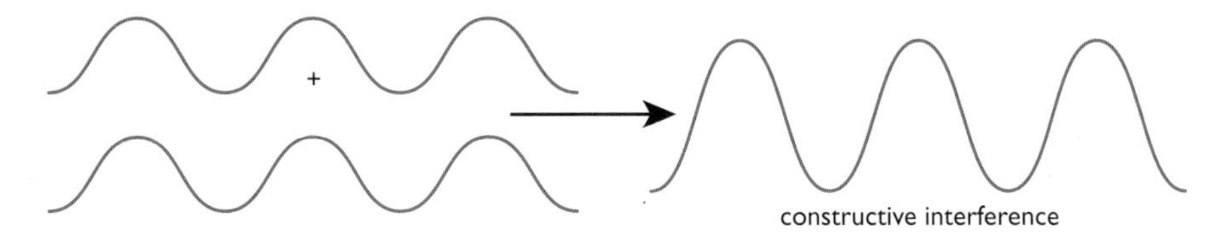

Figure 15.20 *Where two waves arrive with their crests together the waves increase in size.*

In those places where the waves arrive out of step (that is, crests and troughs arrive together) the two wave motions destroy each other. This is known as **destructive interference**.

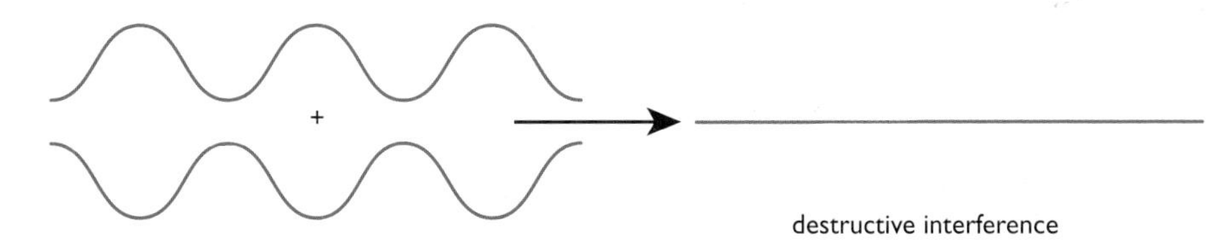

Figure 15.21 *Where the two waves arrive with crests meeting troughs the waves cancel each other out.*

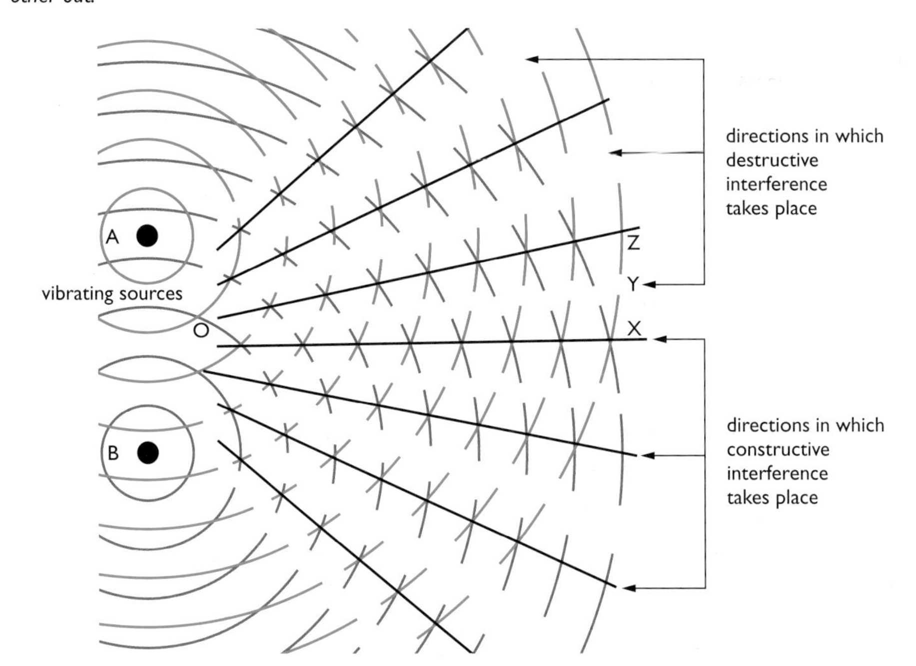

Figure 15.22 *Creating a pattern of constructive and destructive interference.*

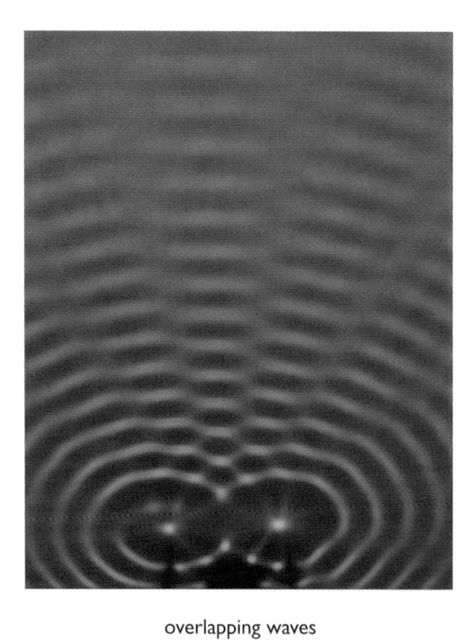

Figure 15.19 *Where waves overlap a pattern of interference can be seen.*

In the direction OX, in Figure 15.22, the waves from each of the sources have travelled the same distance and so meet in step – that is, the crests of the two waves meet simultaneously and the troughs of the two waves meet simultaneously. Constructive interference takes place in this direction.

In the direction OY, the waves from B have travelled half a wavelength ($\frac{1}{2}\lambda$) further than those from A. The waves have a path difference of $\frac{1}{2}\lambda$. As a result, the crests from A arrive at the same time as the troughs from B, and vice versa. Destructive interference takes place in this direction.

In the direction OZ, the path difference between the waves is λ. The waves will again meet in step and constructive interference will take place. This is true for all directions where the path difference between the waves is a whole number of wavelengths.

Destructive interference will take place in all those directions where the path difference is $(n + \frac{1}{2})\lambda$,where n is a whole number.

End of Chapter Checklist

If you haven't got a copy of your specification, read the introduction on page iv.

You will need to be able to do some or all of the following. Check your Awarding Body's specification (syllabus) to find out exactly what you need to know.

- Know that waves carry energy from place to place, but not matter.
- Describe the difference between transverse and longitudinal waves.
- Recall examples of transverse and longitudinal waves.
- Recall the meaning of amplitude, wavelength, frequency and period of a wave.
- Use the relationship $T = 1/f$.
- Use the relationship $v = f\lambda$.
- Know that all waves can be reflected, refracted, diffracted and create interference patterns.
- Know that the change in speed of water waves as they cross the boundary between two different depths of water may cause them to change direction (refract).
- Know that diffraction of a wave is most pronounced when the wavelength of the wave and the size of the aperture through which it is travelling are the same.
- Know the conditions under which constructive and destructive interference occur.

Questions

More questions on wave properties can be found at the end of Section C on page 205.

1 ***a)*** Explain the difference between a transverse wave and a longitudinal wave.

b) Give one example of each.

c) Draw a diagram of a transverse wave. On your diagram, mark the wavelength and amplitude of the wave.

2 Plane water waves in a ripple tank pass through a narrow gap in a barrier.

a) Draw the appearance of the waves when the gap is much larger than the wavelength of the water waves.

b) Draw the appearance of the waves when the gap is the same size as the wavelength of the water waves.

c) Explain why designers must choose the size of a harbour entrance carefully if boats moored there are to be in calm water.

3 The diagram below shows the displacement of water as a wave travels through it.

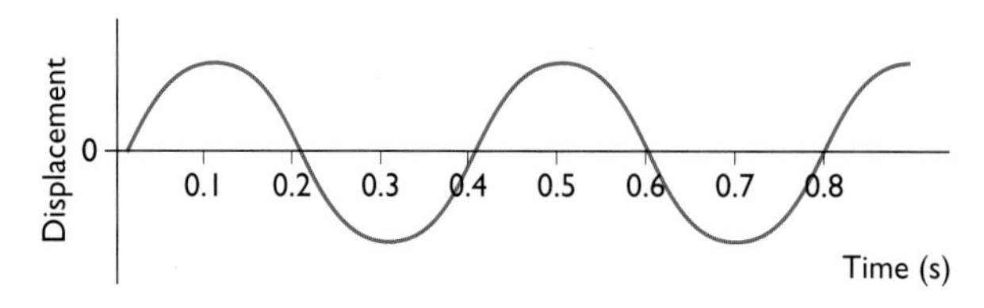

From the diagram calculate:

a) the period of the wave

b) the frequency of the wave.

Chapter 16: Light Waves and Reflection

We see objects because they emit or reflect light. In this chapter you will learn how light behaves when it reflects from different surfaces, and you'll find out about the shadows and images it can create.

Seeing the light

The patient shown in Figure 16.1 has a cataract. The front of one of his eyes has become so cloudy that he is unable to see. Nowadays it is possible to remove this infected part of the eye and replace it with a clear plastic that will again allow light to enter the eye.

There are many sources of light, including the Sun, the stars, fires, light bulbs and so on. Objects such as these that emit their own light are called **luminous** objects. When the emitted light enters our eyes we see the object. Most objects, however, are **non-luminous**. They do not emit light. We see these non-luminous objects because of the light they **reflect**.

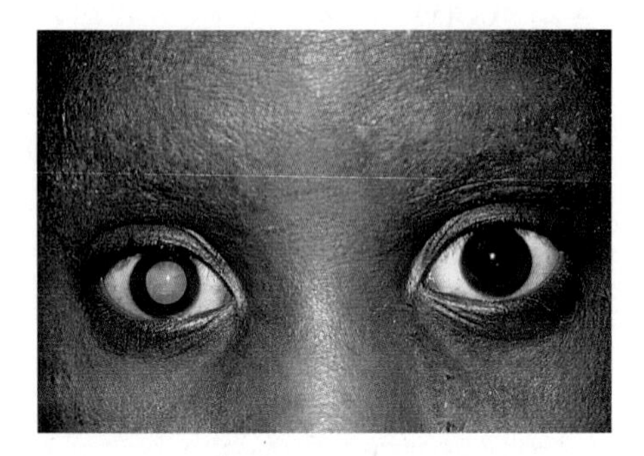

Figure 16.1 *Cataracts mean that light cannot enter the eye correctly.*

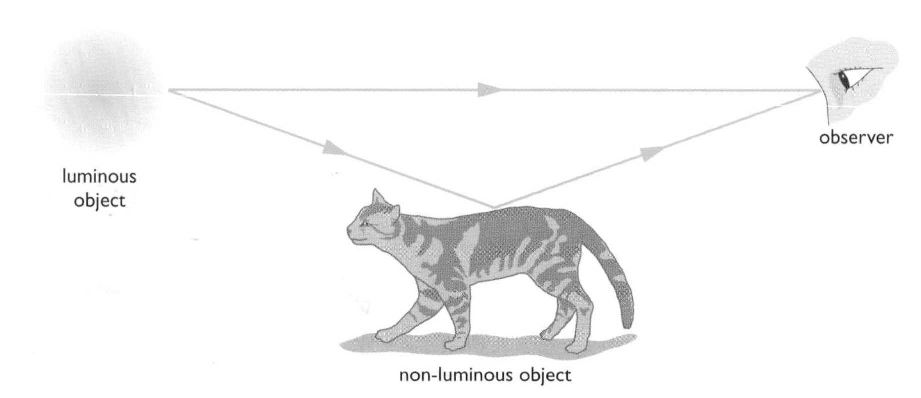

Figure 16.2 *Luminous objects, such as the Sun, give out light. Non-luminous objects only reflect light.*

Travelling light

Photographs like Figure 16.3 suggest that light travels in straight lines. We can confirm that this is the case by doing a simple experiment.

Figure 16.3 *Light travels in straight lines.*

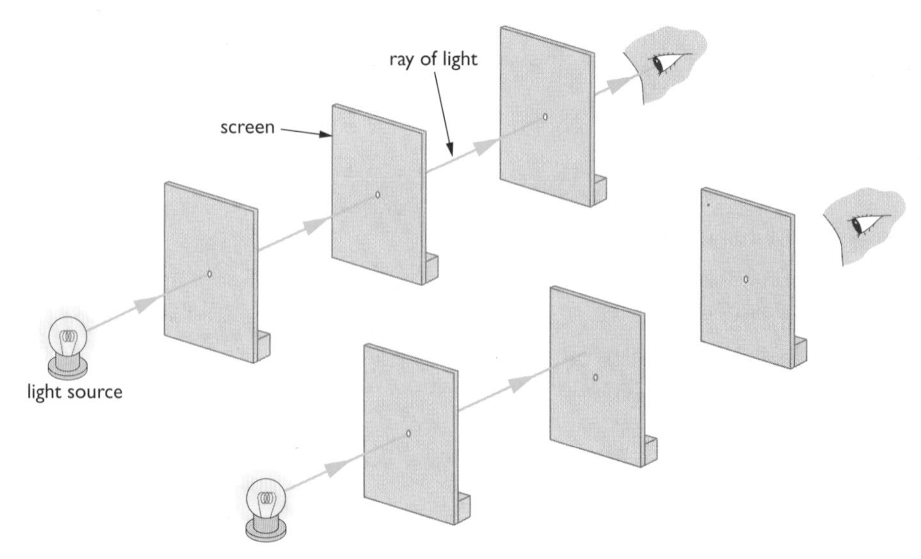

Figure 16.4 *The screens must be aligned exactly if the light is to pass through the holes. This demonstrates that light travels in straight lines.*

Place three screens with small holes in their centres in front of a luminous object such as a filament bulb. Thread a piece of cotton through the holes and pull it taut so that the holes are all in a straight line. When the cotton is removed you can look through all of the holes and see the light from the bulb. If any of the screens are moved so that the holes are no longer in a straight line, light from the bulb is unable to reach your eye. Light must therefore travel in straight lines.

As light travels in straight lines, we often draw ray diagrams to explain its behaviour. The arrowheads drawn on the lines show the direction in which the light is moving.

Shadows

Figure 16.5 *This shadow-puppet show uses opaque objects to cast a shadow on the screen.*

The photograph in Figure 16.5 shows a shadow-puppet show. Opaque objects, which do not allow light to pass through them, are placed between a light source and a screen. The shadow cast on the screen has the same shape as the object that is creating it. This is further proof that light travels in straight lines.

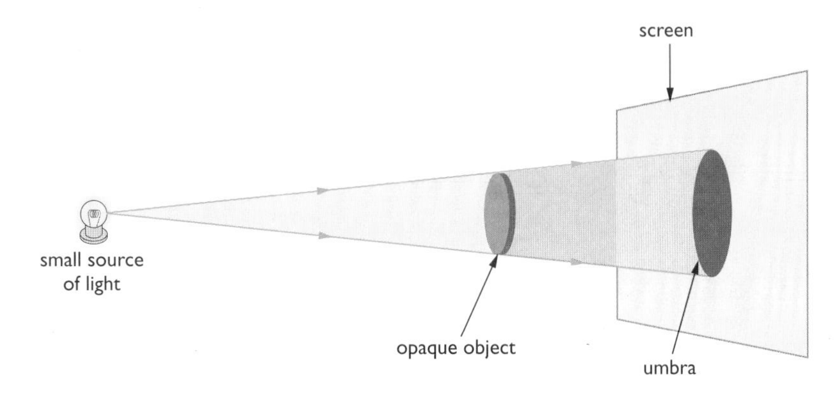

Figure 16.6 *A small source of light produces a dark shadow called an umbra.*

If we place an opaque object in front of a small source of light, a sharp dark shadow called an **umbra** is created.

ideas evidence

The dual nature of light

It is well known that light travels as waves ... or does it? Sir Isaac Newton suggested that light travelled as small particles called "corpuscles". His theory was accepted by many scientists until in 1801 Thomas Young produced an interference pattern by passing light through two narrow slits. Interference patterns are created by waves and not by particles, so Newton's theory was abandoned and replaced with the wave theory of light.

In the early 1900s an experiment was carried out by scientists investigating the emission of electrons by some metals when exposed to light (this is called the "photoelectric effect"). This effect could only be explained if they regarded light as being made up of particles. They called these particles "quanta" and the theory was called the quantum theory of light.

Nowadays we describe light as having a dual nature. Sometimes it behaves as if it is a wave and sometimes it behaves as if it consists of particles.

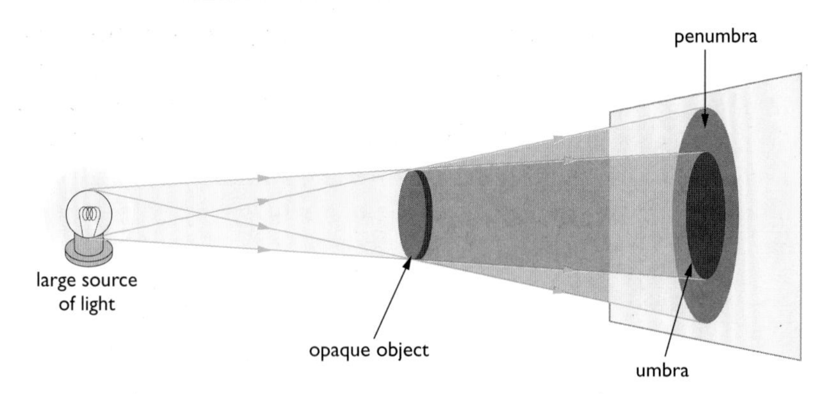

Figure 16.7 *If a large source of light is used, the shadow has a fuzzy edge called the penumbra because some light gets round the object.*

If the source of light is large, a second lighter shadow called a **penumbra** is also formed. The penumbra is less dark than the umbra as only part of the light from the source is blocked off.

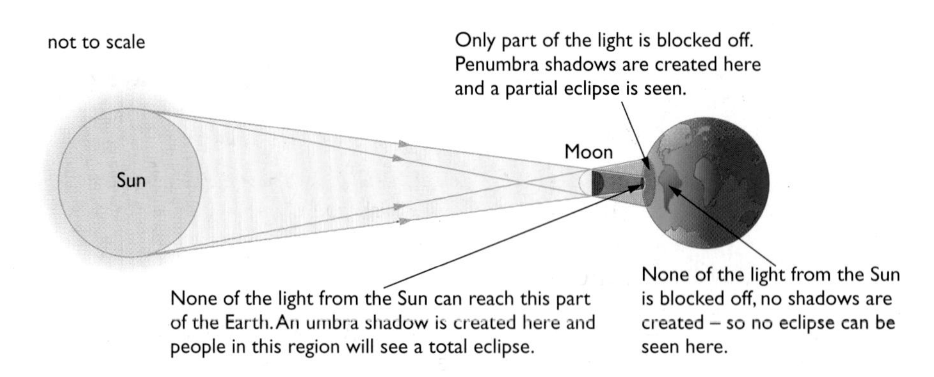

Figure 16.8 *A solar eclipse produces both umbra and penumbra shadows.*

Reflection

When a ray of light strikes a plane (flat) mirror, it is reflected so that the **angle of incidence** is equal to the **angle of reflection**.

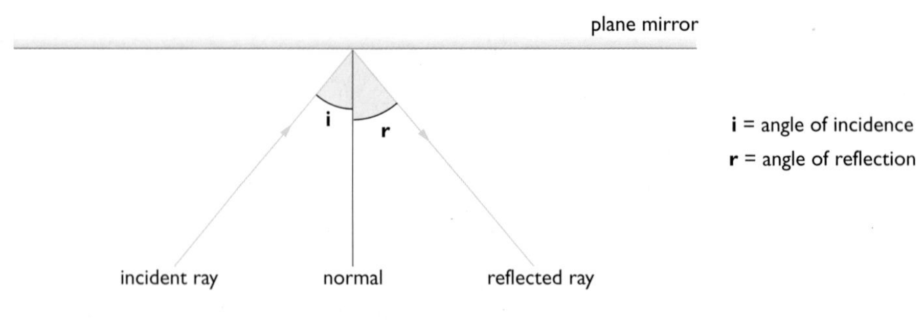

Figure 16.9 *Light is reflected from a plane mirror. The angle of incidence is equal to the angle of reflection.*

Mirrors are often used to change the direction of a ray of light. One example of this is the simple periscope, which uses two mirrors to change the direction of rays of light.

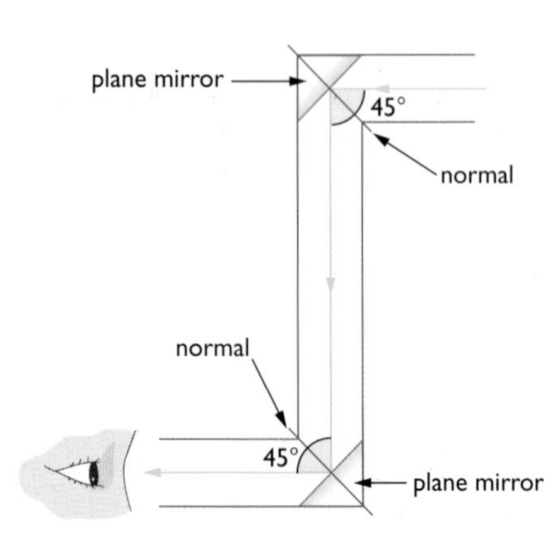

Figure 16.10 *A periscope is used to see over or around objects.*

Rays from the object strike the first mirror at an angle of 45° to the normal. The rays are reflected at 45° to the normal and so are turned through an angle of 90° by the mirror. At the second mirror the rays are again turned through 90°. Changing the direction of rays of light in this way allows an observer to use a periscope to see over or around objects.

Plane and diffuse reflection

If a beam of parallel rays strikes a plane surface, all of the rays will be reflected in the same direction. A large number of rays will enter the eyes of an observer and so the surface appears shiny. This is **plane reflection**.

If a beam of parallel rays strikes a surface that is not flat, the rays will be scattered in all directions. Only a small number of rays will enter the eyes of an observer and therefore the surface will appear dull or matt. This is **diffuse reflection.**

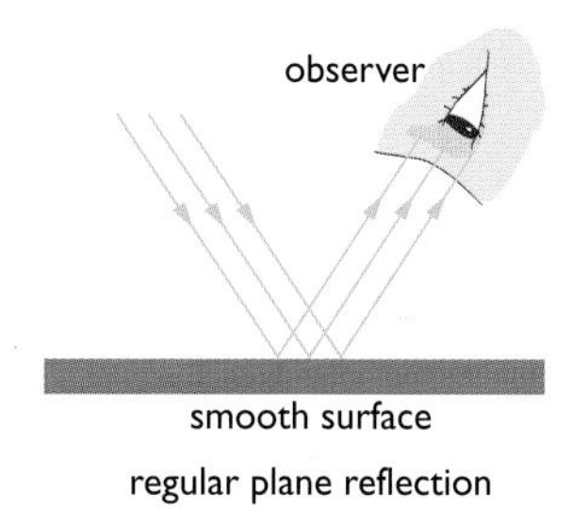

Figure 16.11 *A plane surface reflects all the light rays that hit it, in the same direction.*

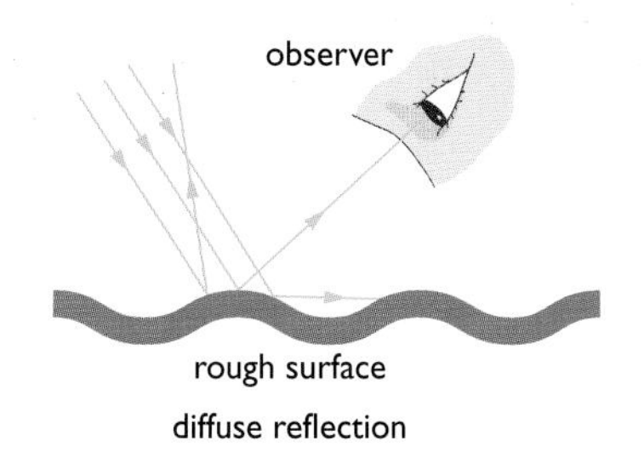

Figure 16.12 *A surface that is not flat will reflect light rays in many different directions.*

Before you polish your shoes they are likely to look dull because their surfaces are rough. After you have polished them the surfaces are smooth, so your shoes look shiny.

Images created by a plane mirror

When you look into a plane mirror, you see images of the room that appear to be behind the mirror. These images are created by rays of light from objects inside the room striking the mirror and being reflected into your eyes. Figure 16.13 shows how these images are created.

Because rays of light normally travel in straight lines, your brain interprets the rays as having come from **I** (that is, an image of the object is seen at **I**). Images like these, through which rays of light do not actually pass, are called **virtual images**. Images created with rays of light actually passing through them (for example, on a cinema screen) are called **real images**.

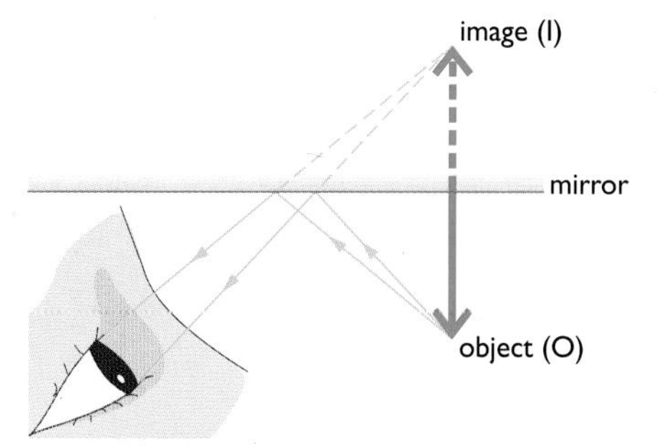

Figure 16.13 *Virtual images look like they are behind the surface of the mirror.*

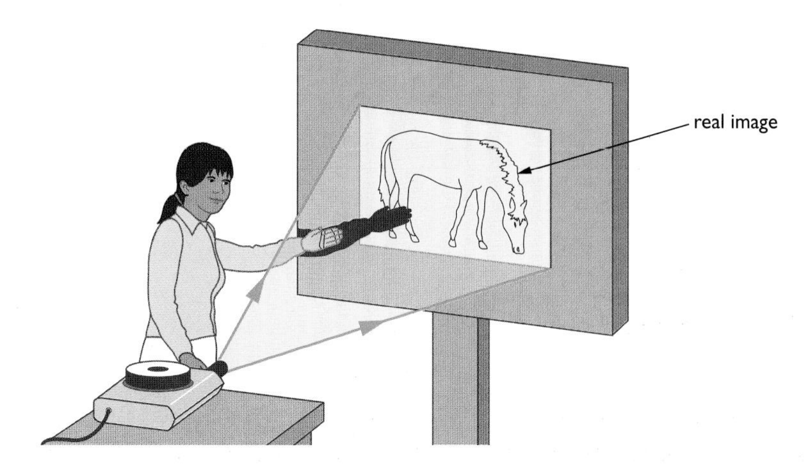

Figure 16.14 *Real images are produced when light passes through them.*

If you can produce an image on a screen, it is real.

Properties of an image in a plane mirror

- The image is as far behind the mirror as the object is in front.
- The image is the same size as the object.
- The image is virtual – that is, it cannot be produced on a screen.
- The image is **laterally inverted** – that is, the left side and right side of the image appear to be interchanged.

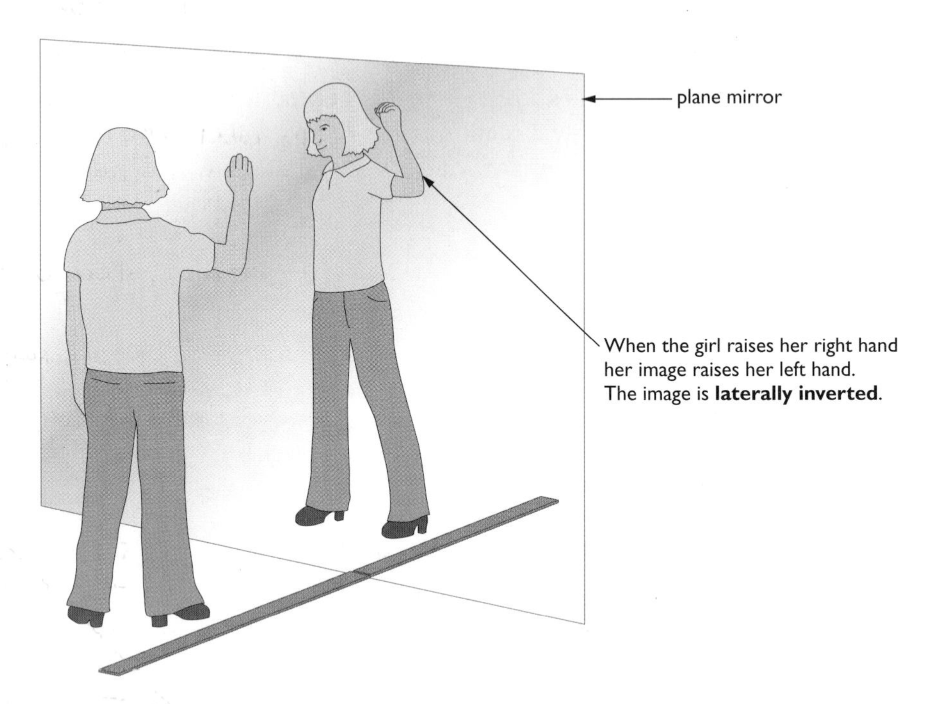

Figure 16.15 *The image in a mirror is the same size as the object, is laterally inverted and is virtual. The image also appears to be the same distance from the mirror as the object.*

Figure 16.16 *Emergency vehicles such as ambulances often have mirror writing on their bonnets so that drivers in front can read the name when they look in their mirrors.*

End of Chapter Checklist

If you haven't got a copy of your specification, read the introduction on page iv.

You will need to be able to do some or all of the following. Check your Awarding Body's specification (syllabus) to find out exactly what you need to know.

- Know that luminous objects are seen by the light they emit and non-luminous objects are seen by the light they reflect.
- Explain with the help of ray diagrams how shadows are created with both small and large sources of light.
- Know that light is reflected from a plane surface so that the angle of incidence is equal to the angle of reflection.
- Explain the difference between a virtual image and a real image.
- Draw ray diagrams to show how an image is created in a plane mirror.
- Describe the properties of images created by a plane mirror.

Questions

More questions on light waves and reflection can be found at the end of Section C on page 205.

1 ***a)*** Explain the difference between luminous objects and non-luminous objects.

b) Give one example of each.

c) Draw a labelled diagram to show how an observer sees a non-luminous object.

2 **a)** What is an "opaque" object?

b) What is a shadow?

c) Draw a labelled diagram to show how a large source of light can produce two types of shadow – that is, both umbra and penumbra.

3 Draw a ray diagram to show how a ray of light can be turned through 180° using three plane mirrors. Mark on your diagram a value for the angle of incidence at each of the mirrors.

4 ***a)*** Draw an accurate diagram to show how a plane mirror creates an image of an object.

b) Explain the difference between a real image and a virtual image.

c) State five properties of an image created in a plane mirror.

d) A man stands 5 m in front of a plane mirror. How far is he from his image?

e) The man walks towards the mirror at a speed of 1 m/s. At what speed are the man and his image approaching each other?

5 Explain with diagrams how a beam of parallel rays is reflected from:

a) a surface that appears dull

b) a surface that appears shiny.

Chapter 17: Refraction

When light travels through several transparent materials, as it reaches a boundary it may be refracted, dispersed or reflected. In this chapter you will learn why light behaves like this, and read about some useful applications of such effects.

Figure 17.1 *This rainbow is caused by refraction.*

A **medium** is a material – such as glass or water, through which light can travel. The plural of medium is media.

Light does travel more slowly in air than in a vacuum but the difference is negligible.

Rays of light can travel through many different transparent media, including vacuum, air, water and glass. In a vacuum and in air light travels at a speed of 300 000 000 m/s. In other media it travels more slowly. For example, the speed of light in water is approximately 200 000 000 m/s. When a ray of light travels from air into glass or water it slows down as it crosses the boundary between the two media. The change in speed may cause the ray to change direction. This change in direction of a ray is called **refraction**.

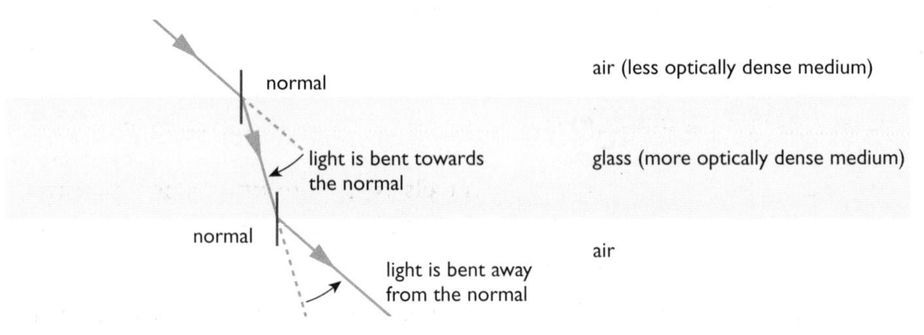

Figure 17.2 *Light rays bend as they travel from air into glass and out again. This is called refraction.*

As a ray enters a glass block it slows down and is refracted towards the normal. As the ray leaves the block it speeds up and is refracted away from the normal (see page 157).

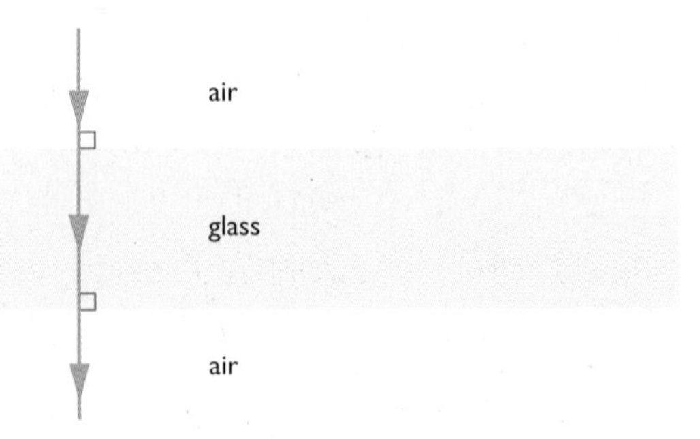

Figure 17.3 *If the light hits the boundary at 90° the ray does not bend.*

If the ray strikes the boundary between the two media at 90°, the ray continues without change of direction.

Why are rays of light refracted?

It is the change in the speed of a ray of light that causes it to be refracted as it crosses the boundary between two media. Figure 17.4 shows a ray of light of width AB passing through a glass block.

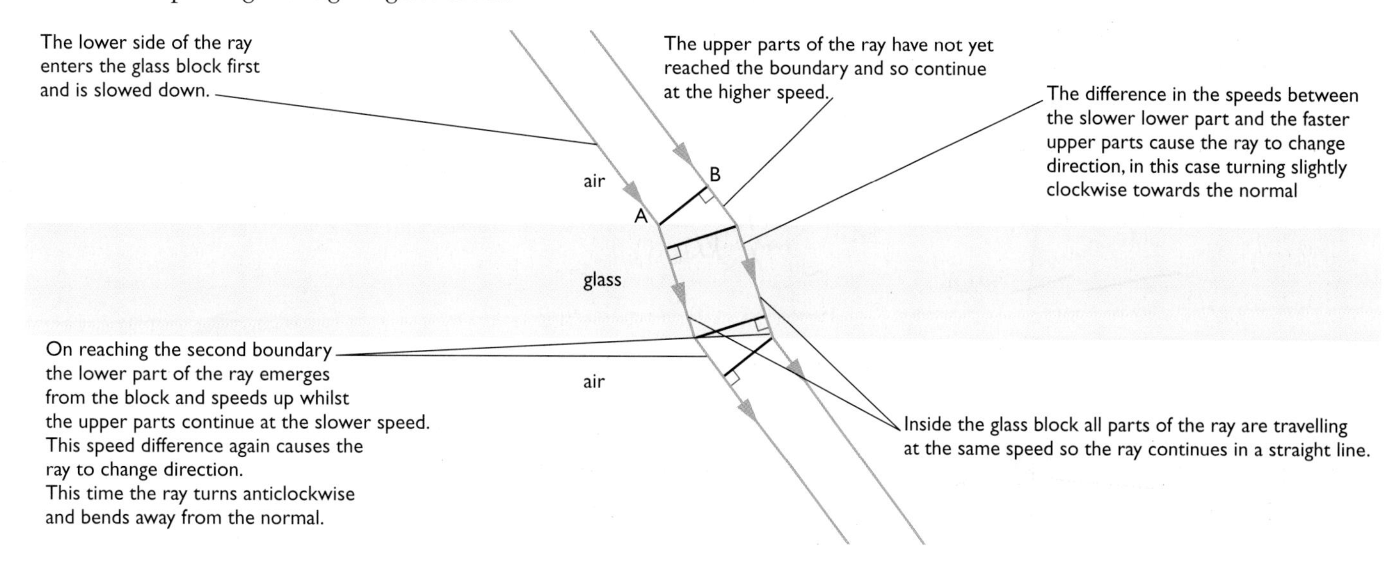

Figure 17.4 Light rays are refracted because they slow down as they enter the glass block.

If the ray strikes either boundary at 90°, all parts of the ray change speed at the same time so refraction does not occur.

Effects of refraction

We are used to the idea that light travels in straight lines. Consequently, when the direction in which a ray is travelling changes by refraction we sometimes see some unusual effects.

Figure 17.5 shows a pencil half immersed in water. Although we know the pencil to be straight, it appears to be bent. This effect is caused by refraction.

Figure 17.5 *The pencil looks as if it is bent at the water surface.*

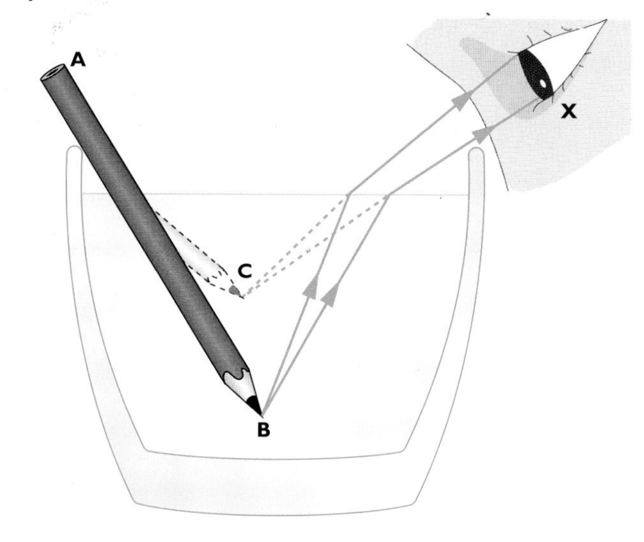

Figure 17.6 *The pencil looks bent because the light rays bend away from the normal at the surface of the water.*

When rays of light from the bottom of the pencil reach the surface of the water they bend away from the normal. If you put your eye in position X the rays therefore appear to have come from C rather than from B and this is where you will see the bottom of the pencil. Similarly all parts of the pencil between A and B will, because of refraction, appear to have come from positions between A and C. The pencil therefore looks bent.

Swimming pools always look shallower than they really are. This again is an effect of refraction. Rays of light from the bottom of the pool are refracted away from the normal as they cross into the air. To an observer the bottom of the pool therefore appears to be at X, and not at Y, so the pool appears to be shallower than it really is.

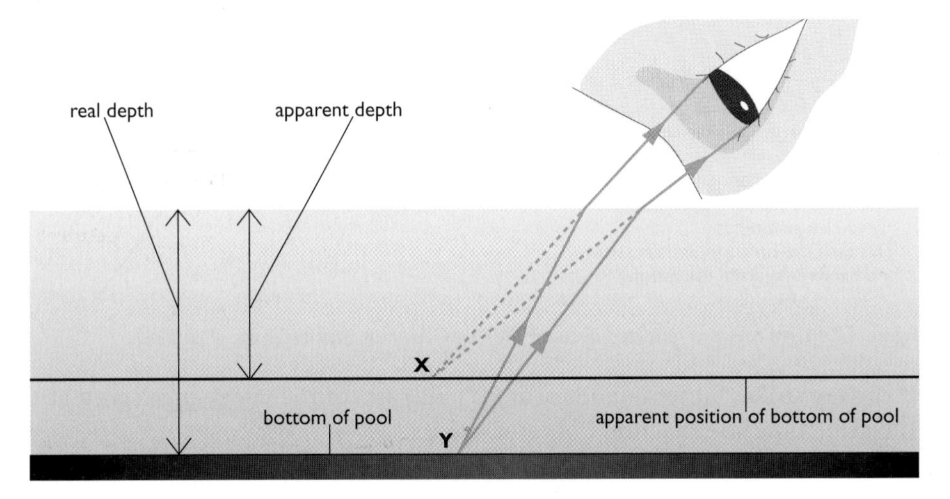

Figure 17.7 Swimming pools look shallower than they really are.

Refractive index

Different materials refract light by different amounts. The amount of refraction depends upon the speed of the light in the different media. The **refractive index** is a way of comparing these speeds.

$$\textbf{refractive index of a medium} = \frac{\textbf{speed of light in a vacuum}}{\textbf{speed of light in the medium}}$$

Note that because the refractive index is a ratio it has no units.

worked example

Example 1

Calculate the refractive index of water if the speed of light in a vacuum is 3×10^8 m/s and in water is 2×10^8 m/s.

$$\text{refractive index of a medium} = \frac{\text{speed of light in a vacuum}}{\text{speed of light in the medium}}$$

$$\text{refractive index of water} = \frac{3 \times 10^8 \text{ m/s}}{2 \times 10^8 \text{ m/s}}$$

$$\text{refractive index of water} = 1.5 \quad \text{(no units)}$$

Light travels more slowly in an optically more dense medium.

Dispersion

When white light passes through a prism, it emerges as a band of colours called a **spectrum**. The spectrum is formed because white light is a mixture of colours and each colour travels through the prism at a slightly different speed – that is, the prism has a different refractive index for each of the colours. Because of this, each of the colours emerges from the prism travelling in a slightly different direction. The process is called **dispersion**. The speed of red light is changed least and so red has the smallest deviation. The speed of violet light is changed the most and so violet has the largest deviation.

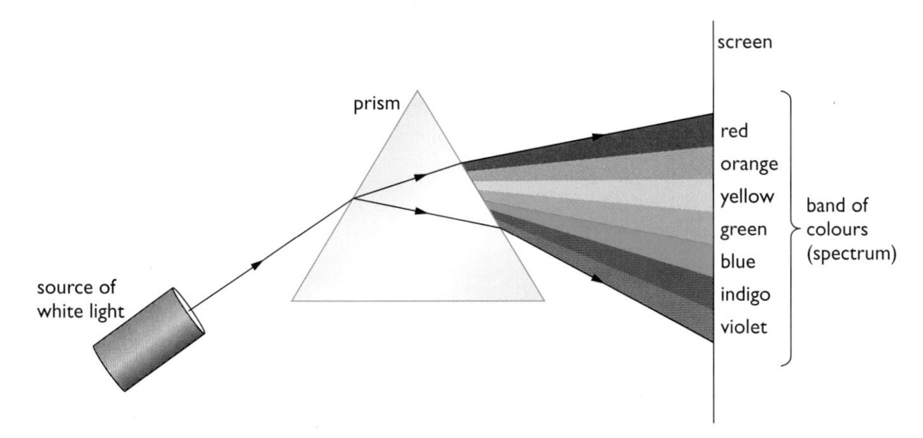

Figure 17.8 *A prism separates white light into its colours because the prism has a slightly different refractive index for each of the colours.*

If a second inverted prism is placed behind the first, the coloured lights recombine to produce white light. This confirms that white light is a mixture of coloured lights.

Most colours of light can be made by mixing other colours. There are three which cannot be produced by mixing. They are red, green and blue. These are the **primary colours** of light.

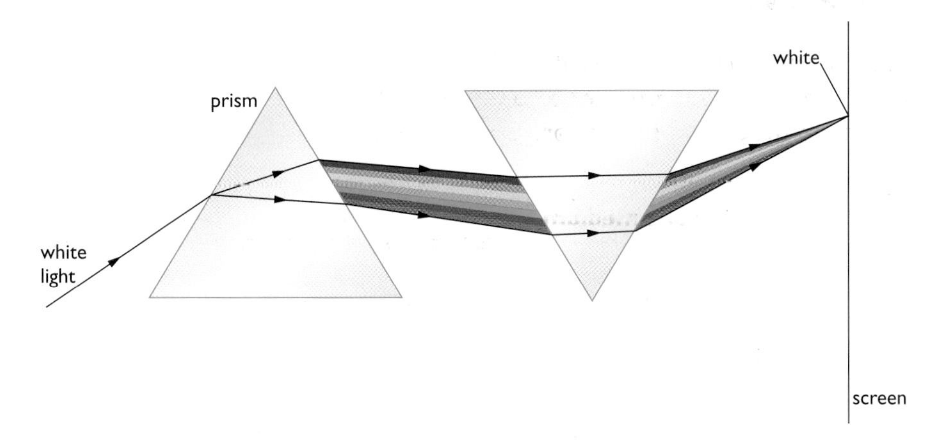

Figure 17.9 *Adding another prism causes the light to be recombined.*

Total internal reflection

When a ray of light passes from an optically more dense medium into an optically less dense medium – for example, from glass into air – the majority of the light is refracted away from the normal but there is a small amount that is reflected from the boundary.

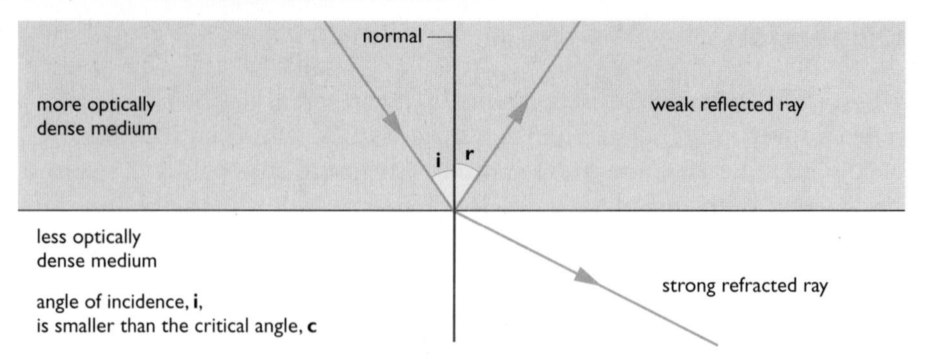

Figure 17.10 *A ray of light is refracted as it passes from a more dense to a less dense medium, but a little ray of light is reflected.*

As the angle of incidence in the more dense medium increases the angle of refraction also increases until, at a special angle called the **critical angle**, the angle of refraction is 90°.

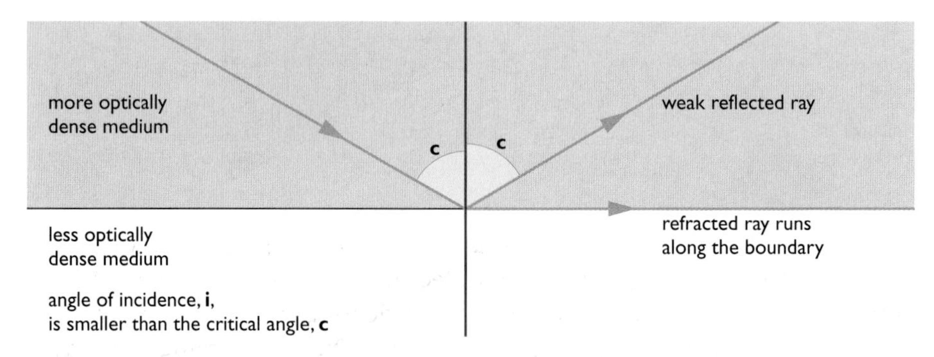

Figure 17.11 *At the critical angle the light is refracted at 90° to the normal.*

If the angle of incidence in the glass is further increased, *all* of the light is reflected from the boundary. The ray is said to have undergone **total internal reflection.**

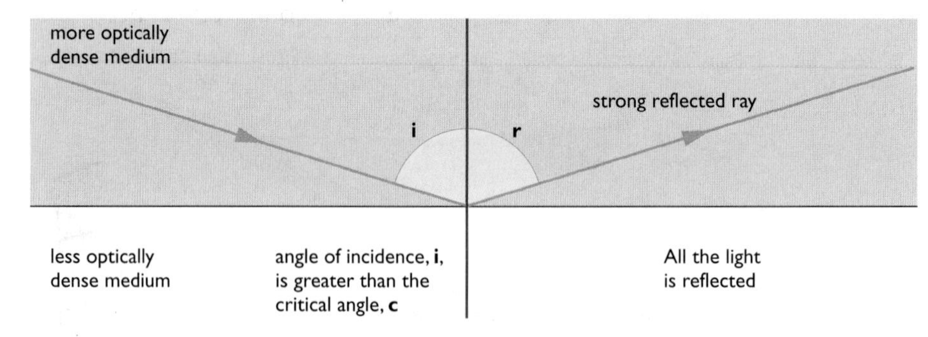

Figure 17.12 *Above the critical angle, the light is reflected back into the more dense medium.*

The value of the critical angle depends upon the media on either side of the boundary. Assuming that the less dense medium is air, then the critical angle for glass is typically 42° and the critical angle for water is 49°.

Using total internal reflection

If we look carefully at the image of an object created by a plane mirror we may see several faint images around the main central image. These multiple images are due to several partial reflections and refractions at the non-silvered glass surface of the mirror.

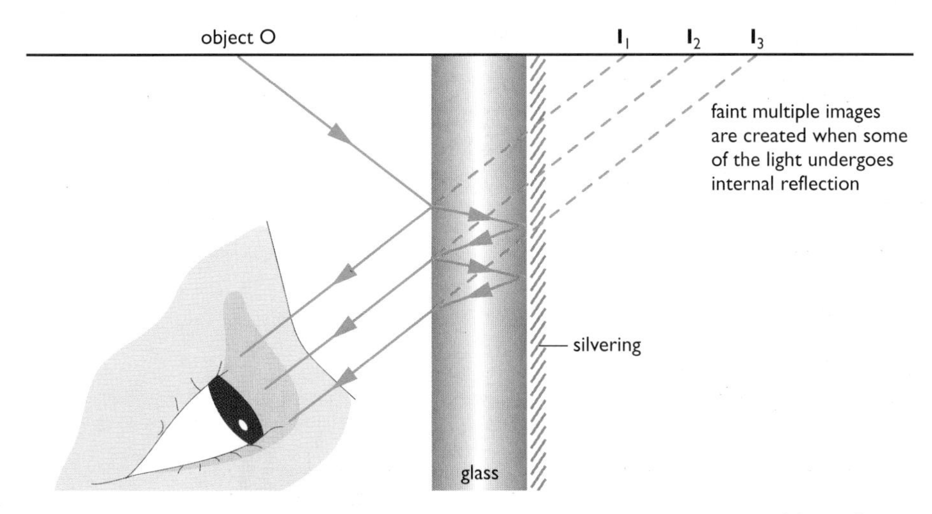

Figure 17.13 *The faint multiple images we can sometimes see in a mirror are caused by total internal reflection.*

To avoid this problem, particularly when high quality images are required, glass prisms are often used to alter the direction of the light rather than mirrors.

The prismatic periscope

Light passes normally through the surface AB of the first prism (that is, it enters the prism at 90°) and so is undeviated. It then strikes the surface AC of the prism at an angle of 45°. The critical angle for glass is 42° so the ray is totally internally reflected and is turned through 90°. On emerging from the first prism the light travels to a second prism which is positioned such that the ray is again totally internally reflected. The ray emerges parallel to the direction in which it was originally travelling.

The final image created by this type of periscope is likely to be sharper and brighter than that produced by a periscope that uses two mirrors, as no multiple images are created.

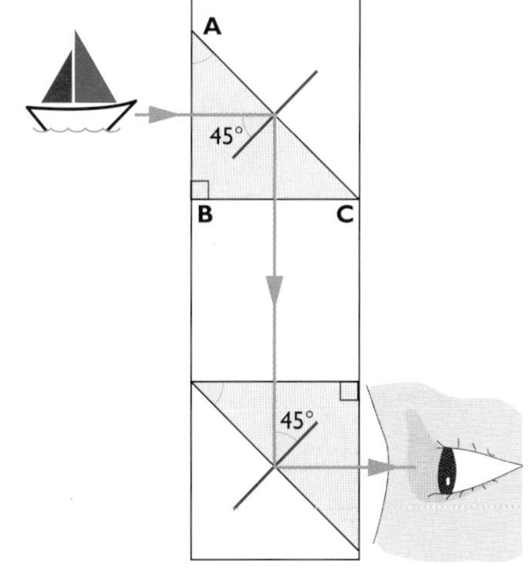

Figure 17.14 *Prisms can be used in periscopes instead of plane mirrors.*

Reflectors

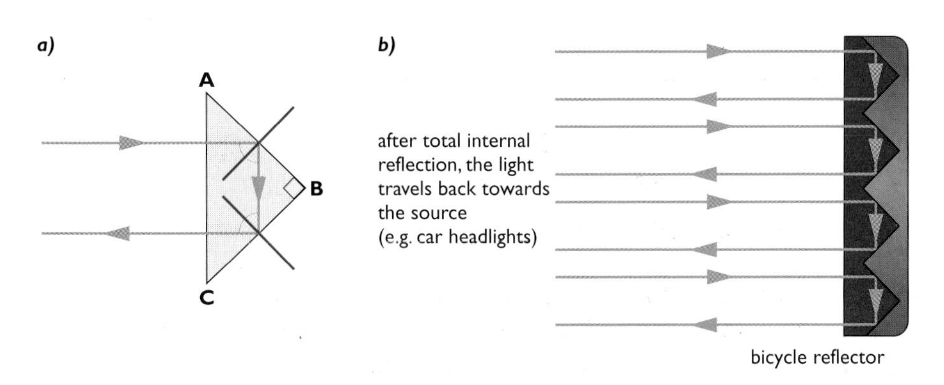

Figure 17.15 *Prisms can also be used as reflectors.*

Light entering the prism in Figure 17.15 undergoes total internal reflection twice. It emerges from the prism travelling back in the direction from which it originally came. This arrangement is used in bicycle reflectors and binoculars.

Each side of a pair of binoculars contains two prisms to reflect the incoming light. Without the prisms, binoculars would have to be very long to obtain large magnifications and would look like a pair of telescopes.

Optical fibres

One of the most important applications for total internal reflection is the **optical fibre**. This is a very thin strand composed of two different types of glass. There is a central core of optically dense glass (high refractive index) around which is a "cladding" or "coat" of optically less dense glass.

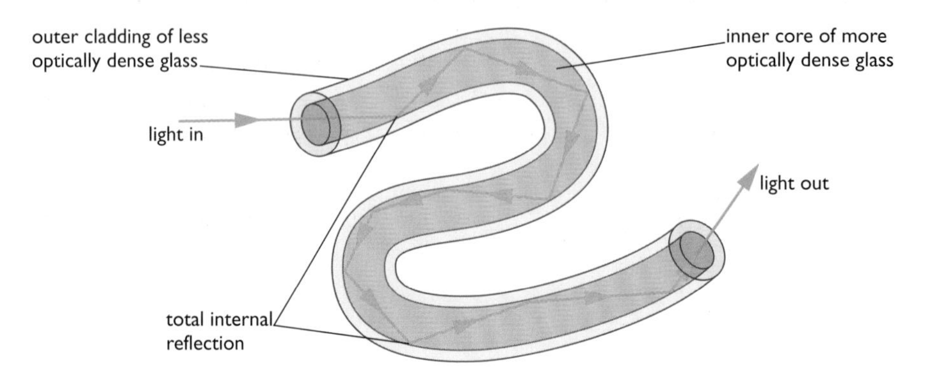

Figure 17.16 *In an optical fibre light undergoes total internal reflection.*

As the fibres are very narrow, light entering the inner core always strikes the boundary of the two glasses at an angle that is greater than the critical angle. No light escapes across this boundary. The fibre therefore acts as a "light pipe" providing a path that the light follows even when the fibre is curved.

Large numbers of these fibres fixed together form a **bundle**. Bundles can carry sufficient light for images of objects to be seen through them. If the fibres are tapered it is also possible to produce a magnified image.

The endoscope

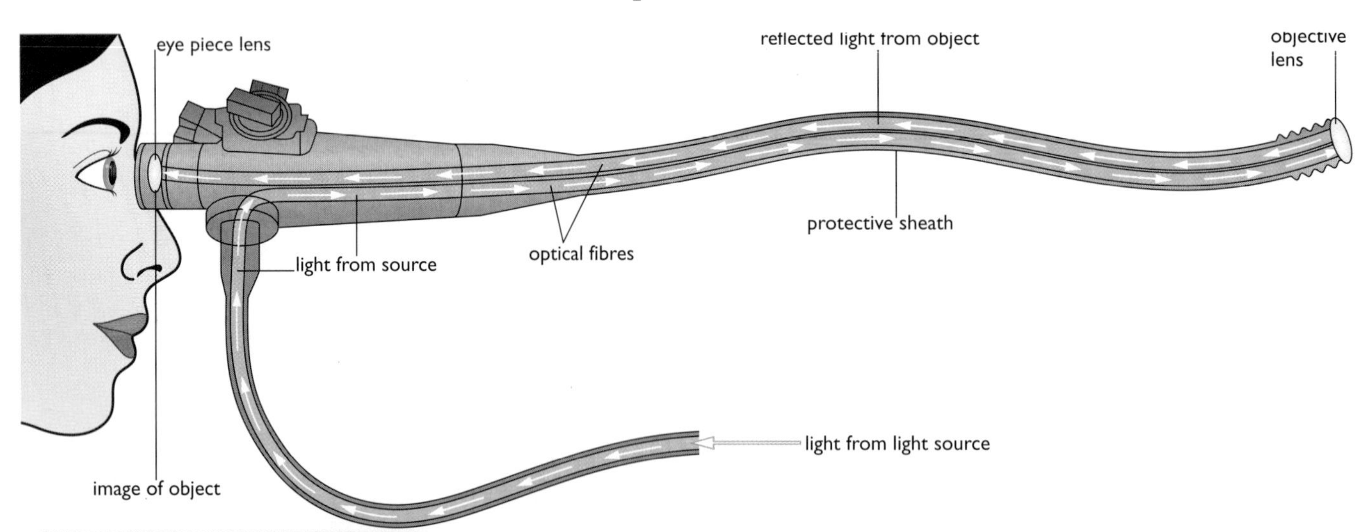

Using optical fibres to see what they are doing, surgeons can carry out operations through small holes made in the body, rather than making large incisions. This is called 'keyhole surgery'. It causes less distress to patients and usually leads to a more rapid recovery.

Figure 17.17 *Optical fibres are used in endoscopes to see inside the body.*

Figure 17.17 shows optical fibres in an endoscope. The endoscope is used by doctors to see the inside the body – for example, to examine the inside of the stomach. Endoscopes can also be used by engineers to see inaccessible parts of machinery.

Light travels down one bundle of fibres and illuminates the object to be viewed. Light reflected by the object travels up a second bundle of fibres. An image of the object is created by the eyepiece.

Optical fibres in telecommunications

Modern telecommunications systems use optical fibres rather than copper wires to transmit messages. Electrical signals from a telephone are converted into light energy by tiny lasers, which send pulses of light into the ends of optical fibres. A light-sensitive detector at the other end changes the pulses back into electrical signals, which then flow into a telephone receiver (ear piece).

The table shows a comparison of the properties of copper cables and optical fibres used to transmit signals.

Copper cables	Optical fibres
expensive	low cost
quite heavy	low weight
high signal loss – repeater stations needed every 4 or 5 km to amplify signal	very little signal loss – repeater stations needed approximately every 100 km
cables capable of carrying large numbers of signals at the same time	fibres can carry very large numbers of signals at the same time
signal can be corrupted by electric and magnetic fields	signal is unaffected by electric and magnetic fields

End of Chapter Checklist

If you haven't got a copy of your specification, read the introduction on page iv.

You will need to be able to do some or all of the following. Check your Awarding Body's specification (syllabus) to find out exactly what you need to know.

- Explain how the change in speed of a ray of light causes it to be refracted.
- Know that light bends towards the normal as it enters an optically more dense medium and bends away from the normal when it enters an optically less dense medium.
- Recall the equation:

$$\text{refractive index of a medium} = \frac{\text{speed of light in a vacuum}}{\text{speed of light in the medium}}$$

- Explain, with examples, how refraction can create virtual images.
- Explain that dispersion occurs because waves of different frequencies (waves of light of different colours) travel at different speeds in transparent materials – that is, the material has a slightly different refractive index for each frequency.
- Recall the conditions under which total internal reflection takes place.
- Describe several applications of total internal reflection including prismatic periscopes, reflectors and optical fibres.

Questions

More questions on refraction can be found at the end of Section C on page 205.

1 ***a)*** Draw a diagram to show the path of a ray of light travelling from air into a rectangular glass block at an angle of about 45°.

b) Show the path of the ray as it emerges from the block.

c) Explain why the ray changes direction each time it crosses the air/glass boundary.

d) Draw a second diagram showing a ray that travels through the block without being deviated.

2 Explain why the water in a pond always looks shallower than it really is. Include a diagram with your explanation.

3 Calculate the refractive index of a block of glass if light travels through the glass at a speed of 1.9×10^8 m/s. The speed of light in a vacuum is 3.0×10^8 m/s.

4 ***a)*** Draw a diagram to show how a prism can cause dispersion.

b) Explain why the white light is dispersed.

5 Draw three ray diagrams to show what happens to a ray of light travelling in a more dense medium if it strikes the boundary with a less dense medium at an angle:

a) less than the critical angle

b) equal to the critical angle

c) greater than the critical angle.

6 ***a)*** What is meant by "total internal reflection of light" and under what conditions does it occur?

b) Draw a diagram to show how total internal reflection takes place in a prismatic periscope.

c) Give one advantage of using prisms in a periscope rather than plane mirrors.

d) Draw a second diagram to show how a prism could be used to turn a ray of light through 180°. Give one application of a prism used in this way.

7 ***a)*** Explain why a ray of light entering an optical fibre is unable to escape through the sides of the strand. Include a ray diagram in your explanation.

b) What would happen to a ray of light inside an optical fibre if the outer glass had a higher optical density than the inner glass?

c) Explain how doctors use optical fibres to see inside the body.

8 ***a)*** Explain how optical fibres are used in telecommunications systems.

b) Give four reasons why optical fibres are being used to replace copper wires in telecommunications systems.

Chapter 18: Lenses and Optical Instruments

Lenses refract light rays in a predictable way. In this chapter you will learn about different kinds of lenses, and see how lenses help to create sharp images in all kinds of optical instruments, including your own eyes.

Lenses

Lenses are specially shaped pieces of glass or plastic that refract light in a predictable and useful manner. There are two main types of lenses:

1 **convex** lenses, which refract rays of light so that they converge

2 **concave** lenses, which refract rays of light so that they diverge.

Convex lenses

When rays travelling parallel to the **principal axis** pass through a convex lens they are made to converge to a point on the axis called the **focal point**. The distance from the focal point to the centre of the lens is called the **focal length**. Lenses that are quite thick converge the rays quite quickly and so have short focal lengths. Lenses that are thin converge the rays more gently and so have quite long focal lengths.

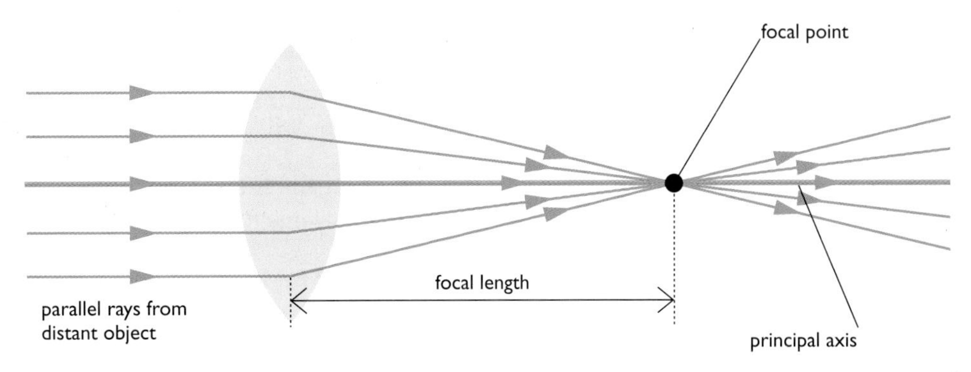

Figure 18.1 *The distance from the centre of the lens to the focal point is known as the focal length.*

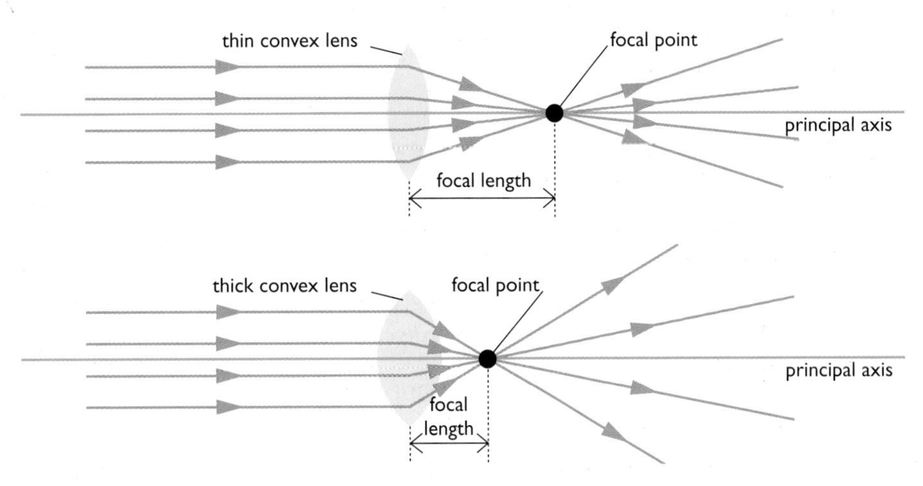

Figure 18.2 *A convex lens bends light towards a focal point. The thicker the lens, the closer the focal point is to the lens.*

Convex lenses can be used to create a range of different kinds of images – for example, real, virtual, magnified, diminished, inverted and so on. The type of image produced by a lens depends upon the position of the object and the focal length of the lens. We can predict the position, nature and size of an image created by a lens using diagrams drawn to scale.

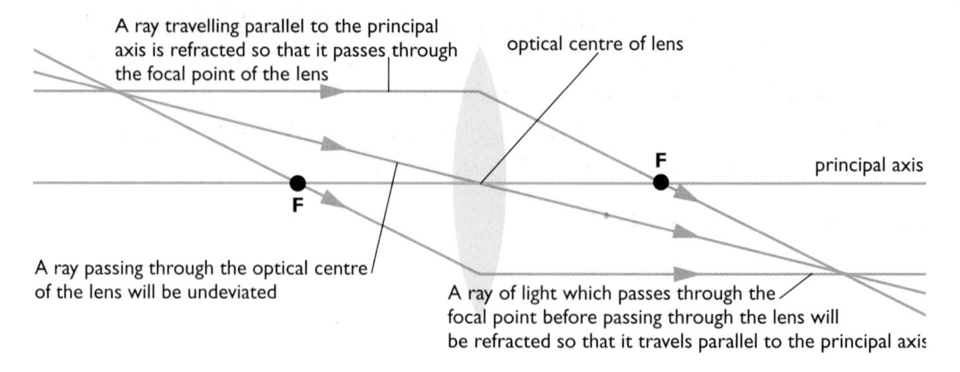

Figure 18.3 *We can predict the behaviour of these three rays.*

There are three rays whose behaviour we can predict as they pass through a convex lens, as shown in Figure 18.3.

1 A ray of light travelling parallel to the principal axis is refracted so that it passes through the focal point of the lens.

2 A ray of light that passes through the focal point before passing through the lens will be refracted so that it travels parallel to the principal axis when it leaves the lens.

3 A ray of light that passes through the optical centre of the lens will be undeviated – that is, it does not change direction.

We can locate the image produced by a lens using any two of these rays. It is where the rays meet.

Object placed beyond 2F

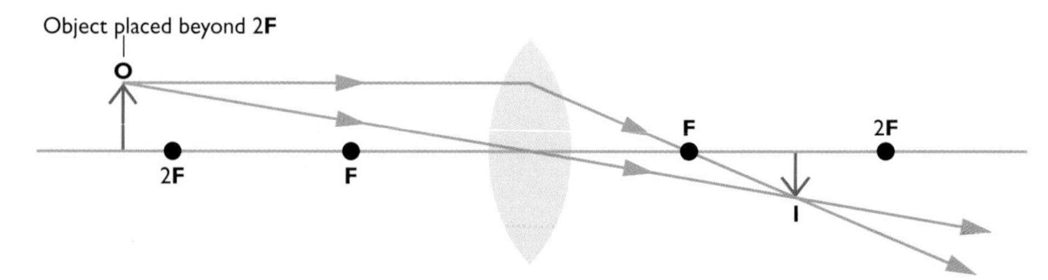

Figure 18.4 *If the object is beyond **2F**, the image is inverted and smaller than the object.*

If an object is placed a distance more than twice the focal length from the lens (beyond 2F), the image created is upside down (inverted), smaller than the object (diminished), real (rays of light actually pass through it) and between F and 2F on the opposite side of the lens to the object. This is what happens in the eye and in cameras.

Object placed at 2F

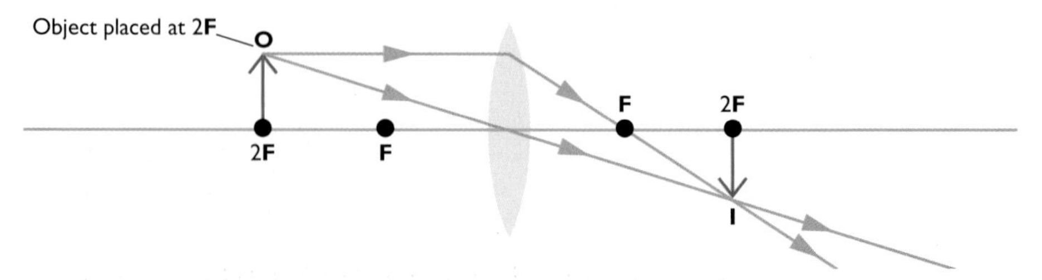

Figure 18.5 *If the object is placed at **2F**, the image is inverted and the same size as the object.*

If an object is placed at a distance equal to twice the focal length of the lens (at 2F), the image created by the lens is inverted, the same size as the object, real and at 2F on the opposite side of the lens to the object. This arrangement is used in optical telescopes where it is necessary to invert the image without altering its size – for example, in terrestrial telescopes.

Object placed between F and 2F

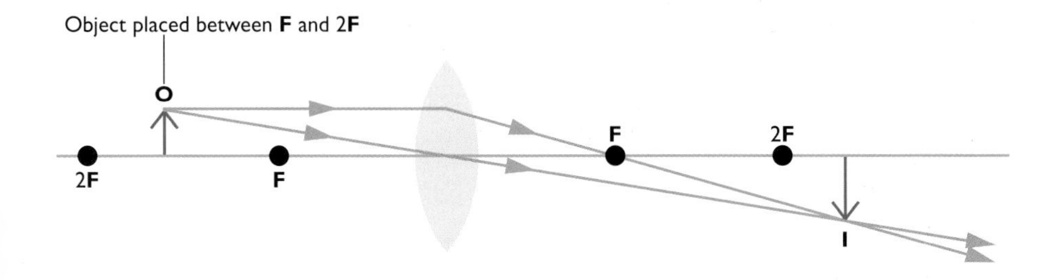

Figure 18.6 *If the object is placed between* **F** *and 2***F***, the image is inverted and larger than the object.*

When an object is placed between F and 2F, the image created is inverted, larger than the object (magnified), real and is further away than 2F on the far side of the lens. Slide projectors use this arrangement of lenses (so remember to put the slides in upside down!).

Object placed at F

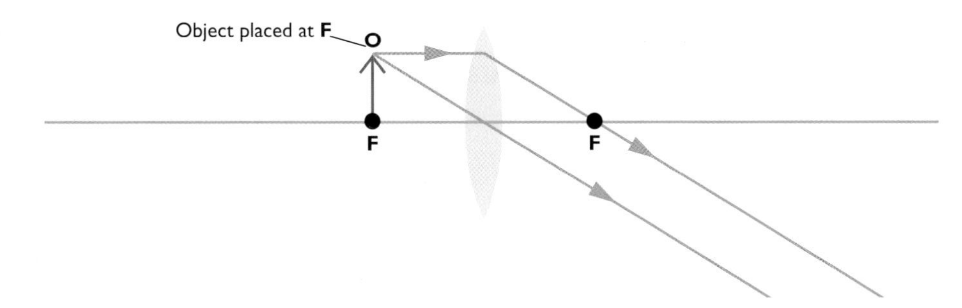

Figure 18.7 *If the object is placed at* **F***, the light is refracted as a parallel beam.*

When an object is placed at F, the light is refracted as a parallel beam. This arrangement might be used in spotlights or searchlights.

Object placed between F and the lens

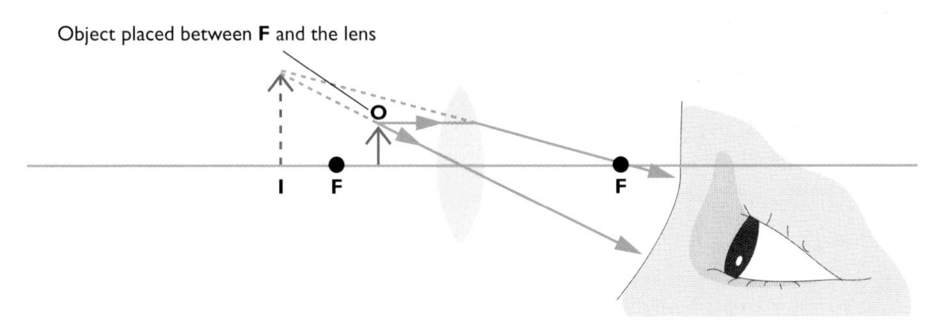

Figure 18.8 *If the object is placed between* **F** *and the lens, the image is upright and larger than the object.*

A virtual image has no rays of light passing through it and so cannot be projected onto a screen. The image is created by our eyes, which perceive light diverging from this point.

If an object is positioned at a distance smaller than the focal length of the lens, the image created is upright, magnified, virtual and on the *same* side of the lens as the object. Used in this way, a convex lens is acting as a **magnifying glass**.

Concave lenses

When rays travelling parallel to the principal axis pass through a concave lens they are made to diverge as if they had come from the principal focus.

An object placed in front of a concave lens will produce an image that is virtual, upright and smaller than the object regardless of its position and the focal length of the lens.

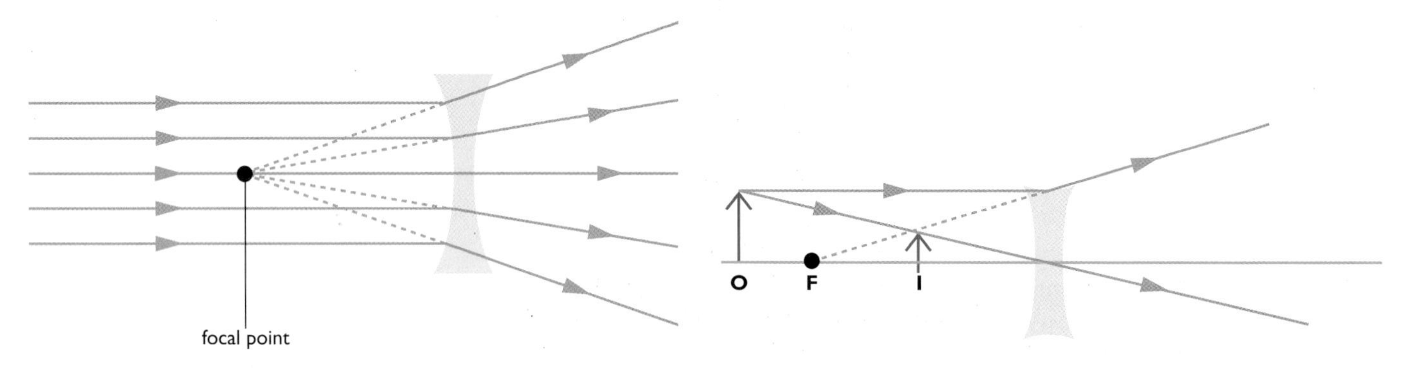

Figure 18.9 *A concave lens makes light rays diverge.*

Figure 18.10 *A concave lens produces an image that is upright and smaller than the object.*

Optical instruments that use lenses

The projector

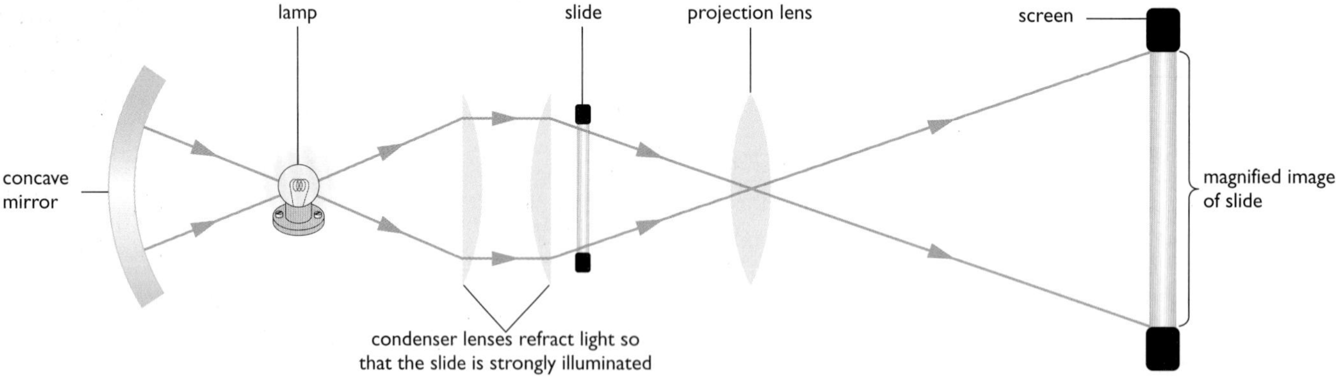

Figure 18.11 *A projector uses a combination of lenses.*

A projector is designed to produce a real, magnified image of a slide on a screen. The slide is strongly lit using a combination of lenses called condenser lenses and a concave mirror placed around the light source. The slide is positioned so that it is just beyond the focal point of the projection lens. This lens is usually held in a screw mounting. Turning the mounting alters the distance between the lens and the slide and focuses the image on the screen.

The lens camera

The camera is designed to produce a real, diminished image of an object on photographic film at the back of the camera.

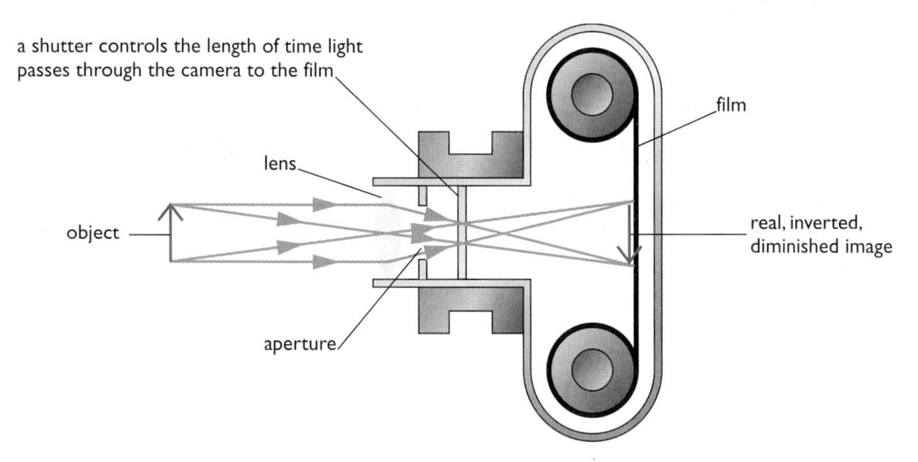

Figure 18.12 *A camera produces an image that is inverted and smaller than the object.*

Light from an object enters the camera through a small hole called the **aperture**. This light is focused by a lens onto the film, creating a sharp image. The images of objects at different distances are brought to focus by altering the distance between the lens and the film.

The eye

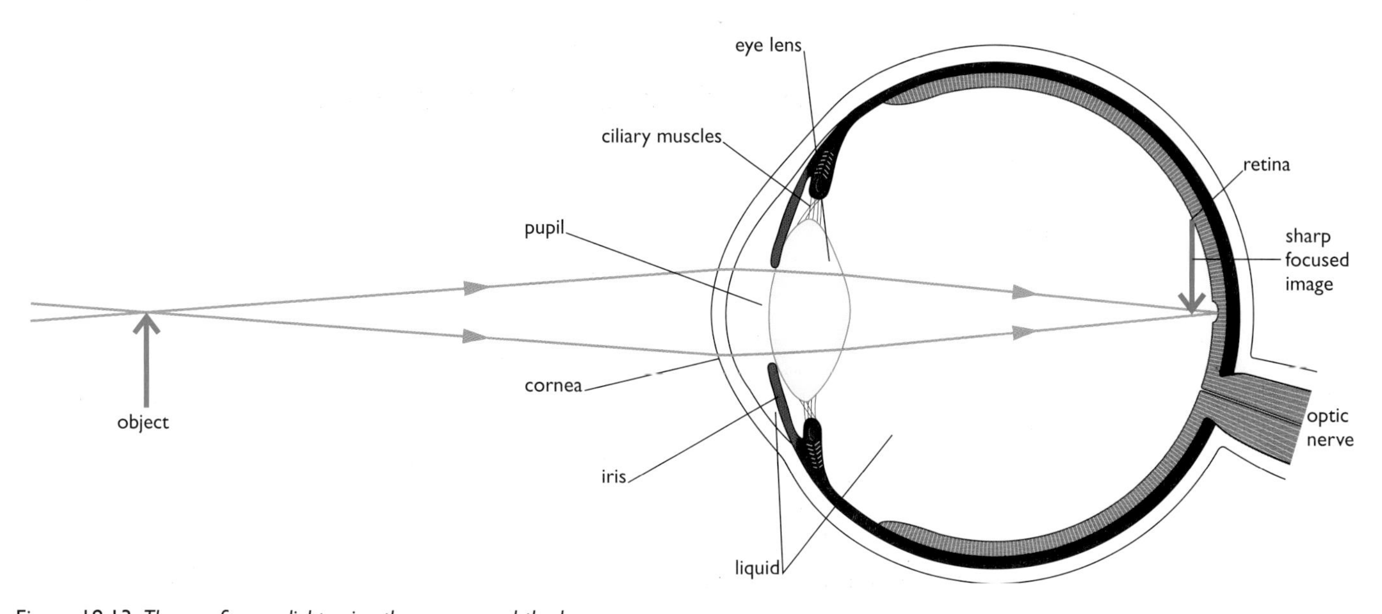

Figure 18.13 *The eye focuses light using the cornea and the lens.*

Light enters the eye through a transparent window called the **cornea**. The cornea has a convex shape and begins to converge the rays. They then travel through a small circular aperture called the **pupil**. Behind the pupil is the **eye lens**. This refracts the light in order to produce a sharp focused image on the **retina** at the back of the eye. The retina is sensitive to light and sends electrical messages along the **optic nerve** to the brain. Although the image created on the retina is inverted our brains are able to interpret this correctly.

An image created on the retina of the eye persists for about one fifteenth of a second. This is why it is possible to create the illusion of motion by presenting a rapid series of still pictures, each showing a slightly different scene, on a film or in a flicker book.

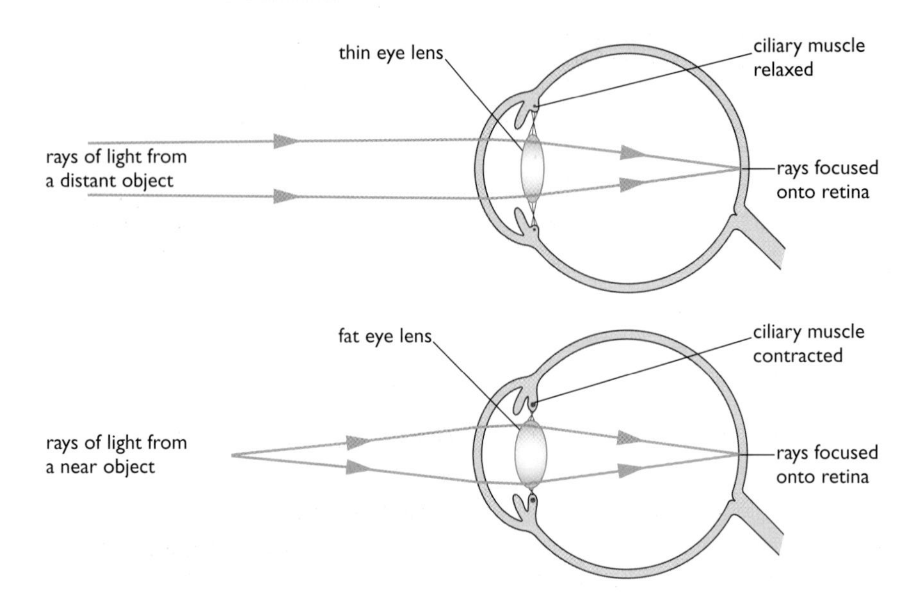

Figure 18.14 *To focus on a near object the lens is thick. To focus on an object further away the lens is thinner.*

The images of objects at different distances are brought to focus on the retina by altering the shape of the eye lens. When an object is close, the **ciliary muscles** surrounding the lens contract, the lens becomes fatter and produces a sharp image of the near object on the retina. When an object is far away, the ciliary muscles relax and the lens becomes thinner again, producing a sharp image on the retina.

Eye defects

Many people have an eye defect that prevents them from producing sharp images of objects both near and far.

If you can see near objects clearly but not distant objects, you are said to be **short-sighted**. This inability to see distant objects is likely to be caused by your eyeballs being too long or the ciliary muscles in each eye being unable to relax sufficiently to make the lens thin enough. The result of either of these problems is that the rays of light from a distant object are brought to focus *in front of* the retina – that is, short of the retina.

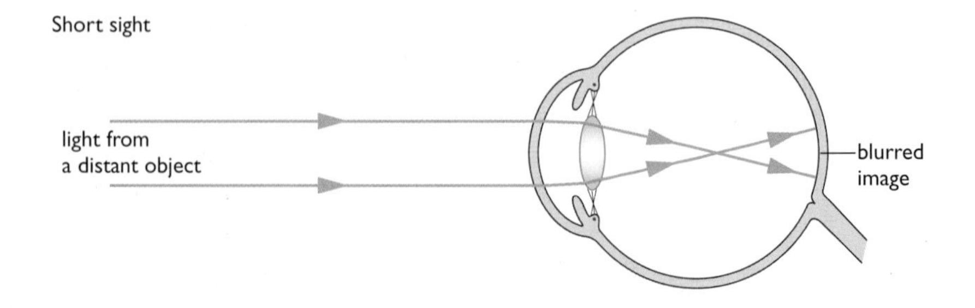

Figure 18.15 *Being short sighted means that the image is focused in front of the retina.*

To correct short-sightedness, a diverging (concave) lens of the appropriate focal length is placed in front of the eye. This helps to diverge the rays before they enter the eye, so that the lens can then focus them and form an image on the retina of the eye.

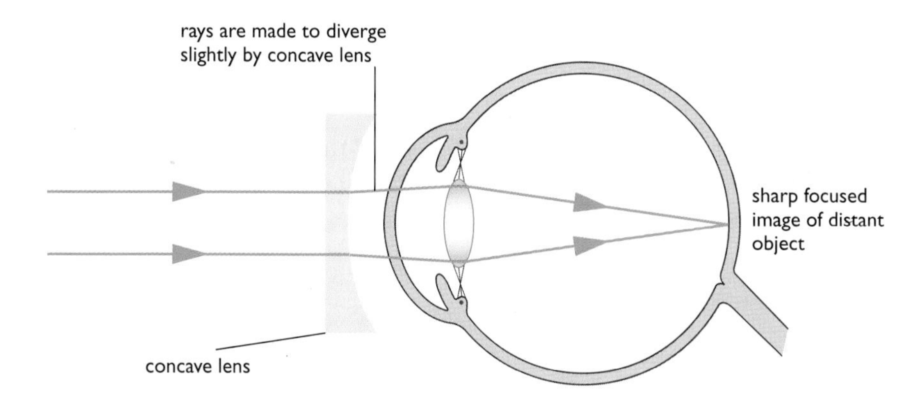

Figure 18.16 *Short sight is corrected with a concave lens.*

If you can see distant objects clearly but cannot see near objects, you are said to be **long-sighted**. This may be due to your eyeballs being too short or the ciliary muscles being unable to produce a fat enough lens. The result of either of these problems is that light from near objects is brought to focus *behind* the retina.

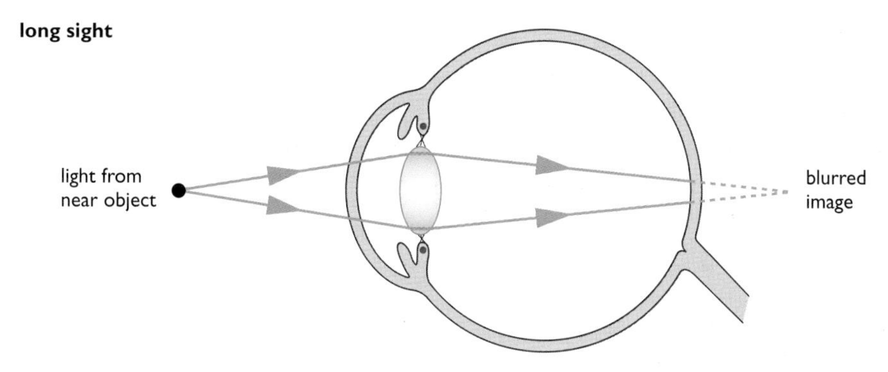

Figure 18.17 *Being long sighted means that the image is focused behind the retina.*

To correct long-sightedness, a converging (convex) lens of appropriate focal length is placed in front of the eye to help bring the rays together on the retina, producing a sharp image.

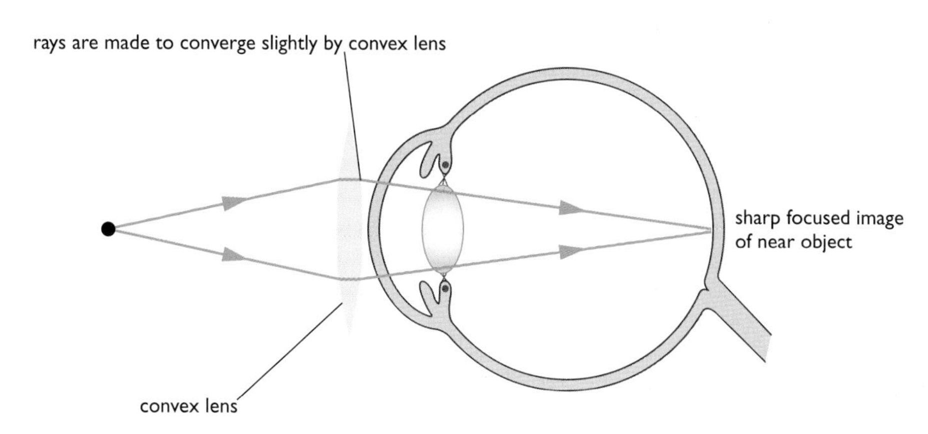

Figure 18.18 *Long sight is corrected with a convex lens.*

End of Chapter Checklist

If you haven't got a copy of your specification, read the introduction on page iv.

You will need to be able to do some or all of the following. Check your Awarding Body's specification (syllabus) to find out exactly what you need to know.

- Distinguish between the action of a convex (converging) lens and a concave (diverging) lens.
- Draw ray diagrams to show how images are produced by convex and concave lenses.
- Explain, using ray diagrams, how the magnifying glass, the projector and the simple camera work.
- Recognise the different ways in which a camera and the eye produce sharp images of objects at different distances.

Questions

More questions on lenses can be found at the end of Section C on page 205.

1 ***a)*** Why do lenses change the directions of rays of light?

b) Describe what happens to a beam of parallel rays as they pass through a convex lens.

c) Describe what happens to a beam of parallel rays as they pass through a concave lens.

d) Explain in your own words the terms *focal point* and *focal length*.

e) Compare the effects of passing a parallel beam of rays through a convex lens with a long focal length, and a convex lens with a short focal length.

2 Draw a ray diagram to show how a convex lens can be used to produce a virtual image. Name one application for this arrangement of object and lens.

3 ***a)*** Draw a ray diagram to show how a projector produces an image of a slide on a screen.

b) What is the purpose of the condenser lenses and the curved mirror?

c) Where is the slide placed in relation to the focal point of the projection lens?

d) How can the image of a slide be re-focused if the screen on which it is seen is moved further away from the projector?

4 How do a simple camera and the eye change from creating an image of a near object to creating an image of a distant object?

5 ***a)*** Draw a diagram of the eye showing the paths of rays of light for someone who is short-sighted when they are looking at a distant object.

b) Draw a second diagram to show how this eye defect can be corrected using a concave lens.

Chapter 19: Sound and Vibrations

Sound waves are longitudinal waves, rather than transverse waves like light, but they can be reflected and diffracted in just the same way. In this chapter you will learn about the nature and behaviour of sound waves, and find out about some properties of vibrations.

Figure 19.1 *The sound produced by the speakers must be loud but also of good quality.*

The photograph in Figure 19.1 shows just part of the sound system used by a pop group playing at a concert. This equipment must produce sounds that are loud enough to be heard by all the audience and the sound quality must be good enough for the music to be appreciated. In this chapter, we are going to look at how sounds are made, how they travel as waves and some of the uses of sound waves.

Sound waves

Sounds are produced by objects that are vibrating. We hear sounds when these vibrations, travelling as sound waves, reach our ears.

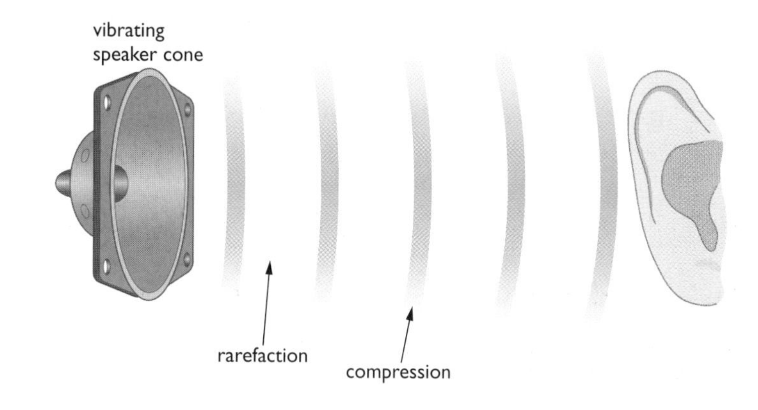

Figure 19.2 *The loudspeaker vibrates and produces sound waves.*

As the speaker cone moves to the right, it pushes air molecules closer together, creating a **compression**. These particles then push against neighbouring particles so that the compression appears to be moving to the right. Behind the compression is a region where the particles are spread out. This region is called a **rarefaction**. After the cone has vibrated several times, it has created a series of compressions and rarefactions travelling away from it. This is a **longitudinal sound wave** (see page 142).

When the waves enter the ear, they strike the eardrum and make it vibrate. These vibrations are changed into electrical signals, which are then detected by the brain.

Sound waves in different materials

Sound waves can travel through:

1 solids – this is why you can hear someone talking in the next room, even when the door is closed

2 liquids – this is why whales can communicate with each other when they are under water

3 gases – the sound waves we create when we speak travel through gases (in the air).

Sound waves cannot travel through a vacuum because there are no particles to carry the vibrations. We can demonstrate this with the experiment shown in Figure 19.3.

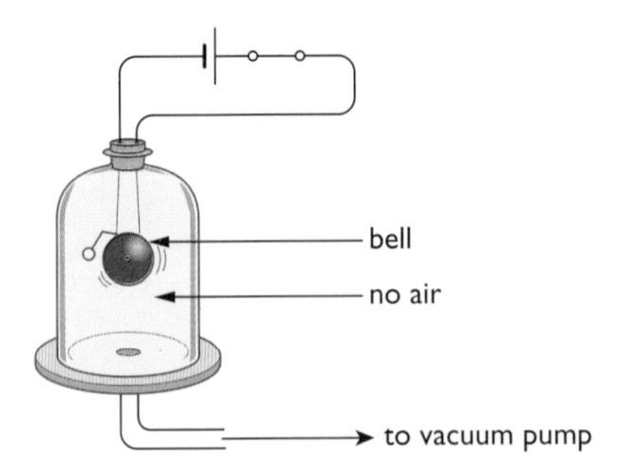

Figure 19.3 *We can only hear the bell when there is air in the jar.*

When there is air in the bell jar, the bell can be seen and heard to ring. However, if the air is removed using a vacuum pump, we can see that the bell is still ringing but it cannot be heard. This simple experiment demonstrates that light waves can travel through a vacuum (because we can still see the bell) but sound waves cannot (because we cannot hear the bell any more).

Sound travels faster in denser materials.

The speed of sound in air is approximately 340 m/s, although this value does vary a little with temperature and pressure. In liquids and solids the particles are much closer together. This means that they are able to transfer sound energy more quickly. The speed of sound in seawater is approximately 1500 m/s. The speed of sound in a solid such as concrete or steel is approximately 5000 m/s.

Reflection and diffraction

Sound waves behave in the same way as any other wave.

When a sound wave strikes a surface it may be **reflected**. Like light waves, sound waves are reflected from a surface so that the angle of incidence is equal to the angle of reflection. A reflected sound wave is called an **echo**.

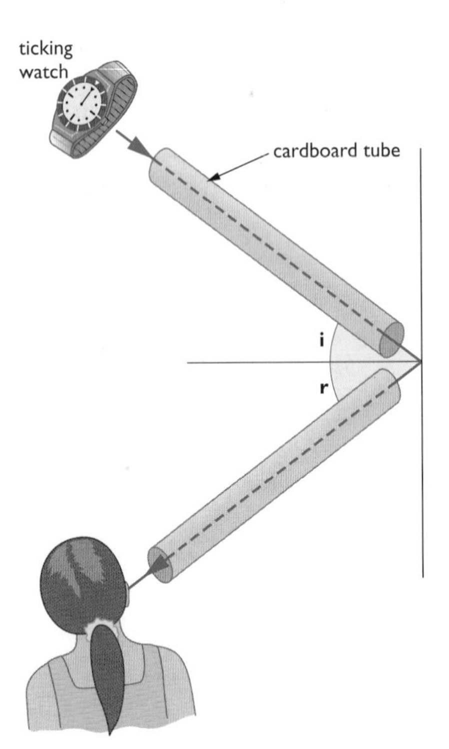

Figure 19.4 *Sound waves are reflected in the same way that light rays are reflected.*

Ships often use echoes to discover the depth of the water beneath them. This is called **echo sounding**. Sound waves are emitted from the ship and travel to the seabed. Equipment on the ship detects some of the sound waves that are reflected by the seabed. The depth of the sea can be calculated from the time between sending the sound wave and detecting the echo. The system of using echoes in this way is called "sonar" (SOund, Navigation And Ranging).

Figure 19.5 *Reflected sound can be used to tell ships about the depth of the sea beneath them.*

We can often hear sounds even when the sound waves cannot travel in a straight line from the source to our ears. Sound can be **diffracted**.

Figure 19.6 *Sound waves can be diffracted.*

The wavelength of some sound waves is approximately the same as the width of a doorway. These waves will therefore spread out as they pass through the door. If the waves did not diffract there would be "sound shadows" where no sound waves would reach.

Pitch and frequency

Small objects, such as the strings of the violin in Figure 19.7, vibrate quickly and produce sound waves with a high **frequency**. These sounds are heard as notes with a high **pitch**.

Larger objects, such as the strings of the cello, vibrate more slowly and produce waves with a lower frequency. These sounds have a lower pitch.

The frequency of a source is the number of complete vibrations it makes each second. We measure frequency in **hertz** (Hz). If a source has a frequency of 50 Hz this means that it vibrates 50 times each second and therefore produces 50 waves each second.

Figure 19.7 *The violin produces notes that are higher pitched than those from the cello.*

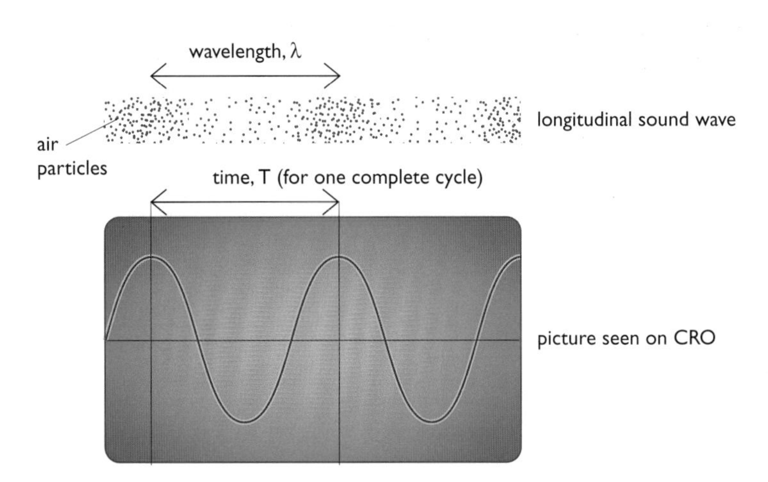

Figure 19.8 *Although we cannot see an actual sound wave we can see a representation of it using a piece of apparatus called a* ***cathode ray oscilloscope*** *(CRO).*

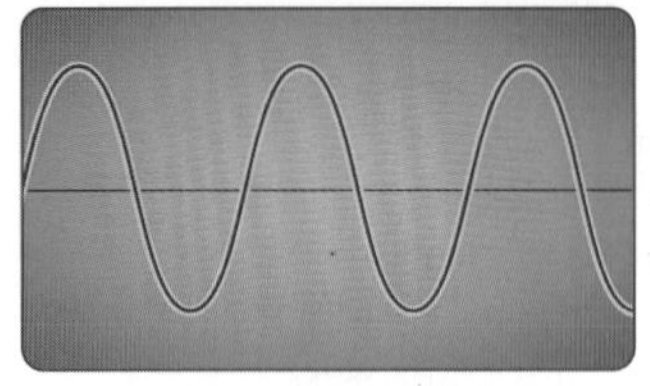

low pitched sound

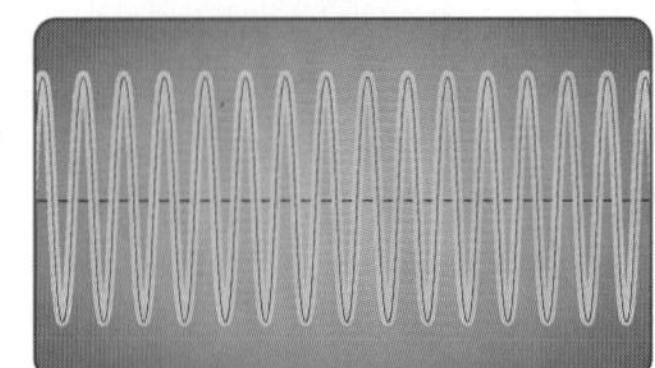

high pitched sound

CRO settings are the same for each wave

Figure 19.9 *High pitched sounds and low pitched sounds seen on a CRO. A low pitched sound has fewer complete waves per second – so fewer complete waves are seen on the CRO.*

The relationship between the frequency (f), wavelength (λ) and speed of a sound wave (v) is described by the equation:

$$v = f\lambda$$

worked example

Example 1

Calculate the wavelength of a sound wave that is produced by a source vibrating with a frequency of 85 Hz. The speed of sound in air is 340 m/s.

Using $v = f\lambda$

$340\,\text{m/s} = 85\,\text{Hz} \times \lambda$

$\lambda = 340\,\text{m/s} / 85\,\text{Hz}$

$\lambda = 4\,\text{m}$

The wavelength is 4 m.

Audible range

The average person can only hear sounds that have a frequency higher than 20 Hz but lower than 20 000 Hz. This spread of frequencies is called the **audible range** or **hearing range**. The size of the audible range varies slightly from person to person and usually becomes narrower as we get older.

Some objects vibrate at frequencies greater than 20 000 Hz. The sounds they produce cannot be heard by human beings and are called **ultrasounds**. Some objects vibrate so slowly that the sounds they produce cannot be heard by human beings. These are called **infrasounds**.

Figure 19.10 *Some animals can make and hear sounds that lie outside the human audible range. Dolphins can communicate using ultrasounds. Elephants can communicate using sounds that have frequencies too low for us to hear.*

Uses of ultrasound

Ultrasounds are often used in hospitals to monitor the condition of a fetus in the womb. The waves are generated in special electrical circuits and then emitted by a probe held against the pregnant woman's abdomen. A sensor within the probe detects waves that have been reflected from different surfaces within the womb. A computer processes these reflections creating an image of the fetus on a screen. X-rays were once used to monitor unborn babies but are now thought to be too dangerous.

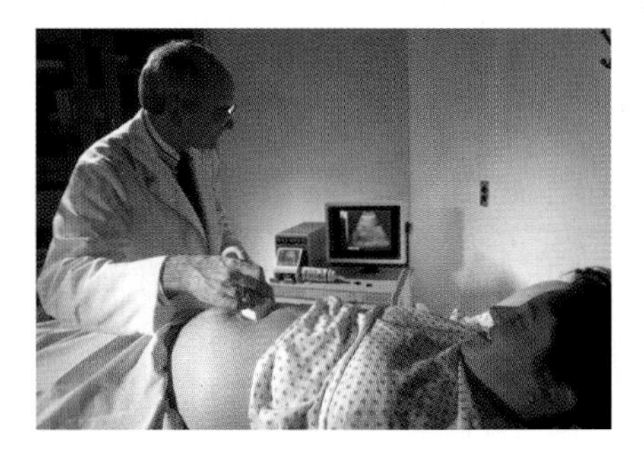

Figure 19.11 *Ultrasound can be used to look at a fetus in the womb.*

Figure 19.12 *Bats use ultrasound to help them navigate and locate prey.*

Ultrasound is also used in industry to clean very delicate pieces of equipment without having to take them apart. It can also be used to check metal castings for internal cracks and flaws. The flaws and cracks are detected by ultrasound that is reflected from them.

Modern sonar equipment used on ships (see page 174) does not emit sound waves in our audible range but uses ultrasounds instead. This is because the ultrasonic waves can be emitted as a narrow beam, which does not spread out as much as audible sounds. Ultrasounds are also absorbed less by the water and so produce stronger echoes.

Bats emit ultrasounds and then listen for echoes. Their brains process these returning signals so that they can navigate around objects as well as hunt for prey.

Loudness

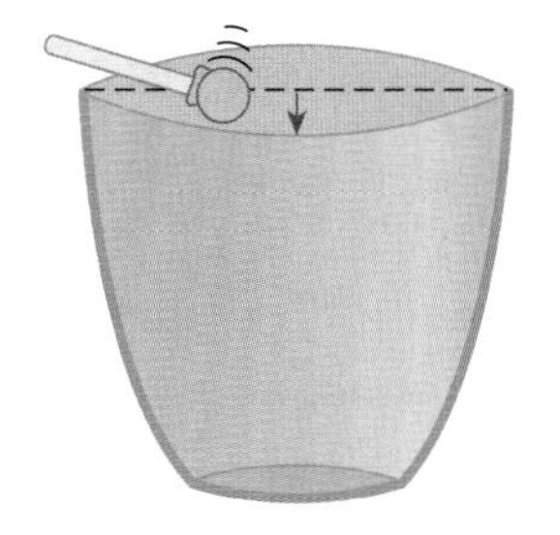

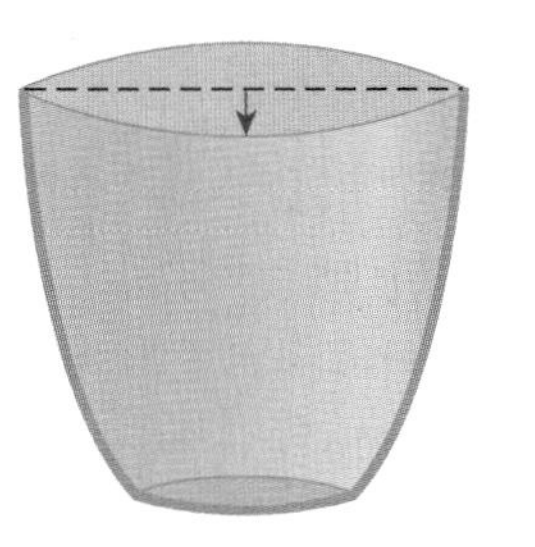

Figure 19.13 *If you strike a drum hard, you get a louder sound than if you beat it gently.*

If the drum in the Figure 19.13 is struck hard, lots of energy is transferred to it from the hammer. The drum skin vibrates up and down with a large amplitude and creates regions of very compressed air molecules. As these compressions move away from the source they carry lots of energy, which we hear as a loud sound. If the drum is struck more gently the compressions created are less dense and less energy is transferred by the sound waves. We hear these as quieter sounds.

It is important to remember that sound waves are longitudinal. CROs display all waves as if they are transverse, because they show the amplitude of the wave against **time** on the horizontal scale.

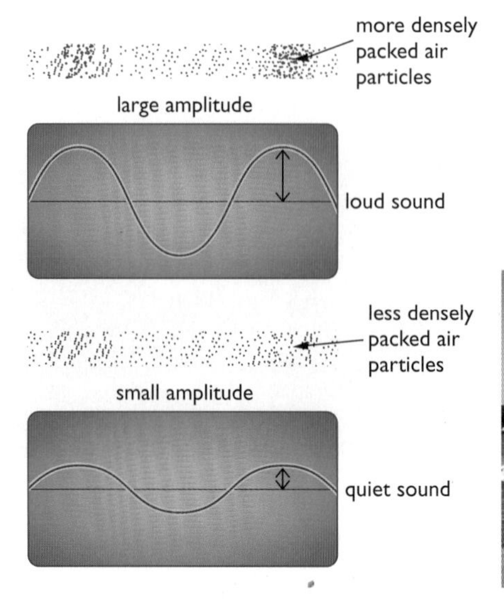

Figure 19.14 *Loud sounds and quiet sounds on a CRO.*

Figure 19.15 *The photo above shows a scientist measuring the sound level as a speedboat passes him.*

The amplitude of a wave gives us some idea of the energy it is transferring. For a sound wave, its amplitude is a measure of its loudness.

Figure 19.16 *Traffic produces noise at about 80 dB.*

The decibel scale

Noise is unwanted sound. It may be sound from a nearby road or airport. It may be noise from the machinery people use at work. It might even be the music someone else wants to listen to but you do not. Noise can make it difficult for you to concentrate, it may cause you stress and it can cause physical harm to your body. It is therefore important that we measure and monitor noise levels.

We measure noise levels in **decibels**. The table below gives some examples of the loudness of everyday events on the decibel scale.

Source	Loudness on decibel scale (dB)
sound that is just audible	0 dB
whisper	25 dB
normal conversation	60 dB
traffic next to a busy road	80 dB
noisy factory	90 dB
disco loudspeaker (near by)	110 dB
jet aircraft taking off (near by)	120 dB
noise causing pain	130 dB

The decibel scale is not a linear scale. It is a logarithmic scale. This means a rise of 10 dB on the decibel scale corresponds to a ten-fold increase in the intensity of the sound, and a 20 dB rise corresponds to a one-hundred-fold increase in the intensity of sound. The values in the table tell us that the intensity of sound from a disco is 100 000 times that of a normal conversation!

Persistent exposure to loud sounds can cause permanent damage to our hearing. Workers who use noisy machinery should wear ear defenders. When we listen to our music using headphones, we must be aware of the loudness of the sounds we are listening to.

Figure 19.17 *It is easy to damage your hearing if the music on your headphones is too loud.*

Vibrations

Figure 19.18 *A swing will vibrate back and forth. The frequency of the vibration depends on the length of the chains holding the swing.*

All objects have their own **natural frequency** of vibration. A swing or pendulum given a single push will vibrate, or oscillate, back and forth at a frequency that depends upon its length. The longer the swing or pendulum, the lower its natural frequency.

When we pluck one string of a guitar, it will vibrate at its own natural frequency. We can alter the frequency of vibration by:

- altering the length of the string that is vibrating – making the string shorter will increase its frequency of vibration
- choosing a string of different thickness (that is, different mass) – using a string of greater mass will decrease the frequency of vibration
- altering the tension of the string – the lower the tension, the lower the natural frequency of the string.

Figure 19.19 *We can change the frequency of the note produced by a guitar by altering the length of the vibrating part of the string.*

If a swing or a guitar string is given a single push or displacement each will then vibrate at its own natural frequency. These are examples of **free vibrations**.

This is not the case for the vibrating cone of a loudspeaker. When we turn on a radio or television, its speaker receives lots of pulses or displacements created by the currents passing through the speaker coils. The frequency of these pulses controls the frequency of vibration of the loudspeaker cone. To produce high-pitched notes the cone is made to vibrate quickly. To produce lower-pitched notes it is made to vibrate more slowly. These are examples of **forced vibrations**.

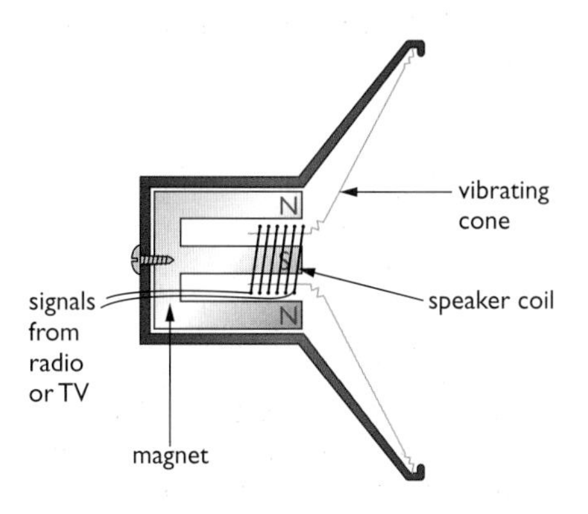

Figure 19.20 *The loudspeaker is made to vibrate at the frequency of the current passing through its coils.*

We make use of resonance when pushing someone on a swing. If we give the swing small pushes at a frequency equal to its natural frequency the amplitude of its vibration gets larger and larger.

Resonance

If the frequency of a driving vibration (the pulses that cause an object to vibrate) is the same as the natural frequency of the object, the amplitude of vibration of the object will gradually increase. This effect is called **resonance**.

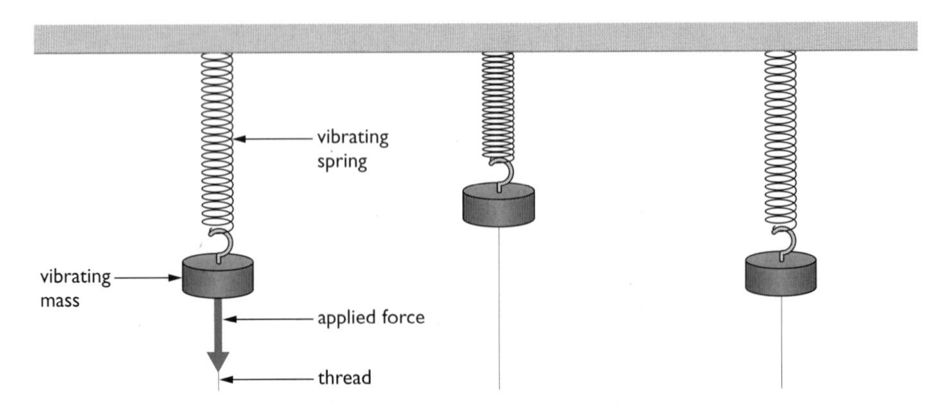

Figure 19.21 *The spring vibrates when the mass is pulled and released.*

If the thread on the end of the mass in Figure 19.21 is pulled and then released, the mass and spring will vibrate vertically. If the thread is pulled each time the mass is moving downwards the amplitude of vibration of the mass will gradually increase. This is an example of resonance. If the thread is pulled at a frequency that is not the natural frequency of the system, the mass may be moving upwards and the amplitude of vibration is likely to decrease.

We can also demonstrate resonance using **Barton's** pendulum, as shown in Figure 19.22.

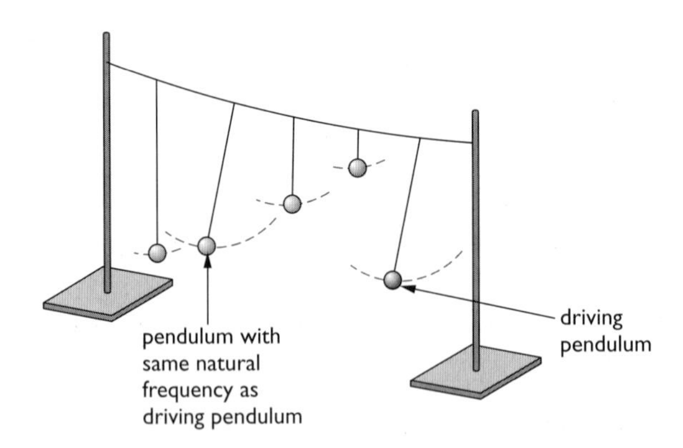

Figure 19.22 *Barton's pendulum demonstrates resonance – the pendulum with the same natural frequency as the driving pendulum vibrates with the largest amplitude.*

When the driving pendulum is displaced and then released it vibrates or oscillates at its own natural frequency. As it does so, small pulses of energy are sent to the other pendulums through the supporting thread. For any pendulum that has the same natural frequency (same length) as the driving pendulum these pulses of energy add together and the amplitude of oscillation of the pendulum increases. For pendulums that do not have this natural frequency (that is, they are of a different length) the pulses of energy do not add together constructively and there is no increase in the amplitude of their vibration. Resonance does not occur.

Figure 19.23 *Tacoma Bridge, USA.*

When cross-winds of the right speed blew across the Tacoma Narrows Bridge in the USA they caused resonant vibrations. The amplitude of these vibrations became so large that the bridge collapsed. Nowadays models of new bridges are tested in wind tunnels to ensure that the wind will not make them vibrate at their natural frequency.

Vibrations in musical instruments

When a string is made to vibrate, interference occurs between travelling waves that are reflected from the fixed ends of the string. As a result a **standing** or **stationary wave pattern** is created.

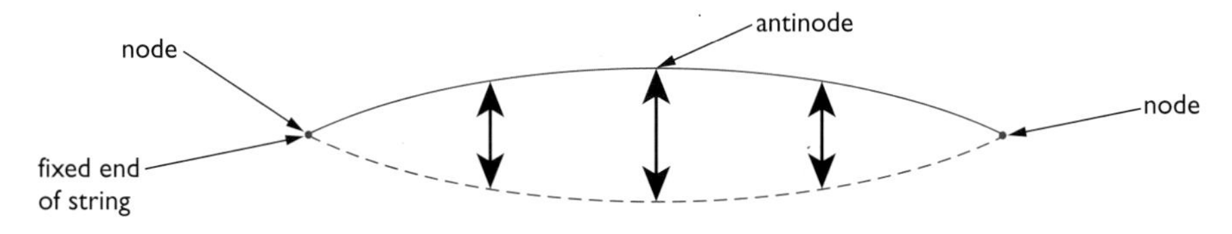

Figure 19.24 *The fundamental mode of vibration for a string.*

At certain places along the string there is no vibration. These places are called **nodes**. In some places the amplitude of vibration is a maximum. These are called **antinodes**.

It is possible for a string to vibrate in several different ways or **modes** at the same time. The diagrams in Figure 19.25 show some of these modes of vibration. The quality of the note produced by a stringed instrument depends upon which of these modes of vibration are present, and upon their relative strengths.

a)

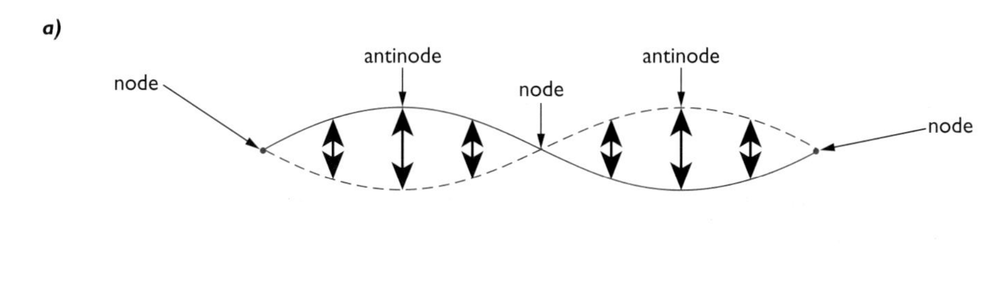

b)

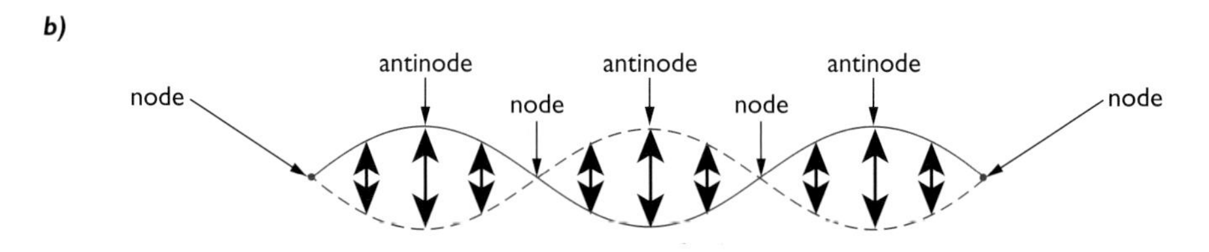

Figure 19.25 a) *This mode of vibration is called the first overtone,* b) *this mode of vibration is called the second overtone.*

Resonance plays an important role in many musical instruments. Without resonance the amplitude of the vibrations would be so small we would hardly be able to hear the sounds. We can demonstrate this using the apparatus shown in Figure 19.26.

We can change the frequency of vibration of the loudspeaker using a signal generator. At certain frequencies we will hear sudden increases in the loudness of the note. These increases in loudness are caused by resonance between the frequency of the vibration of the loudspeaker and the natural frequency of the air column inside the tube.

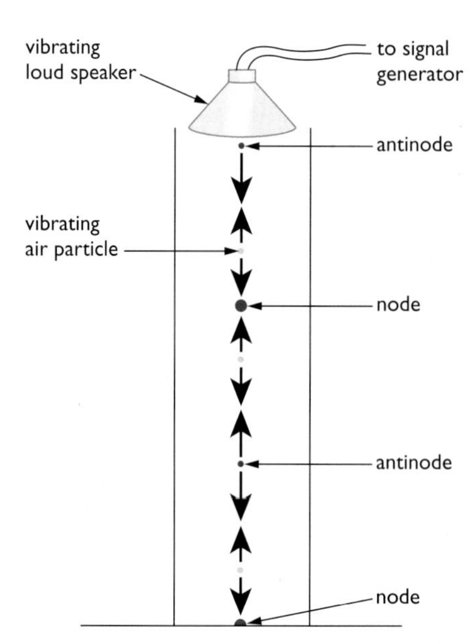

Figure 19.26 *When resonance occurs, a standing wave is created in the air column.*

These waves are called standing or stationary waves because the positions of the nodes and antinodes do not move – that is, their shape appears to be stationary.

End of Chapter Checklist

If you haven't got a copy of your specification, read the introduction on page iv.

You will need to be able to do some or all of the following. Check your Awarding Body's specification (syllabus) to find out exactly what you need to know.

- Recall that sound waves are longitudinal waves that can travel through solids, liquids and gases. Like other waves, they can be reflected and diffracted.
- Understand that sounds are produced when objects vibrate. The greater the amplitude of vibration, the louder the sound. The faster the vibration, the higher the frequency and pitch of the sound produced.
- Describe the amplitudes and frequencies of sounds as shown on a cathode ray oscilloscope.
- Understand that humans have an audible range, or hearing range. Sounds with frequencies higher than this range are called ultrasounds.
- Describe several uses of ultrasound including prenatal scanning, sonar and quality control.
- Recall that the amplitude of a sound wave is a measure of its loudness.
- Understand the conditions necessary for resonance.
- Describe the effects of resonance in a variety of systems including musical instruments.

Questions

More questions on sound and vibrations can be found at the end of Section C on page 205.

1 ***a)*** Name a musical instrument that is used to produce high-pitched notes.

b) Explain why the musical instrument you have named in part ***a)*** produces high-pitched notes.

c) Explain how you would produce loud sounds from this musical instrument.

d) Draw the trace you might expect to see on a CRO when this instrument is producing a loud, high-pitched note.

2 ***a)*** What is an echo?

b) Explain how echoes are used by ships to find the depth of the ocean beneath them.

c) A ship hears the echo from a sound wave 4 s after it has been emitted. If the speed of sound in water is 1500 m/s, calculate the depth of the water beneath the ship.

3 ***a)*** What is meant by the phrase "a person's audible range is 20 Hz to 20 000 Hz"?

b) What are ultrasonic sounds?

c) Explain how ultrasonic sounds are used to monitor the well-being of a fetus in the womb.

d) This monitoring used to be done using X-rays. Explain why ultrasonic waves are now used instead.

e) Calculate the wavelength of ultrasonic waves whose frequency is 68 000 Hz. Assume that the waves are travelling through air at a speed of 340 m/s.

4 ***a)*** Two astronauts stand 2 m apart on the surface of the Moon. Their radio link is broken. Both men shout as loud as they can. But neither of them can hear what the other is saying. Explain why the astronauts are unable to hear each other. (**Hint**: What is the atmosphere like on the Moon?)

b) The two astronauts move closer together so that their helmets touch. Explain why they can now hear each other.

5 ***a)*** Sound waves are emitted from a source that is vibrating with a large amplitude, and from a source that is vibrating with a small amplitude. Explain, using diagrams, the difference between the two sets of sound waves.

b) Draw two diagrams to show how these waves would appear on a CRO.

c) What is the decibel scale?

d) Suggest what values on the decibel scale the following might have:

i) a scream

ii) a whisper

iii) a small bird singing.

e) Suggest one way in which you could protect your hearing from permanent damage caused by loud sounds.

6 ***a)*** Explain the difference between free vibrations, forced vibrations and resonance. Give one example of each.

b) Describe three ways in which the pitch of a sound produced by a string on a violin could be increased.

Chapter 20: Using Waves

The electromagnetic spectrum is a family of waves, varying in wavelength and frequency. Although it is continuous, it is helpful to consider groups of waves within the entire spectrum that have a smaller range of wavelengths and frequencies. Each group has distinct properties and therefore can be used in different ways. In this chapter, you will learn about many ways in which we use electromagnetic waves, from cooking to communication.

Figure 20.1 *Lasers are a source of concentrated light waves. They can be used to carry large amounts of energy and coded information over long distances.*

In this chapter, we shall be looking at the ways in which we make use of waves.

The electromagnetic spectrum

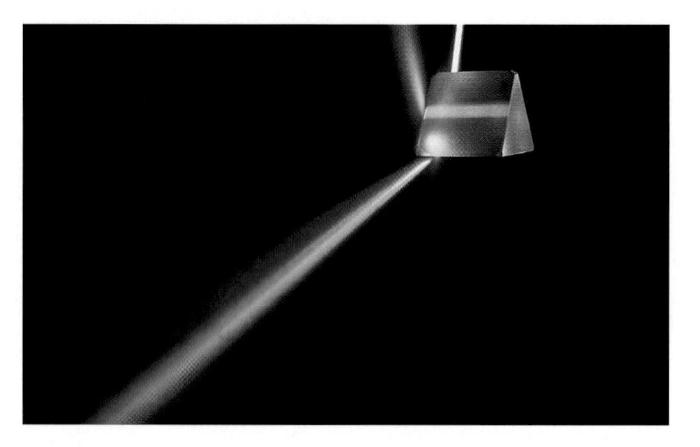

Figure 20.2 *The visible part of the electromagnetic spectrum.*

The **electromagnetic spectrum** is a continuous spectrum of waves, which includes the visible spectrum. At one end of the spectrum the waves are of very long wavelength (low frequency), while at the other end the waves have a very short wavelength (high frequency). All the waves have the following properties:

1 they all transfer energy

2 they are all transverse waves

3 they all travel at the speed of light in a vacuum (300 000 000 m/s)

4 they can all be reflected, refracted and diffracted.

Although the spectrum is continuous – that is, the wavelengths vary gradually and smoothly from one end to the other – it is easier to study if we divide it into groups. Different groups of waves have different properties because they have different frequencies and wavelengths. The table opposite shows the different groups of waves in order, and gives some of their uses.

	Typical frequency (Hz)	Typical wavelength (m)	Sources	Detectors	Uses
Radio waves	10^5–10^{10}	10^3–10^{-1}	radio transmitters, TV transmitters	radio and TV aerials	long, medium and short wave radio TV (UHF)
Microwaves	10^{10}–10^{11}	10^{-1}–10^{-3}	microwave transmitters and ovens	microwave receivers	mobile phone and satellite communication, cooking
Infra-red	10^{11}–10^{14}	10^{-3}–10^{-6}	hot objects	skin, blackened thermometer, special photographic film	infra-red cookers and heaters, TV and stereo remote controls, night vision
Visible light	10^{14}–10^{15}	10^{-6}–10^{-7}	luminous objects	the eye, photographic film, light-dependent resistors	seeing and communication (optical fibres)
Ultraviolet (UV)	10^{15}–10^{16}	10^{-7}–10^{-8}	UV lamps and the Sun	skin, photographic film and some fluorescent chemicals	fluorescent tubes and UV tanning lamps
X-rays	10^{16}–10^{18}	10^{-8}–10^{-10}	X-ray tubes	photographic film	X-radiography
Gamma rays	10^{18}–10^{21}	10^{-10}–10^{-14}	radioactive materials	Geiger–Müller tube	sterilising equipment and food, radiotherapy

Radio waves

Radio waves have the longest wavelengths in the electromagnetic spectrum. They are used mainly for communication.

You do not need to remember the values of frequency and wavelength given in the table but you do need to know the order of the groups and which has the highest frequency or longest wavelengths. Most importantly, you need to realise that it is these differences in wavelength and frequency that give the groups their different properties – for example, gamma rays have the shortest wavelengths, the highest frequencies, and carry the most energy.

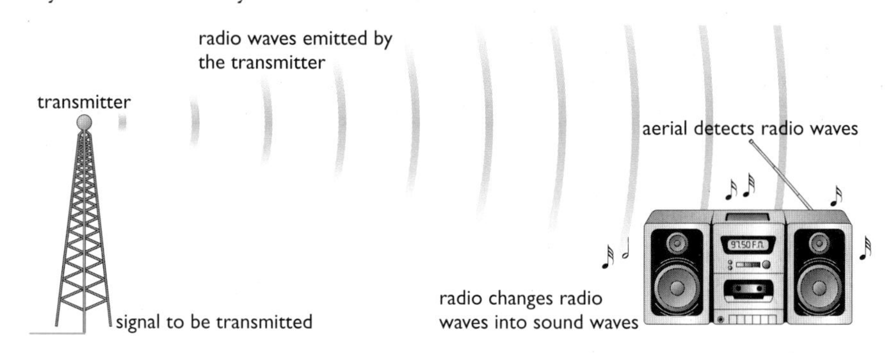

Figure 20.3 *Radio waves are emitted by a transmitter and detected by an aerial.*

Radio waves are given out (emitted) by a transmitter. As they cross an aerial, they are detected and the information they carry can be received (see page 191). Television and FM radio use radio waves with the shorter wavelengths (see page 190) to carry their signals.

ideas
evidence

- **1844** First public demonstration of Samuel Morse's electric telegraph
- **1861** Coast to coast telegraph communications established in USA
- **1866** First telegraph communication between USA and Europe
- **1876** Alexander Graham Bell speaks the first full sentence transmitted by telephone
- **1895** Guglielmo Marconi sends his first words by radio transmission
- **1901** Marconi transmits his first message across the Atlantic
- **1926** John Logie Baird transmits first television pictures
- **1935** Commercial television available in the UK (broadcasting to the 100 televisions in the coutry)
- **1955** Colour television commercially available in USA
- **1962** The satellite Telestar 1 is put into orbit allowing transmission between continents but only when satellite is on the proper position
- **1963** Satellites put into geostationary orbit allowing continual intercontinental communications.

The BBC (www.bbc.co.uk) website contains lots of information on the history of communication

Figure 20.4 *Timeline of telecommunications.*

Microwaves

Microwaves are used for cooking foods, communications and radar.

Figure 20.5 *Food cooks quickly in a microwave oven because water molecules in the food absorb the microwaves.*

Food placed in a microwave oven cooks more quickly than in a normal oven. This is because water molecules in the food absorb the microwaves and become very hot. The food therefore cooks throughout rather than just from the outside.

Microwaves are used in communications. The waves pass easily through the Earth's atmosphere and so are used to carry signals to orbiting satellites. From here, the signals are passed on to their destination or to other orbiting satellites. Messages sent to and from mobile phones are also carried by microwaves. The fact that we are able to use mobile phones almost anywhere in the home and at work confirms that microwaves can pass through glass, brick and so on.

Infra-red

All objects, including your body, emit infra-red radiation. The hotter an object is, the more energy it will emit as infra-red. Special cameras designed to detect infra-red waves can be used to create images even in the absence of visible light. These cameras have many uses, including searching for people trapped in collapsed buildings, tracking criminals and checking for heat loss from buildings.

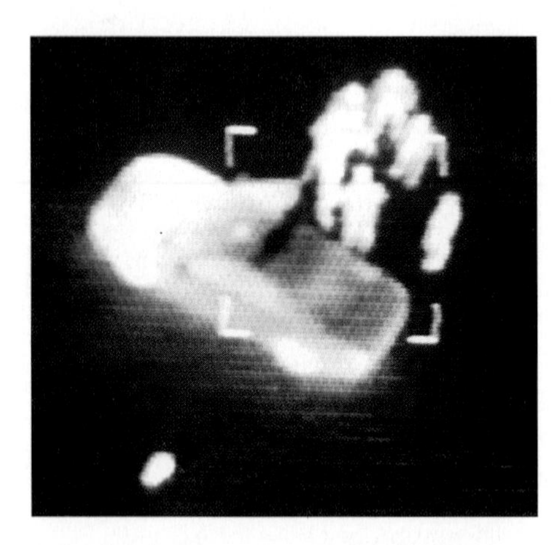

Figure 20.6 *This picture was taken using the infra-red waves being emitted by the people and the car.*

Infra-red radiation is also used in remote controls for televisions, videos and stereo systems. It is very convenient for this purpose because the waves are not harmful, they have a low penetrating power and will therefore operate only over small distances. So they are unlikely to interfere with other signals or waves.

Visible light

This is the part of the electromagnetic spectrum that is visible to the human eye. We use it to see. Visible light from lasers is used to read compact discs and barcodes. It can also be sent along optical fibres, so it can be used for communication or for looking into inaccessible places such as inside the body of a patient.

Ultraviolet

Part of the light emitted by the Sun is ultraviolet (UV) light. UV radiation is harmful to human eyes and can cause damage to the skin.

UV causes the skin to tan but overexposure will lead to sunburn and blistering. Ultraviolet radiation can also cause skin cancer. Protective goggles or glasses and skin creams can block the UV rays and will reduce the harmful effects of this radiation.

Figure 20.7 *UV light can cause sunburn so we need to protect our skin.*

Ozone in the Earth's atmosphere absorbs large quantities of the Sun's UV radiation. There is real concern at present that the amount of ozone in the atmosphere is decreasing due to pollution. This may lead to increased numbers of skin cancers in the future.

Some chemicals glow, or fluoresce, when exposed to UV light. This property of UV light is used in security markers. The special ink is invisible in normal light but becomes visible in UV light.

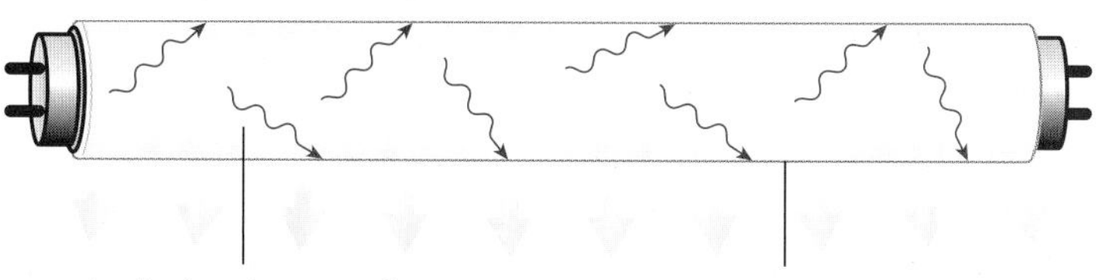

Figure 20.8 *Fluorescent tubes glow when UV light hits the fluorescent coating in the tube.*

Fluorescent tubes glow because the UV light they produce strikes a special coating on the inside of the tube, which then emits visible light.

ideas evidence

X-rays

X-rays pass easily through soft body tissue but cannot pass through bones. As a result, radiographs or X-ray pictures can be taken to check a patient's bones.

Over-exposure to X-rays can cause cancer. Workers such as radiographers who are at risk of overexposure therefore stand behind lead screens or wear protective coating.

X-rays are also used in industry to check the internal structures of objects – for example, to look for cracks and faults in buildings or machinery, and at airports as part of the security checking procedure.

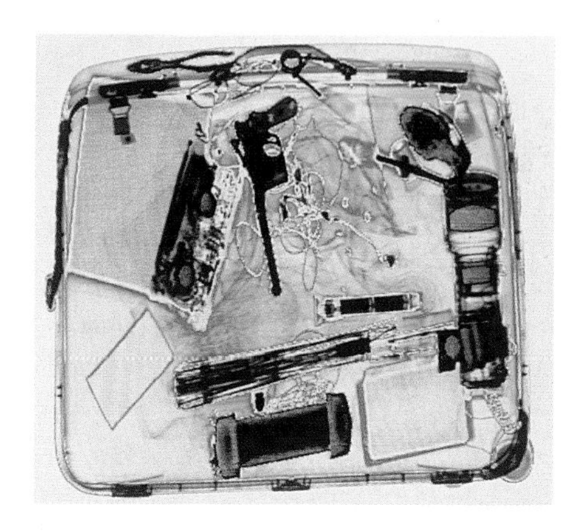

Figure 20.9 *X-rays were used to see what was in this suitcase.*

Gamma rays

Gamma rays, like X-rays, are highly penetrating rays and can cause damage to living cells. They are used to sterilise medical instruments, to kill micro-organisms so that food will keep for longer, and to treat cancer using radiotherapy.

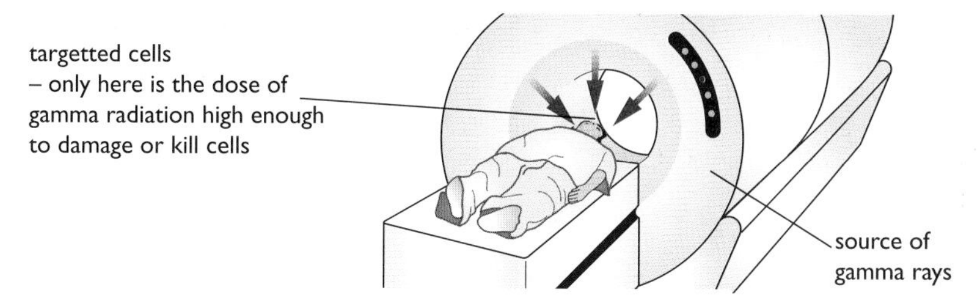

Figure 20.10 *The gamma rays are aimed carefully so that they cross at the exact location of the cancerous cells.*

Communicating using waves

Morse code can be transmitted using light, sound, electricity or radio waves. It uses combinations of long and short signals to represent letters and numbers. The sequence . . . _ _ _ . . . represents the letters "SOS", which is the international distress signal.

When we talk to the person next to us, we use sound waves. If we want to communicate with someone who is further away we might use light waves – for example, semaphore (signalling with flags) or beacons. With modern radio and other telecommunication systems, we can use waves to communicate with each other over large distances and very quickly.

Figure 20.11 *Semaphore being used to communicate over distance.*

Figure 20.12 *These machines all allow us to communicate through a telephone line.*

Telephones, fax machines and Internet-linked computers can all be used to transmit information. All three use the telephone system but before their information can travel down the telephone lines, it must be converted into a stream of electrical pulses or light pulses. These pulses may carry the information as **digital signals** or **analogue signals**.

Digital signals

To send a message using a digital signal, the information is converted into a sequence of numbers called a **binary code**. The code uses just two digits (0 and 1) rather than the ten digits we normally use. These numbers are then converted into a series of electrical pulses and sent down the telephone lines.

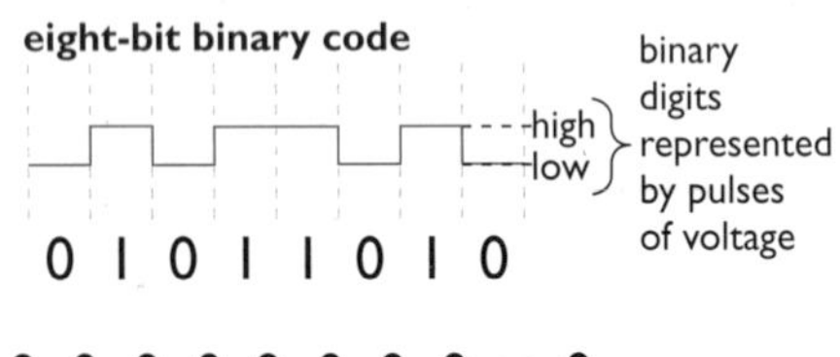

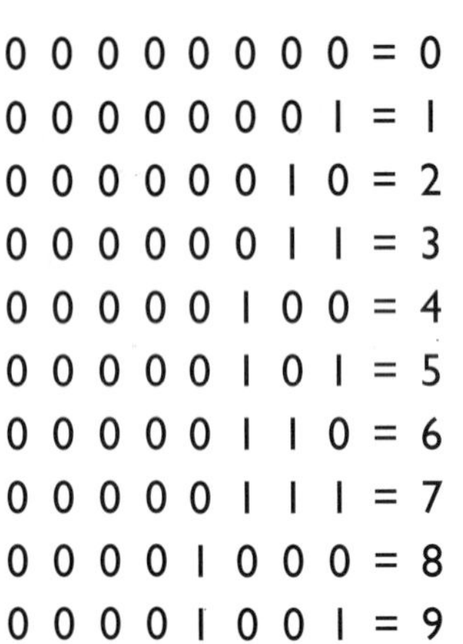

0 0 0 0 0 0 0 0 = 0
0 0 0 0 0 0 0 1 = 1
0 0 0 0 0 0 1 0 = 2
0 0 0 0 0 0 1 1 = 3
0 0 0 0 0 1 0 0 = 4
0 0 0 0 0 1 0 1 = 5
0 0 0 0 0 1 1 0 = 6
0 0 0 0 0 1 1 1 = 7
0 0 0 0 1 0 0 0 = 8
0 0 0 0 1 0 0 1 = 9

Figure 20.13 *An eight-bit pattern of 1s and 0s can be used to represent any of the letters of the alphabet, numbers 0 to 9 and other important symbols such as punctuation marks.*

Analogue signals

In the analogue method, the information is converted into electrical voltages or currents that vary continuously.

Figure 20.14 shows that a microphone converts sound waves into continuous electrical signals. These signals are then amplified (made stronger) and fed into a loudspeaker.

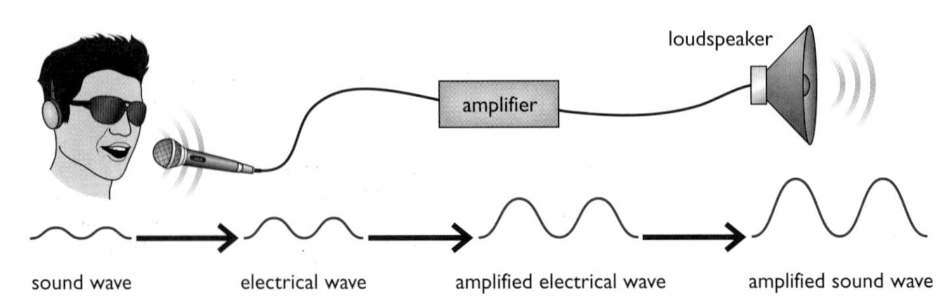

Figure 20.14 *An analogue signal follows the actual pattern of the original information.*

Advantages of using digital signals

There are several advantages in transmitting information in a digital form rather than as analogue signals.

All signals become weaker during transmission and need to be amplified or regenerated. Regeneration of digital signals creates a clean, accurate copy of the original signal. But when analogue signals are amplified, any accompanying noise is also amplified. Eventually the level of noise may "drown out" the original signal or introduce errors in the information being carried.

Digital systems are generally easier to design and build than analogue systems. They also deal with data that is easy to process.

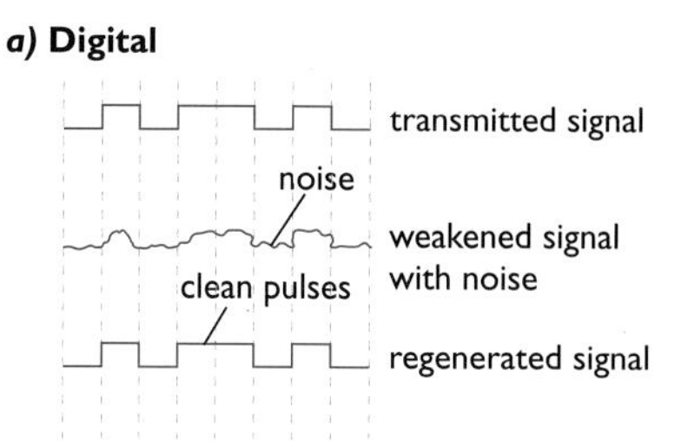

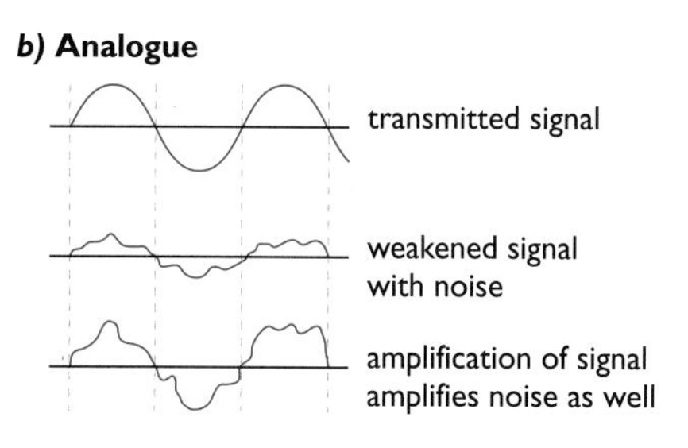

Figure 20.15 *Digital signals are easy to regenerate, analogue signals are easily distorted.*

We normally think of noise as being unwanted sounds but in this context noise is stray unwanted voltages or currents that distort the waveform of the signal.

Converting analogue signals to digital signals

Analogue signals can be converted into digital signals using a **pulse code modulator (PCM)** before they are transmitted.

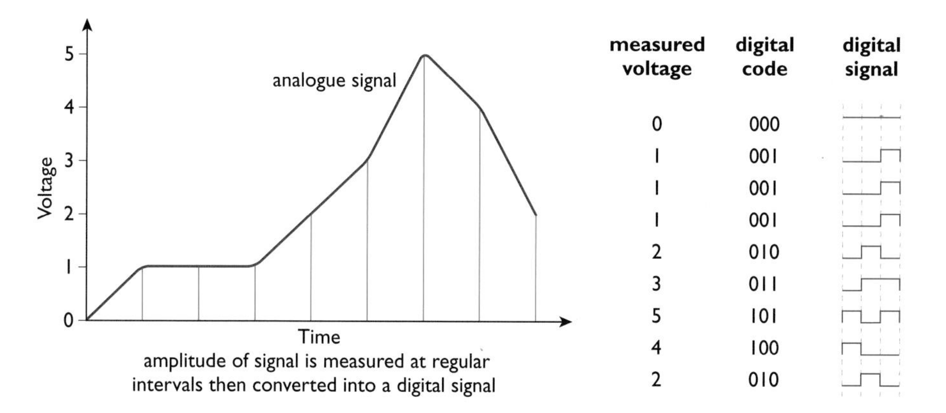

measured voltage	digital code	digital signal
0	000	
1	001	
1	001	
1	001	
2	010	
3	011	
5	101	
4	100	
2	010	

Figure 20.16 *How analogue signals are converted into digital signals.*

The PCM measures, or **samples**, the amplitude of the signal. The amplitude is then converted into binary code, which can be transmitted digitally. Typically, telecommunications companies sample signals approximately 8000 times each second.

Sending the signals down the line

Digital signals may pass through a telecommunications network as:

1 pulses of electricity travelling through copper wires, or

2 pulses of light travelling through optical fibres.

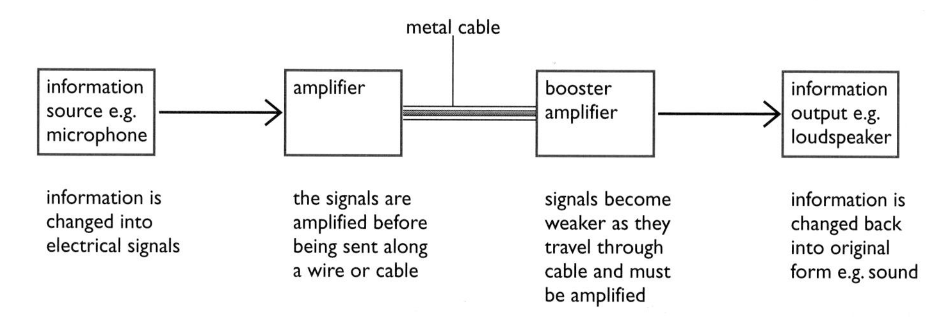

Figure 20.17 *An electrical telecommunications system.*

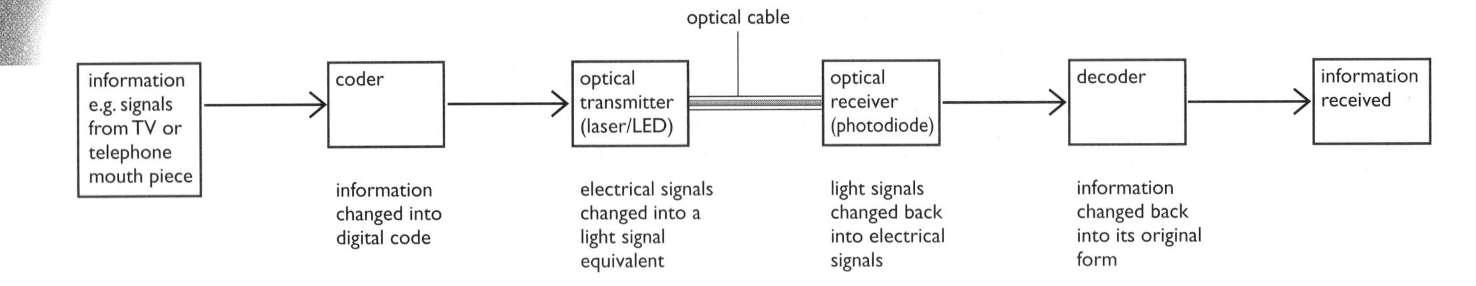

Figure 20.18 *An optical telecommunications system.*

In the optical system digital electrical signals are converted into a *light signal equivalent* using light-emitting diodes (LEDs) or lasers. Having travelled through the cables, the light is converted back into electrical signals using **photodiodes**. These signals are then decoded and converted back into the original form of the data.

Optical fibres are rapidly replacing copper wires and cables. Optical fibres have several important advantages over copper wires:

- optical fibres are capable of carrying far more information – just one fibre can carry as many as 30 000 phone calls
- optical fibres are free from noise
- signals can travel far greater distances along optical cables before needing regeneration (100 km in optical fibres compared with 4 km in copper cables)
- optical fibres are much lighter and smaller
- the "cross talk" (leakage of information) between channels using a single cable is much less with optical fibres
- greater security is possible – it is much more difficult to tap into or bug a conversation that is being transmitted through optical cables.

Communicating using radio waves

Radio waves can be sub-divided into three or more smaller groups depending upon their frequency and wavelength. These different radio bands have different transmission properties and therefore have different uses. The table below gives some examples of the uses of these different bands.

Frequency band	Uses
low frequency (LF): 30 kHz to 300 kHz	long-wave radio – capable of communicating over large distances
medium frequency (MF): 300 kHz to 3 MHz	medium-wave radio – local and distant radio communication
high frequency (HF): 3 MHz to 30 MHz	short-wave radio – CB radio
very high frequency (VHF) and ultra high frequency (UHF): 30 MHz to 3 GHz	FM radio, TV
microwaves: 3 GHz and upwards	satellite communication, mobile phones, radar

Radio waves may travel from a transmitter to a receiver in three different ways – as ground waves, sky waves or space waves.

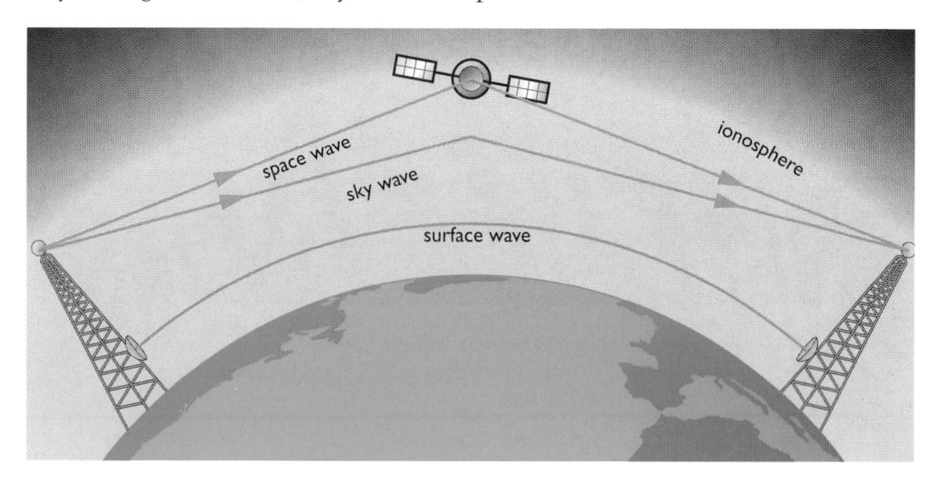

Figure 20.19 *Radio waves can be transmitted as space waves, sky waves or ground waves.*

Ground waves

- Range is limited because of absorption of the waves' energy by the ground. The higher the frequency, the greater the absorption.
- Low frequency, long wavelength radio waves are diffracted by large obstacles such as hills and buildings. As a result, there are few radio shadows where the signals cannot be received.

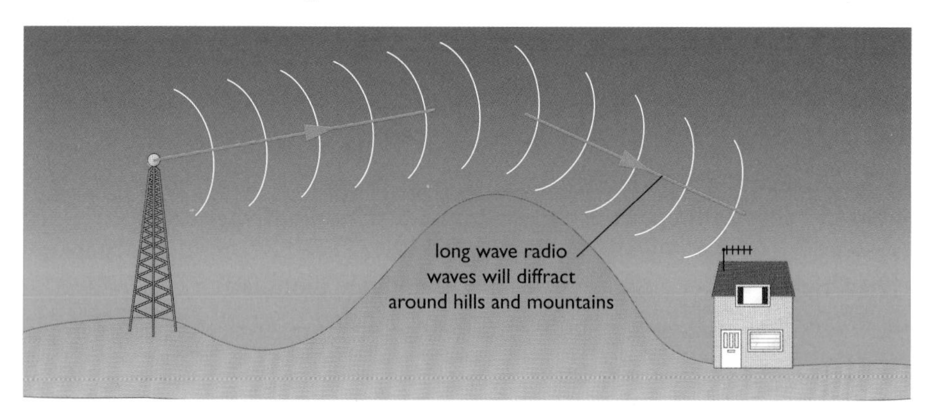

Figure 20.20 *Long wavelengths diffract around mountains.*

Sky waves

- Radio waves with frequencies less than 30 MHz may be directed upwards to part of the atmosphere called the **ionosphere**, which is typically between 80 km and 500 km above the Earth. The ionosphere reflects these waves back down to the Earth's surface (see Figure 20.21).
- It is possible for these waves to "bounce" all around the world until they eventually become too weak to be detected.
- If ground waves and sky waves transmitted from the same source are received at the same place, they will have travelled different distances and interference may take place between the two sets of waves. This may cause the received signal to alternately grow and fade in strength. (This is an example of constructive and destructive interference.)

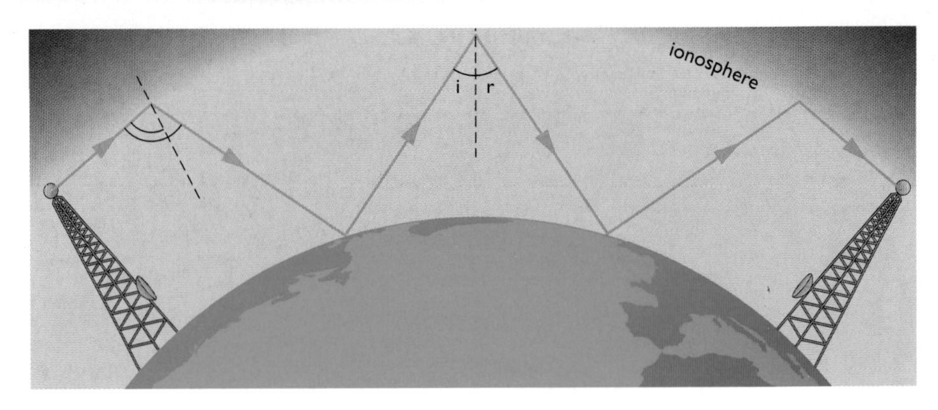

Figure 20.21 *Sky waves bounce off the ionosphere and the Earth.*

Space waves

- Radio waves with high frequencies (UHF and VHF), and microwaves, pass through the ionosphere to orbiting **geostationary satellites**. These are satellites that are over the same part of the Earth's surface all the time. From here the transmissions can be directed back down to a ground station on the Earth or passed on to other satellites.
- It is possible to use space waves to communicate with any part of the Earth's surface using a network of geostationary satellites.

Transmitting radio waves

Sound waves have a very limited range in air. It is impossible to use sound waves alone to communicate over large distance. Radio waves travel much greater distances (and much faster) with relatively small losses in intensity and therefore are used to carry sound waves.

Radio waves are emitted when alternating currents flow in an aerial. These radio frequency (rf) currents are combined with audio frequency (af) currents. There are two ways in which the waves can be combined. These are called amplitude modulation (AM) and frequency modulation (FM). The modulated signal is then amplified and transmitted as a radio wave.

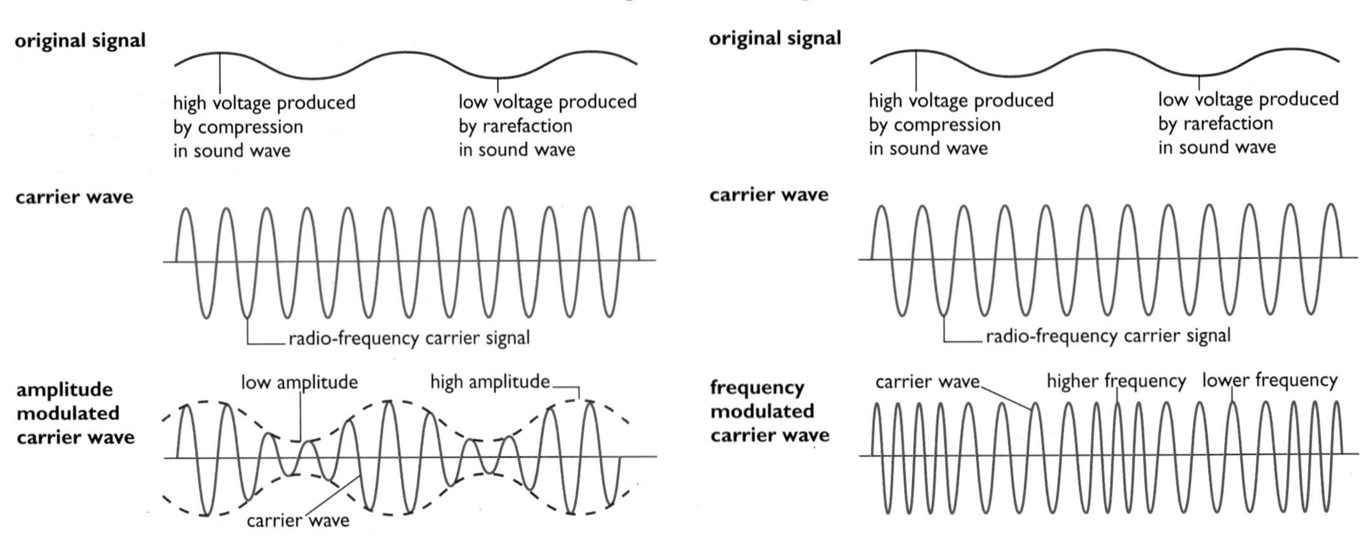

Figure 20.22 *In amplitude modulation, the amplitude of the carrier wave is altered.*

Figure 20.23 *In frequency modulation, the frequency of the carrier wave is altered.*

A radio receiver detects radio waves and turns them into sound waves. It works like a transmitter in reverse. Radio waves passing through an aerial produce electrical signals, which are demodulated – that is, the carrier wave and the signal it is carrying are separated. The signal is amplified and fed into a loudspeaker or ear piece to produce sounds similar to those that entered the microphone.

Radio signals that travel as FM suffer much less interference and produce better quality sounds. Their range however is much shorter than that of most AM waves.

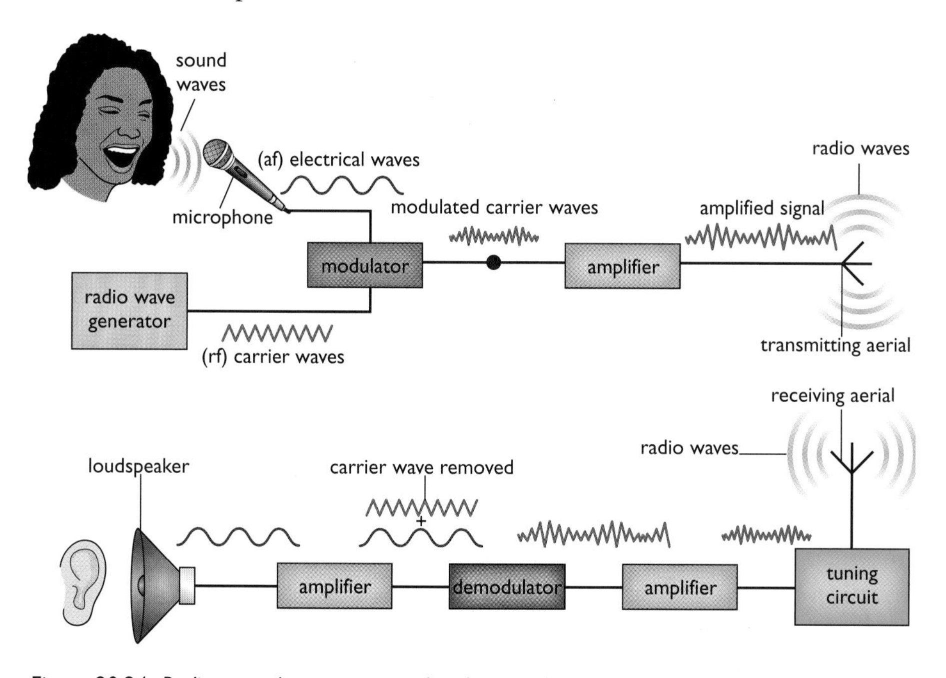

Figure 20.24 *Radio waves being transmitted and received.*

Storing data

Vinyl discs and record players

One of the earliest methods of storing data was the vinyl disc, or record. Information, usually music, was stored on the disc in analogue form. The depth and contours of a single spiral groove corresponded to the frequency and loudness of the recorded sound. As the disc rotated, a needle followed the spiral groove. The groove was not smooth so the needle moved up and down and from side to side. These movements were detected and changed into electrical signals that were then amplified and fed into loudspeakers or headphones.

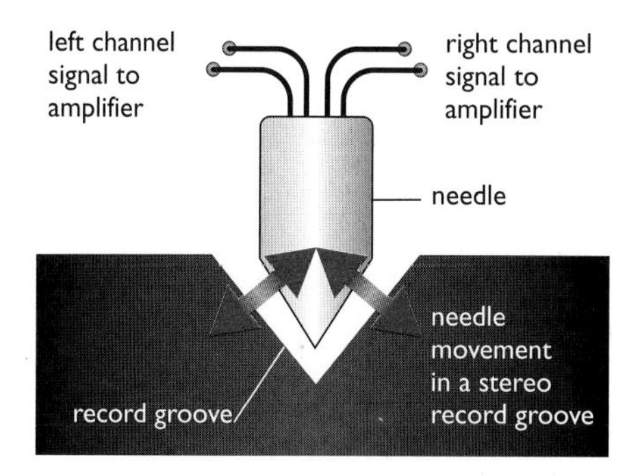

Figure 20.25 *The needle moves in a groove in the vinyl. The movements are changed into electrical signals.*

Magnetic tapes

When the tape recorder was developed, it offered the opportunity for everyone to record their own music or messages.

Recording

The plastic tape used for recording is coated with a layer of magnetic material such as iron oxide or chromium oxide. Before any recording these magnetic particles are arranged randomly and there is no overall magnetic effect. When a recording is made electrical signals from a source pass through electromagnets in the record head and a permanent magnetic pattern is imprinted on the tape.

Replaying the tape

When the tape is played back, the reverse process occurs. As the magnetic pattern passes the replay head, electrical signals are generated that are identical to those fed into the record head. These are amplified and passed out through a loudspeaker.

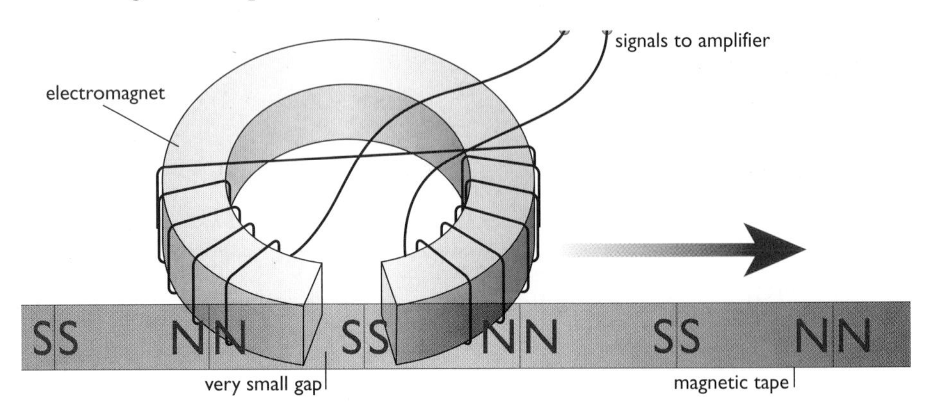

Figure 20.26 *Replaying the signal that has been recorded onto a magnetic tape.*

Figure 20.27 *These days, compact discs have replaced vinyl records and magnetic tapes as the most common way to store recorded music.*

CD-Recordable discs are coated with a photosensitive dye. A write-laser creates a series of opaque non-reflecting spots in place of the bumps found on a conventional CD.

Compact discs (CD)

Information is stored on a compact disc in digital form on a spiral track. The track is formed from a series of raised bumps. A set of lenses and mirrors direct a laser beam at the disc as it spins.

If the beam hits a flat part of the disc – that is, a place where there is no bump – the light is reflected with no phase difference across it. Constructive interference takes place and a photodiode produces an electrical signal that represents the digit 1 in the binary system.

If the beam strikes a bump a phase difference is created between that part of the beam reflected from the background surface and that reflected from the bump. Destructive interference takes place between the two halves and the photodiode records this as a 0.

This sequence of 1s and 0s – the digital output – is fed into a digital-to-analogue converter and is amplified, before being fed into a loudspeaker.

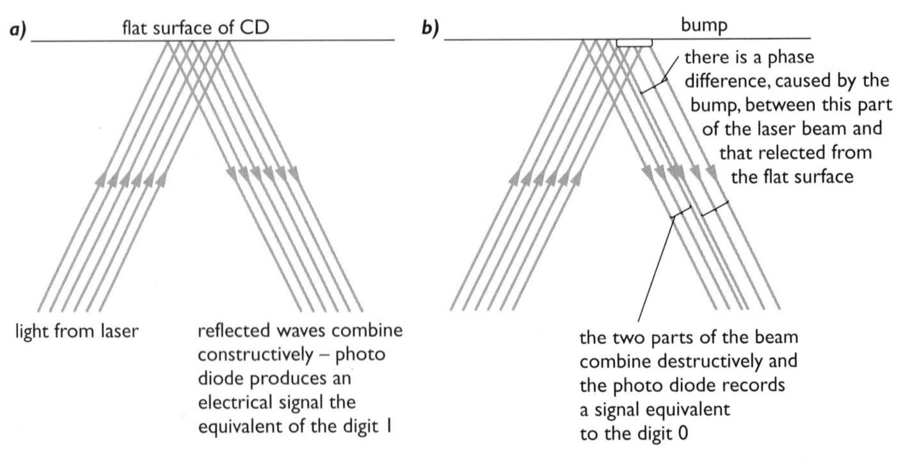

Figure 20.28 When the laser hits the flat bit of the disc, it is reflected uniformly. When the laser hits a bump, waves reflecting from the background and from the bump interfere with each other destructively.

Advantages of the storing data on a CD

- A large amount of music or other information can be stored in a small space. The bumps on a CD are very small, about one-hundredth of the breadth of a human hair in length. So a 12 cm disc will play music for over an hour or store approximately 100 million words of text (about 10 books).
- There is no damage to the surface of the disc when it is played. The information is read by a laser, which causes no wear and tear to the surface. The surfaces of vinyl discs and magnetic tapes do deteriorate with time and usage. Consequently there is a decrease in the quality of the sound they reproduce.
- Dust and imperfections on vinyl discs cause background noise. This is not a problem with CDs.

End of Chapter Checklist

If you haven't got a copy of your specification, read the introduction on page iv.

You will need to be able to do some or all of the following. Check your Awarding Body's specification (syllabus) to find out exactly what you need to know.

- Understand that the electromagnetic spectrum is a family of waves, which includes radio waves, microwaves, infra-red waves, visible light, ultraviolet waves, X-rays and gamma rays.
- Recall those properties that are common to all waves in the spectrum.
- Recall the order in which these waves appear in the spectrum and relate this to their wavelengths and frequencies.
- Recall some of the uses of these waves.
- Recall some of the dangers associated with exposure to certain types of radiation.
- Know that waves can be used to carry information.
- Explain the advantages and disadvantages of light and radio waves for communication.
- Describe the advantages of using digital signals rather than analogue signals.
- Describe AM and FM.
- Describe how information can be stored and the advantages of storing information digitally.

Questions

More questions on using waves can be found at the end of Section C on page 205.

1 ***a)*** Name four wave properties that are common to all members of the electromagnetic spectrum.

b) Name three types of wave that can be used for communicating.

c) Name two types of wave that can be used for cooking.

d) Name one type of wave that is used to treat cancer.

e) Name one type of wave that might be used to "see" people in the dark.

f) Name one type of wave that is used for radar.

2 Explain why:

a) microwave ovens cook food much more quickly than normal ovens

b) X-rays are used to check for broken bones

c) it is important not to damage the ozone layer around the Earth

d) food stays fresher for longer after it has been exposed to gamma radiation.

3 ***a)*** What is binary code?

b) Explain the difference between analogue signals and digital signals.

c) Explain the advantages of sending messages in digital form.

d) Explain how an analogue signal can be converted into a digital signal.

4 ***a)*** Draw a labelled block diagram to show the different processes applied to a sound wave that is transmitted and received as an AM radio wave.

b) Explain with diagrams the differences between ground waves, sky waves and space waves.

c) Give one advantage of transmitting a signal as FM radio waves rather than AM radio waves.

5 ***a)*** Draw a labelled block diagram to show how optical cables can be used in a telephone network.

b) Explain why optical fibres are replacing copper wires in modern telephone systems.

6 Compare the way in which data is stored on a compact disc and on a magnetic tape.

Chapter 21: Tectonic Plates and Seismic Waves

The Earth is not a solid, uniform ball of static rock. It has a layered structure made up of different materials with different properties, and the thin outer layer consists of a number of plates that are in constant motion over the Earth's surface. In this chapter you will learn about the evidence that has helped us to understand the Earth's structure, in particular the evidence from earthquake shock waves.

Early ideas, continental drift

ideas
evidence

If we look carefully at a map of the world we can see clearly that the edges of some land masses seem to have shapes that fit together like a jigsaw puzzle.

At the beginning of the twentieth century, most scientists believed that the Earth was a solid body whose land masses were fixed and motionless. In 1912 a German meteorologist called Alfred Wegener put forward a theory to explain why land masses have shapes that suggest they are parts of a jigsaw puzzle. Wegener proposed that all the continents were at one time joined together to form a single land mass he called Pangaea. He suggested this single land mass subsequently split up into the various continents, which then drifted apart. The theory was known as **Wegener's theory of continental drift**.

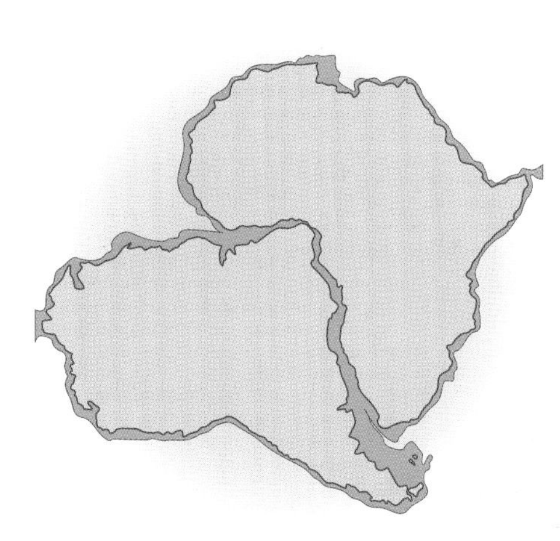

Figure 21.1 *The continents of South America and Africa seem to fit together.*

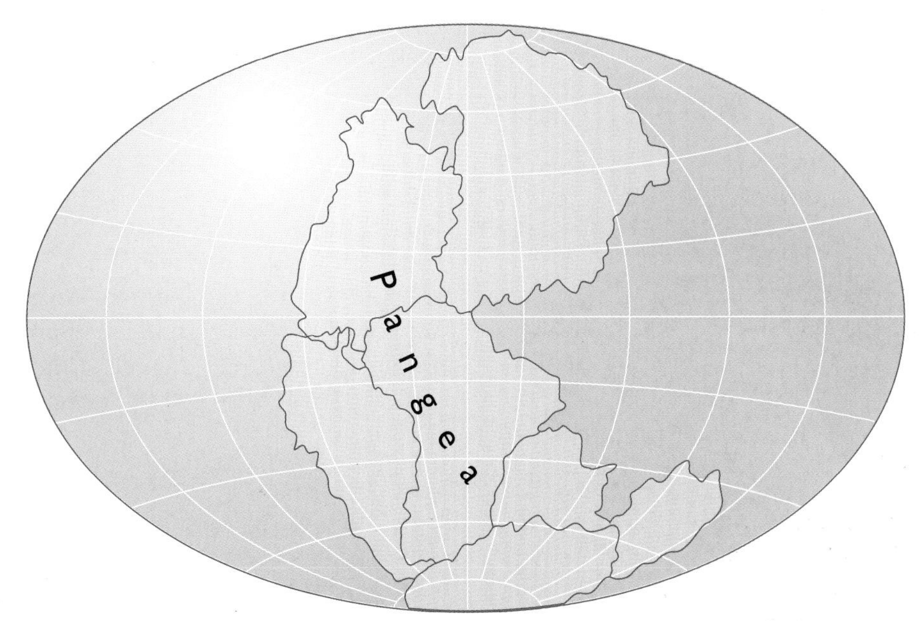

Figure 21.2 *Pangaea was a "supercontinent" that existed about 200 million years ago.*

Wegener's theory was not based solely on the jigsaw shapes of the continents. There was other evidence that appeared to support his theory.

- There are similarities between the plants and reptiles found on continents that are separated by large distances.
- There are similarities between the coastline rocks of the different continents as if at one time they had not been separated by oceans.
- There are geological features that suggest that the climates experienced by the land masses were very different to the climates they have today. For instance, there is evidence of the effects of glaciers in the tropical rain forests of Brazil, suggesting that the South American continent may once have been closer to the South Pole.

Even as early as 1620 Francis Bacon noted that the shapes of the east coast of South America and the west coast of Africa would fit together like two pieces of a jigsaw.

Many scientists of the time were sceptical of Wegener's theory, especially as he was unable to explain a mechanism for the drift.

After Wegener put forward his theory new evidence emerged. Some of this evidence supported his ideas but some seemed to suggest that his model was not accurate and needed some modification.

Sea floor spreading

In 1948, while investigating islands in the Atlantic, William Maurice Ewing discovered a mountain range below the sea that ran most of the length of the Atlantic Ocean – from the Arctic to as far south as the southern tip of South Africa. He noted that the rocks of this range were volcanic and very young. This suggested that the size of the sea floor between the continents was increasing.

Further examination of these rocks in the 1950s showed the presence of some iron-rich minerals that had aligned themselves along the Earth's magnetic field when they were formed after volcanic eruptions. It was already known that the Earth's magnetic field periodically reverses (that is, its north magnetic pole becomes a south magnetic pole, and vice versa). After each reversal, the iron-rich minerals in newly-formed rock align themselves according to the new polarity – that is, in the opposite direction to those in rocks created before the reversal. As time goes on, and the Earth's magnetic field reverses again and again, this creates a striped pattern of alternating magnetic alignment in the new rock. A pattern of this kind was found on the Atlantic sea floor, and was almost symmetrical either side of the Mid-Atlantic Ridge (the mountain range discovered by Ewing). This supported the idea that the sea floor was spreading as new rock was created in the middle, along the line of the ridge, and the land masses on either side of the Atlantic were getting further apart.

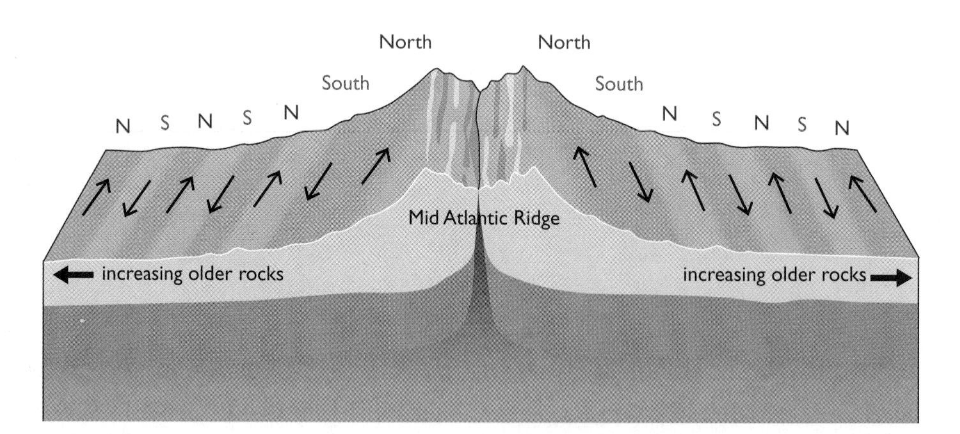

Figure 21.3 *Symmetrical magnetic stripes suggest that new crust is being created in the Mid-Atlantic and the sea floor is spreading.*

In 1966 Harry Hess studied these seafloor rocks and confirmed that the age of the rocks did increase from the centre of the sea floor towards the North American coast and that new rocks are still being formed. His measurements suggested that the Atlantic sea floor was spreading by up to 5 cm a year. Although this may seem very slow, it is fast enough to account for the present-day gap of thousands of kilometres between the coasts of Europe or Africa, and the Americas.

One major difficulty with the idea of sea floor spreading was the obvious conclusion that the surface area of the Earth must be getting larger. This was known not to be true. The only other possible explanation was that parts of the Earth's crust were being destroyed elsewhere. These areas were later discovered and include the fringes of the Pacific Ocean. This led to a new theory called **plate tectonics**, which replaced Wegener's theory of continental drift.

Plate tectonics

One of the major flaws with Wegener's theory of continental drift was that he thought that each continent was "floating" on a plate and that these plates were drifting on a never changing ocean. We now believe that the lithosphere (the crust of the Earth and the solid part of the mantle – see Figure 21.4) is cracked and divided into a number of moving pieces called **tectonic plates**. Unlike Wegener's drift theory, these plates are not composed of land masses alone – some are just land, but some are just ocean (and the rocks beneath the ocean) and some consist of both land and ocean.

There are two types of crust – **continental** and **oceanic**. These terms do not refer to continents or oceans but to the type of rock of which the crust consists. Continental crust is composed of older, lighter rocks such as granite and can be up to 40 km thick. Oceanic crust is composed of much younger, denser rocks such as basalt and is 5–10 km thick. It is at the boundaries of these moving plates that volcanic, earthquake and mountain-forming zones are found.

The mantle is made up of rocks rich in iron and magnesium. The upper layer of the mantle is fairly rigid but the remainder is in a semi-molten state (see Figure 21.5).

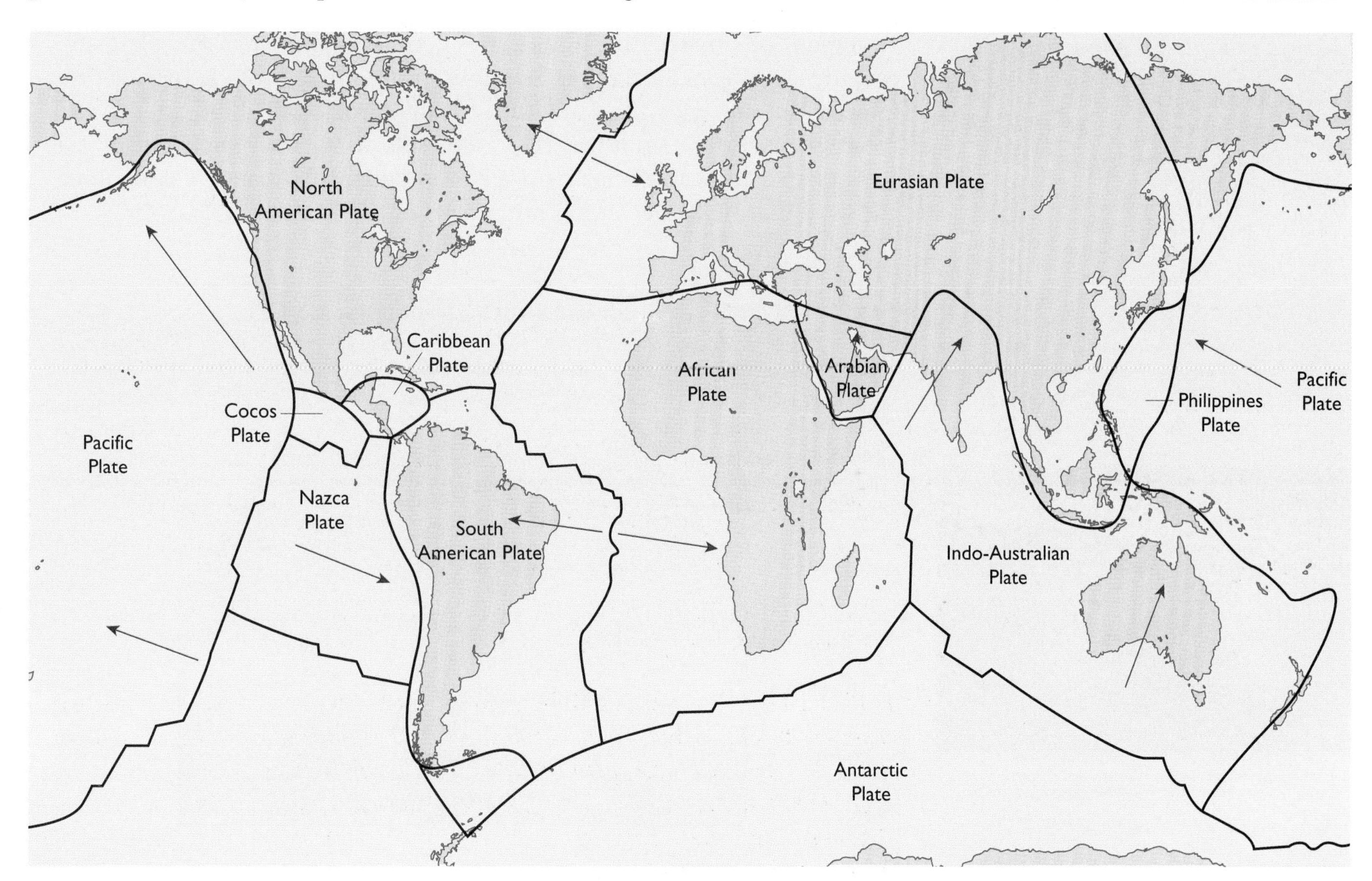

Figure 21.4 *Plate boundaries.*

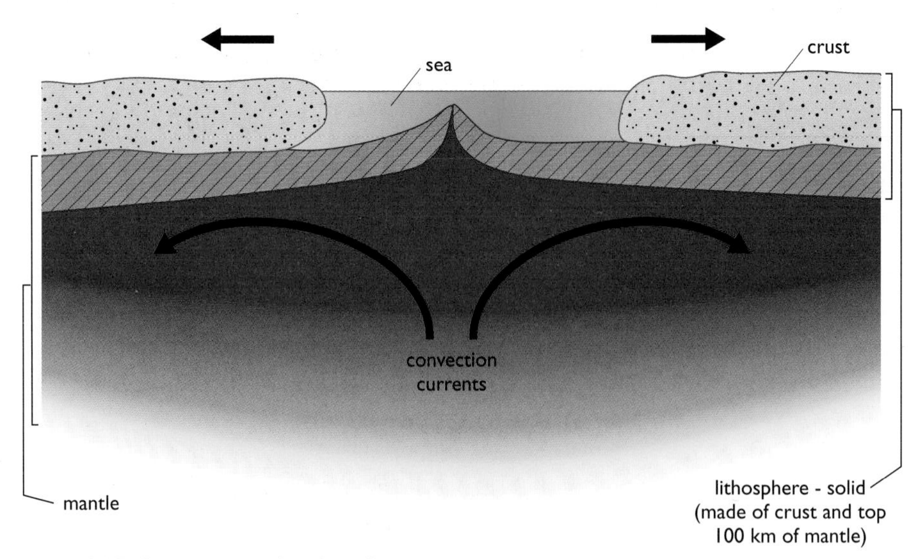

Figure 21.5 *A constructive plate boundary.*

Below the lithosphere at some depth (approximately 100 km) the mantle is able to flow very slowly. This mobile rock beneath the rigid plates is believed to be moving due to convection currents. The driving force for these movements is thought to be the heat energy released by elements such as uranium, thorium and potassium, which are undergoing radioactive decay.

Divergent boundaries

The Great Rift Valley starts in Jordan and runs through Africa to Mozambique. As the plates move apart a new sea may eventually form in the middle of Africa.

Divergent or **constructive boundaries** occur where the tectonic plates are moving away from each other. Fractures in the surface, also known as rifts, appear. The rifts widen and are filled by liquid rock, called **magma**, which creates new crust when it becomes solid. This, as we have already seen, leads to floor spreading and is taking place deep under the Atlantic Ocean along the Mid-Atlantic Ridge, and in other parts of the world.

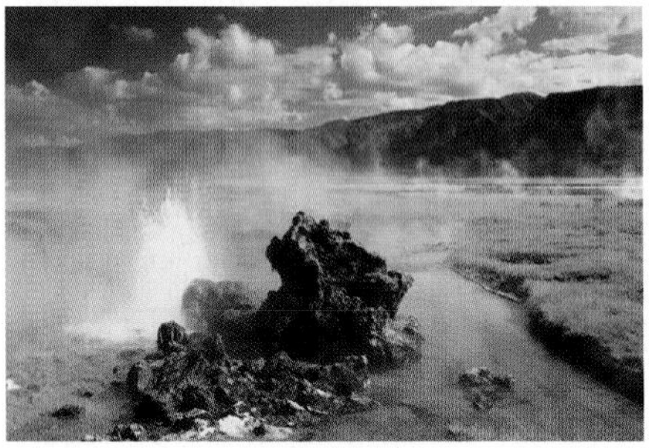

Figure 21.6 *Hot springs are common in rift valleys because the magma is closer to the surface.*

cross-section of the Mid-Atlantic Ridge

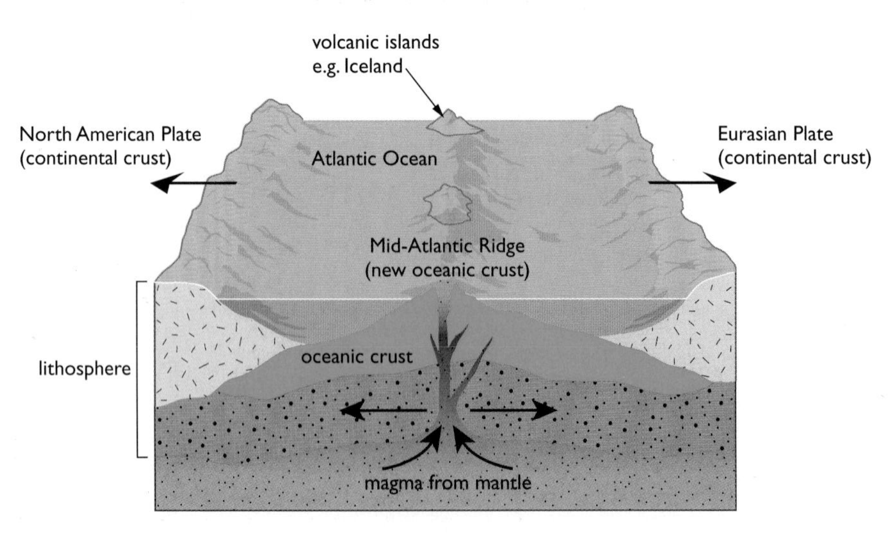

Figure 21.7 *A divergent or constructive plate boundary results in the widening of an ocean, or even the formation of a new sea.*

Convergent boundaries

Convergent or **destructive boundaries** occur where two plates are moving towards each other. If the plates are both continental plates, the crust tends to be pushed upwards and sideways. The collision between the Eurasian plate and the Indian plate resulted in the formation of the **fold mountains** we now call the Himalayas.

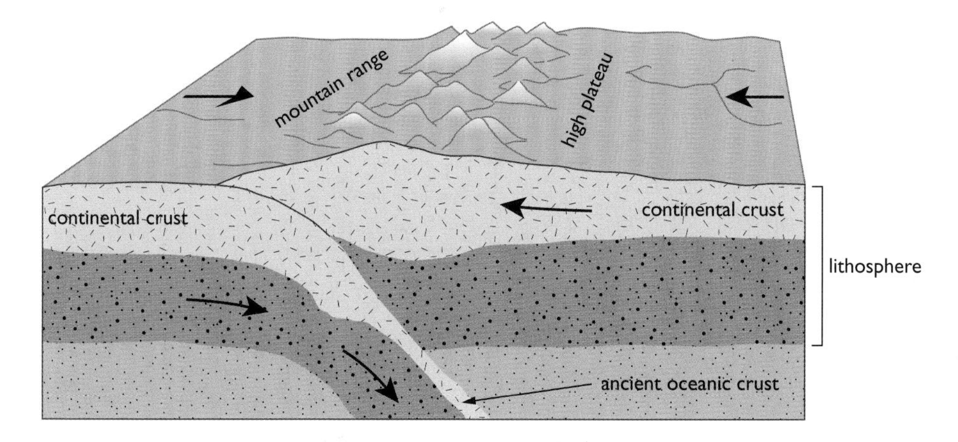

Figure 21.8 *A convergent or destructive plate boundary between two continental plates can result in mountain-building.*

If one of the converging plates is an oceanic plate and the other is a continental plate, the lighter continental plate is crumpled and lifted upwards often resulting in a mountain chain. The heavy oceanic lithosphere is forced downwards into the mantle. The beginning of the down-buckling is often marked by a deep trench running parallel to the chain of mountains created by the buckled continental plate. As the lithosphere descends, the increase in pressure can trigger earthquakes and the heat produced by friction may be sufficient to convert part of the plate back into magma. The area where the plate descends into the mantle is called the **subduction zone**. The newly formed magma may rise to the Earth's surface, where it causes volcanic activity. This is happening along the west coast of South America, as the Nazca plate is being subducted below the South American plate.

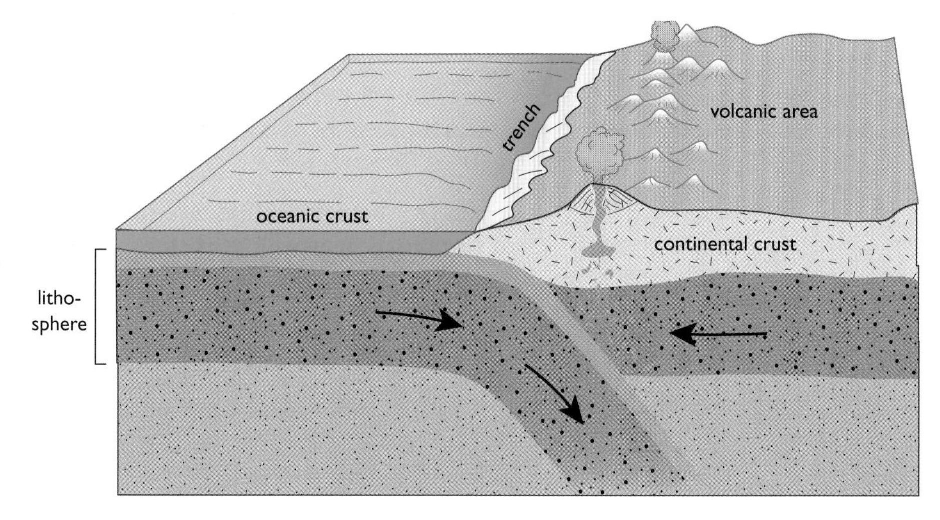

Figure 21.9 *An oceanic plate and a continental plate converge, resulting in a trench parallel to the plate boundary.*

Approximately 80% of the world's active volcanoes lie above subduction zones.

ideas
evidence

Predicting when a volcano is going to erupt can make the difference between life and death. If scientists can pin-point the likely time of an eruption, people living in the danger area can be evacuated. However, predicting volcanic eruptions is very difficult – there are lots of variables involved, and every volcano is different.

There are certain warning signs that vulcanologists (people who study volcanic activity) can look for that indicate when an eruption is likely to occur.

- Liquid rock called magma moves and collects in large chambers or reservoirs beneath the ground. This movement is often accompanied by small seismic tremors. Monitoring the area for these tremors can provide some clues about whether an eruption is close.
- As the magma comes nearer to the surface it releases gases. Monitoring the concentration of these gases (CO_2 and SO_2 in particular) may provide some information about a likely eruption.
- As the magma comes near to the surface, there is a strong possibility that the land around the volcano may deform – for example, it may swell. Monitoring the shape of this land may be another means of predicting if an eruption is about to take place.

Accurate predictions are still very difficult in volcanology. Magma entering a reservoir may cool below the surface rather than erupting. The most accurate predictions are made where volcanoes have been studied for some time and their past behaviour is well documented.

The US Geological Survey has lots of information about volcano monitoring – search the web for "USGS" to find out more.

Figure 21.11 *The rows in this field have been offset by an earthquake.*

When two oceanic plates converge one of them is usually subducted and a deep trench is formed. The descending plate gets hotter as it moves down into the mantle and molten rock rising from it can form volcanoes. Where these rise above sea level volcanic islands are created. This is happening in the Pacific Ocean where the Pacific plate is being subducted below the Indo-Australian plate.

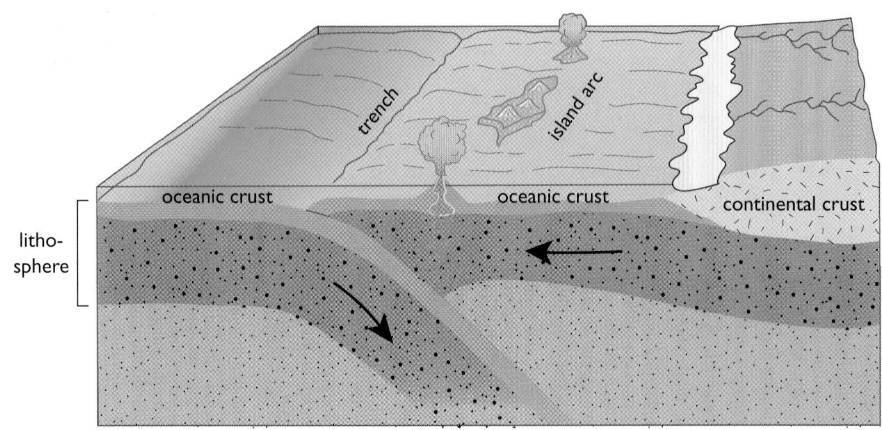

Figure 21.10 *Two oceanic plates colliding can result in an "arc" of volcanic islands.*

Transform boundaries

Transform boundaries occur when two plates slide past each other. Crust is neither created nor destroyed, as the tectonic plates are moving neither away nor towards each other. The San Andreas Fault near the coast of California is an example of a transform fault.

The plates do not slide smoothly. The plates become stuck, and enormous stresses and strains build up until suddenly the plates force their way past each other. This movement of the plates and sudden release of stress can cause earthquakes.

Figure 21.12 *Buildings in San Francisco after an earthquake.*

Seismic waves

Seismic waves are shock waves created in the Earth by the energy released by earthquakes. Much of our knowledge of the structure of the Earth comes from studying these seismic waves. Before we can understand how conclusions were drawn from this information we need to understand some of the basic properties of seismic waves.

Although surface waves are mentioned here, the Awarding Bodies focus on P and S waves.

There are three main types of seismic waves. These are called surface waves, P-waves (primary waves) and S-waves (secondary waves).

- **Surface waves** provide little information about the Earth's structure because they are only found in the crust. It is the rippling motion of these waves that causes most of the damage to buildings and other structures on the Earth's surface.
- **P-waves** are longitudinal waves. They travel slightly faster than S-waves and can pass through solids and liquids.
- **S-waves** are transverse waves. They travel slightly slower than P-waves and are unable to pass through liquids.

Seismic stations around the world monitor earthquakes and the shock waves they produce. Information from seismic waves indicates that the speeds of both P- and S-waves increase as they travel deeper into the mantle. As a result the waves undergo refraction – that is, they change direction.

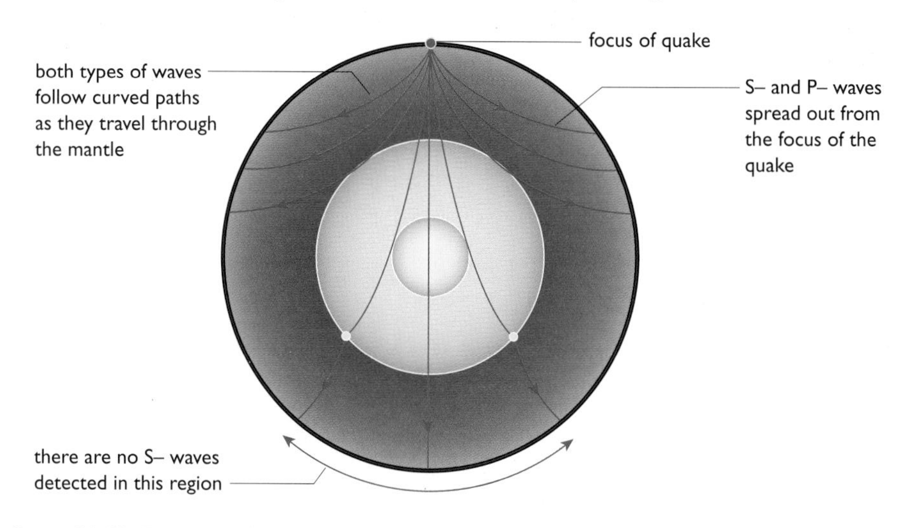

Figure 21.13 *The paths of the waves can be worked out by measuring when they arrive at different places on the surface after they have travelled through the centre of the Earth.*

To help you remember the difference between P-waves and S-waves you can think of it this way. P-waves are **P**ressure waves (like sound waves) but S-waves move from **S**ide to side.

If the paths they follow are gently curved this is accepted as indicating a gradual change in the properties of the material through which they are travelling. Where there is an abrupt change in a wave's direction this is interpreted as a sudden change in the properties of the material through which it is travelling. For example, the wave may have moved into a totally new material or perhaps is travelling from a solid to liquid or liquid to solid.

Conclusions drawn from seismographic studies

The **crust** of the Earth is a thin outer layer of relatively low-density rock. Its depth varies from just a few kilometres in some areas under the oceans, to over 70 km under continental mountain ranges.

Beneath the crust is the **mantle**. This is a hot solid but it has some liquid properties – it can flow very slowly. The mantle becomes more dense towards the centre of the Earth. Both P- and S-waves follow curved paths as they travel through the mantle, showing that the changes in the properties of the material are gradual.

Half way to the centre of the Earth, there are abrupt changes to the paths followed by both types of wave. This suggests that they have entered into a material with a completely different structure to that of the mantle. We call this region the **core** and believe it consists of an outer and inner part.

The outer core will not allow S-waves to pass through it, suggesting that it is liquid. Abrupt refractions of P-waves at the boundary with the inner core suggest that the inner part is solid.

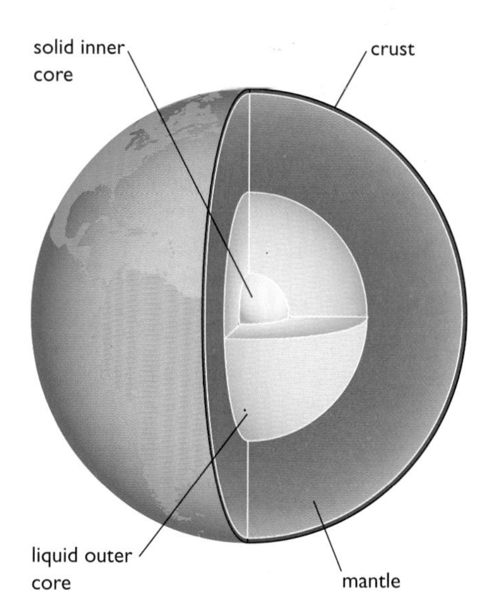

Figure 21.14 *The Earth is not a uniform solid ball of rock. It has a layered structure.*

End of Chapter Checklist

If you haven't got a copy of your specification, read the introduction on page iv.

You will need to be able to do some or all of the following. Check your Awarding Body's specification (syllabus) to find out exactly what you need to know.

- Explain the theory of continental drift and the evidence supporting it.
- Explain why the theory of continental drift was replaced by the theory of plate tectonics.
- Understand the different processes that may take place at the boundaries of tectonic plates.
- Recall that earthquakes and volcanic eruptions occur at the boundaries of tectonic plates.
- Understand how the study of seismic waves has led to a greater understanding of the structure of the Earth.

Questions

More questions on plate tectonics can be found at the end of Section C on page 205.

1 ***a)*** What is Wegener's theory of continental drift?

b) Give three pieces of evidence that supported Wegener's theory.

c) Why were some scientists sceptical of this theory?

2 ***a)*** Explain why the discovery of the Mid-Atlantic Ridge and the subsequent investigation of its rocks lead to the idea that the sea floor must be spreading.

b) Why does sea floor spreading not result in an increase in the surface area of the Earth?

3 ***a)*** What are tectonic plates?

b) Explain why tectonic plates move.

c) Explain why fold mountains are created at the boundaries between two tectonic plates that are moving in to each other.

4 ***a)*** What is a subduction zone?

b) Explain why there may be volcanic activity above subduction zones.

5 ***a)*** Explain how an earthquake may be caused by two plates sliding past each other. Name one example of a place where earthquakes have been caused by this kind of motion.

b) What is a seismic wave?

c) List the different properties of P-waves and S-waves.

d) Explain how studying these shock waves has helped scientists to understand the structure of the Earth.

e) Look at this diagram of the Earth showing its internal structure. Write out the correct words for the parts of the Earth labelled A to D.

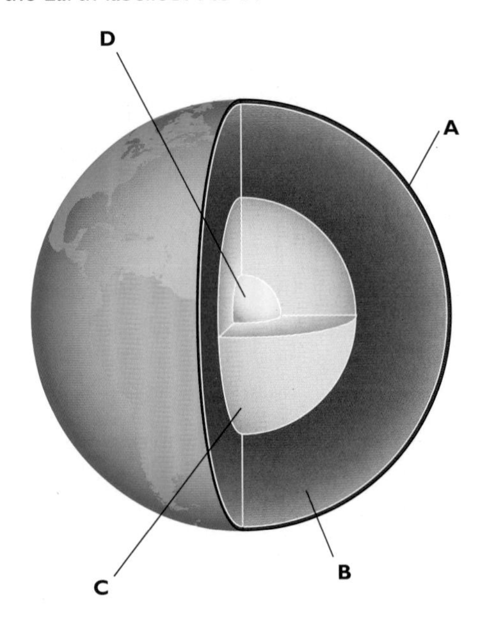

End of Section Questions

1 The diagram below shows water waves passing through the entrance of a model harbour.

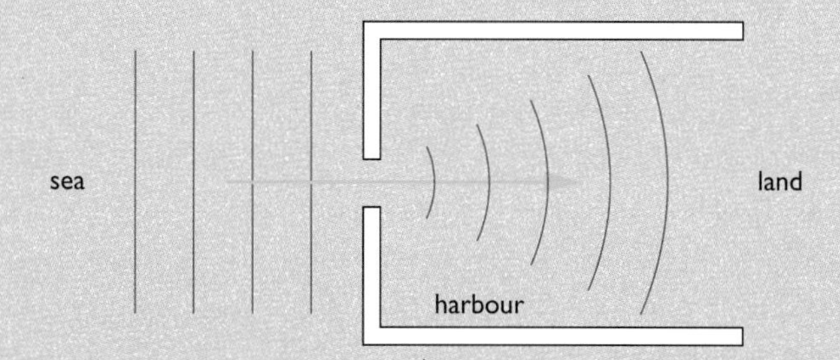

a) Describe what happens to the waves as they leave the gap between the harbour walls. *(1 mark)*

b) What is this process called? *(1 mark)*

c) Describe one change that could be made to the above arrangement in order to reduce this effect. *(1 mark)*

The diagram below shows a cross-section of the water waves.

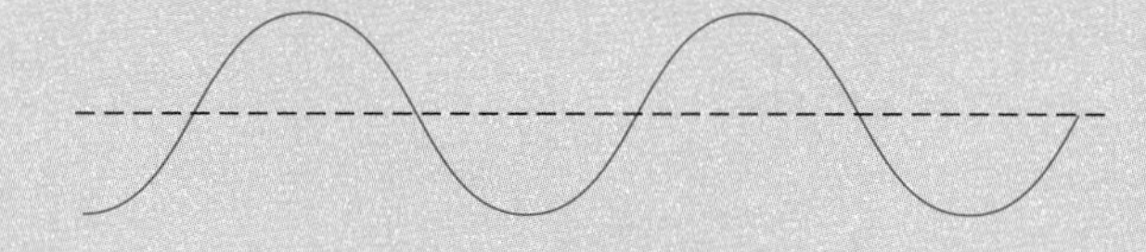

d) Copy this diagram and mark on it:

i) the wavelength of the wave (λ)

ii) the amplitude of the wave (A). *(2 marks)*

e) A water wave travelling at 20 m/s has a wavelength of 2.5 m. Calculate the frequency of the wave. *(3 marks)*

Total 8 marks

2 The diagram below shows a ray of light travelling down an optical fibre.

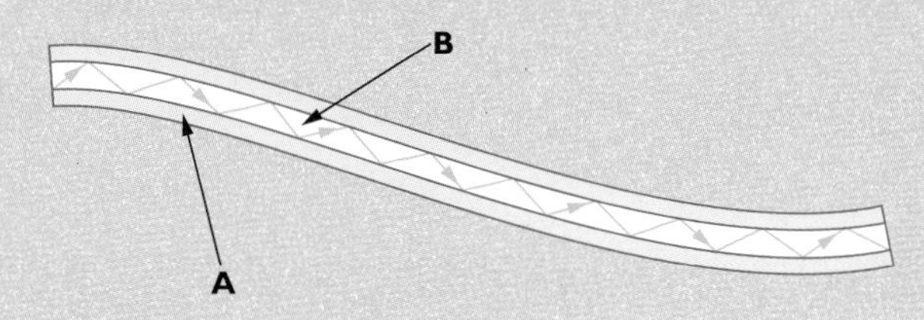

a) What is A? *(1 mark)*

b) What is B? *(1 mark)*

c) Why is light reflected from the boundary between A and B? *(2 marks)*

d) Optical fibres are gradually replacing copper cables used in telecommunications systems. Give three advantages of using optical fibres instead of copper cables to carry messages. *(3 marks)*

e) Describe one medical use for optical fibres. *(1 mark)*

Total 8 marks

3 **a)** *i)* Explain the difference between a longitudinal wave and a transverse wave.

ii) Give one example of each type of wave. *(4 marks)*

A girl stands 500 m from a tall building and bangs two pieces of wood together. At the same instant her friend starts a stopwatch. The sound waves created by the two pieces of wood strike the building and are reflected. When the two girls hear the echo they stop the stopwatch and note the time. The girls repeat the experiment four more times. The results are shown in the table below.

Experiment	Time in seconds
1	2.95
2	3.00
3	2.90
4	3.20
5	2.95

b) Why did the girls repeat the experiment five times? *(1 mark)*

c) Calculate the speed of sound using the above results. *(6 marks)*

d) One of the girls thought that their answer might be affected by wind. Was she correct? Explain your answer. *(2 marks)*

Total 13 marks

4 The electromagnetic spectrum contains the following groups of waves: infra-red, ultraviolet, X-rays, radio waves, microwaves, visible spectrum and gamma rays.

a) Put these groups of waves in the order they appear in the electromagnetic spectrum starting with the group that has the longest wavelength. *(2 marks)*

b) Write down four properties that all of these waves have in common. *(4 marks)*

c) Write down one use for each group of waves. *(7 marks)*

d) Which three groups of waves could cause cancer? *(3 marks)*

e) Which three groups of waves can be used to communicate? *(3 marks)*

Total 19 marks

5 The block diagram below shows how information, initially in the form of sound waves, is transmitted as radio waves.

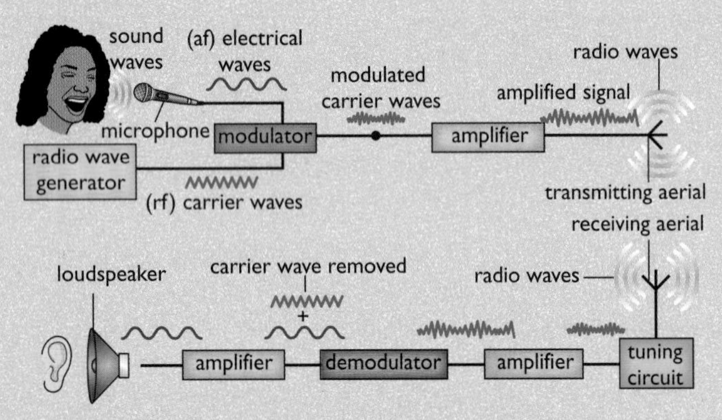

a) What does the microphone do to the sound waves? *(1 mark)*

b) What is the purpose of the modulator? *(1 mark)*

c) Explain with diagrams the meaning of
i) AM and
ii) FM. *(2 marks)*

d) Give one advantage of FM over AM. *(1 mark)*

e) What is a demodulator? *(1 mark)*

f) Give one advantage of transmitting information in a digital form rather than in analogue form. *(1 mark)*

Total 7 marks

6 The diagram below shows two tectonic plates moving towards each other.

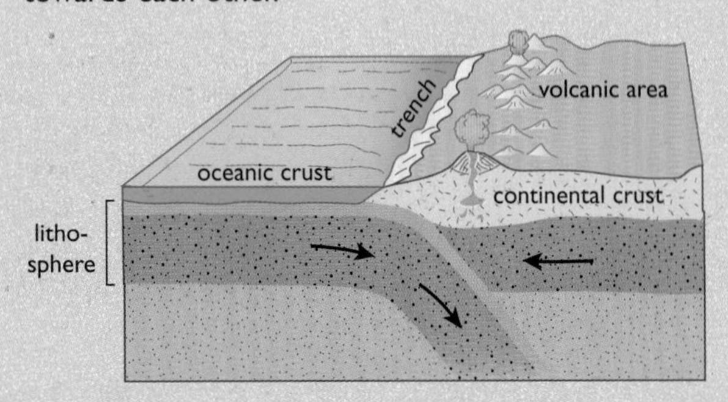

a) Explain in detail why tectonic plates move. *(3 marks)*

b) Explain what happens in the collision zone, where two plates moving towards each other meet, to:

i) the oceanic plate

ii) the continental plate. *(6 marks)*

c) Explain why volcanoes are often found above collision zones. *(2 marks)*

Total 11 marks

7 The movement of tectonic plates can create seismic waves that then travel through the Earth. The diagram below shows examples of the paths taken by P-waves and S-waves as they travel through the Earth.

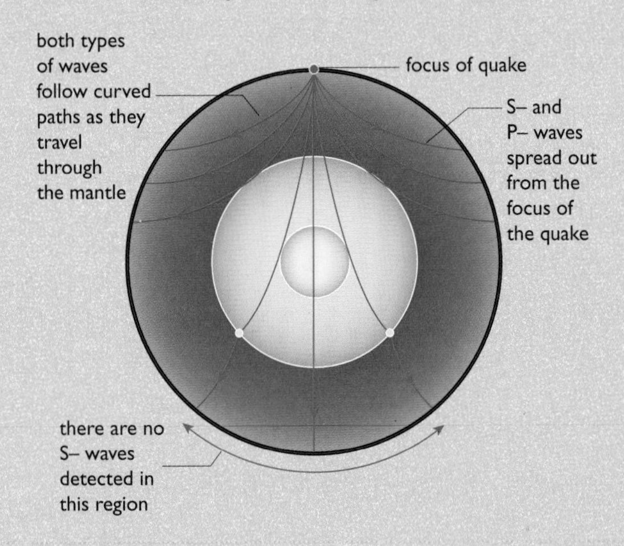

a) Give three differences between P-waves and S-waves. *(3 marks)*

b) Why do the waves travel in curved paths? *(1 mark)*

c) Why do some of the waves suddenly change direction? *(1 mark)*

d) There are no S-waves on the side of the Earth furthest from the earthquake. What does this suggest about the core of the Earth? *(2 marks)*

Total 7 marks

Chapter 22: The Evolution of the Solar System, the Stars and the Universe

Our Earth is just one planet orbiting the Sun. The Sun is one of billions of stars in a galaxy we call the Milky Way. This galaxy is just one of billions in the Universe, which are all speeding away from one another. In this chapter you will learn something about the huge distances between the stars and galaxies, about the "life cycle" of a star, about how the Universe began and how it may end. You will also learn about attempts to find out if there is life elsewhere in our Universe.

The Milky Way

Our nearest star is the Sun. It is approximately 150 million kilometres from the Earth and is a very average star. Its surface temperature is approximately 6000°C whilst temperatures within its core are about 15 000 000°C. Gravitational forces between stars cause them to cluster together in enormous groups called **galaxies**. Our Galaxy is a **spiral galaxy** called the **Milky Way**. We are approximately two thirds of the way out from the centre of our Galaxy along one of the arms of the spiral.

The Universe is mainly empty space within which are scattered large numbers of clusters of stars – astronomers believe that there are billions of galaxies in the Universe. The distances between galaxies are millions of times greater than the distances between stars within a galaxy. The distances between the stars in a galaxy are millions of times greater than the distances between planets and the Sun.

Figure 22.1 *A spiral-shaped galaxy, something like our own.*

Distances in space are so large that it would be very awkward to use our normal units (kilometres). Instead, we use **light years** as our measure of distance. One light year is the distance travelled by a ray of light in one year. The distance to our nearest star after the Sun, Proxima Centauri, is 4.3 light years. The distance to the centre of the Milky Way from Earth is approximately 26 000 light years. The nearest galaxies to the Milky Way are approximately 2 000 000 light years away.

Formation of a solar system

Stars are formed when large amounts of dust and gases are drawn together by gravitational forces. As this matter is pulled together, very high temperatures are created (as high as 15 000 000°C). At such high temperatures, **nuclear fusion reactions** begin. In these reactions the nuclei of small atoms, such as hydrogen, join or fuse together to produce new larger nuclei, like that of helium. These reactions release large amounts of energy and a star begins to form.

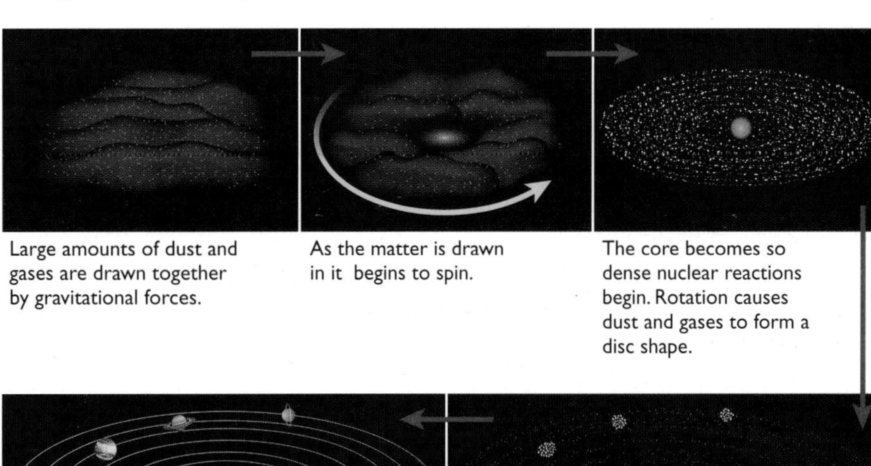

Figure 22.2 *How a solar system is formed.*

As the clouds of dust and gas initially collapse, they may cause the cloud to begin to spin. As the cloud becomes more dense, the spin rate increases. When the core becomes hot it may separate from the spinning cloud, which then takes the shape of a flat spinning disc. It is likely that this is how our Solar System was formed, with the hotter core forming the Sun and the outer gases condensing to form the planets and other orbiting bodies.

Life of a star

When a star is forming, there are strong gravitational forces pulling the dust and gases closer together. At the same time, the energy released by nuclear reactions is creating forces that are trying to make the matter expand. At some stage during the development of a star, these two forces will balance. The star is then said to be in its **main stable period**. This is the stage that our star, the Sun, has reached. The length of time a star spends in this period depends upon its mass but it could be billions of years. As a rule, smaller stars stay in this period for longer than bigger stars. Our Sun is about halfway through this period.

Eventually the supply of hydrogen nuclei within a star is used up. The forces trying to make the star expand become smaller and the star begins to collapse. This contraction, however, causes a rise in the temperature of the core of the star and a second nuclear reaction begins. Helium nuclei now fuse together and again release energy, which causes the star to expand again. As it does so the outer layers become cooler and redder in colour. The star is now a **red giant**.

Stars that have a mass similar to the Sun will shine steadily for approximately 10 000 million years. Stars with much greater masses convert their hydrogen fuel more quickly and therefore have shorter lives.

What happens next depends upon the mass of the star. As the nuclear fuel is used up, stars like our Sun will gradually contract and under the forces of gravity change into a **white dwarf** containing matter millions of times more dense than any found on Earth. A handful of this matter would have a mass greater than 1000 tonnes! As a white dwarf cools it becomes duller and duller and eventually it stops giving out light. It is now a **black dwarf**.

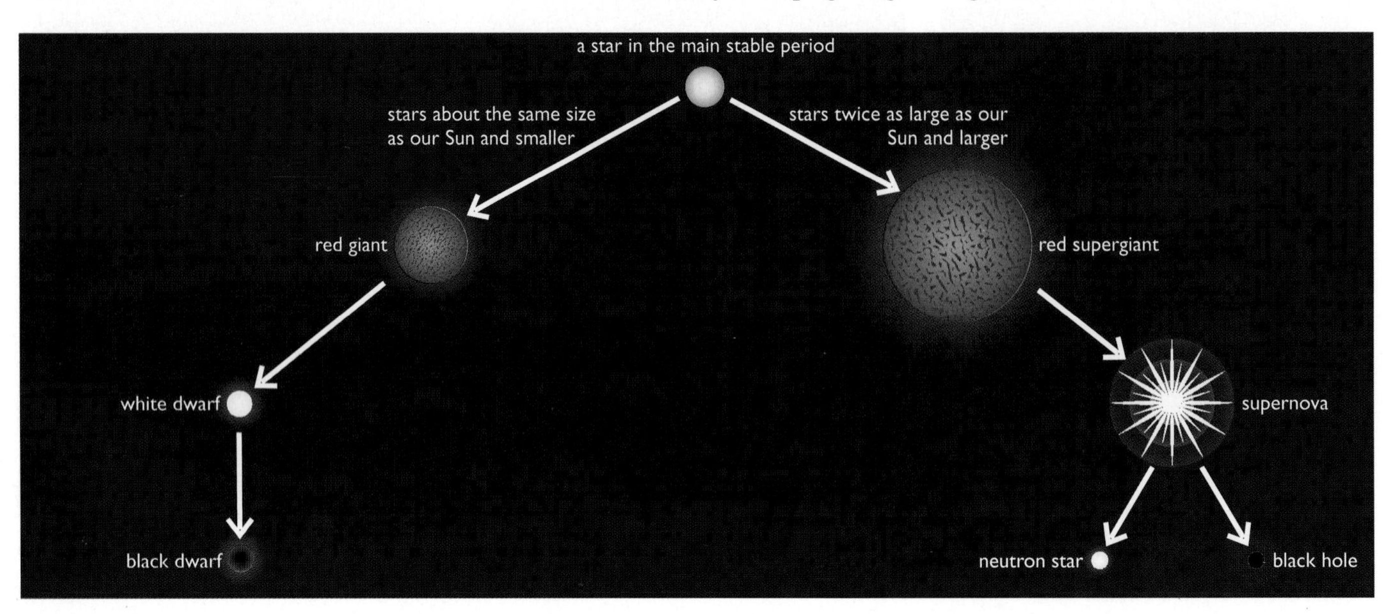

Figure 22.3 *The life of a star.*

If a star has a mass that is much larger than that of our Sun, it has a different future. After the main stable period, its collapse inwards releases enough energy to cause carbon nuclei, created in earlier nuclear reactions, to fuse together. This in turn causes the star to expand to a size hundreds of times bigger than our Sun. These stars are called **red supergiants**.

When these fusion reactions stop, the star collapses. As it collapses, the star heats up again and new fusion reactions start once more. This process can repeat itself for several cycles. Eventually, however, the release of energy causes the star to become unstable and it explodes, throwing dust and gases into space. This exploding star is called a **supernova**. What is left behind is a star just a few kilometres across, which emits no light. It is called a **neutron star** and is so dense that a handful of the material would have a mass of about 10 000 million tonnes!

When extremely large stars collapse they may instead leave behind a **black hole** – a place in space where matter is so densely concentrated that its gravity will allow nothing to escape from it, not even light.

There are nuclei of heavy elements present in our Solar System. These were probably formed from material released when earlier stars exploded.

Origins and future of the Universe

No one knows exactly how the Universe began or what its future will be. We can, however, put forward some ideas about both its past and its future based on observations and measurements of what we see today.

The Doppler effect – red shift

If you stand at the side of a race track with your eyes shut you can tell if a car is coming towards you or moving away by listening to the pitch of the sounds produced by its engine. The pitch (frequency) appears higher as the car approaches and lower as it moves away. This change in pitch is called the **Doppler effect**, and it is a property of all waves – not just sound waves. By observing the light from distant galaxies we can determine which way they are moving.

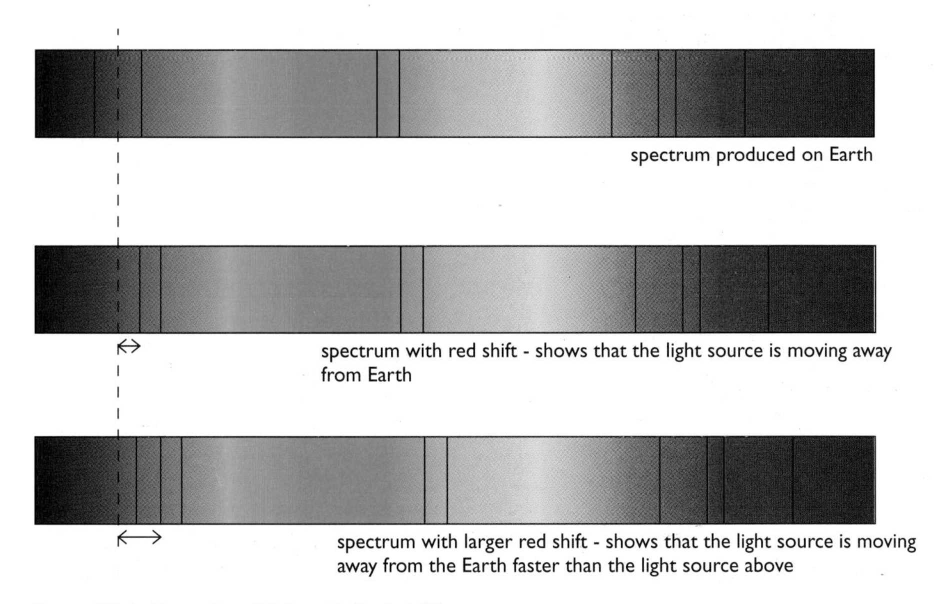

Figure 22.4 *Examples of light with "red shift".*

ideas
evidence

If we observe light from the Sun through an instrument called a spectrometer, we can see the visible spectrum from the Sun. Across these coloured bands are a few dark lines caused by elements in the outer layers of the Sun absorbing some of the light (Figure 22.4). When we look at light that has come from a distant galaxy, these lines are not in the same place. They appear at lower frequencies – that is, they have moved towards the red part of the spectrum. This is called **red shift**. Red shift indicates that the source of light – in this case, the distant galaxy – is moving away from us.

If the light from lots of different galaxies is compared, some patterns emerge.

1 Galaxies are moving away from us, and away from each other.

2 The largest red shift comes from the most distant galaxies. This means that the most distant galaxies are moving away from us at the greatest rate.

Both of these patterns support the idea that the Universe is expanding, and that when the Universe was very young all the matter it contains might have been in one place.

Scientists now believe that the Universe began with a large explosion, scattering mass outwards. This theory of how the Universe began is called the **Big Bang theory**.

ideas
evidence

It is uncertain what will happen to the Universe in the future. There are three suggested possibilities. The size of the gravitational forces between the matter in the Universe will determine which possibility comes true.

1 If there is enough matter in the Universe, forces of attraction will gradually slow down the expansion of the Universe and eventually pull all the matter back together again. Scientists refer to this possible event as the Big Crunch. At this point, there may be another explosion and the Universe may begin to expand again. This model is known as the **pulsating universe**.

2 If there is not enough matter in the Universe, it may simply continue to expand.

3 Finally, there is the possibility that the Universe may reach a certain size, then stop expanding and remain static.

Figure 22.5 *Is there anybody out there?*

Search for extraterrestrial life

ideas
evidence

Figure 22.5 shows a radio telescope being used to search for signs of **extraterrestrial** life. People have wondered if there is life on other planets for many years. Serious attempts to answer this question began in 1959 when scientists realised that it would be possible to send radio messages between stars. If they could detect radio messages between stars then this would confirm the existence of other advanced civilisations. Astronomers used radio telescopes to search for these signals. They looked initially at nearby Sun-like stars, gradually spreading their search to much more distant stars.

The SETI Institute was founded in 1984 to continue the Search for Extra Terrestrial Intelligence in the Universe.

Although as yet there have been no positive results, this is not surprising because the number of stars to be investigated is huge. In addition, the radio telescopes may be looking for signals with the wrong frequency. Another problem is that distances in space are so vast that answers to the radio transmissions we have been sending since the 1960s could only have come from near worlds. This means that we have probed only a very small section of the Universe. It is also possible that other advanced civilisations use a completely different method to communicate and therefore do not use radio waves.

It may be that the proof that life exists or existed elsewhere in the Universe can be found much closer to home than the stars of distant galaxies. It could be found by investigating the planets of our own Solar System. The most likely candidate to have been host to extraterrestrial life is Mars. It has conditions that are closest to those required to sustain life as we know it. Of all the planets in the Solar System Mars has an average surface temperature closest to that on Earth and it has a thin atmosphere. Also water once flowed over its surface and it has ice caps of frozen water.

The NASA Viking missions carried the first unmanned probes that landed on the surface of Mars. The landers carried instruments that monitored the weather and analysed the atmosphere for chemical changes that may have been caused by the presence of life. They carried out experiments on the Martian soil to detect the presence of life now or in the distant past. The tests looked for carbon-based life similar to our own but the results were inconclusive.

Strangely enough, the strongest evidence for the existence of life on Mars came from a rock found on Earth. The rock was found in the Antarctic in 1984. It was recognised as being a meteorite that had come from Mars and was more than 3.6 billion years old. Within the meteorite, scientists found what they thought were organic molecules, several mineral features that indicated that there had been some biological activity, and possible microscopic fossils of primitive bacteria-like life.

We still do not know if life exists elsewhere in the Universe. More missions are being planned to send probes to Mars and Europa (one of the moons of Jupiter where surface conditions may be favourable for supporting life) and the search of the skies for alien radio signals continues.

Figure 22.6 *Surface features on Mars have led scientists to believe that there was once flowing water.*

Figure 22.7 *This is believed to be a fragment of a meteorite from Mars.*

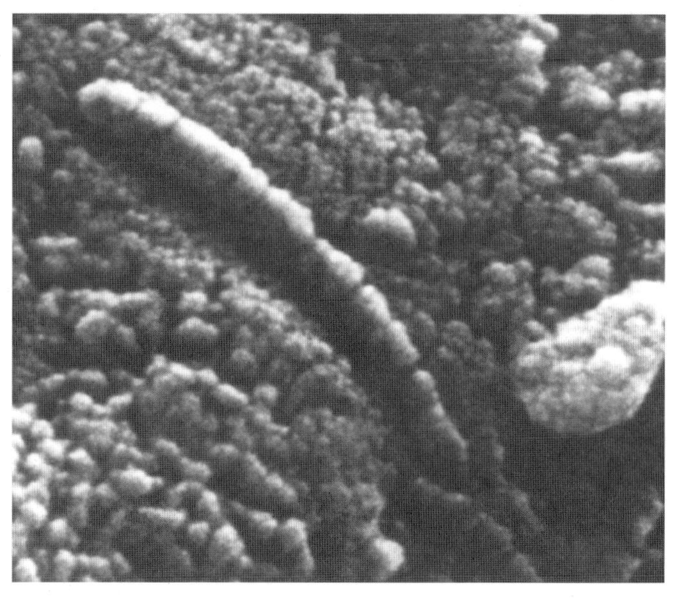

Figure 22.8 *This is possibly a fossilised Martian cell found inside the meteorite.*

End of Chapter Checklist

If you haven't got a copy of your specification, read the introduction on page iv.

You will need to be able to do some or all of the following. Check your Awarding Body's specification (syllabus) to find out exactly what you need to know.

- Understand that our Sun is just one of many billions of stars in a galaxy called the Milky Way, and that there are billions of galaxies in the Universe.
- Understand that the distances between stars within our Galaxy are millions of times greater than the distances between planets in our Solar System and that the distances between galaxies are millions of times greater than the distances between stars in a galaxy.
- Describe the formation of a star and understand how a star may change during its lifetime.
- Describe the Big Bang theory for the formation of the Universe and recall the three possible futures for our Universe.
- Understand that there is the possibility of extraterrestrial life and that there are difficulties in confirming this.

Questions

More questions on stars, the Solar System and the Universe can be found at the end of Section D on page 225.

1 ***a)*** From what types of materials do scientists believe our Solar System was formed?

b) What forces pulled these materials together?

c) Explain in detail what happened to these materials as the Solar System formed. Include diagrams with your explanations.

2 ***a)*** Explain why a star gives out large amounts of energy as it forms.

b) The nearest star to our Solar System is 4.3 light years away. If light travels at 300 000 km/s, calculate the distance of this star from us in kilometres.

3 ***a)*** Explain with diagrams the life cycle of a star that has a mass similar to our Sun.

b) What is a supernova?

c) What is a black hole?

4 ***a)*** What is the Doppler effect?

b) Explain how the Doppler effect can be used to determine if distant galaxies are moving towards us or away from us.

5 ***a)*** What is the Big Bang theory of the origin of the Universe?

b) What evidence is there to support this theory?

c) What are the three possible futures of the Universe? Explain how the amount of matter in the Universe will determine which of these possible futures will occur.

6 ***a)*** Why did the real search for extraterrestrial life begin in 1959?

b) Describe two ways in which scientists are searching for proof of extraterrestrial life.

Chapter 23: The Earth and the Solar System

Our Solar System is held together by gravitational forces, acting between bodies. These forces hold planets, asteroids and comets in orbit around the Sun, and keep moons in orbit around planets. In this chapter, you will read about the planets of our Solar System and their orbits, and learn about how the motion of the Earth gives rise to our days and nights, and seasonal changes. You'll also see how the height of orbit of an artificial satellite around Earth depends on the job the satellite has to do, and find out how to calculate its orbital speed.

Many years ago travellers, and in particular sailors, believed that the Earth was flat and if they travelled too far, they would fall over the edge. Figure 23.1 is a photograph that was taken from a Lunar Orbiter 1 orbiting the Moon in 1966. It clearly shows that the ancient travellers had incorrect ideas about the size and shape of the Earth.

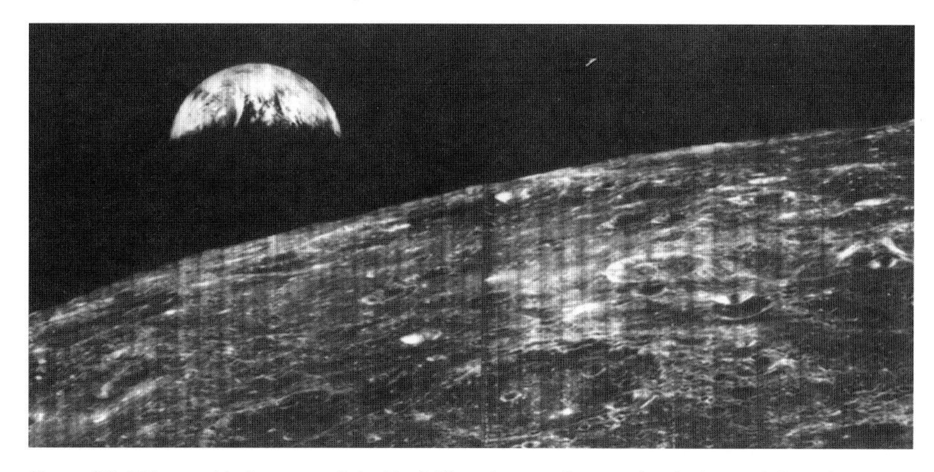

Figure 23. I *The worlds first view of the Earth. This photograph was taken by Lunar Orbiter I orbiting the Earth in 1966.*

Our explanations of the behaviour of astronomical bodies, such as moons, planets and stars are based upon the evidence we have before us today. In the future new evidence may be found, which may cause us to change our present-day ideas and models.

The Earth

We usually have 365 days in a year but to account for the quarter day we add an extra day to the year every fourth year. This "leap year" therefore has 366 days.

The Earth is a ball of rock spinning around an axis, whilst at the same time travelling around the Sun. It takes the Earth 24 hours (1 day) to make a complete rotation around its axis, and $365\frac{1}{4}$ days (1 year) to orbit the Sun.

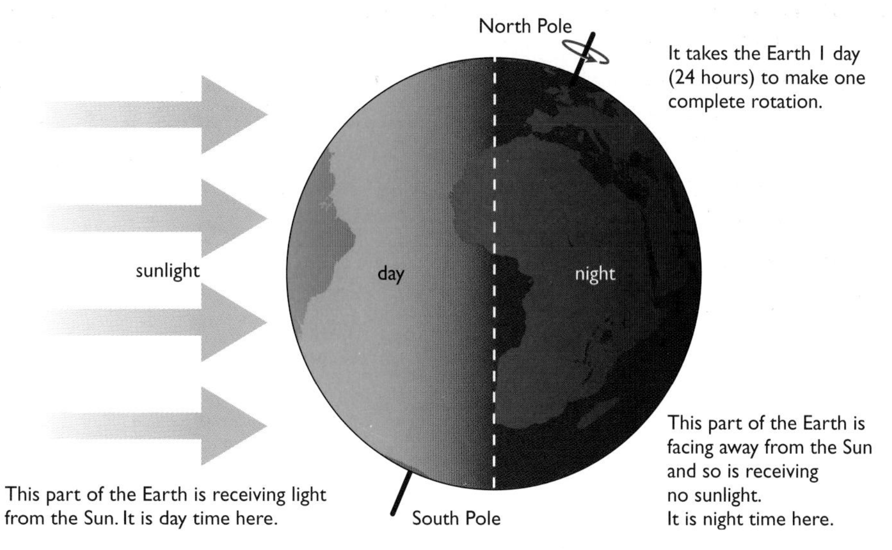

Figure 23.2 *The Earth spins once around its axis every 24 hours.*

The seasons

The Earth's axis is tilted as shown in Figure 23.3. This tilting causes **seasonal changes**.

1 The number of hours of daylight is greater in the summer than in the winter – that is, "the days are longer" in the summer.

2 The Sun follows a higher path across the sky in the summer.

3 The temperatures in the summer are higher than those in the winter.

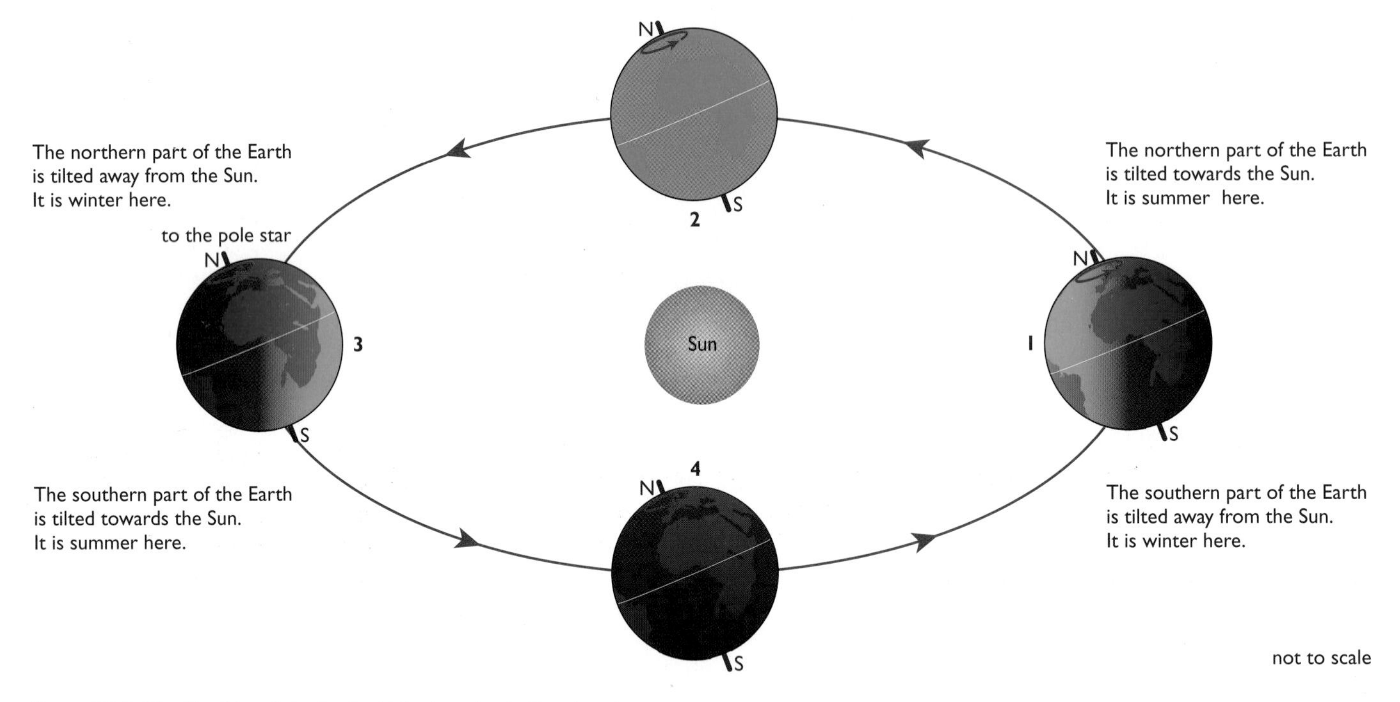

Figure 23.3 *The Earth's axis is tilted, which gives rise to the changing seasons.*

The Solar System

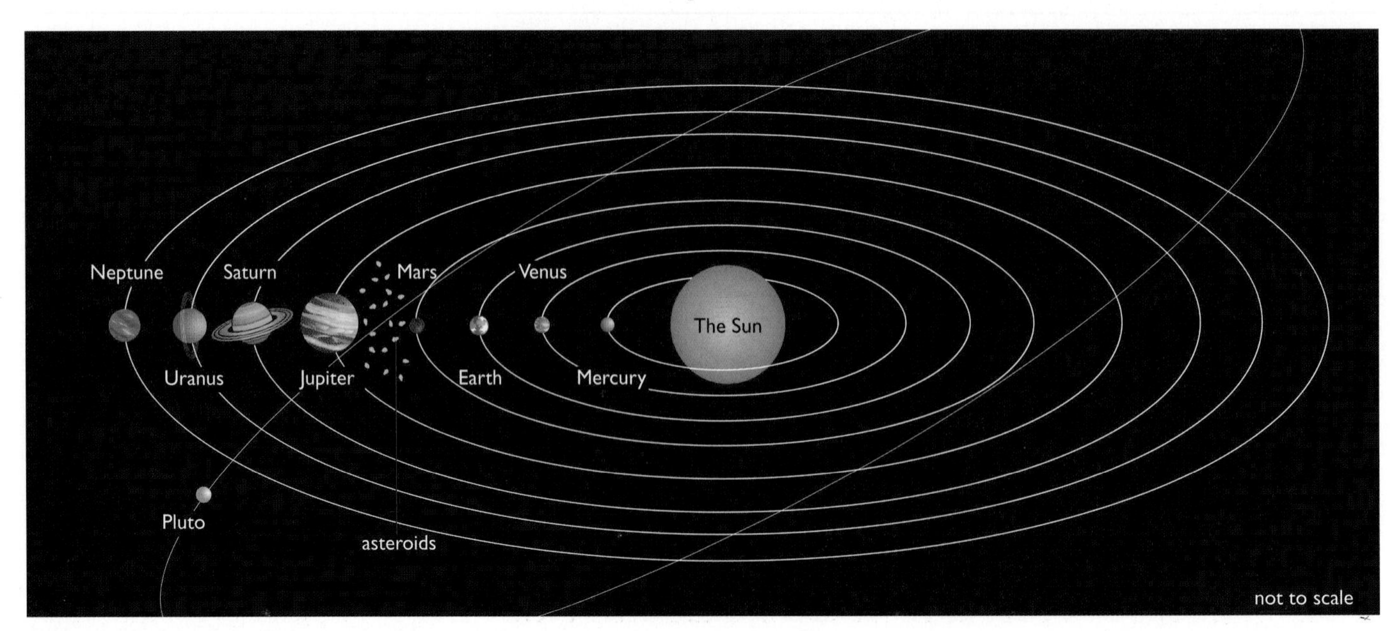

Figure 23.4 *Our Solar System.*

The Earth is one of nine planets that orbit the Sun. The orbits of the planets are **elliptical** (like squashed circles) with the Sun close to the centre.

The table gives some information about the planets, compared to Earth.

Planet	Average distance from Sun compared with the Earth	Time for one orbit of the Sun in Earth years	Diameter compared with the Earth	Average surface temperature (°C)	Surface gravity compared with the Earth
Mercury	0.4	0.2	0.4	+350	0.4
Venus	0.7	0.6	0.9	+470	0.9
Earth	1.0	1.0	1.0	+15	1.0
Mars	1.5	1.9	0.5	–30	0.4
Jupiter	5.0	12	11	–150	2.4
Saturn	9.5	30	9	–180	1.0
Uranus	19	84	4	–210	0.8
Neptune	30	165	4	–220	1.2
Pluto	39	248	0.2	–230	0.4

Planets are **non-luminous** objects – they do not emit light. We see them because they *reflect* light from the Sun. The five planets closest to the Sun (not including the Earth) are all visible to us with the naked eye. The last three planets – Uranus, Neptune and Pluto – are so far away that they were only discovered after the invention of the telescope.

You do not need to remember the information in this table for your exam but you should be able to spot trends and suggest reasons for them.

Figure 23.5 *We can see Saturn because it reflects light from the Sun, but does not emit light itself.*

As the figures in the table show, the planets vary enormously in size and distance from the Sun. There are, however, some trends we can identify from the table.

1 The time it takes for a planet to orbit the Sun increases as its distance from the Sun increases.

2 The further a planet is from the Sun, the colder the average temperature of its surface.

Although Mercury is the closest planet to the Sun, it does not have the highest average surface temperature. Venus has the highest average temperature. This is because Venus has an atmosphere that "traps" the heat from the Sun. Mercury has no atmosphere to trap the heat and so is not as hot.

Planets are held in orbit by the **gravitational pull** of the Sun. The gravitational pull on Mercury is quite large compared with that exerted on Pluto. This is the reason Mercury follows a much more tightly curved path than Pluto.

Figure 23.6 *These photographs, taken over several days, show how the planet Mars moves across the background of stars, the Paleiades. The position in the sky in which we see a planet depends upon where it and the Earth are in their orbits.*

Planets and stars can look very similar in the night sky. But if you watch them over a period of several nights the planets will change their positions against the background of the distant stars. The word "planet" comes from the ancient Greek word meaning *wanderer*.

Moons

Moons are natural objects that orbit a planet. Their motions, like those of the planets, are determined by gravitational forces. Moons are non-luminous objects. We see them because they reflect light from the Sun.

The Earth has just one moon. It is approximately 400 000 km from the Earth and has a mass and surface gravity just one sixth that of the Earth. The Moon has no atmosphere and has a surface that is covered with craters caused by the impact of meteorites.

It takes the Moon 28 days (1 **lunar month**) to orbit the Earth. The Moon, like the Earth, spins on its axis, but much more slowly than the Earth turns. It completes one full rotation every 28 days. Because the time it takes to complete one orbit around the Earth is the same as the time for one rotation, the Moon always keeps the same part of its surface facing the Earth.

Figure 23.7 *Gibbous moon. There are many craters on the surface of the Moon.*

Figure 23.8 *Cresent moon.*

The shape of the Moon looks different to us at different times of the lunar month depending upon how much of the illuminated side we can see. If we can see all of the illuminated surface, it is called a full moon. The photograph in Figure 23.8 was taken when only part of the illuminated surface could be seen. This is a crescent moon. These different shapes are called the **phases of the Moon**.

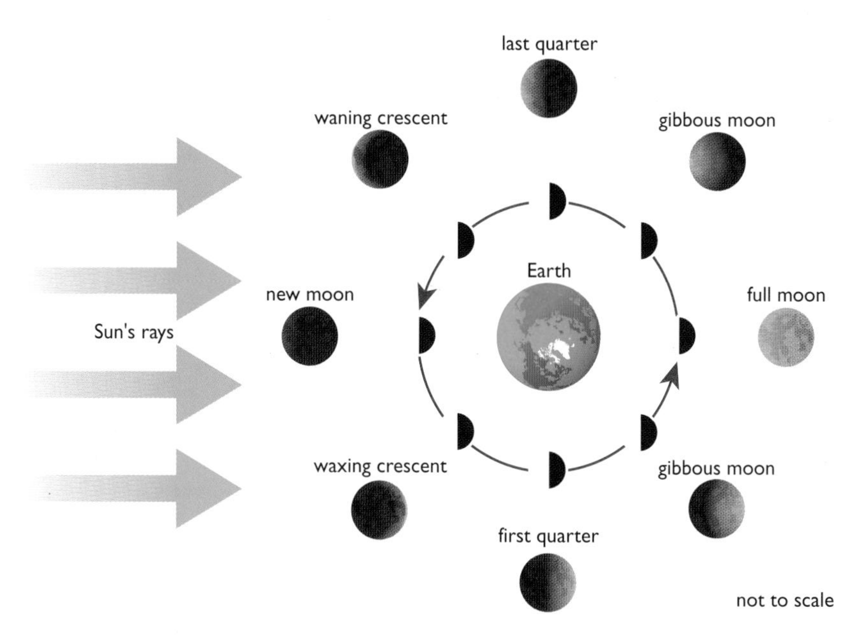

Figure 23.9 *The phases of the Moon.*

Some planets have no moons and some have more than one. For instance, Mars has two moons, Jupiter has 16 moons and Uranus has 20 moons. Some moons are very small and not easily seen from the Earth. It is possible that in future years we will discover new moons orbiting some of the outermost planets in our Solar System.

Comets

Comets orbit the Sun. They are approximately 1–30 km in diameter and made of dust and ice. Their orbits are very elongated. At times they are very close to the Sun, while at other times they are found at the outer reaches of the Solar System.

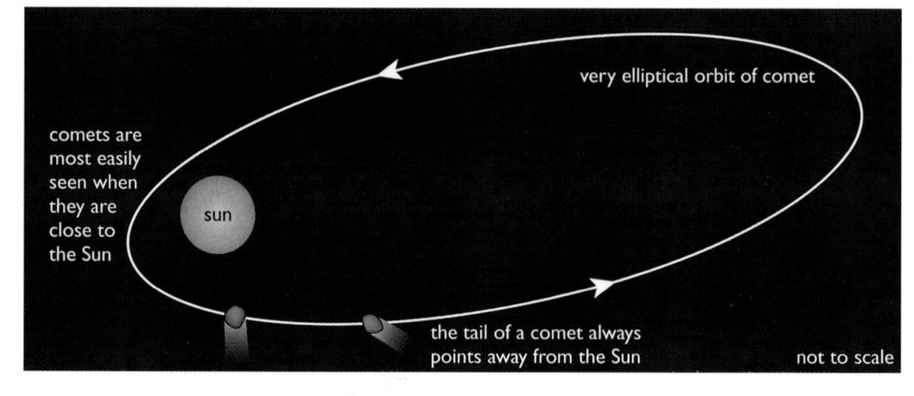

Figure 23.10 *As a comet gets closer to the Sun, the gravitational forces acting upon it increase and it speeds up. At the opposite end of its orbit, a long way from the Sun, the gravitational forces are smaller, so the comet travels at its slowest speed.*

Figure 23.11 *Halleys comet could be last viewed from Earth in 1986.*

Close to the Sun, some of a comet's frozen gases evaporate, forming a long tail that shines in the sunlight. These tails can be millions of kilometres in length. Perhaps the most famous of the comets is Halley's Comet, which visits our part of the Solar System every 76 years. It was last visible from the Earth in 1986.

Asteroids

Asteroids are minor planets or rocks that orbit the Sun. There is a belt of asteroids between the orbits of Mars and Jupiter. They vary greatly in size from just a few metres to several hundreds of kilometres across. The first asteroid to be discovered was called Ceres. It has a diameter of 933 km. It is believed that asteroids were formed at the same time as the rest of the Solar System and possibly are the rocky remains of a planet that broke apart or failed to form.

ideas evidence

Gravitational forces

The movements of all astronomical bodies – for example, planets, comets and asteroids – are determined by gravitational forces. It was careful observation and measurement of the movements of planets that lead to developments in the theory of gravity.

ideas evidence

Astronomical models

Initially, it was thought that the Earth was at the centre of the Universe and other astronomical bodies moved around it. This was known as the Ptolmaic System, after the Egyptian astronomer Ptolemy who put forward this idea in approximately 150 BC. Later, however, careful observation of the movements of the stars and planets lead to this model being abandoned and new models with the Sun at the centre of the Solar System (heliocentric systems) were suggested.

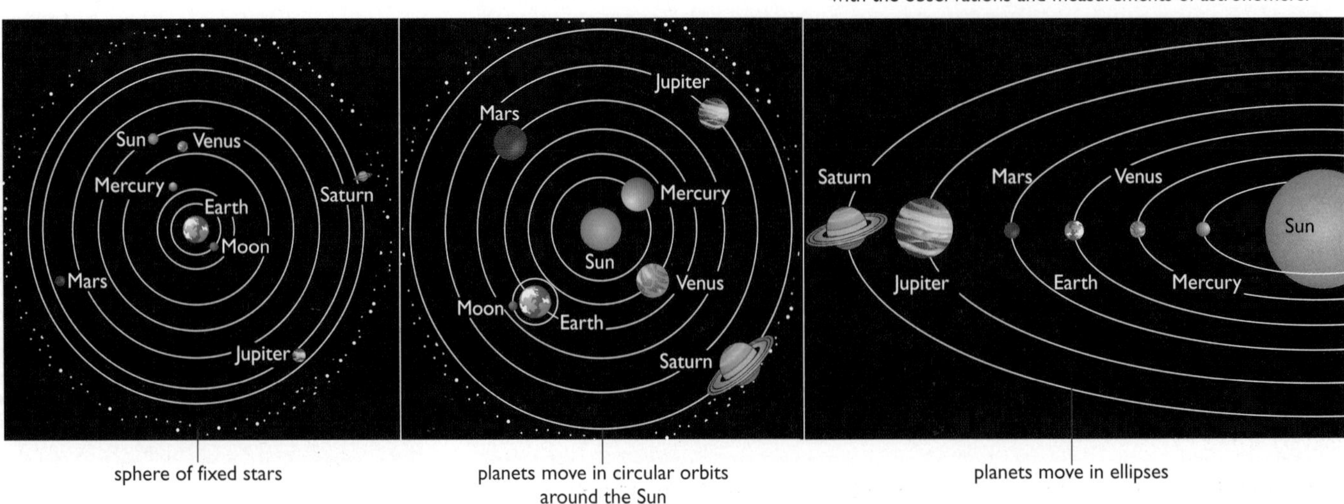

Figure 23.12 *The development of astronomical models.*

In 1687, a scientist called Isaac Newton put forward a theory of gravity. This theory not only explained the movements of the planets in the Solar System but also eventually provided an explanation of how the stars and the Solar System were formed.

Newton's Laws of Gravity

Newton suggested that between any two objects there is a force of attraction. This attraction is due to the masses of the objects. He called this force **gravitational force**. The size of this force depends upon:

1 the masses of the two objects

2 the distance between the masses.

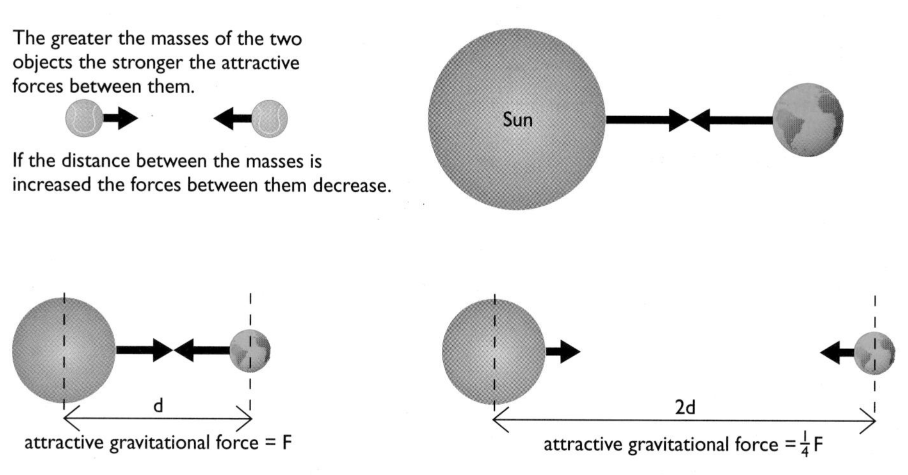

Figure 23.13 *Gravitational forces obey an inverse law – that is, if the distance between the masses is doubled, the forces between them are quartered; if the distance between them is trebled, the forces become one ninth of what they were.*

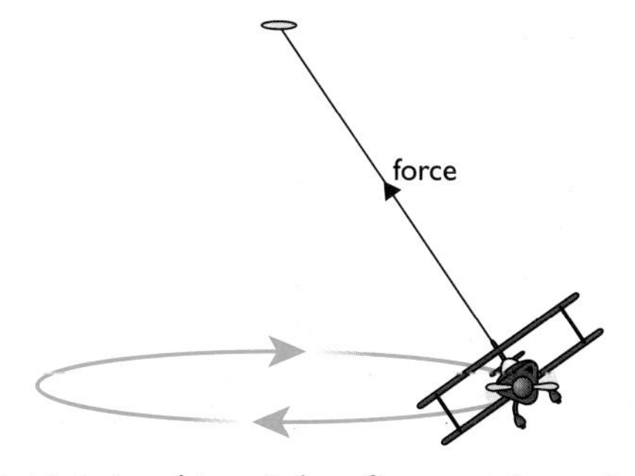

Figure 23.14 *If an object is to travel in a circle, a force must be applied to it.*

The gravitational attraction between two objects with small masses is extremely small. Only when one or both of the objects has a very large mass – for example, a moon or a planet – is the force of attraction noticeable.

Moving in a circle

The model aircraft in Figure 23.14 flies in a circle because a force is being applied to it through the wire (see page 135). If the wire breaks the aircraft will fly away in a straight line.

Planets and moons travel in orbits around the Sun. They too, therefore, must have a force applied to them. This force is the gravitational attraction between two masses. Our Sun contains over 99% of the mass of the Solar

System. It is the gravitational attraction between this mass and each of the planets that holds the Solar System together and causes the planets to follow their curved paths.

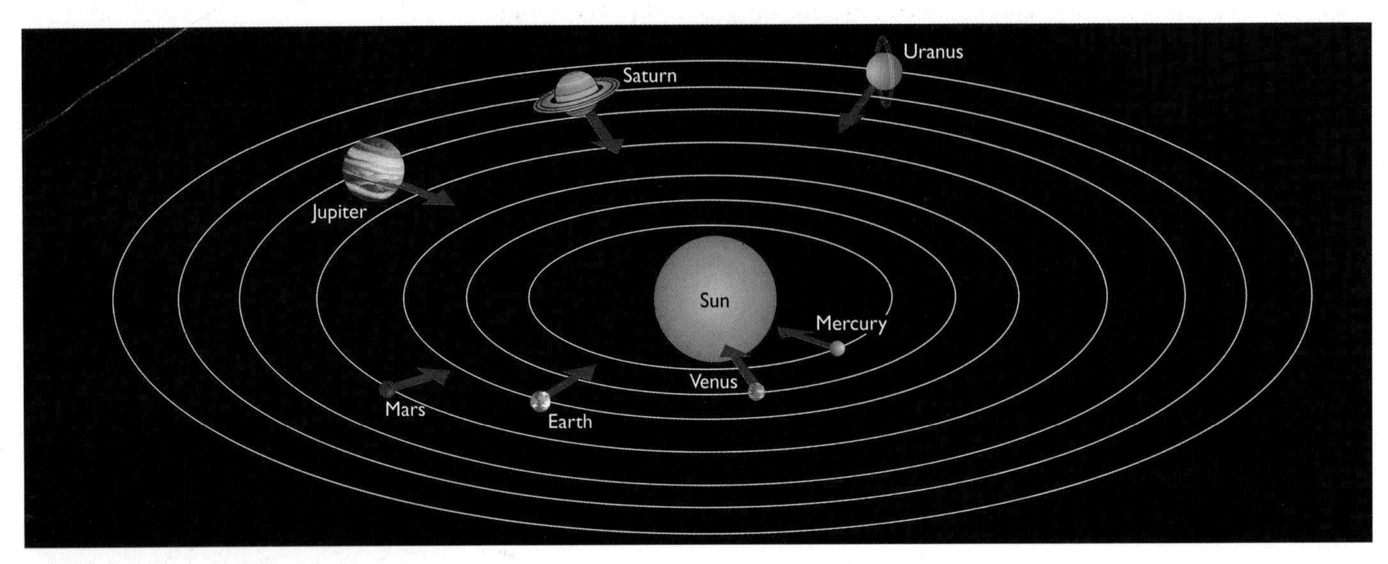

Figure 23.15 *Gravitational forces make the planets follow curved paths.*

Figure 23.16 *This is an artist's impression of NASA's Quikscat satellite. The dish at the bottom measures winds over the ocean's surface.*

Astronauts in orbit around the Earth feel weightless. This is not because there are no forces acting upon them. If this were the case, they would be moving in a straight line. Gravitational forces do act upon astronauts. They cause them to fall, but the space vehicle they are travelling in is also falling at the same rate. As a result, the astronauts appear to be weightless.

Satellites

Satellites are objects that orbit a planet. They are held in orbit by gravitational forces. Moons are examples of **natural satellites**. Some objects that orbit the planets are manufactured objects. They are **artificial satellites**.

Since 1957, when Sputnik 1 was launched, thousands of artificial satellites have been put into orbit around the Earth. They relay telephone and TV signals, they monitor the Earth's surface and they observe the Universe beyond the Earth. The orbit given to a satellite is chosen to match its purpose.

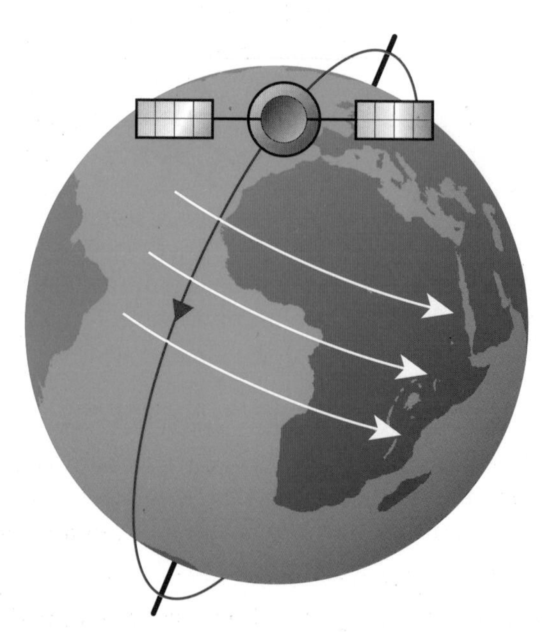

Figure 23.17 *A satellite in low polar orbit can be used to monitor the whole of the Earth's surface.*

Satellites that are used for monitoring the whole of the surface of the Earth are usually put into fast low orbits over the poles. The Earth spins beneath them and they can therefore scan the whole of the surface once every day. Typically, they are about 400 km or more above the surface of the Earth and complete approximately 14 orbits each day. These satellites have many uses including monitoring the Earth for nuclear tests and observing variations in ocean temperature and plankton concentration.

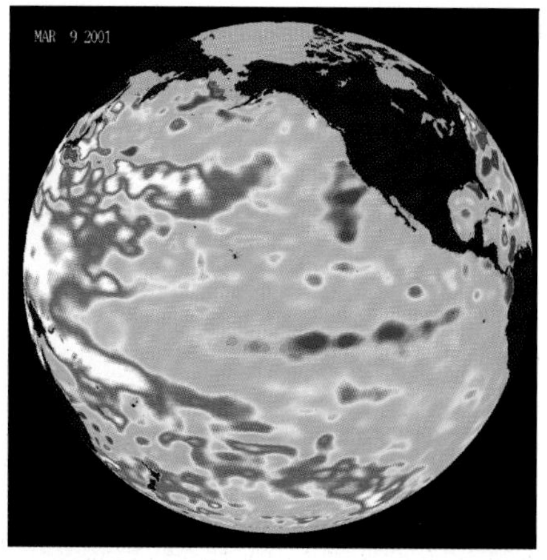

Figure 23.18 *An infra-red photograph taken from an orbiting satellite, showing the temperature of the Earths oceans.*

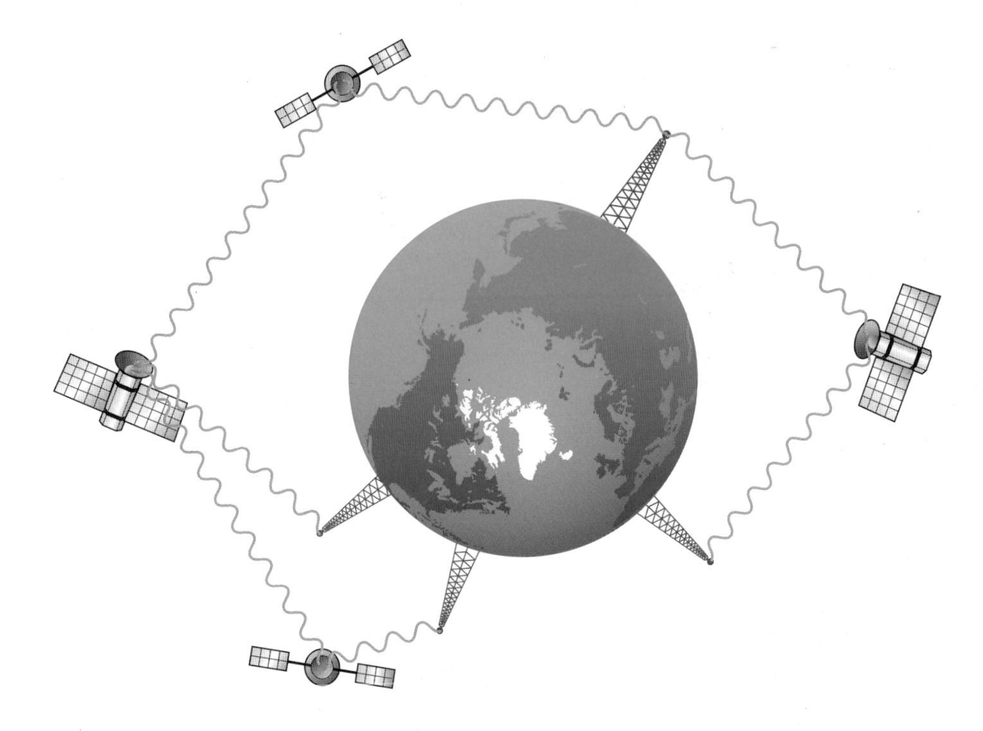

Figure 23.19 *Communication satellites allow transmitted signals to be received anywhere around the world.*

Satellites may carry passive or active sensors. Passive sensors will record radiation emitted or reflected from the Earth – for example, infra-red or visible light. The disadvantage of this type of monitoring is that it can be affected by clouds. Active sensors record reflected radio signals that have been emitted from the satellite. The frequencies of these emissions are chosen so that they can penetrate any cloud cover.

Communication satellites are usually placed in high orbits above the equator – that is, approximately 36 000 km above the Earth's surface. Their heights and speeds are usually adjusted so that the rate at which they orbit the Earth is the same as the rate at which the Earth rotates. As a result, a satellite stays above the same part of the Earth. The satellite is said to be in a **geostationary orbit** (see page 136). A string of such satellites allows communication using microwaves between all places on the Earth, as shown in Figure 23.19.

There is, however, a limit to the number of communication satellites that can be put into geostationary orbit. If more than 400 were put into such orbits, the signals from one satellite would interfere with the signals from another.

Figure 23.20 *Mount Haleakala Observatory, Hawaii.*

When astronomers use ground-based telescopes to observe the Universe, the images they obtain are not perfectly clear because they have to look through the Earth's atmosphere. This problem can be minimised by building observatories high on mountain tops. There is then less atmosphere to affect the light arriving from distant stars and galaxies.

An even better solution to this problem is to take telescopes above the atmosphere and install them on space stations or satellites orbiting high above the Earth. Figure 23.22 was taken by an orbiting telescope called the Hubble space telescope. The photographs from this telescope allow us to see features of distant space that we have never been able to see before.

Figure 23.21 *The Hubble space telescope being launched from the cargo bay of the Space Shuttle.*

Figure 23.22 *The Hubble space telescope captured this image of the giant galacatic nebula NGC 3603. It shows various stages in the life cycle of stars in this one view.*

Satellites are also used to investigate the planets in our Solar System and their moons. Once in orbit around the planet or moon to be studied, the satellites can observe these bodies from above and release landing craft that descend to the surface and then relay the information that they gather back to Earth.

Figure 23.23 *This Viking lander was released by a satellite put into orbit around Mars. The lander sent detailed information about the Martian atmosphere and soil back to Earth. Here it is being prepared for launch*

Orbital speeds

The speeds of satellites vary greatly depending on the tasks they are performing. For example, communication satellites are put in high orbits and travel at approximately 3 km/s, while those monitoring the whole surface of the Earth are put into low polar orbits with speeds of about 8 km/s.

We can calculate the speed of a satellite using the equation:

$$\text{speed} = \frac{\text{distance}}{\text{time}}$$

In orbital terms, the distance travelled is the circumference of the orbit:

$$\text{distance} = 2 \times \pi \times \text{radius of orbit}$$

The time taken is the time for one complete orbit (T). So:

$$\textbf{speed of satellite} = \frac{2\pi r}{T}$$

The gravitational attraction experienced by a satellite depends on the height of its orbit. The relationship between the gravitational force that keeps the satellite in this orbit and its speed is given by the equation:

force = mass × (orbital speed)²/r

$$F = \frac{mv^2}{r}$$

Remember to be consistent with your units. Distances are likely to be given in kilometres and so must be converted into metres. In the example, you must multiply both 8.2 km and 6600 km by 1000.

worked example

Example 1

a) Calculate the speed of a satellite that is orbiting 200 km above the Earth's surface and completes one orbit in 1 h 24 min. The radius of the Earth is 6400 km.

Using speed $= 2\pi \times \frac{r}{T}$

$$\text{speed} = 2\pi \times \frac{6600}{1.4} \times 60 \times 60$$

$= 8.2$ km/s

b) Calculate the gravitational force exerted on the satellite. The mass of the satellite is 1000 kg.

Using $F = \frac{mv^2}{r}$

$$F = 1000 \times \frac{(8.2 \times 1000)^2}{6600} \times 1000$$

$= 10\,000$ N

End of Chapter Checklist

If you haven't got a copy of your specification, read the introduction on page iv.

You will need to be able to do some or all of the following. Check your Awarding Body's specification (syllabus) to find out exactly what you need to know.

- Explain day and night in terms of the rotation of the Earth.
- Explain the length of a year and the seasonal changes in terms of the movement of the Earth around the Sun.
- Know that the Earth is one of several planets in the Solar System.
- Recall how the planets move relative to one another, and be able to interpret physical data about them particularly with regard to their masses and their orbital distances from the Sun.
- Understand the differences in the orbits of planets, moons and comets.
- Understand that the motions of all bodies are affected by the gravitational forces between them.
- Understand how the masses of, and distance between, two bodies affect the gravitational attraction between them.
- Explain the different uses of artificial satellites and how these different uses affect the orbits into which they are placed.

Questions

More questions on the Earth and Solar System can be found at the end of Section D on page 225.

1 Using the table on page 215, answer the following questions.

a) Name two planets closer to the Sun than the Earth.

b) Name two planets that have a higher surface gravity than the Earth.

c) Name one planet that is approximately the same size as the Earth.

d) Name the planet that has the highest average surface temperature.

e) One "Mars year" is 1.9 Earth years. How long is a year on Pluto?

2 ***a)*** Draw a diagram to show how the tilting of the Earth's axis gives rise to the seasons.

b) Write down three seasonal changes that are experienced in the UK.

c) Explain the difference between a year on Earth and a year on Mars.

3 ***a)*** Name four astronomical bodies that are kept in orbits by gravitational forces.

b) What two factors determine the size of the gravitational force between two objects?

c) What major change did Kepler make to the model of the Solar System proposed by Copernicus?

d) Why did Kepler suggest these changes?

4 ***a)*** What is an asteroid?

b) What is a comet?

c) Describe two differences between an asteroid and a comet.

d) In which direction does the tail of a comet point?

e) Describe how the speed of a comet changes as it orbits the Sun.

f) Name two non-luminous objects we might see in our Solar System.

g) Find out what a meteorite is.

5 ***a)*** What is a natural satellite?

b) What is an artificial satellite?

c) Give three uses for an artificial satellite.

d) What is a geostationary orbit?

e) What forces keep a satellite in orbit?

f) Suggest one reason why satellites put into low orbits may gradually lose height and occasionally need to use their rocket motors to take them back up. (**Hint**: You may need to refer back to Chapter 14.)

End of Section Questions

1 The diagram below shows the Earth. Part of its surface is receiving sunlight, while part of it is in shadow.

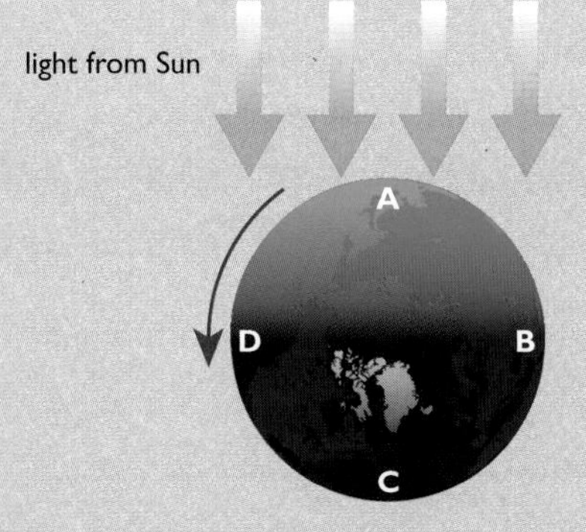

a) Identify one place on the Earth (A, B, C or D) where:

i) it is dawn *(1 mark)*

ii) it is the middle of the night *(1 mark)*

iii) it is early evening *(1 mark)*

iv) it is midday. *(1 mark)*

b) Explain the meaning of:

i) one day *(1 mark)*

ii) one year *(1 mark)*

iii) summer *(1 mark)*

iv) winter. *(1 mark)*

Total 8 marks

2 The table below contains information about the planets in our Solar System.

Planet	Surface gravity compared with the Earth	Distance from Sun compared with the Earth	Period in Earth years
Mercury	0.4	0.4	0.2
Venus	0.9	0.7	0.6
Earth	1.0	1.0	1.0
Mars	0.4	1.5	1.9
Jupiter	2.6	5.0	12
Saturn	1.1	9.5	30
Uranus	0.9	19	84
Neptune	1.2	30	165
Pluto	0.2	39	248

a) Name three planets that have a stronger gravitational pull on their surface than there is on Earth. *(1 mark)*

b) How long is a year on Saturn? *(1 mark)*

c) If the distance from the Earth to the Sun is 150 million kilometres, calculate the distance of Saturn from the Sun. *(2 marks)*

d) Assuming that the orbital path for Saturn is circular, calculate its orbital speed in km/year. (**Hint**: circumference of a circle = $2\pi r$.) *(3 marks)*

e) Plot a graph of period (the time for one orbit) against distance from the Sun (compared with the Earth) for all the planets in the Solar System. *(5 marks)*

f) What trends or conclusions can you draw from your graph? *(2 marks)*

Total 14 marks

3 *a)* Draw a diagram to show the shape of the orbit of a comet around the Sun. *(2 marks)*

b) Describe and explain how the speed of a comet changes as it travels around its orbit. *(4 marks)*

c) Many comets have tails. What are these tails made of and in which direction do they point? *(2 marks)*

Total 8 marks

4 Stars are not unchanging features of the night sky. They form, they follow a life cycle and then they die.

a) Stars are made of matter that is mostly dust and gases. What pulls this matter together to begin the formation of a star? *(1 mark)*

b) What kind of reactions begin when this matter is subject to high temperature and very large pressures? *(1 mark)*

c) Which element takes part in these reactions when a star is first formed? *(1 mark)*

d) In which stage of its life cycle is our Sun? *(1 mark)*

e) Explain why stars eventually expand and become red giants or red supergiants. *(2 marks)*

f) Explain the different futures of red giants and red supergiants. *(4 marks)*

Total 10 marks

5 The diagram below shows two satellites of the same mass in orbit around the Earth.

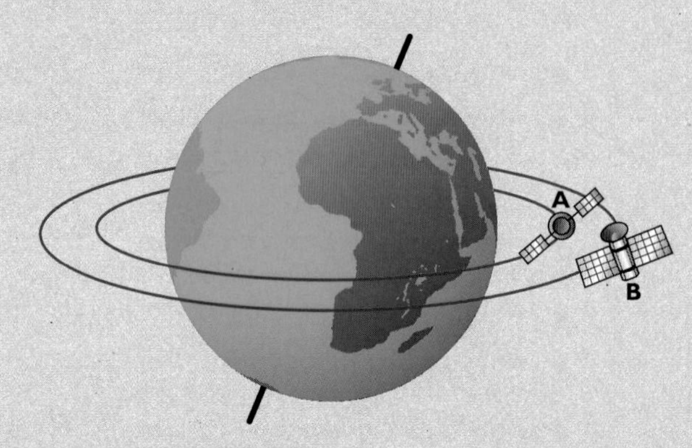

a) Which satellite is experiencing the strongest gravitational force? Explain your answer. *(2 marks)*

b) Which satellite has the highest orbital speed? *(1 mark)*

c) What is a geostationary orbit? *(2 marks)*

d) Give one use for a satellite that is put into a geostationary orbit. *(1 mark)*

e) Give one use for a satellite that is put into a low polar orbit. *(1 mark)*

f) Explain the advantages of putting a telescope into an orbiting satellite. *(2 marks)*

Total 9 marks

6 Choose two planets in our Solar System, one closer to the Sun than the Earth, the other further from the Sun than the Earth. Describe in as much detail as you can why life is unlikely to exist on these planets.

Total 6 marks

7

visible spectrum
red | yellow | green | blue | indigo | violet

radio | microwave | infra-red | UV | X-rays | gamma-rays

a) What is red shift? *(1 mark)*

b) If scientists discovered in the future that the light from stars in distant galaxies showed much less red shift than it does today, what conclusions could they draw from this information? *(2 marks)*

c) What conclusion could they draw if they discovered some galaxies emitting light with "violet shift"? *(2 marks)*

d) What would be the likely final fate of the Universe if lots of galaxies with violet shift were discovered? *(1 mark)*

Total 6 marks

Section E: Energy Resources and Energy Transfer

Chapter 24: The Need for Energy

Whenever anything happens, energy is transferred from one form to another – indeed, without energy things simply can't happen! In this chapter, you will learn that energy can be transferred into many different forms, including sound, light, movement, heat and potential energy. You will also find out that, although energy is never destroyed, in every energy conversion some energy is "lost" to the surroundings, often as heat.

Figure 24.1 *Energy comes in different forms – as sound, movement, light and heat, for example.*

For things to happen we need energy! Energy is used to transport people and goods from place to place, whether it is by train, boat or plane or on the backs of animals or people. Energy is needed to lift objects, make machinery work and run all the electrical and electronic equipment we take for granted in our modern world. Energy is needed to make heat and light. The demand for energy increases every day because the world's population is increasing. People consume energy in the form of food and need energy for the basics of life, like warmth and light. As people become wealthier they demand much more than the basics, so the need for energy grows!

Sound energy can be used to break up small stones that can form inside a person's body, without the need for an operation. Sound energy is also used in medicine to examine the inside of the body, as an alternative to X-rays (see page 187).

Types of energy

Energy comes in many different forms. We get our energy from the food we eat. Food provides stored **chemical energy** that we can burn to turn it into other types of energy. We use the energy from food to generate **heat energy** (thermal energy) to help to keep us warm. Our muscles convert the chemical energy into **movement energy** (kinetic energy). Some of this movement energy is used as we speak producing energy in the form of **sound**.

We need heat for our homes, schools and workplaces. We also need to convert energy to **light** for our buildings, vehicles and roads. Most of the energy needed for these purposes is converted from **electrical energy**. We shall see later (see page 229) that electrical energy can be made from other forms of energy, like chemical or nuclear energy. Some electrical energy is produced from the **gravitational potential energy** stored in water kept in reservoirs in mountainous areas. We can also use the energy stored in the hot core of the Earth. We see the evidence of this huge supply of heat in volcanoes and thermal springs. Energy from the heat underground is called **geothermal energy**.

Energy can also be stored in springs as **elastic potential energy**. This type of stored energy is used in things like clocks, toys and catapults.

The main source of energy for the Earth is our Sun. This provides us with heat, light and other forms of energy.

Energy transfer and conversion

For energy to be useful, we need to be able to transfer it from place to place and to be able to convert it into whatever form we require. Unfortunately when we try to do either of these things there is usually some energy converted to unwanted forms. We often refer to these unwanted forms as "wasted" energy because it is not being used for a useful purpose.

Here are some examples of wasted energy.

- An electric heater may be used to heat water in a house. The hot water will be stored in a tank for use when required. Although the tank may be well insulated, some energy will escape from the water and some will be used to heat up the copper that the tank is made from. Both processes mean that some of the energy that is converted to heat from electricity is lost because it is not doing what we want it to do – it is not making the water hot.
- When we fill a car with petrol, the main purpose is to convert the chemical energy stored in the fuel into movement energy, with a small amount doing other things like providing electrical energy for the lights or radio. But the process of converting the energy is not perfectly **efficient**, as not *all* of the energy is used to do what we want. (A formal definition of efficiency is given on page 241.) A considerable amount of the energy supplied by the fuel is converted to heat, most of which is lost or wasted by heating up the surroundings. Some of the energy is converted to sound, which can be unpleasant for people both inside and outside the car. A lot of energy is used to overcome the various friction forces that oppose movement of parts within the car and the car's movement along the road. Friction causes energy conversion to occur, too, usually producing unwanted heat energy.

Unwanted energy conversions during energy transfer reduce efficiency. This problem is the same whether the system is a small one like a car, or a large system like the nation-wide electricity generation and distribution industry. We need to be aware of where our precious energy is going if we are to find ways of using it to best effect.

Energy conversions

We have many ways of converting energy from one form to another.

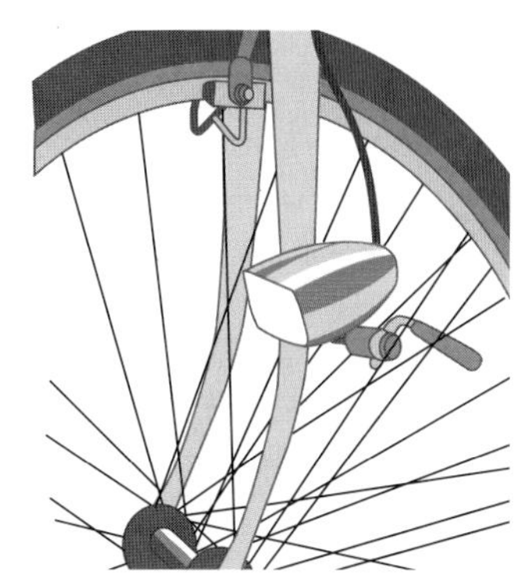

Figure 24.2 *Energy is converted from one form to another, and to another, and so on.*

In Figure 24.2, stored chemical energy in the food is converted into a range of other chemicals, some of which – like carbohydrates, for example – are used to produce heat to maintain our body temperature, and energy for movement through muscle activity. Having eaten a meal, the cyclist in Figure 24.2 is converting the chemical energy stored in his body into movement energy. The movement is initially in the cyclist's legs and is then transferred to the machine (the bicycle). The cyclist is also producing additional heat energy, which is lost to the surroundings. Friction in various parts of the bicycle will also result in energy being converted from movement into heat and sound.

The dynamo or generator fitted to the bicycle's wheel converts some of the movement energy of the wheel into electrical energy. The lamp then converts the electrical energy into light and heat. As the electric current flows in the dynamo–lamp circuit, heat will be produced in the conducting wires, too.

Examples of other energy conversions are given in the questions at the end of the chapter, on page 232.

Conservation of energy

The Law of Conservation of Energy is a very important rule. It states that:

Energy is not created or destroyed in any process.

(It is just converted from one type to another.)

We often hear about the "energy crisis": as our demand for more energy increases our reserves of energy in the form of fuels like oil and gas are rapidly being used up. The Law of Conservation of Energy makes it seem as if there is no real problem – that energy can never run out. We need to understand what the law really means.

Physicists believe that the amount of energy in the Universe is constant – energy can be changed from one form to another but there is never any more or any less of it. This means we cannot use energy up. However, if we

Even though the amount of energy in the Universe is constant it is becoming more spread out and so less usable. Some scientists think that all the energy in the Universe is ultimately going to be converted into heat and that everywhere in the Universe will end up at the same, very low temperature. This "end of the world" scenario is sometimes referred to by the dramatic name of "heat death". If the Universe does end up this way, it is not expected to do so for some time yet, so carry on working for those exams!

consider our little piece of the Universe, the problem becomes more obvious. As we make energy do useful things – for example, move a car – some of it will be converted to heat. Some of this heat energy will be radiated away from the Earth and "lost" (not destroyed or used up). "Lost" in this sense means not available for us to use any more. This is just like a badly insulated house; if heat energy escapes, it is lost from the system we call our home, and is no longer available to keep us warm.

Energy flow diagrams (Sankey diagrams)

We use different ways to show how energy is transferred and converted. Energy transfer diagrams show the energy input, the energy conversion process and the energy output. The system may be a very simple one with just one main energy conversion process taking place. An example of a simple system with its energy transfer diagram is shown in Figure 24.3a.

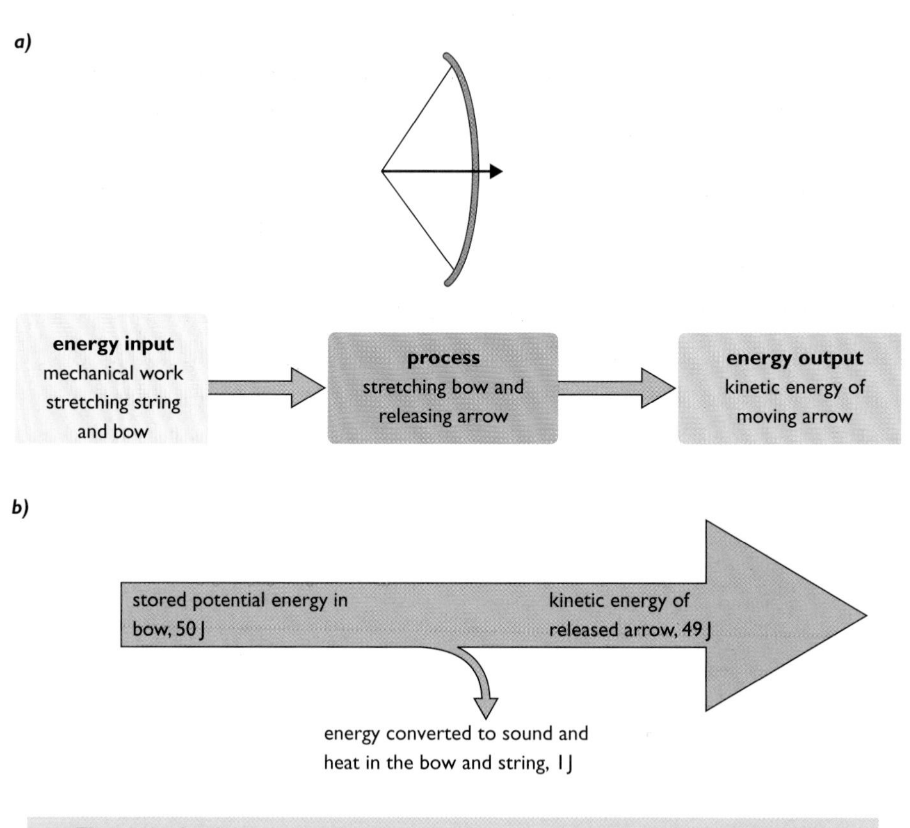

The width of each arrow must be proportional to the amount of energy represented: if you draw the main arrow 50 mm wide, then the kinetic energy end should be 49 mm wide and the arrow showing the energy converted to heat and sound is just 1 mm wide.

Figure 24.3 a) *Energy transfer diagram for an arrow being fired from a bow,* b) *Sankey diagram for an arrow being fired from a bow.*

Energy flow diagrams, sometimes called Sankey diagrams, are a simpler and clearer way of showing what becomes of an energy input to a system. The energy flow is shown by arrows *whose width is proportional to the amount of energy involved*. Broad arrows show large energy flows, narrow arrows show small energy flows. A Sankey diagram for a simple process is shown in Figure 24.3b.

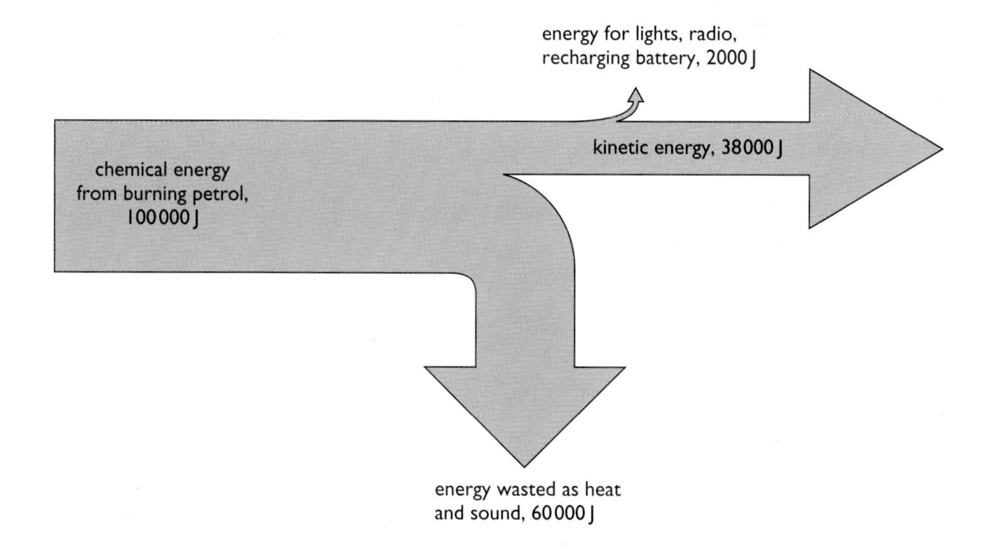

Figure 24.4 *Sankey diagram showing the energy flow in a typical car.*

Figure 24.4 shows a Sankey diagram for a more complex system – the energy flow for a car. Chemical energy in the form of petrol is the input to the car. The energy outputs from the car are:

- electrical energy (from the alternator) to drive lights, radio and so on, to charge the battery (conversion to chemical potential energy) and provide ignition
- movement (kinetic) energy from the car engine
- wasted energy as electrical heating in wiring and lamp filaments, as frictional heating in various parts of the engine and alternator, and as noise.

End of Chapter Checklist

If you haven't got a copy of your specification, read the introduction on page viii.

You will need to be able to do some or all of the following. Check your Awarding Body's specification (syllabus) to find out exactly what you need to know.

- Recognise that energy exists in the following forms: heat (thermal), light, electrical, sound, movement (kinetic), chemical, nuclear, and potential (stored) as both gravitational and elastic potential energy.
- Understand how energy may be converted from one form to another.
- Know that energy is conserved.
- Appreciate that energy conversions usually involve some energy being converted to unwanted forms or being transferred to places other than where it is needed.
- Draw energy flow diagrams (Sankey diagrams) for a variety of everyday and scientific situations.

Questions

More questions on the need for energy can be found at the end of Section E on page 291.

1 Describe the main energy conversions taking place in the following situations:

a) turning on a torch

b) lighting a candle

c) rubbing your hands to keep them warm

d) bouncing on a trampoline.

2 Copy and complete the following Sankey diagrams. Remember that the width of the arrows must be proportional to the amount of energy involved. This has been done for you in example ***a)***.

a) For an electric lamp:

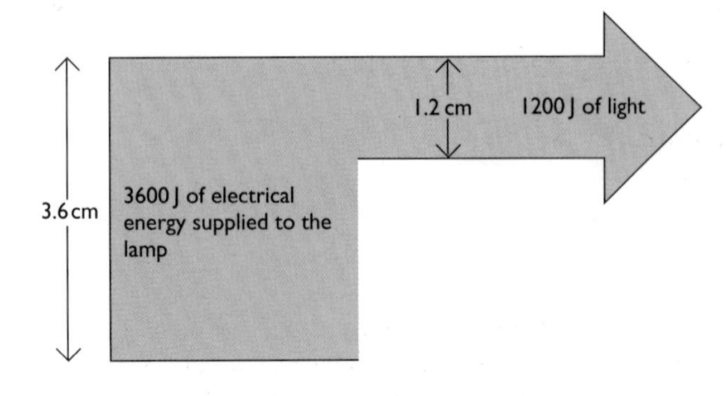

b) For a washing machine:

1.2 MJ of kinetic energy to rotate the drum

6 MJ of heat energy

0.8 MJ of energy wasted as heat and sound

3 Draw a Sankey diagram for the following situation. An electric kettle is used to heat some water. 350 kJ of energy are used to heat the water, 10 kJ raise the temperature of the kettle and 40 kJ escape to heat the surroundings.

4 Draw a Sankey diagram to show the energy flow in a 200 MW transformer. Its purpose is to transform electrical energy at high voltage (input) into electrical energy at low voltage (output). It wastes 15% of its energy input. A third of the energy that is wasted goes to heat the iron core of the transformer (label this as the "iron loss") and nearly all the remainder is lost as heat in the copper windings (label this as the "copper loss").

5 A ball is dropped. It hits the ground with 10 J of kinetic energy and rebounds with 4 J of kinetic energy.

a) What happens to 6 J of the energy during the bounce?

b) Draw a Sankey diagram for the energy flow that takes place during the bounce.

Section E: Energy Resources and Energy Transfer

Chapter 25: Work, Power and Efficiency

Work is calculated by multiplying the force applied by the distance through which the force moves – the bigger the force, or the longer the distance through which it moves, the more work is done. Work always involves an energy transfer. Power is the rate at which energy is transferred, and efficiency is a measure of how much of the input energy to a system is converted to *useful* output energy. In this chapter you will learn how to calculate the work done in a system, its power and its efficiency, as energy is transferred.

Figure 25.1 *James Joule (1818–1889) was the son of a wealthy Manchester brewer. He was tutored by James Dalton and carried out scientific research in his own laboratory, built in the basement of his father's home.*

The unit of energy is named after James Joule. It was Joule who realised that heat was a form of energy. He showed that kinetic energy could be converted into heat. At that time heat was measured in calories. In 1843, Joule worked out "the mechanical equivalent of heat", or, more simply, he found the "exchange rate" between calories and the unit then used for work and energy (the foot-pound weight).

Calories are still used when we talk about the energy content of food. If you look at dietary information on the side of a food product packet, like the one shown in Figure 25.2, you will notice that the energy content is given in kilojoules as well as the kilocalorie equivalent.

GUIDELINE DAILY AMOUNTS

EACH DAY	WOMEN	MEN
Calories	2000	2500
Fat	70g	95g

NUTRITIONAL INFORMATION

	Typical value per 100 g	30 g serving with 125 ml of semi-skimmed milk
ENERGY	1550 kJ 370 kcal	700 kJ* 170 kcal
PROTEIN	8g	7g

* Energy contribution of 125 ml of semi-skimmed milk 250 kJ (60 kcal)

Figure 25.2 *Dietary information panel from a packet of cornflakes.*

Energy and work

Energy is the ability to do work.

This statement tells us what energy *does* rather than what energy *is*. We know that energy comes in a wide variety of different forms but we are really interested in what energy can *do* – the answer is that energy does **work**.

We need to define work in a way that is measurable. Some types of work are not easy to quantify! Mechanical work, like lifting heavy objects, is easy to measure: if you lift a heavier object, you do more work; if you lift an object through a greater distance, again, you do more work. The definition of work in physics is:

work done, WD = force applied, F × distance through which it is applied, d
(in joules) (in newtons) (in metres)

$$WD = Fd$$

The unit of work and, therefore, energy is the **Joule**.

1 J of work is done when a force of 1 N is applied through a distance of 1 m in the direction of the force.

worked example

Example 1

height lifted
2 m
weight 500 N

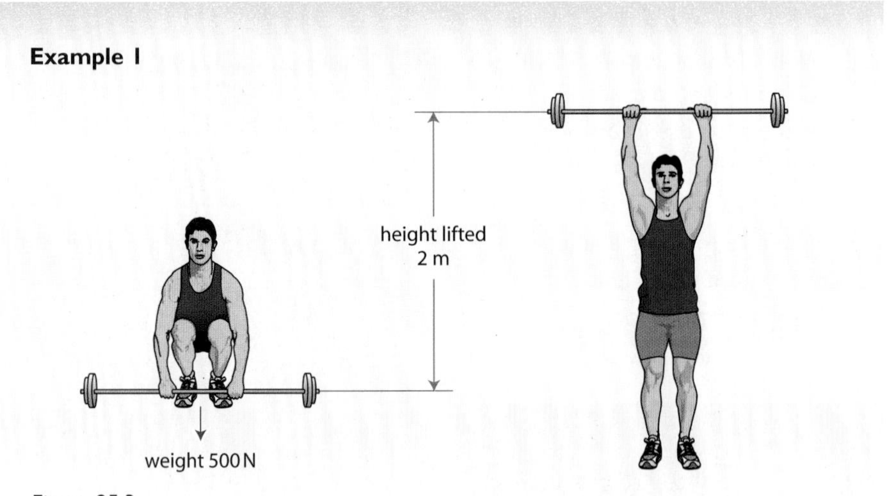

Figure 25.3.

Figure 25.3 shows a weight lifter raising an object that weighs 500 N through a distance of 2 m. To calculate the work done we use:

$$WD = Fd$$
$$= 500\,N \times 2\,m$$
$$= 1000\,J$$

This work done on the weight has increased its energy. This is explained opposite, under the section on gravitational potential energy.

worked example

Example 2

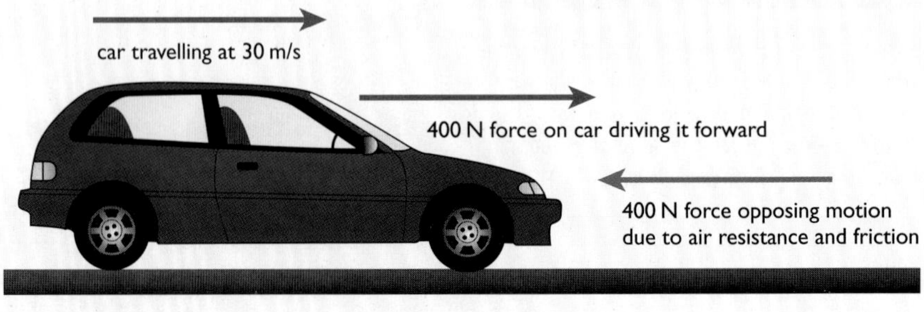

car travelling at 30 m/s
400 N force on car driving it forward
400 N force opposing motion
due to air resistance and friction

Figure 25.4.

In the example shown in Figure 25.4 the force acting on the car is not accelerating it – instead, it is being used to balance the forces opposing its motion. The resultant force on the car is zero, so it keeps moving in a straight line at constant speed. To work out the work done on the car in one second we substitute the force required, 400N, and the distance through which it acts in one second, 30m, in the equation:

WD = Fd

so WD = 400 N × 30 m

= 12 000 J or 12 kJ

Gravitational potential energy (GPE)

In the weight lifting example given opposite, the weight lifter has used some chemical energy to do the work. We know that energy is conserved so what has happened to the chemical energy that the weight lifter used? Some has been converted to heat in the weight lifter's body. The remainder has been transferred to the weight because he has increased its height in the gravitational field of the Earth. The energy that the weight has gained is called **gravitational potential energy** or GPE.

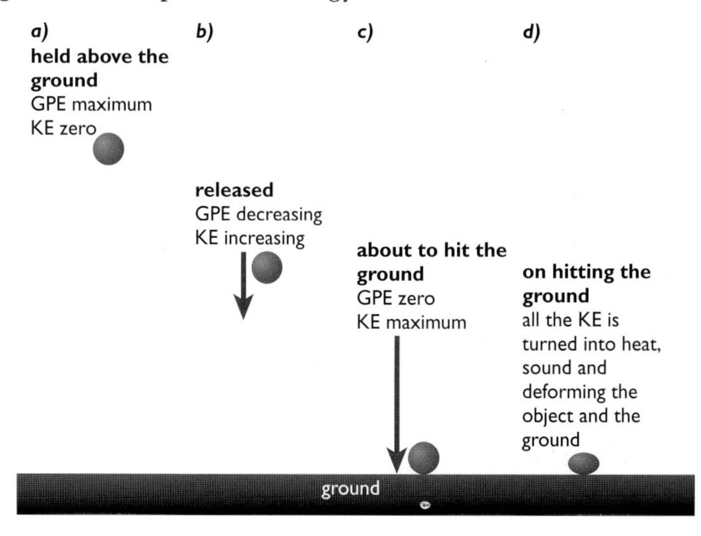

Figure 25.5 *When a raised object falls, its gravitational potential energy is converted first to kinetic energy and then to heat and sound.*

work done = Fd
= mg × h

so, lifting the object through a distance of **h** m involves doing **mgh** J of work

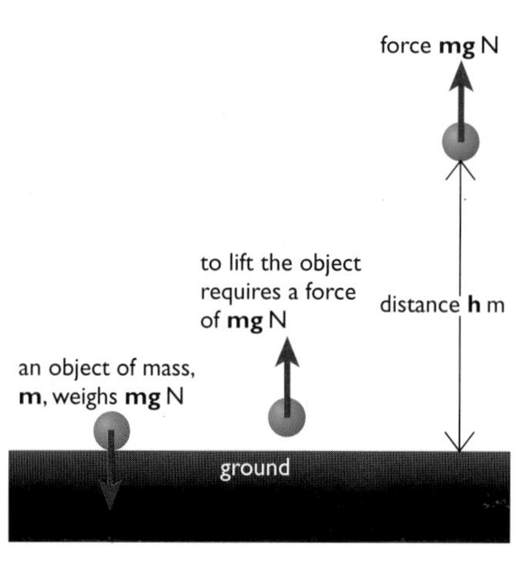

Figure 25.6 *The work done to lift an object is equal to the GPE the object has at its new height.*

In Figure 25.5, we can see the GPE possessed by the weight being converted into other forms as the weight falls. The weight accelerates because of the force of gravity acting on it, so it gains kinetic energy. When it reaches the ground all the initial GPE is converted to kinetic energy. When it hits the ground all the movement energy is then converted into other forms, mainly heat and sound.

The *change* in GPE of an object is defined as:

change in GPE (in joules) = mass of object, m (in kilograms) × gravitational field strength, g (in newtons/kilogram) × distance raised against gravitational force, h (in metres)

GPE = mgh

Reminder: Gravitational field strength is the force acting per kilogram on a mass affected by gravity. The gravitational field strength, g, on the surface of the Earth is approximately 10N/kg. Since the weight of an object is mg, increase in GPE is a special version of the equation WD = Fd, with F = mg and d = h.

This change in the GPE of an object will be an *increase* if we apply a force on it in the *opposite direction* to the pull of gravity – that is, if we lift it off the ground. For the sake of simplicity, we usually assume that an object has no GPE before we do work on it.

Kinetic energy (KE) is the energy possessed by moving objects.

Meteorites and kinetic energy

The amount of energy possessed by a moving object depends on its speed and its mass. As the Earth travels through space, orbiting the Sun, it runs the risk of colliding with chunks of matter that are drawn into the gravitational field of the Solar System. In fact, this is a common occurrence. If you have ever seen a shooting star – or, to give it its proper name, a **meteor** – you have seen the streak of light produced as a small piece of space debris burns up on entering our atmosphere. This is an example of kinetic energy being converted to heat and light by the friction produced between the air and the object passing through it.

The meteorite that caused the Arizona crater is thought to have hit the Earth travelling at 11 000 m/s and to have had a mass of 10^9 kilograms. It hit the ground with an energy equivalent to a 15 megaton hydrogen bomb, 1000 times greater than the atomic bomb dropped on Hiroshima at the end of the Second World War.

Figure 25.7 *Meteors burn up on entering our atmosphere – we see this as shooting stars.*

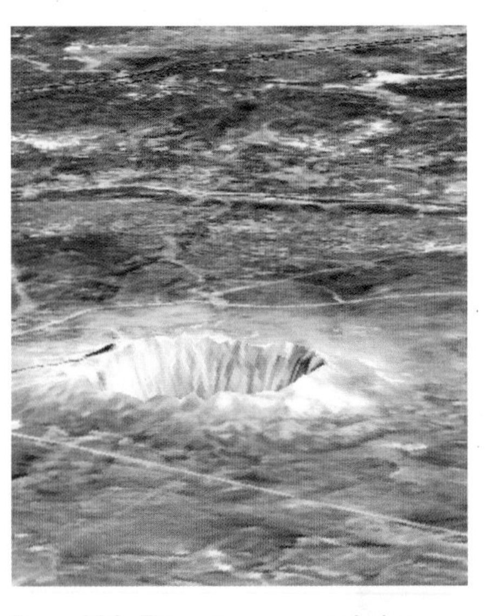

Figure 25.8 *This crater was created when a meteorite collided with Earth in Arizona.*

Small pieces of space debris do not present a threat to life on this planet. Friction against the atmosphere converts their kinetic energy into so much heat that they are vaporised into minute, harmless particles. Bigger pieces of matter are not completely burned up before reaching the Earth's surface, so they hit the ground still carrying kinetic energy.

Meteors are called **meteorites** when they hit the ground.

Small meteorites cause little damage and most impacts are in uninhabited regions or in the oceans. Large meteorites, however, can carry enormous amounts of kinetic energy and the energy conversions on impact with the Earth are devastating.

The kinetic energy of a moving object is calculated using the formula:

$$\textbf{KE (in joules)} = \tfrac{1}{2}\,\textbf{mass of object, m (in kilograms)} \times \textbf{(speed of object, v (in metres/second))}^2$$

$$\mathbf{KE = \tfrac{1}{2}mv^2}$$

worked example

Example 3

Calculate the kinetic energy carried by a meteorite of mass 500 kg (less than that of an average sized car) hitting the Earth at a speed of 1000 m/s.

$$KE = \frac{1}{2}mv^2$$
$$= \frac{1}{2} \times 500 \text{ kg} \times (1000 \text{ m/s})^2$$
$$= 250\,000\,000 \text{ J (or 250 MJ)}$$

Calculations using work, GPE and KE

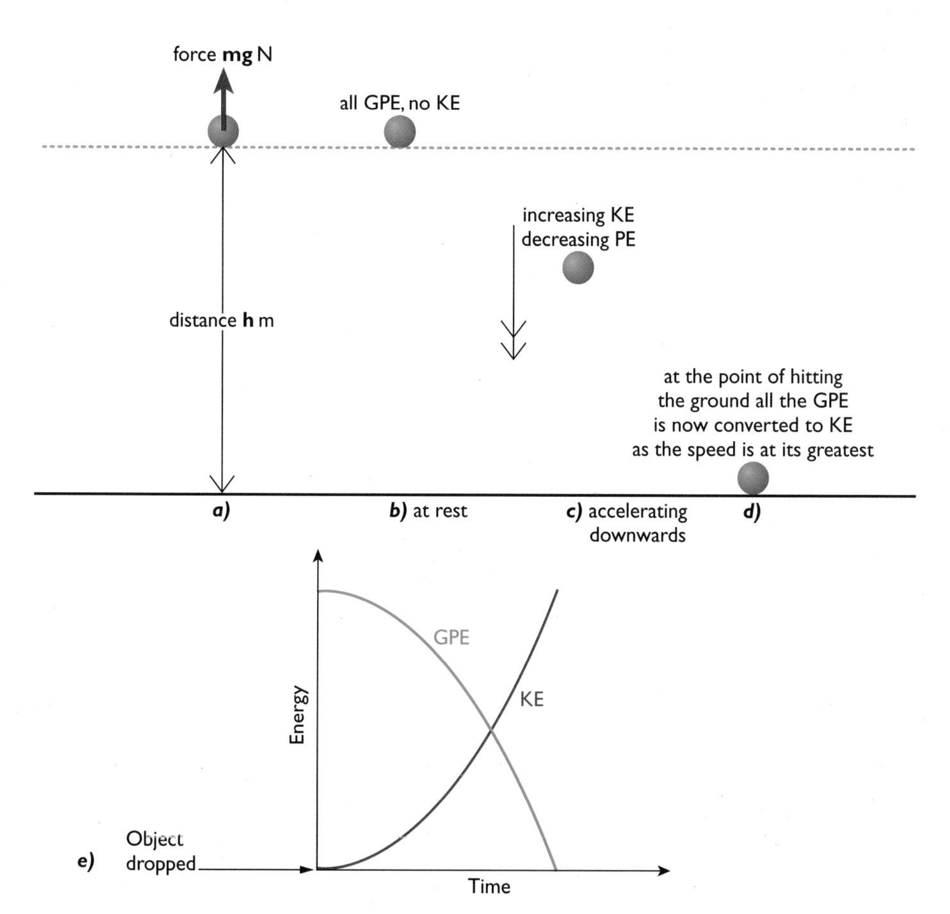

Figure 25.9 *GPE and KE of a falling object:* a) *doing work to lift an object,* b) *gain in GPE,* c) *GPE converting to KE during fall,* d) *all KE at end of fall,* e) *graph showing relationship between GPE and KE as the object falls.*

Work transfers energy to an object: WD = Fd. An object of mass, m, weighs m × g newtons so the force, F, needed to lift it is mg (Figure 25.9). If we raise the object through a distance h, the work done on the object is mg × h. This is also the gain in GPE.

When the object is released, it falls – it loses GPE, but gains speed and so gains KE. At the end of the fall, all the initial GPE of the stationary object has been converted into the KE of the moving object. The graphs in Figure 25.9 show how the GPE of the object is changing into KE as it falls. The sum of the two graphs is always the same, as the loss of GPE is equal to the gain in KE.

work done lifting object = gain in GPE = gain in KE of the object just before hitting the ground, when dropped

***worked* example**

This calculation has assumed that the effect of air resistance is very small. In practice, air resistance will reduce the height that an object will reach when thrown in the air. Neglecting air resistance, we can use $h = (v^2)/2g$ to work out the height reached.

Example 4

If you throw an object vertically upwards with an initial speed of, say, 20 m/s you can work out how high it will reach before falling back to Earth. At the highest point of its flight all the initial KE you have given it will be converted to GPE, according to the formula:

$$\frac{1}{2}mv^2 = mgh$$

(You can see that you don't need to know the mass of the object for this calculation. But the mass will affect how much energy you must use to throw it at a particular speed.)

Substituting v = 20 m/s and g = 10 N/kg:

$$\frac{1}{2} \times m\ \text{kg} \times (20\ \text{m/s})^2 = m\ \text{kg} \times 10\ \text{N/kg} \times h\ \text{m}$$

$$\frac{1}{2} \times m\ \text{kg} \times 400\ \text{m}^2/\text{s}^2 = m\ \text{kg} \times 10\ \text{N/kg} \times h\ \text{m}$$

$$m\ \text{kg} \times 200\ \text{m}^2/\text{s}^2 = m\ \text{kg} \times 10\ \text{N/kg} \times h\ \text{m}$$

$$h = \frac{m\ \text{kg} \times 200\ \text{m}^2/\text{s}^2}{m\ \text{kg} \times 10\text{N/kg}}$$

$$h = 20\ \text{m}$$

Elastic potential energy

Reminder: An elastic material is one that changes shape when a force is applied to it but returns to its original shape when the force is removed.

Mechanical energy can be stored in elastic materials. We call this **elastic potential energy** (EPE).

Springs store elastic potential energy. They come in a wide variety of shapes and sizes. Some are used to store energy to drive mechanical devices. The energy is stored by doing work on the spring, and is released when the spring is allowed to return to its original shape. Elastic and rubber bands are used in the same way to store energy to power toys. Springs are also used to absorb movement energy to smooth out the bumps when travelling in wheeled vehicles.

Figure 25.10 The bow stores EPE when the string is pulled back.

Power

Power is the measure of how fast energy is transferred or transformed.

James Watt is remembered as the inventor of the steam engine and is said to have been inspired by watching the lid on a kettle being forced up by the pressure of the steam forming inside. Neither story is accurate, but what is true is that Watt, working in partnership with Matthew Boulton, patented improvements to the steam engine that made it a commercial product and revolutionised industry and transport. Potential customers wanted to know just how fast these engines could do work, so Watt calculated the rate of working of his engines in comparison with the rate at which a typical horse could work – the unit was the "horsepower".

Figure 25.11 *James Watt (1736–1819) was a Scottish engineer who improved the performance of the steam engine, and can be said to have started the Industrial Revolution – the beginning of the machine age.*

The horsepower is only used in the car industry these days. The modern unit of power is named in honour of James Watt. The watt (W) is the rate of transfer or conversion of energy of one joule per second (1J/s).

$$\textbf{power, P (in watts)} = \frac{\textbf{work done (in joules)}}{\textbf{time taken, t (in seconds)}}$$

$$P = \frac{WD}{t}$$

1 horsepower is equivalent to 746 watts.

Some power calculations

You may have done a simple experiment involving running up a flight of stairs, to measure your output power. You do work as you raise your GPE, and to find your power output in watts you divide the work done by the time taken. The experiment is described in Figure 25.12. Notice that calculating the work you do against gravity using force × distance works just as well as using the formula for GPE (mass × gravitational field strength × height).

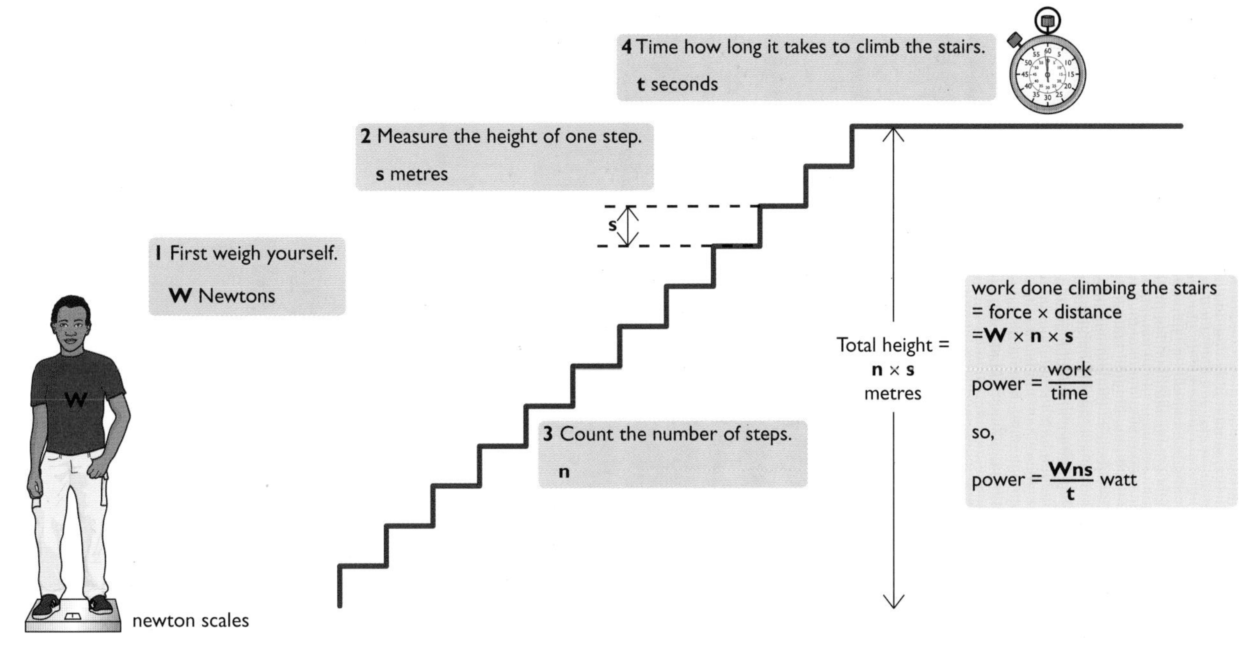

Figure 25.12 *An experiment to measure your output power.*

A more convenient way of raising your GPE and getting to a higher floor in a building is to take a lift. The lift will convert its energy input, usually electrical, into kinetic energy and then, if you are going up, into GPE. As usual, unwanted energy conversions are inevitable – sound and heat will be produced. If we know the weight of the lift and its contents and the height through which it moves, we can calculate the work done in the usual way. If we measure the time that the lift journey takes we can then calculate the power output of the lift motor. (Strictly this will be the *useful* power output – it will not take account of the wasted power due to unwanted energy conversions.)

If we know the speed, v, at which a lift moves, we can do the power calculation in a slightly different way, using:

$$\underset{\text{(in watts)}}{\textbf{power, P}} = \underset{\text{(in newtons)}}{\textbf{force acting, F}} \times \underset{\text{(in metres/second)}}{\textbf{speed in direction of force, v}}$$

$$\mathbf{P = Fv}$$

Notice that force × speed is equivalent to force × $\dfrac{\textbf{distance}}{\textbf{time}}$

worked example

Example 5

Suppose a lift and passengers have a combined weight of 4000 N and the lift moves upwards with an average speed of 3 m/s. What is the useful power output of the lift motor?

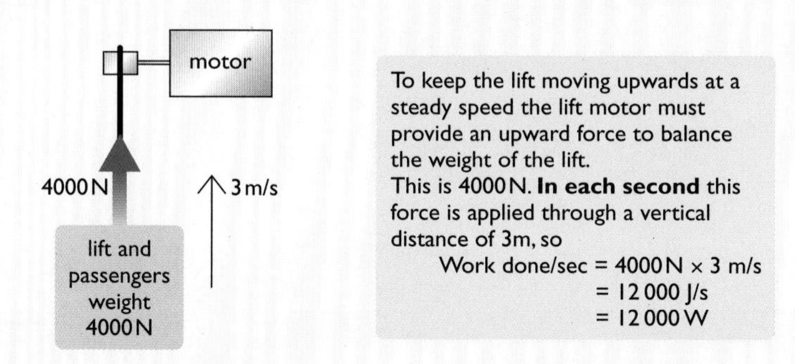

Figure 25.13.

To keep the lift moving upwards at a steady speed, the lift motor must provide an upward force to balance the weight of the lift. This is 4000 N. In each second, this force is applied through a vertical distance of 3 m, so:

work done/second = 4000 N × 3 m/s
= 12 000 J/s
= 12 000 W

Efficiency

In all the examples where energy has been converted, we have had to remember that a proportion of the energy input has been wasted. Remember that "wasted" means converted into forms other than the useful form required. We would like our energy conversion systems to be perfect with *all* the input energy being converted to the form that we want – for example, all the input energy for an electric lamp being converted into light rather than some being converted to heat. Real systems do not achieve this level of complete, or 100%, **efficiency**. The efficiency of an energy conversion system is defined as:

$$\textbf{efficiency} = \frac{\textbf{useful energy output from the system}}{\textbf{total energy input to the system}}$$

Efficiency does not have a unit because it is a number between 0 and 1. This number represents the fraction of the energy input that is converted to the form of energy that we want. Sometimes efficiency is given as a percentage:

$$\textbf{efficiency} = \frac{\textbf{useful energy output from the system}}{\textbf{total energy input to the system}} \times \textbf{100\%}$$

The efficiency of a system can also be expressed in terms of power input and useful power output:

$$\textbf{efficiency} = \frac{\textbf{useful power output from the system}}{\textbf{total power input to the system}}$$

If you calculate the efficiency of a system and get an answer greater than 1 (or 100%) then you have put the numbers into the formula the wrong way round!

worked example

Example 6

a) A 60 W tungsten filament lamp is 5% efficient. How much useful light energy does it emit per second?

$$\text{efficiency} = \frac{\text{useful energy output from the system}}{\text{total energy input to the system}} \times 100\%$$

Total electrical energy input *per second* = 60 J (Remember that 1 W = 1 J/s)

$$\text{So, } 5\% = \frac{\text{useful light energy output from the lamp}}{60\text{ J}} \times 100\%$$

$$\therefore \text{ useful light energy output from the lamp} = \frac{5 \times 60}{100}$$

$$= 3\text{ J per second}$$

b) Fluorescent lamps are five times more efficient than tungsten filament lamps. What would the electrical power rating of a fluorescent lamp giving out the same amount of useful light energy be?

The efficiency of a fluorescent lamp = 5 × 5%

= 25%

$$\text{efficiency} = \frac{\text{useful power output as light from lamp}}{\text{electrical power input to lamp}} \times 100\%$$

$$25\% = \frac{3\text{ W}}{\text{electrical power input to lamp}} \times 100\%$$

$$\therefore \text{ electrical power input to lamp} = \frac{3 \times 100\text{ W}}{25}$$

$$= 12\text{ W}$$

End of Chapter Checklist

If you haven't got a copy of your specification, read the introduction on page iv.

You will need to be able to do some or all of the following. Check your Awarding Body's specification (syllabus) to find out exactly what you need to know.

- Recall and apply the relationship: work done = force × distance.
- Understand that doing work involves the transfer of energy.
- Know that the gravitational potential energy (GPE) of an object is increased when it is moved against the pull of gravity.
- Calculate the increase in GPE using the formula: GPE = mgh.
- Know that moving objects possess kinetic energy (KE) and calculate KE using: $KE = \frac{1}{2}mv^2$.
- Appreciate that, as an object falls, gravity does work on the object converting its GPE into KE.
- Know that power is the rate of doing work or converting energy, and carry out power calculations using: power = work done/time, or power = Fv.
- Carry out efficiency calculations using:

$$\text{efficiency} = \frac{\text{useful energy output from the system}}{\text{total energy input to the system}}$$

Questions

More questions on work, power and efficiency can be found at the end of Section E on page 291. In the questions below, where necessary, take the strength of the Earth's gravity to be 10 N/kg.

1 James Joule showed that heat is a form of energy. He did this by showing that heat can be produced by using mechanical energy.

a) Describe how you would convert mechanical energy into heat energy.

b) How could you convert heat energy into mechanical energy?

2 **a)** What is the unit of work?

b) Define the unit of work.

c) How much work is done in each of the following situations?

i) A bag of six apples each weighing 1 N is lifted through 80 cm.

ii) A rocket with a thrust of 100 kN travels to a height of 200 m.

iii) A weight lifter raises a mass of 60 kg through a height of 2.8 m.

iv) A lift of mass 200 kg lifts three people of mass 50 kg each through a distance of 45 m.

3 Water from a hydroelectric power station reservoir is piped from a reservoir at a height of 800 m above sea level to turbines in the power station itself. The power station is 250 m above sea level. The reservoir holds 200 million (2×10^8) litres of water. If a litre of water has a mass of 1 kg, how much gravitational potential energy is stored in the water in the reservoir?

4 **a)** Explain how to calculate the kinetic energy possessed by a moving object.

b) Work out the kinetic energy of the following:

i) a man of mass 80 kg running at 9 m/s

ii) an air rifle pellet of mass 0.2 g travelling at 50 m/s

iii) a ball of mass 60 g travelling at 24 m/s.

5 A catapult fires a stone of mass 0.04 kg vertically upwards. If the stone has an initial kinetic energy of 48 J, how high will it travel before it starts to fall back to the ground?

6 If a coin is dropped from a height of 80 m, how fast will it be travelling when it hits the ground? State any assumptions you may need to make.

7 Define *power* and state its unit.

8 A person with a mass of 40 kg runs up a flight of stairs in 12 s. The flight of stairs has 20 steps and the height of each step is 20 cm.

a) How much does the person weigh, in newtons?

b) What is the total height that the person has climbed?

c) How much work is done in climbing the stairs?

d) What is the power output of the person running up the stairs?

9 A drag car, of mass 500 kg, accelerates from rest to a speed of 144 km/h in 5 s.

a) What is its final speed in:

i) m/h (metres per hour)

ii) m/s?

b) What is the increase in KE of the drag car?

c) What is the average power developed by the drag car's engine?

10 Define the efficiency of an energy conversion system. If the drag car in the above example is 30% efficient, how much chemical energy has it used in the 5 s of acceleration? What has become of the rest of the energy supplied to the drag car in the form of fuel?

Chapter 26: Thermal Energy

This chapter is about the ways in which thermal (heat) energy is transferred from a hotter place to a cooler place.

Figure 26.1 *Sometimes we want to prevent the transfer of thermal energy – to keep our bodies warm, for example. In other circumstances, the movement of thermal energy from place to place can be useful – for example, rising warm air creates 'thermals' that can carry a glider up to great heights.*

Thermal or heat energy is a form of energy that is possessed by "hot" matter. We shall see that "hot" is a relative term. There is a temperature called **absolute zero** that is the lowest possible temperature. Any matter that is above this temperature has some thermal energy. The kinetic energy of the minute particles that make up all matter produces the effect we call heat.

Thermal energy travels from a place that is hotter (that is, at a higher temperature) to one that is colder (at a lower temperature). In this chapter, we shall look at the different ways in which thermal energy is transferred between places that have different temperatures.

Conduction

Thermal conduction is the transfer of thermal (heat) energy through a substance without the substance itself moving.

Figure 26.2 *Metal skewers allow heat to be transferred to parts which are away from the heat.*

If you have ever cooked kebabs on a barbecue with metal skewers, as shown in Figure 26.2, you will have discovered conduction! The metal over the burning charcoal becomes hot and the heat energy is transferred along the skewer by conduction. In metals, this takes place quite rapidly and soon the handle end is almost as hot as the end over the fire. Metals are good **thermal conductors**. If you use skewers with wooden handles you can hold the wooden ends much more comfortably because wood does not conduct thermal energy very well. Wood is an example of a good thermal **insulator**.

The process of energy transfer by conduction is explained in terms of the behaviour of tiny particles that make up all matter. In a hot part of a substance, like the part of the skewer over the glowing charcoal, these particles have more kinetic energy. The more energetic particles transfer some of their energy to particles near to them. These therefore gain energy and then pass energy on to particles near to them. The energy transfer goes on throughout the substance. This process takes place in all materials.

In metals, the process takes place much more rapidly, because metals have **free electrons** that can move easily through the structure of the metal, speeding up the transfer of energy.

"Free electrons" are electrons that are not bound to any particular atom in the structure of a substance. Metals usually have huge numbers of free electrons per unit volume. Copper, for example, has about 10^{29} free electrons in each cubic metre. As these free electrons carry electric charge as well as energy, it is no coincidence that good thermal conductors are usually good conductors of electricity too.

Convection

Convection is the transfer of heat through fluids (liquids and gases) by the upward movement of warmer, less dense regions of fluid.

You may have seen a demonstration of convection currents in water, like the one shown in Figure 26.3. The water is heated just under the purple crystal – the crystal colours the water as it dissolves, which lets you see the movement in the water. The heated water expands and becomes less dense than the colder surrounding water, so it floats up to the top of the beaker. Colder water sinks to take its place, and is then heated too. At the top, the warm water starts to cool, becomes more dense again and will begin to sink, so a circulating current is set up in the water. This is called a **convection** current.

Lava lamps work by convection (Figure 26.4). Two liquids that do not mix are used in a lava lamp – one clear and one coloured. As the coloured liquid at the bottom is heated, it expands. As a result of it expanding, the coloured liquid becomes less dense so it floats up to the top. When it eventually cools it sinks back to the bottom, demonstrating convection beautifully!

Substances tend to expand when heated because the particles of which they are made have more kinetic energy. As they move around more, the average distance between the particles increases.

Figure 26.3 *Demonstration of convection currents, using a potassium manganate(VII) crystal in water.*

Figure 26.4 *The 'blobs' in a lava lamp move up and then down again in convection currents.*

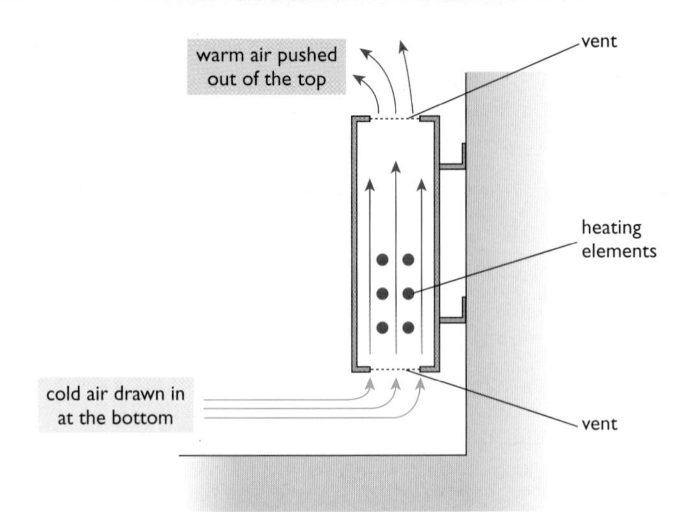

Figure 26.5 *A convector heater relies on the effects of convection.*

Convection occurs in any **fluid** substance – that is, in liquids and gases. Convector heaters (Figure 26.5) heat air, which then floats out of the top of the heater to the top of the room. Cold air is drawn in at the bottom and this in turn is heated. In this way, heat energy is eventually transferred to all parts of the room.

Air and water both allow heat transfer to take place readily by convection as they are both fluids, but neither are good thermal conductors (they are insulators). This insulating property of both water and air is put to good use in situations where they are not able to circulate easily. For example, woollen clothing keeps you warm because air gets trapped in the fibres. The trapped air is heated by your body and forms a warm insulating layer that helps to stop you losing heat. In the same way, a wetsuit keeps a diver warm because a thin layer of water is trapped next to the diver's skin.

Radiation

Remember that we are talking about heat transfer – this is *thermal* radiation not *nuclear* radiation!

Figure 26.6 *Heat is transferred from a heater by radiation.*

When you turn on a bathroom heater, as shown in Figure 26.6, you will feel the effect almost instantly. If you consider the two ways that heat can be transferred discussed so far, you will see that neither can explain how heat is getting from the hot element to your hands. Conduction does not occur that rapidly, even in good thermal conductors, and air is a poor thermal conductor. Convection results in heated air floating *upwards* on colder, denser air.

There are two things you should notice about this example.

1 The heat that you feel so quickly is travelling from the heater *in a straight line*.

2 The design of the electric fire includes a specially shaped, very shiny *reflector* similar to the reflector behind a fluorescent light or in a torch.

The reflector is a special shape, called a **parabola**. You can see parabolic reflectors in torches and radio telescopes, for example. Perhaps you have seen parabolic reflectors used to direct your spoken words across a large distance in the "Launchpad" section of the Science Museum in London. You may also have seen a reflector of this shape used in ripple tank experiments with waves (see page 144).

In this example, heat is travelling in the form of *waves*, like visible light. Heat waves are called **infra-red** (IR) **waves** or IR radiation. The army and the emergency services use special cameras, called *thermal imaging cameras*, that can "see" objects giving out IR waves. These cameras show images of people because of the heat radiation from their bodies, even when there is not enough visible light to actually see them. Thermal imaging is also an important tool in the diagnosis of certain illnesses.

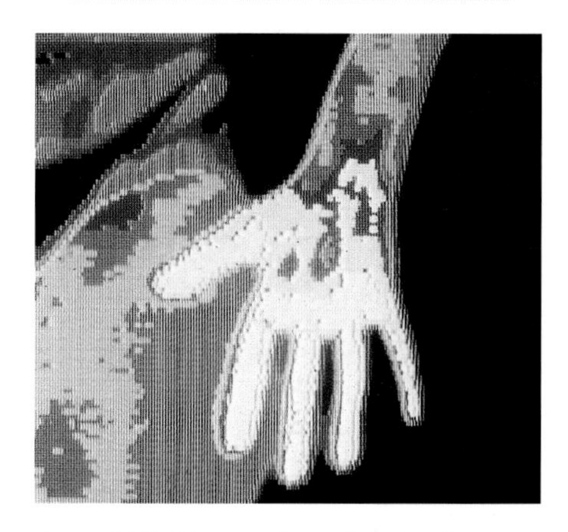

Figure 26.7 *This is a thermal image of a patient showing areas of different temperatures. The hand is white because rheumatoid arthritis gives rise to a temperature increase.*

Thermal radiation is the transfer of energy by infra-red (IR) waves.

IR waves are part of the same "family" of waves as light, radio waves, ultraviolet and so on, called the **electromagnetic** (EM) **spectrum** (see page 184). IR waves, therefore, have the same properties as all the other waves in the EM spectrum. In particular, IR can travel through a vacuum and does so at the speed of light (3×10^8 m/s). It is important that heat can travel in this way, without the need for matter, otherwise we would not receive heat, as well as light, from the Sun.

IR waves can also be reflected and absorbed by different materials, just like visible light. Highly polished, shiny surfaces are good reflectors of thermal radiation. White surfaces also reflect a lot of IR. Matt black and dark surfaces are *poor* reflectors or, to put it more positively, *good* absorbers of heat radiation. Figure 26.8 shows this fact being put to good effect.

Figure 26.8 *Shiny and white surfaces reflect thermal radiation, while matt black surfaces absorb it.*

Figure 26.9 *a) A shiny kettle stays warmer longer.* *b) The heat sink needs to be matt black to "lose" heat to the surroundings quickly, and so stop the transistor overheating.*

If a surface is a good reflector of IR then it is a poor radiator of IR. This means that a hot object with a shiny surface will emit less heat energy in the form of IR than another object at the same temperature with a dull black surface. The kettle in Figure 26.9a has a shiny surface to reduce the rate of heat loss. The heat sink in Figure 26.9b, designed to stop the transistor overheating, has a matt black surface so that it will radiate heat well.

The "greenhouse effect"
Infra-red waves from the Sun can pass through ordinary greenhouse glass. The IR waves heat up the ground, which re-radiates the heat. However, the re-radiated heat is in the form of IR waves with much longer wavelengths. The longer wavelength IR waves cannot pass through glass, so they are trapped inside the greenhouse. Carbon dioxide in our atmosphere acts in the same way as the glass in a greenhouse. It traps the Sun's heat.

Energy efficient houses

We pay for the energy we use in our homes, schools and places of work. Heating is the main use of energy in our homes and – since most domestic heating systems work by burning fuels like coal, oil and gas – it is the main producer of carbon dioxide. (Even if electric heaters are used, most electrical energy is produced by burning fuels in power stations.) Carbon dioxide is a "greenhouse gas" and contributes to global warming. It is, therefore, in everyone's interest that houses are **energy efficient**.

Energy efficiency means using as much as possible of the energy we produce for the desired purpose. So when we turn on the central heating, we want to keep the inside of our homes warm and not allow the heat to escape. If no heat could escape from a house then we would need only heat it until it reached the desired temperature, then never heat it again.

It is worth remembering that, in some countries, the problem is not keeping warm but keeping cool! Keeping cool also requires energy – to run air-conditioning units, fans and other cooling devices. An energy efficient house in a hot climate would stay cool by reducing the rate at which heat *entered* the house.

The key to energy efficient housing is **insulation**. Houses must be designed to reduce the rate at which energy is transferred between the inside and the outside.

In a climate such as that in the UK it is more appropriate to concentrate on keeping heat in, rather than thinking about how to keep it out.

How heat is lost

To insulate a house effectively we must look at *all* the ways in which heat energy can escape. Conduction is the main way heat is transferred between the inside of a building and the outside. Next we need to consider the *places* where conduction occurs: the walls, the windows (and doors) and through the roof.

Heat loss by conduction through the walls can be reduced by using building materials that are good insulators. However, the materials used for building must also have other suitable properties like strength, durability and availability at a sensible price. For walls, bricks are the common building material. Figure 26.10 shows the typical construction of a modern house built to conform to current energy efficiency regulations.

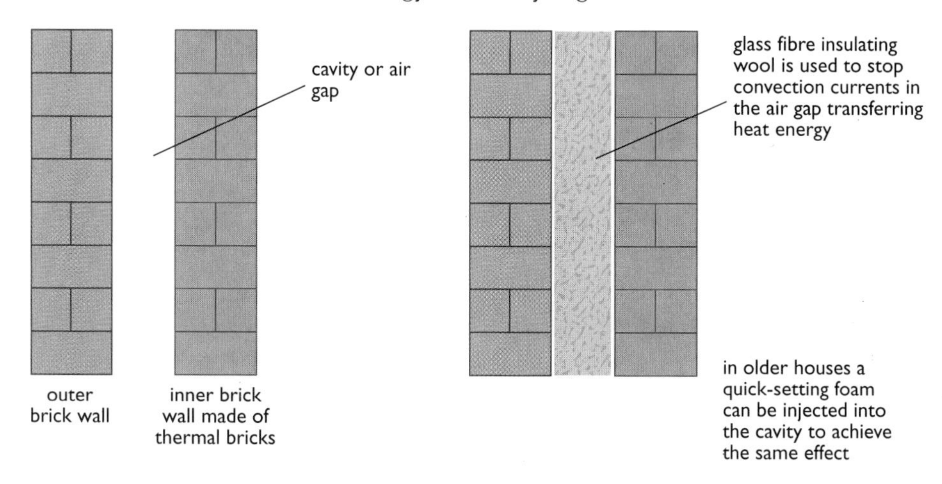

Figure 26.10 *The two-layered wall construction, with the cavity filled with insulation panels, helps to reduce heat loss by conduction, convection and even radiation.*

As you can see, the wall is made with layers of different materials. The outer layer is made with bricks; these have quite good insulating properties, are strong and have good weathering properties. The inner layer is built with thermal bricks with very good insulation properties; they are also light, relatively cheap and quick to work with. The two layers of brick are separated by an excellent thermal insulator in the form of an air cavity or gap.

The walls also stop heat being lost by convection. The cavity or gap between the two walls is wide enough for convection currents to circulate. This means heat is circulated from the warmer surface of one wall to the colder surface of the other. To stop convection currents, the gap in modern houses is filled with insulating panels made of glass fibre matting. This is a lightweight, poor conductor that traps lots of air. The panels are usually surfaced with thin aluminium foil. This highly reflective surface reflects heat in the form of infra-red radiation.

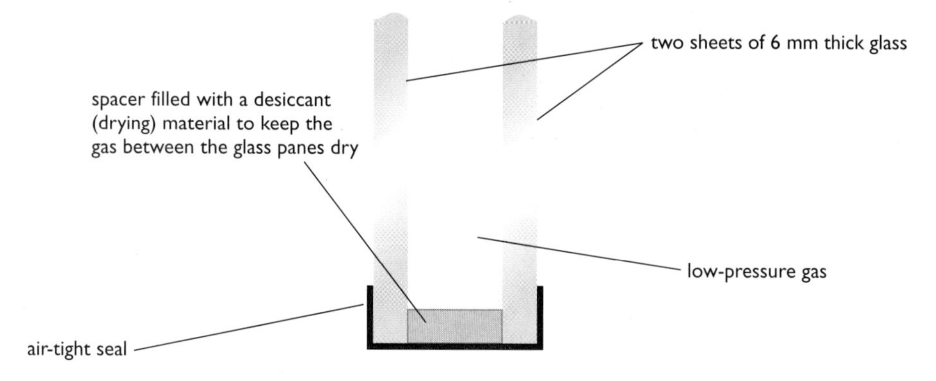

Figure 26.11 *Double glazing helps to stop heat escaping from the home.*

Figure 26.11 shows a cross-section of a typical double glazed window, as used in modern houses. Glass is a poor thermal conductor but is used in thin layers. To improve the insulating properties, two layers of glass are used to trap a layer of air. The thickness of this layer is important. If it is too thin then the

Figure 26.12 *Loft insulation makes a big difference to heat loss through the roof.*

insulation effect is reduced, but if it is too thick then convection currents will be able to circulate and carry heat from the hotter surface to the colder one. In very cold countries triple glazing is used. Modern double glazing uses special glass to increase the greenhouse effect (heat radiation from the Sun can get in but radiation from inside the house is mainly reflected back again).

Roof insulation in modern houses uses similar panels to those used in the wall cavities, trapping a thick layer of air. This takes advantage of the poor conducting property of air, whilst also preventing convection currents circulating. Again, reflective foil is used to reduce radiation heat loss. Figure 26.12 shows houses built before loft insulation was a compulsory building regulation. In some, the owners have installed loft insulation – you should be able to identify which!

There are other measures that can be taken to improve the energy efficiency of houses, which do not relate directly to the mechanisms of heat transfer discussed in this chapter. For example, thermostats and computer control systems for central heating can further reduce the heating needs of a house. They stop rooms being heated too much by switching off the heat when a certain temperature is reached. Another important energy saving measure is the reduction or elimination of drafts from poorly fitting doors and windows.

With your understanding of how heat travels you can save your family money, keep warm *and* reduce global warming.

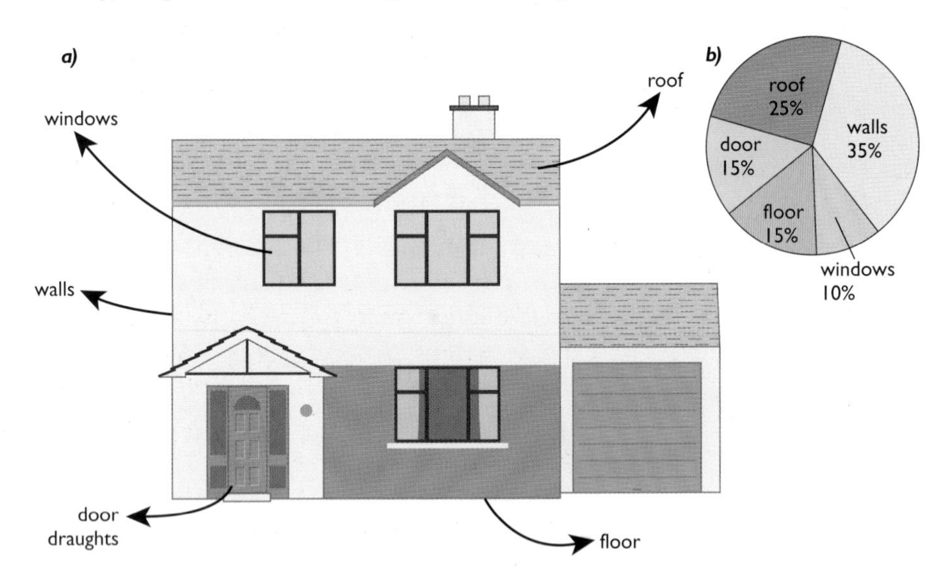

Figure 26.13 a) *How heat energy can be lost from the home.* b) *Percentage of energy.*

The cost effectiveness of insulation

We have discussed how heat can be lost from a house and what things can be done to cut down this heat loss. It is possible to build houses that lose very little heat but each measure that is taken to improve the energy efficiency of a house will increase the building costs. This is also true when we take steps to improve the energy efficiency of existing homes.

Insulation must be **cost effective**. This means that for the price of the insulation we should save a reasonable amount of money by reducing our heating costs.

Suppose the cost of heating a house is £500 per year. Heat loss through the roof might account for 20% of the loss – costing £100 per year. If the loft is insulated this loss might be reduced by half, saving £50 per year. The cost of putting loft insulation in might be £300, in which case the energy saving will pay for the cost of loft insulation in about 6 years. Heat losses through the windows might amount to 15% – a heating cost of £75 per year. Double glazing, costing £10 000, could cut the heat losses through the windows by one third, saving £25 per year. This would mean double glazing takes 400 years to pay for itself! Clearly installing loft insulation is a more cost effective way of improving the energy efficiency of a house.

This example may make it seem crazy to spend money on double glazing, but other benefits may make it worthwhile. Double glazing can cut down condensation on the insides of the windows, reduce unwanted noise from traffic or aircraft, make the house more secure and increase its value.

Insulating people and animals

Earlier in this chapter we saw a picture of a fire fighter in protective clothing designed to reduce the amount of heat getting to their bodies (Figure 26.8). Normally we have the opposite problem and want to keep warm. The obvious method of cutting down heat loss from the body is to wear clothes. Clothes that trap air around the body provide insulation because trapped air cannot circulate and is a very poor conductor. A large proportion of body heat is lost from the head, so hats are the human equivalent of loft insulation.

Wind can cause rapid heat loss from the body. It does this by *forced* convection – that is, making air circulate close to the body surface. It may also cause perspiration to evaporate from the skin more quickly, causing rapid cooling. (The purpose of perspiration is to help the body to lose heat by evaporation, but, if it happens because of strong wind on a cold day, the effect can be life threatening.) These cooling effects of wind contribute to what is called "the wind-chill factor". To reduce the wind-chill effect, a wind-proof outer garment should be worn.

When people do lose body heat at too great a rate they may become **hypothermic**, which means their body temperature starts to fall. If the heat loss is not drastically reduced the condition is potentially fatal. When people are rescued from mountains suffering from the effects of cold they are usually wrapped in thin, highly reflective blankets. The interior reflective surface reflects heat back to their bodies while the outer reflective surface is a poor radiator of heat. Marathon runners are often cloaked in these blankets at the end of the race to keep them warm when their energy reserves are low (Figure 26.1).

Animals keep warm in different ways. You may have noticed birds fluffing up their feathers on cold days in winter. This increases the thickness of the trapped air layer around their bodies, so reducing heat loss by conduction. Some birds, like penguins, will huddle together for warmth (Figure 26.14). Other animals will curl into small balls. This cuts down heat loss by minimising the surface area of their bodies exposed to the cold.

Figure 26.14 *Penguins huddle together for warmth.*

End of Chapter Checklist

If you haven't got a copy of your specification, read the introduction on page viii.

You will need to be able to do some or all of the following. Check your Awarding Body's specification (syllabus) to find out exactly what you need to know.

- Understand that heat energy is transferred from places at high temperature to places at lower temperature.
- Describe how heat can be transferred by conduction, convection and radiation.
- Understand how heat transfer by these processes can be reduced by using insulation.
- Explain what is meant by cost effective insulation.

Questions

More questions on thermal energy can be found at the end of Section E on page 291.

1 Explain the following observations, referring to the appropriate process of heat transfer in each case.

a) Two cups of tea are poured at the same time. They are left to stand for ten minutes. One of the cups has a metal teaspoon left in it. The tea in this cup is cooler than the tea in the other at the end of the ten-minute period.

b) Two fresh cups of tea are poured out. (The others had gone cold!) A thin plastic lid is placed on top of one of the cups. The tea in this cup keeps hot for longer.

2 *a)* Kettles heated on stoves used to be made of copper. Was this a good choice?

b) Copper kettles were usually kept highly polished. If it is not polished, copper turns dull and eventually blackens as it reacts with oxygen in the air. Apart from making the kettle look nice, what is a sound physics reason for keeping a kettle polished?

3 The diagrams below show a physics demonstration about thermal conduction.

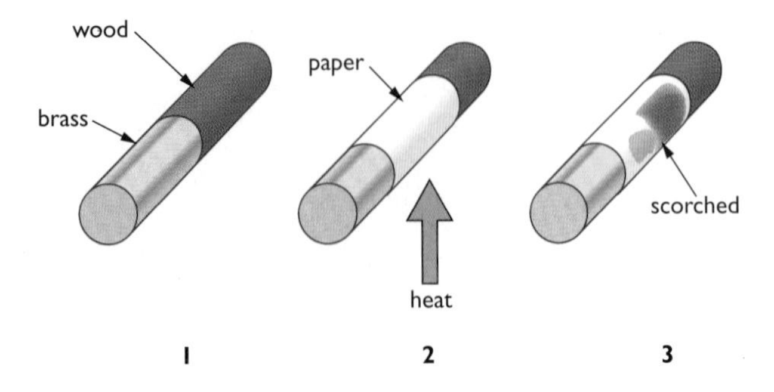

A cylinder is made from a piece of brass fitted to a piece of wood. A strip of paper is glued around the middle. The paper is then heated over a Bunsen burner flame. After a while one end of the paper is noticeably more scorched than the other. Explain why this happens.

4 There are two bench seats in a park, one made of metal, the other made of wood. The metal seat feels much colder to sit on than the wooden one. A student says that it is because the metal seat is at a lower temperature than the wooden one. Explain why this explanation is incorrect, and give a correct explanation of why the metal seat seems colder than the wooden one.

5 Why is the heating element in an electric kettle positioned very close to the bottom of the kettle?

6 One of the latest computers does not use a fan to keep the electronic circuits inside it cool, unlike other PCs. A student notices that the ventilation slots on a PC are positioned on the side, but the slots are on the top and bottom surfaces of this new computer. The designer has applied some physics to the problem of keeping the computer cool. Explain why the new computer does not need a fan.

7 The diagram below shows how Roman mines used to be ventilated.

The mine system had a shaft with a fire lit at the bottom. Explain how this kept the air in the mine system fresh.

8 The diagram below shows the structure of a vacuum flask.

a) Explain how its design makes it more difficult for heat energy to be transferred between the inside of the flask and the outside environment.

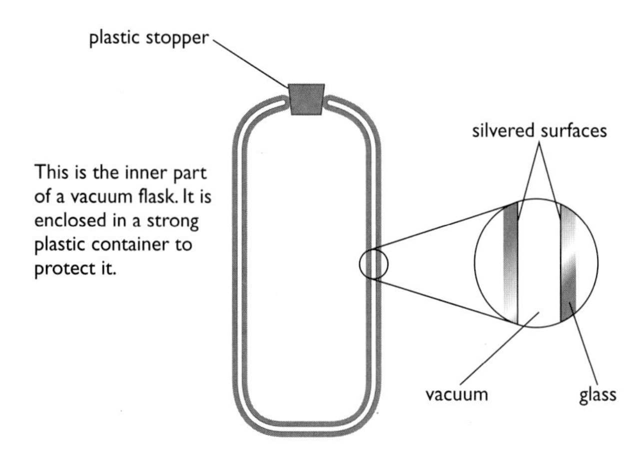

Chapter 27: Energy Resources

In this chapter, you will learn about different sources of energy that are available to us on Earth. You will learn that some resources are renewable, while others cannot be replaced once used. As the demand for energy increases with the human population, there is a danger that non-renewable resources will run out. We therefore need to use fuel efficiently, and to exploit more renewable resources.

Figure 27.1 *Humans consume vast amounts of fuel for transport, heating and cooking.*

The demand for energy increases all the time. The growth of the world population means more people need food and warmth. More people want to be able to travel. The fuels that we use to produce energy are being used up too quickly. We must use our remaining fuel supplies efficiently and look for new sources of energy. In Chapter 26, we saw how we can use energy more efficiently in heating our homes. Clearly we can make the energy supplies we have available last longer by being energy efficient.

We also need to understand what energy resources we have available on the Earth. In this chapter we shall look at different types of energy resource. In particular we shall distinguish between renewable and non-renewable energy resources. New sources of energy are being researched all the time to meet our growing needs. We must also consider the effect that the use of different energy resources has on our environment. Some types of energy resource can cause long-term damage to our environment.

Non-renewable energy resources

ideas
evidence

Fossil fuels

Figure 27.2 *Fossil fuels include coal, oil and natural gas.*

One of the main energy resources available on our planet is its supply of **fossil fuels**. Coal, oil and natural gas are all fossil fuels. They have been formed in the ground from dead vegetation or tiny creatures by a process that has taken millions of years. Once we have used them, it will take millions of years for new reserves of these fuels to be formed. Fossil fuels are, therefore, examples of **non-renewable** energy resources.

A non-renewable energy resource is one that effectively cannot be replaced once it has been used.

Burning fossil fuels has effects on the environment. Burning them releases carbon dioxide into the atmosphere. Carbon dioxide is a "greenhouse" gas. Greenhouse gases trap the Sun's heat in the Earth's atmosphere and cause the average temperature of the atmosphere to rise. This effect is called **global warming** and causes changes in the world's climate and melting of the polar ice caps. Burning coal releases more carbon dioxide into the atmosphere than burning oil or gas. Of the fossil fuels, natural gas produces the least carbon dioxide for the same energy output. There is no practical way of avoiding the release of carbon dioxide into the atmosphere when fossil fuels are burned.

Most types of coal and oil contain some sulphur. When they are burned, this is converted to sulphur dioxide. Sulphur dioxide is then released into the atmosphere where it combines with water to form acid rain. Acid rain causes damage to people, plants and buildings. It is possible to remove the sulphur from these fuels but this increases the cost of the energy produced. It is also possible to remove sulphur dioxide from the waste gases when the fuel is burned, but this also increases the cost.

Figure 27.3 *William Perkin discovered the first synthetic dye in 1856 using substances produced from coal tar.*

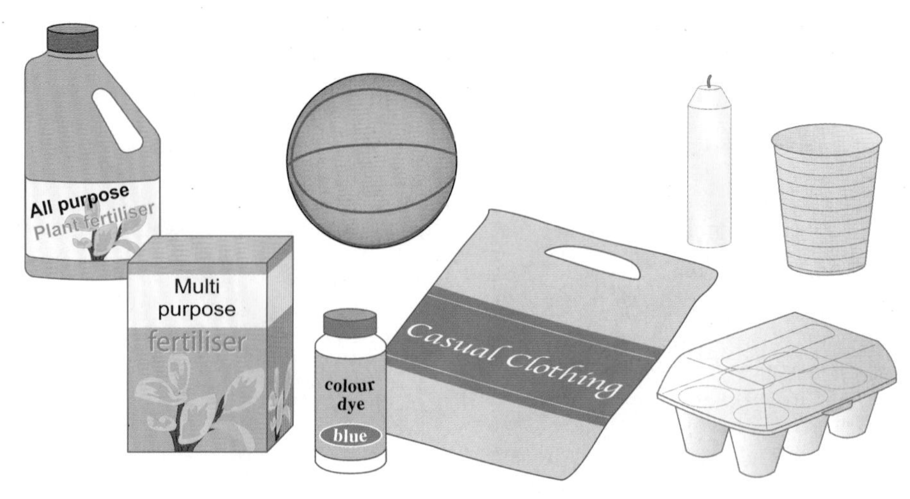

Figure 27.4 *Many products are manufactured using extracts from oil and coal.*

Fossil fuels provide valuable chemicals that can be used in the manufacture of a wide range of useful products (Figure 27.4). Once burned for energy production, these chemical resources are lost permanently. Burning such resources may be a very wasteful way of using them.

Nuclear fuel

"Fast breeder reactors" are so called because they create more nuclear fuel than they consume, but they still require uranium for their operation. The fuel produced is plutonium. This is extremely dangerous to life and is also a material used for the manufacture of nuclear weapons.

Nuclear reactors use uranium to produce energy. For the nuclear process, a particular form or **isotope** (see page 297) of uranium is needed. Although a reactor only needs a small amount of uranium fuel, uranium is in limited supply. Uranium was formed in the unique conditions at the start of the Universe so once it has been used there will be no further supplies. It is, therefore, another example of a non-renewable resource.

Power generated from nuclear processes has the advantage of being "clean". It is clean because the process does not involve the production of greenhouse or other polluting gases. The cost per unit of electricity is very low, but nuclear power stations are expensive to build. The disadvantages of nuclear power are the risk of accidents and the problem of disposal of radioactive material once it is finished with. Accidents that release radioactive materials like uranium and plutonium into the atmosphere cause long-lasting risks to living things. The risk of devastating explosions, like the Chernobyl disaster in 1986, is another significant disadvantage of nuclear fuel.

Renewable energy resources

A renewable energy resource is one that will not run out.

Wood is an example of a **renewable** energy resource. As wood is cut for fuel, new fast-growing trees are planted to replace those cut down. With careful management the supply of wood fuel can be maintained indefinitely. However, burning wood produces pollution and greenhouse gases. Wood is also more valuable if it is not burned because it can be used in building or making furniture.

The demand for fuel and our worries about global warming and pollution have made us search for alternative sources of energy. We need renewable energy supplies that do not pollute the world or contribute to global warming.

Hydroelectric power

Figure 27.5 *Moving water is a renewable energy resource.*

The kinetic energy available in large quantities of moving water has been harnessed for many hundreds of years. Water wheels have been used to convert the energy possessed by water in rivers to grind corn and power industrial machinery. A different kind of "water wheel", called a turbine, is used to turn the generators in a **hydroelectric** power station. These power stations use the stored gravitational potential energy (GPE) of water in high reservoirs built in mountains. The GPE is converted to movement energy (KE) as the water flows down the mountain to the power station below.

The energy produced in this way is renewable. The Sun causes water to evaporate continuously and to be drawn up into the atmosphere. This water then falls as rain to be collected in reservoirs and used again. Moving water is a renewable resource.

Although hydroelectricity is a very clean, renewable resource, building reservoirs and power stations can spoil the landscape. The reservoir may also destroy or alter the natural habitat for wildlife.

Tidal power

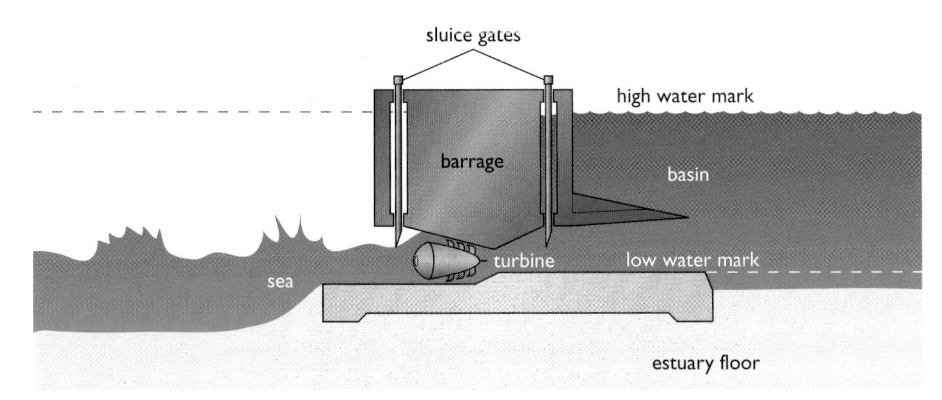

Figure 27.6 a) *Tidal power also involves harnessing the energy in moving water.*

b) *Tidal power station at La Rance in Brittany.*

The tides also involve the movement of huge amounts of water. Tidal power generation schemes, like that at La Rance in Brittany (Figure 27.6b), generate power by turning turbines as the tide flows into a dammed river estuary. As the tide falls the turbines are spun again.

Figure 27.7 *The turbine has angled blades. As water is forced between the blades, they start to spin. The rotating turbine is connected to a generator to make it spin, too.*

The energy for the movement of the tides is provided by the gravitational pull of the Moon and Sun. This is renewable energy using a small fraction of the continuous supply of gravitational energy.

Wave energy

Energy can also be extracted from waves. The continuous movement of the surface of the seas and oceans is the result of a combination of tides and wind. A variety of methods have been developed to make use of the rise and fall of water due to waves. Figure 27.8 shows a system that has been developed to use the energy of waves. Again, this energy is renewable as the movement energy of the waves is continuously available.

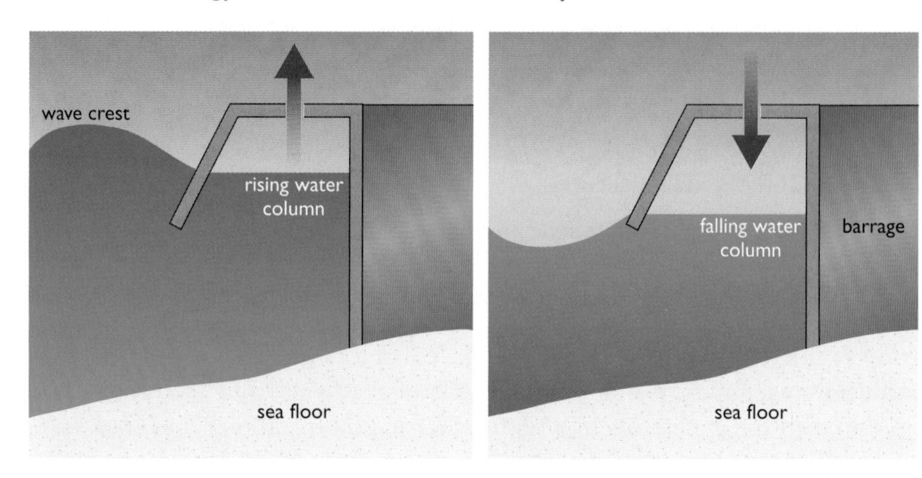

Figure 27.8 *An oscillating water column system for using wave energy.*

Water power is clean, producing no greenhouse gases or unwanted waste products. One disadvantage of harnessing water power is the impact on the environment of "unnatural" features like dams. Another drawback is that the means of converting energy from the movement of water can only be achieved in suitable geographical locations – for example, high mountains for hydroelectric power stations and wide estuaries with a big variation in tide for tidal power.

Wind power

Figure 27.9 *Windmills, old and new.*

Winds are powered by the Sun's heat energy. Wind is a renewable source of energy that has been used for many centuries. Windmills have been used to grind corn and power machinery like pumps to drain lowland areas. Modern windmills (wind turbines) drive generators to provide electrical energy.

The energy produced is clean, but wind power can only be harvested in regions where the wind blows with enough energy for a significant proportion of the year. Wind farms can cause environmental damage, as they change the appearance of the landscape. They also cause some noise pollution.

Solar power

There are two methods of gathering and using energy directly from the Sun.

Figure 27.10 *Photovoltaic cells produce electricity when sunlight falls on them.*

Photovoltaic cells convert light energy directly into electrical energy. They are not very efficient and are expensive. They also require quite bright sunlight to produce useful amounts of electrical energy. These factors limit the usefulness of photovoltaic cells to countries with a high number of sunshine hours. They are only suitable for applications that need small amounts of energy, like calculators and low-power garden lighting. Figure 27.10 shows photovoltaic cells being used to power a communication satellite, and a motorway emergency telephone in southern Europe. Both are remote locations with plenty of sunlight.

Figure 27.11 *Solar heating panels can be used to heat water, even in the UK climate.*

Solar heating panels absorb thermal radiation and use it to heat water. The panels are placed to receive the maximum amount of the Sun's energy. In the northern hemisphere, they must face south and be angled so that light falls on them as directly as possible for as long as possible. The structure of a typical solar heating panel is shown in Figure 27.12.

The design of efficient solar heating panels involves all the methods of heat transfer discussed in Chapter 26. Solar panels are often used in questions about heat transfer.

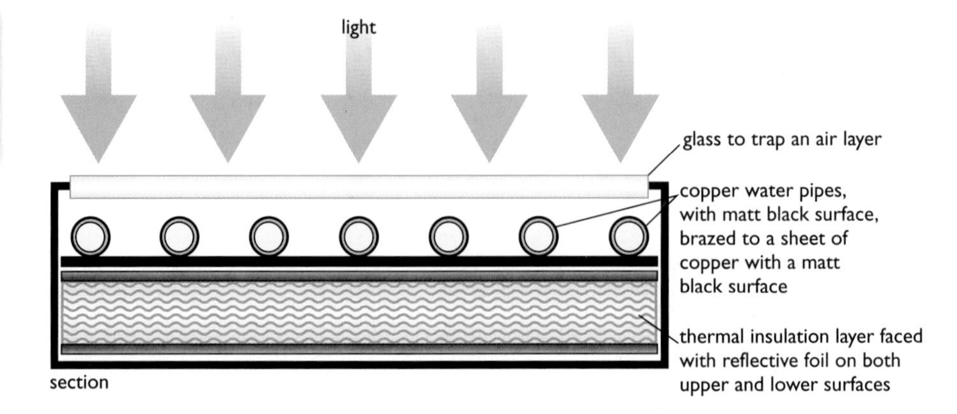

Figure 27.12 *Solar heating panels are designed to absorb as much energy as possible.*

Water is pumped through copper pipes fixed onto a copper sheet. Copper is used because it is an excellent thermal conductor. The surfaces of the sheet and the pipes have a matt black finish as this is the best absorber of heat radiation. The glass traps a layer of air above the copper to help insulate the unit and retain the heat. The backing is also designed to stop heat escaping to the surroundings. This kind of panel is reasonably efficient and the energy produced is more cost efficient than that from photovoltaic cells. Solar heating panels are used widely to provide water heating, but do not provide electrical energy.

Geothermal energy

Figure 27.13 *This picture shows a geothermal electricity generating station in Iceland, with a huge swimming lagoon heated geothermally.*

Geothermal energy is the energy stored deep inside the hot core of the Earth. The heat in regions of volcanic activity was produced by the decay of radioactive elements like uranium. Volcanoes are evidence of the enormous heat and energy beneath the Earth's surface but do not provide a safe or reliable energy resource. However, heat from the ground *can* be used safely. In some areas of the world, like Iceland (Figure 27.13), geothermally heated water is readily available in springs and geysers. This is used to drive the turbines in electricity generation stations. The hot water is also used to provide domestic heating by piping it directly to houses. This resource is renewable, does not produce pollution and does not have a great impact on the environment. There are many areas of the world where geothermally heated springs can, and are, being used to provide energy.

The Sun, our main energy supplier

The Sun has a power output of about 390 billion billion megawatts (3.9×10^{26} W). Every second, about 700 million tons of hydrogen are converted into helium and energy.

Our main source of energy is the Sun. The Sun provides us on Earth with a seemingly endless supply of energy in the form of electromagnetic (EM) waves. The energy reaches us as visible **light** and as **heat**, which are both essential to life on Earth. The Sun's energy supply is not, of course, endless. It is produced by **nuclear fusion**, one of the nuclear processes by which matter is converted into energy. Eventually the Sun will use up all its hydrogen fuel and will undergo a dramatic change that will bring life on Earth to an end. (But don't worry, we are not expecting this particular "end of the world" scenario for another six billion years!).

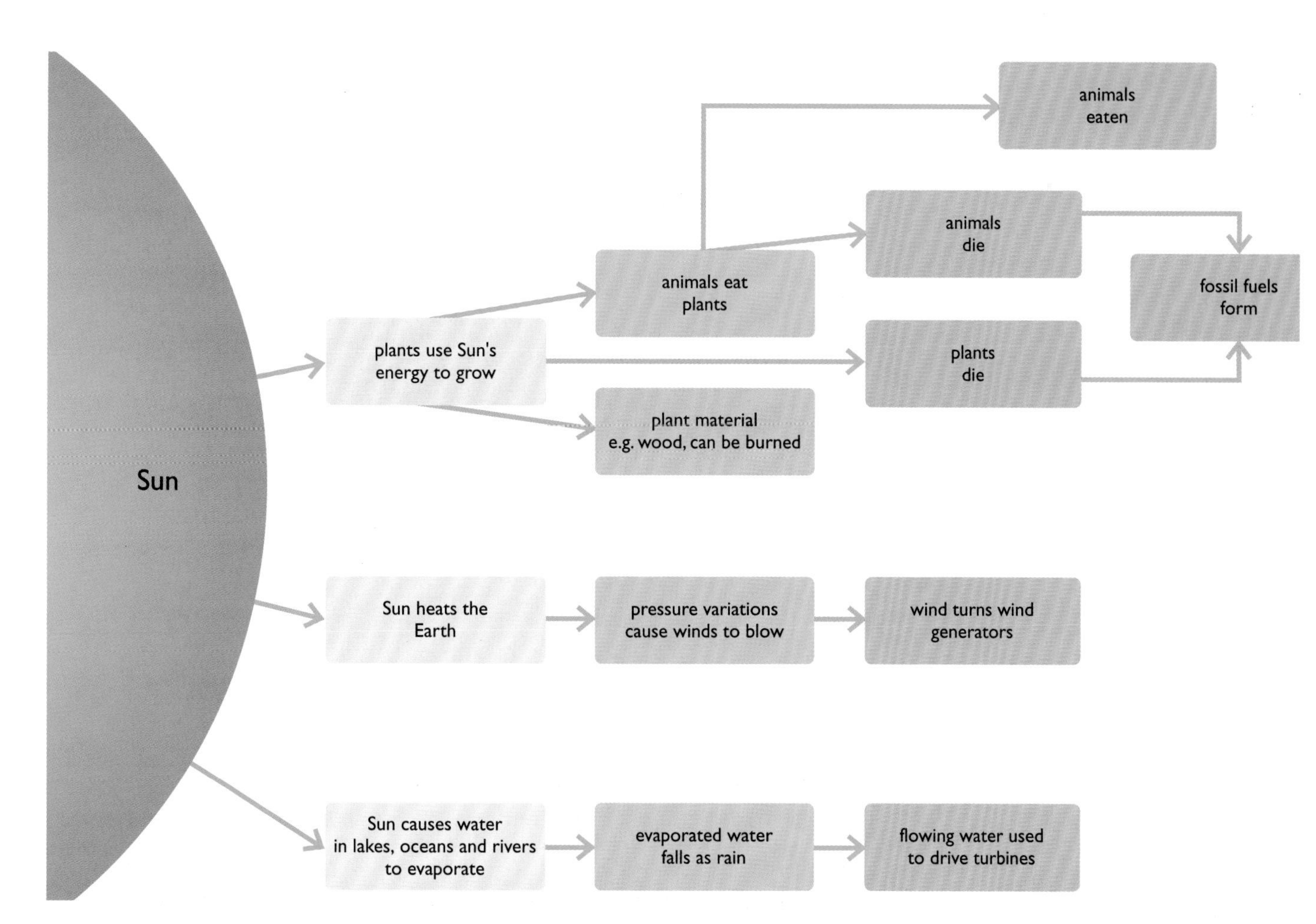

Figure 27.14 *This diagram shows how some of the forms of energy we use on Earth come originally from our Sun.*

End of Chapter Checklist

If you haven't got a copy of your specification, read the introduction on page iv.

You will need to be able to do some or all of the following. Check your Awarding Body's specification (syllabus) to find out exactly what you need to know.

- Understand the difference between renewable and non-renewable energy resources.
- Appreciate that fossil fuels, gas, oil and coal, cannot be replaced once used and all produce greenhouse gases and other pollution when burned.
- Understand what is meant by "clean" energy resources.
- Know of the following sources of renewable energy: wood, wind, the Sun, waves, water and geothermal energy.
- Discuss the advantages and disadvantages of different energy resources.

Questions

More questions on energy resources can be found at the end of Section E on page 291.

1 ***a)*** Explain what is meant by a *non-renewable energy resource*.

b) Give three examples of types of non-renewable energy resources.

2 ***a)*** What is the *greenhouse effect*?

b) How can we reduce the greenhouse effect on the Earth?

3 ***a)*** What are fossil fuels?

b) Explain why fossil fuels are not a clean method of producing energy.

4 Nuclear power generation does not produce greenhouse gases. This is a significant advantage of nuclear power over power stations that burn fossil fuels.

a) What other advantages are there in using nuclear power?

b) What are the disadvantages of nuclear power?

5 Give three examples of renewable energy resources. Compare their advantages and disadvantages.

Chapter 28: Generating Electricity

Electricity is produced in power stations, by generators. In different types of power station, different fuels are used to drive the generators. In this chapter you will learn how some types of power stations are able to react more quickly to fluctuations in demand than others, and how power stations vary in cost efficiency. You will also learn why electricity is transmitted at very high voltages from power stations to the areas where it is needed.

Figure 28.1 *In years gone by, people used many different energy sources to provide light, heat and power for machinery. These days electricity can be used to supply almost any energy requirement.*

In Chapter 27, we looked at energy resources. We need energy for light, heat and a means to make machinery work. In the modern world most of the energy we use is in the form of electricity. In this chapter, we shall look at how electricity is generated and distributed.

Generating electricity

We have already discussed the principle of generating electricity in Chapter 5. When the magnetic field changes around a conductor (such as a length of copper wire), a voltage is **induced** in the conductor. We can make the magnetic field around the conductor change in a number of ways, but it is usually done by moving the conductor within the magnetic field. The conductor is wound into a coil on the **rotor**, which is then spun round in the magnetic field. The magnetic field may be created by permanent magnets, but in a power station generator it is produced by electromagnets. Figure 28.2 shows a typical power station turbine-generator designed to produce a power output of 1200 MW.

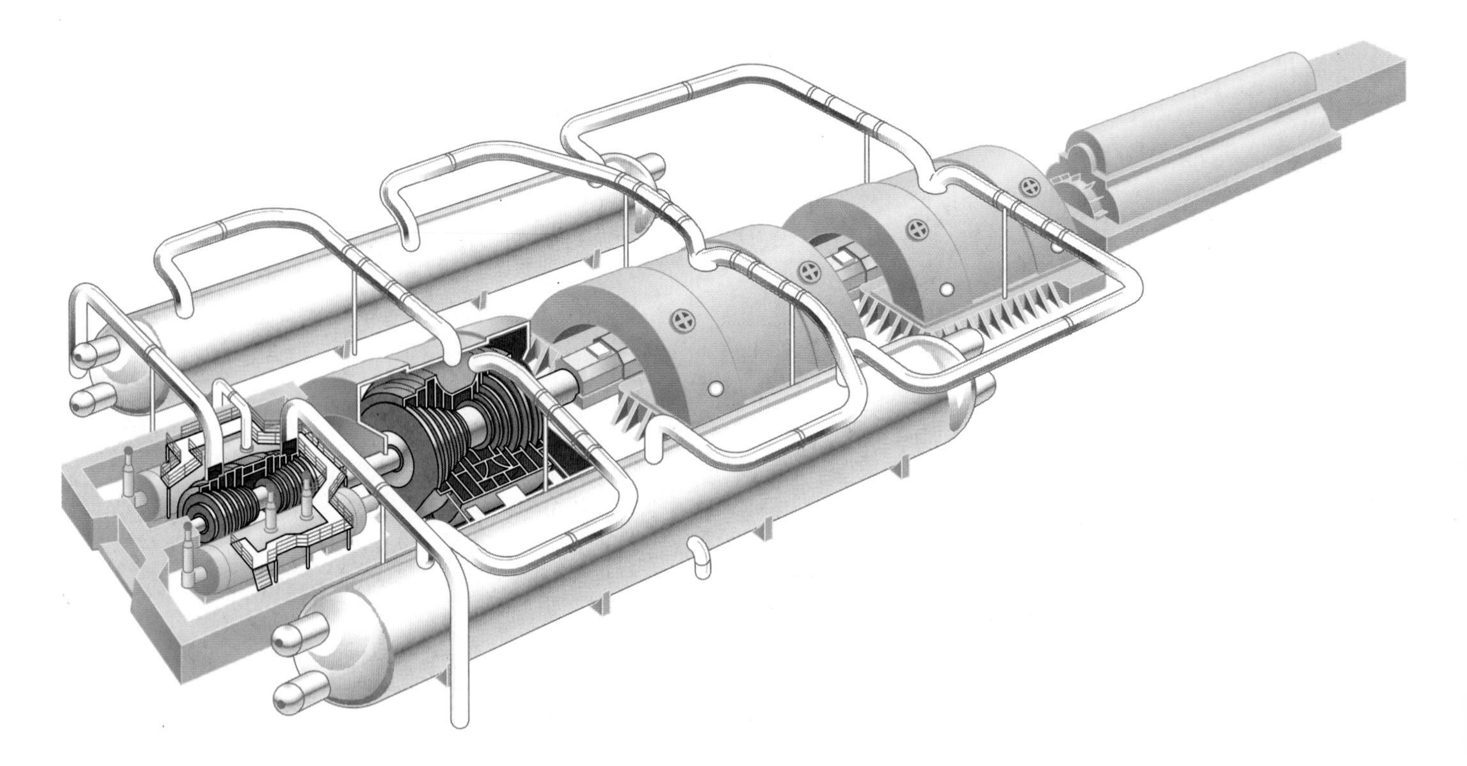

Figure 28.2 *Turbine-generators like this one are used to produce electricity in power stations.*

First, however, we need a source of energy to make the rotor spin. In the UK, the majority of our electricity is made by burning fossil fuels (72% in the year 2000). The fuel is burned to heat water. This produces high pressure steam that is used to make the blades of a turbine spin. A turbine is like a windmill or a fan but with many more blades. The turbine is used to drive the generator. The energy changes involved are shown in Figure 28.3.

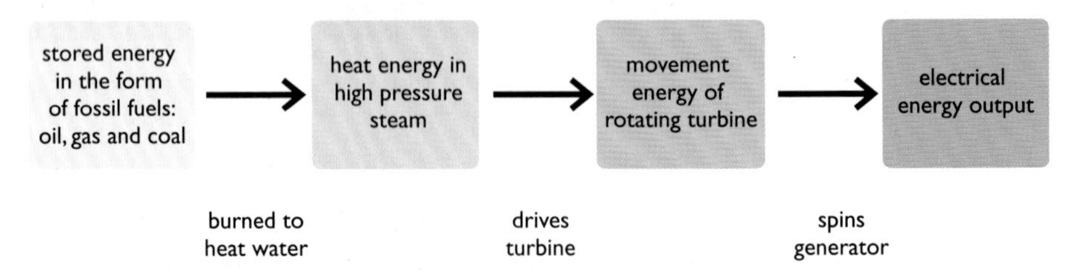

Figure 28.3 *There are several energy changes involved in producing electricity.*

In nuclear power stations, the process is similar. The difference is that the heat is produced by nuclear fission (see page 322) in radioactive fuel, usually uranium or plutonium. This heat is used to make high pressure steam to drive the turbines. Hydroelectric power stations drive the turbines with moving water. Tidal barrage power stations also use water to drive turbines. Wind farms use much smaller generators, each one turned by its own set of windmill blades.

Supply and demand

ideas evidence

We considered some of the environmental problems associated with different ways of producing electricity in Chapter 27. Different types of power stations also differ in the speed with which they can meet changes in the demand for electricity.

The demand for electricity varies from hour to hour, day to day and season to season. The way that the demand varies can be predicted to some extent. For example, there is a surge in demand in the early morning as people wake up, turn on lights and heaters and start to make breakfast, and of course in winter the demand for heat is much greater than in summer. Some sudden surges in demand are less predictable. A popular TV soap with an exciting episode can keep millions of viewers glued to their television sets – if they all decide to make a cup of tea as soon as the adverts come on, electricity consumption will suddenly increase. (Electric kettles use a lot of power.) The companies that supply electricity must be able to cope with these changes in demand, otherwise they are forced to cut off electricity to some consumers. This is not good for customer relations and, of course, means a reduction in the amount of electricity sold.

Nuclear power stations cannot be turned on and off instantly. The process of starting the fission reaction and heating up the core of the nuclear reactor is a lengthy one. Clearly nuclear power stations cannot meet sudden variations in demand.

Power stations that burn fossil fuels can be "fired up" more quickly and can, therefore, deal with changes in demand more rapidly. Coal-fired stations take longer than oil-fired stations to develop the heat required to drive steam through the turbines. Gas-fired stations can respond most quickly to demand surges.

Hydroelectric power stations provide a very reliable energy source with the advantage of being able to respond very quickly to changes in the national demand for electricity. Unlike other types of power station, they are able to operate in reverse. This means that they can use surplus electricity produced by other power stations that cannot be shut down quickly to pump water back up into the high-level reservoirs. This converts the electrical energy back into gravitational potential energy, which can then be reconverted when needed at a later time. This is the only realistic way of 'storing' large amounts of surplus electrical energy.

Wind power is dependent on the strength, direction and frequency of wind. Although wind farms are sited in areas with suitable weather records, they cannot be relied upon to produce electricity at the times when it is needed most.

ideas
evidence

The cost of generating electricity

In Chapter 27, we considered the environmental problems of different energy resources. For example, burning fossil fuels releases polluting gases into the atmosphere and is also a waste of irreplaceable chemical resources. Renewable energy resources, like wind and tidal power, do not produce polluting gases but may damage our environment in other ways. The ideal location for a wind farm is often an area of outstanding natural beauty, and damming a suitable river estuary for a tidal power plant may destroy the habitat of numerous species of wildlife.

The US Department of Energy has proposed the building of a nuclear waste storage facility beneath the Yucca Mountains. High-level nuclear waste will be placed in concrete casks that are supposed to contain the nuclear radiation safely for 300 to 1000 years.

In addition to the environmental costs, planners must also look at the financial costs of electricity generation. Nuclear power uses a relatively cheap fuel. Uranium produces huge amounts of energy, so the cost of energy per unit of fuel used is low. However, building a nuclear power station is very expensive. Nuclear power requires complex technology and very high standards of safety. On top of these "start-up" costs, planners must also consider the expense of decommissioning a nuclear power station at the end of its useful working life. For conventional power stations, this is a routine demolition job, but for nuclear power stations the task is not as straightforward. Radioactive materials must be handled with great care and stored in a way to ensure that none escapes.

We see that, although the running costs of nuclear power are relatively low, the pay-back time is very long. (The pay-back time is how long it takes for the income from selling electricity to cover the cost of building the power station.) The charts in Figure 28.4 show that nuclear electricity production has barely increased over the last decade, but electricity from natural gas has increased enormously.

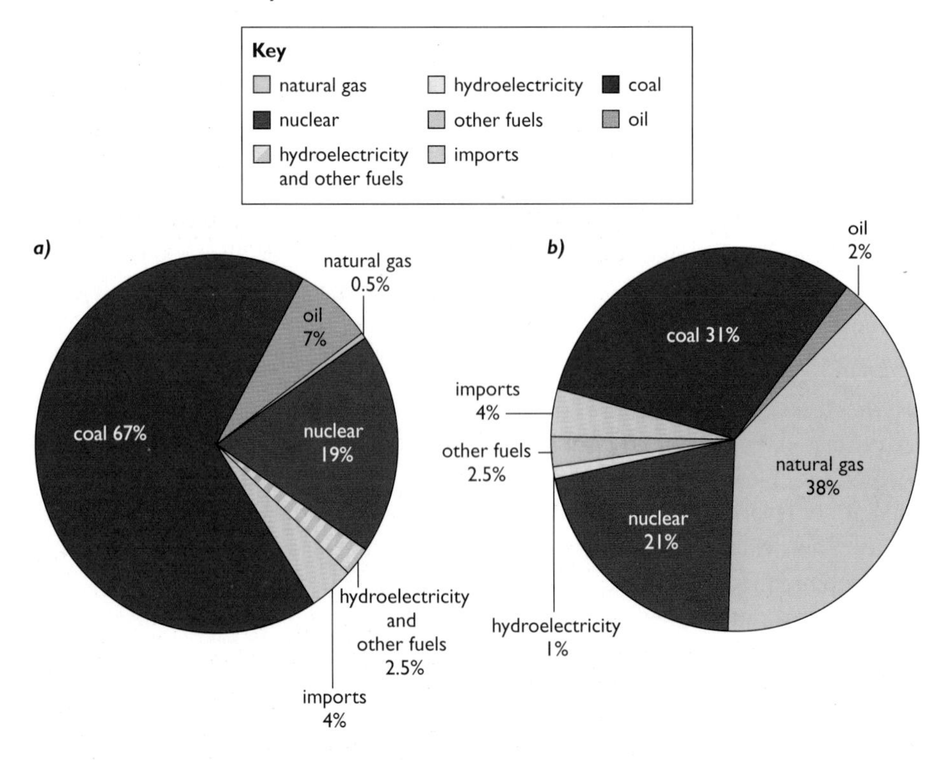

Figure 28.4 *Pie charts to show the proportions of different fuels used to generate electricity* a) *in 1990, and* b) *in 2000.*

The cost of setting up a wind farm is much lower than the cost of building a nuclear power station and there are no fuel costs. However, the amount of energy produced by wind farms is comparatively low. The pay-back time for wind generators is therefore quite long.

Transmission of electricity

Electricity is not necessarily produced where it will be used. Coal-fired power stations are usually situated close to coal fields or ports. The first nuclear power station was built in a remote part of Scotland. Wind farms and hydroelectric power stations are built where the climate and geography are suitable. Electricity is not always needed in the same parts of the country at the same time, as demand varies.

To overcome these problems, electrical energy must be transmitted from the power stations to the consumers.

In Section A, Chapter 2, we saw that the **resistance** of conductors causes energy to be converted from electrical energy to other forms, mainly heat. We also discussed the factors that affect the resistance of a conductor – for example, the longer the conductor, the greater its resistance. You will remember that the power loss, P, in a wire of resistance R ohms carrying a current of I amps is given by:

$$P = I^2R$$

Therefore, the longer the transmission wires are the greater the energy loss will be as electricity travels along them. Other factors affect the resistance, of course. We could reduce the resistance per metre by making the wire much thicker. We could use the very best conductors, like copper and gold. (We could even try to keep the conductors as cold as possible!) Unfortunately, there are practical reasons why we cannot do these things. Thicker wire would make the transmission cables impossibly heavy for pylons to support. Copper is a dense metal with insufficient tensile strength to support its own weight.

Figure 28.5 *High voltage power lines.*

The choice of conductors for overhead power cables is a compromise. Aluminium is used because it has a low density (low weight per cubic metre) and good electrical conductivity. The cables have thin steel cores to give them the necessary strength. A typical overhead cable might have a resistance of 0.5 Ω/km. The example below shows how much energy can be wasted in heating the cables.

worked example

Example 1

Suppose we wish to transmit 9.2 kW of electrical power (P) along a 5 km cable from a small wind generator to a farm. If we did this at the usual mains supply voltage of 230 V the current flow could be calculated using:

$$I = \frac{P}{V}$$

$$= \frac{9200\,\text{W}}{230\,\text{V}}$$

$$= 40\,\text{A}$$

The power loss along the cable can be calculated using:

$$P_{LOSS} = I^2R$$

R is the resistance of the cable = 5 km × 0.5 Ω/km = 2.5 Ω.

This gives a transmission power loss of:

$$P_{LOSS} = (40)^2 \times 2.5$$

$$= 4\,\text{kW}$$

This means that nearly half the generated power is wasted as heat in the transmission cables. Transmitting much higher amounts of power over larger distances would be impossible.

High voltage transmission

The power loss along the transmission cables or lines depends on the square of the current flowing. This means that if the current is *halved*, the power loss is *quartered*. Reducing the current has a big effect on the energy losses due to heating in the transmission lines.

Transformers are used to increase or decrease the voltage of an alternating (ac) supply (see page 44). A well-made transformer has an efficiency of greater than 90%. This means nearly all the electrical power input is available at the output with very little wasted as heat and other non-electrical forms of energy. At GCSE, we assume that transformers are 100% efficient – that is, we assume that the power output is equal to the power input. Using the electrical power formula, this means:

$$V_p \times I_p = V_s \times I_s$$

This can be rearranged:

$$\frac{V_p}{V_s} = \frac{I_s}{I_p}$$

A step-up transformer makes the secondary voltage, V_s, bigger than the primary voltage, V_p. It will, therefore, make the secondary current, I_s, smaller than the primary current, I_p.

Before electrical power is transmitted on the National Grid its voltage is stepped up by a large factor. This means that the transmission current is *reduced* by the same factor. For example, if the voltage is stepped up 10 times then the current will be 10 times smaller. Since power losses through resistance heating depend on the *square* of the current this means that the losses will be reduced by a factor of 100. This is why electricity is transmitted at very high voltages – the lowest voltage used on the National Grid system is 132 kV and the "supergrid", for long distance power transmission, operates at 400 kV. The entire process – from generation at the power station, to transmission across the country and use by the consumer – is illustrated in Figure 28.6.

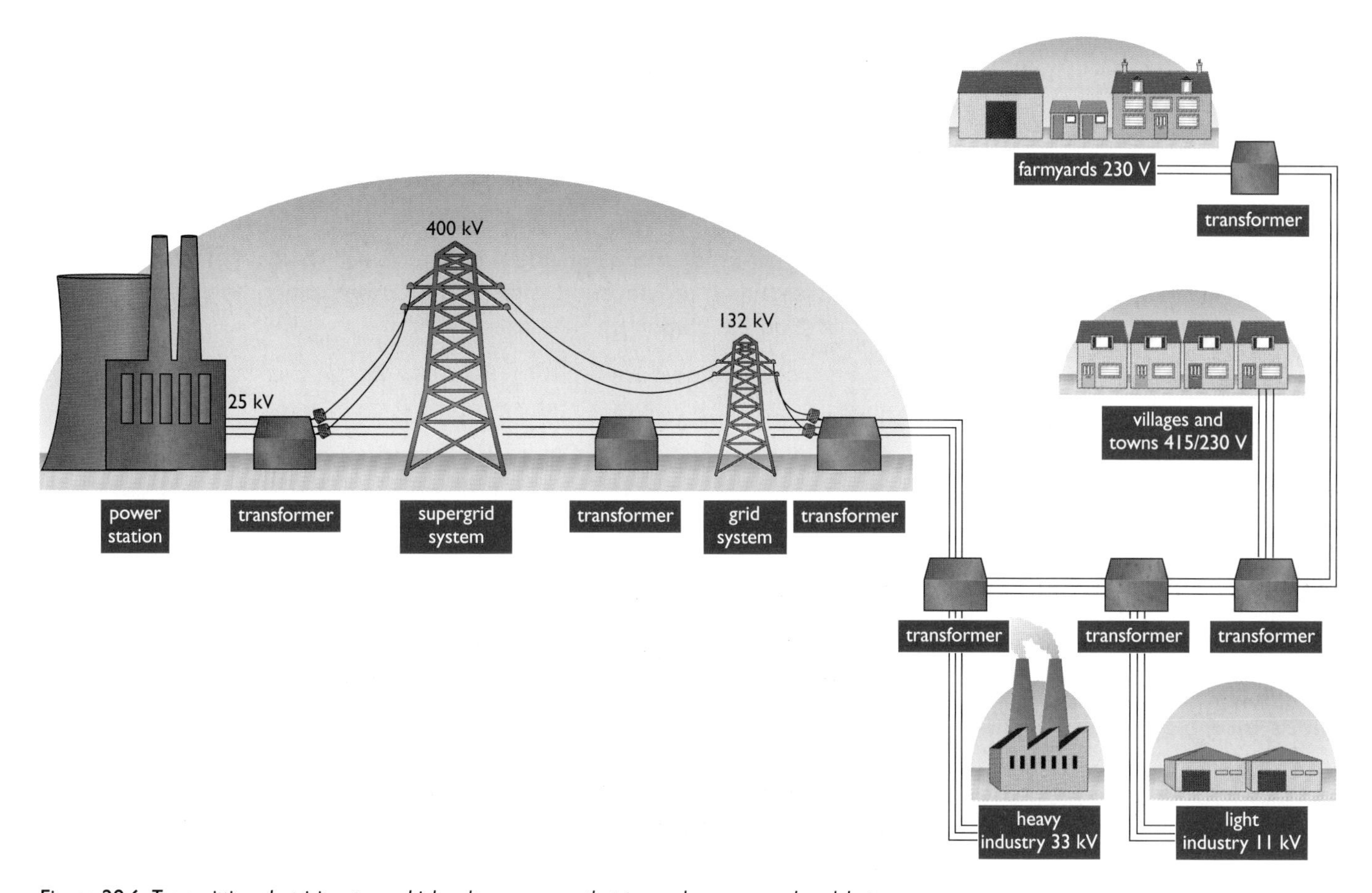

Figure 28.6 *Transmitting electricity at very high voltages means that power losses are reduced, but the voltage must be stepped down again before the supply reaches consumers.*

End of Chapter Checklist

If you haven't got a copy of your specification, read the introduction on page viii.

You will need to be able to do some or all of the following. Check your Awarding Body's specification (syllabus) to find out exactly what you need to know.

- Know that electricity is generated by rotating coils of wire in a strong magnetic field. The machine that makes electricity is called a generator.
- Know that generators are usually rotated by steam-driven turbines. The energy to create the steam is provided by burning chemical fuels or from nuclear reactions.
- Know that generators can also be driven by wind or water power.
- Understand that different types of power station can respond to changes in the demand for electricity at different rates.
- Appreciate the many different cost considerations in the choice of a power station. These include the cost of fuel, the cost of building, maintaining and decommissioning the power station, and the effect of the power station on the environment.
- Understand that transmitting electrical energy over large distances involves energy losses, and explain how these arise.
- Calculate the benefit of high-voltage transmission of energy.

Questions

More questions on generating electricity can be found at the end of Section E on page 291.

1 Copy and complete the following:

Electricity is produced by a machine called a ___________ . This consists of a ___________ that is rotated in a ___________ ___________ . The energy needed to drive the ___________ is usually supplied by a steam-driven ___________ . The steam may be produced by burning fossil fuels like ___________ , ___________ or ___________ .

2 ***a)*** The demand for electricity varies on a hourly and daily basis. Sometimes the changes in demand are predictable, but sometimes they are not. Give two examples of predictable changes in demand and two examples of unpredictable demand variations.

b) Electricity is generated by a number of different types of power station. Give three examples of different types of power station and compare how well they are able to respond to sudden changes in demand.

3 "Tidal power stations provide energy at zero cost and produce no pollution." Give one argument in support of this view, and one argument against it.

4 An electricity substation has a transformer that reduces the voltage from 11 kV to 440 V for use by a factory. The factory uses 132 kW on average during the working day. Calculate the following:

a) the "turns ratio" of the transformer (the value of N_p/N_s)

b) the average current used by the factory

c) the current flowing in the 11 kV transmission cable supplying the substation.

5 Explain the advantages of high-voltage power transmission.

Chapter 29: The Kinetic Theory of Matter

All matter is made up of particles that are continuously moving. The arrangement and movement of the particles determine the properties of the material. In gases, scientists have discovered laws that describe the relationship between pressure, temperature and volume. In this chapter, you will learn what these laws are and how this relationship can be explained in terms of the behaviour of the particles.

All matter is made up of tiny particles. (See Chapter 26, where the transfer of heat by conduction in a solid is explained in terms of these particles.) The **kinetic theory of matter** uses this idea to explain the different properties of matter. We think that all matter is made up of particles that are moving. The way that the particles are arranged and the way that they move determine the properties of the material, such as its state at room temperature or its density.

The states of matter

Some of the properties of a substance depend on the chemicals it is made from. However, substances can exist in different states. The main states of matter are solid, liquid and gas. We are used to finding some substances in each state in everyday life – for example, water is familiar as ice, as water and as steam. There are other substances that we usually see in only one state – for example, we rarely see iron in any other state than solid, or experience oxygen in any other state than as a gas.

Properties of the different states of matter

Most of this chapter will be about the properties of gases, but we shall mention solids and liquids for completeness.

Solids

Solids have a definite rigid structure and they are often very dense. The **density** of a substance is a measure of how tightly packed the particles are. We define density as follows:

$$\textbf{density} = \frac{\textbf{mass}}{\textbf{volume}}$$

Density is measured in kilograms per cubic metre (kg/m^3), or grams per cubic centimetre (g/cm^3). The following table shows the densities of some common solids, liquids and gases.

	Density	
Substance	**in kg/m^3**	**in g/cm^3**
iron	7800	7.8
aluminium	2700	2.7
wood (typical softwood)	600	0.6
oil (cooking)	920	0.92
water	1000	1
helium	0.18	0.00018

Solids have higher densities because the particles that they are made from are very closely packed. There are strong forces between the particles.

These forces between the particles give solid objects their definite shape and, in some materials, a great deal of strength.

Although the particles are held together by strong forces, they can still move. They "vibrate" about their fixed positions in the solid. When we supply energy to a solid, by heating it, the particles vibrate more – they move more quickly. We notice the increase in the kinetic energy of the particles in a substance as an increase in the *temperature* of the substance.

Liquids

Liquids share a property with gases – they have no definite shape. However, the particles that make up liquids tend to stick together, unlike gas particles. Liquids will occupy the lowest part of any container but gases will expand to fill any container that they are in. Liquids have much greater densities than gases. This is because the particles in liquids are still very close together, like they are in solids. Because the particles in liquids are close together, they still attract one another and hold together. In liquids, there is no fixed pattern and the particles can move around more freely than in solids. As we heat liquids, the movement of the particles becomes more energetic.

Gases

In gases the particles are very spread out, with large spaces between them. This means that the forces holding them together are small. Gases have very low densities and no definite shape. Gases can also be squashed into a smaller space (compressed). Solids and liquids are very difficult to compress.

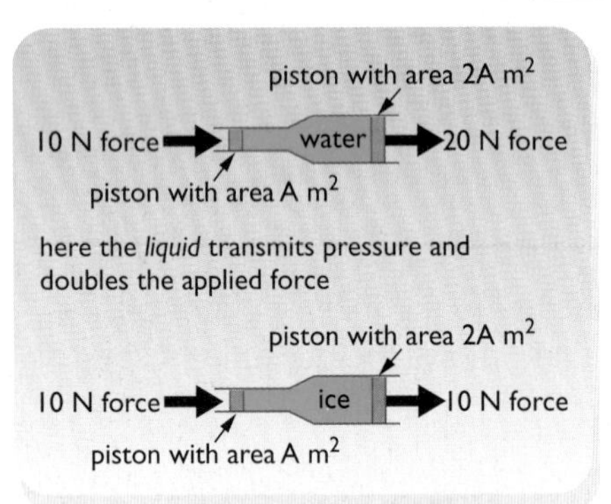

Figure 29.1 *Solids transmit forces but liquids and gases transmit pressure.*

Summary of the properties of solids, liquids and gases

Property	Solids	Liquids	Gases
definite shape	yes	no	no
can be easily compressed	no	no	yes
relative density	high	high	low
can flow (fluid)	no	yes	yes
can transmit forces	yes	no	no
expands to fill all available space	no	no	yes

The particles (molecules) in a solid: (Figure 29.2a)

- are tightly packed
- are held in a fixed pattern or crystal structure by strong forces between them
- vibrate around their fixed positions in the structure.

The particles (molecules) in a liquid: (Figure 29.2b)

- are tightly packed
- are not held in fixed positions but are still bound together by strong forces between them
- move at random.

The particles (molecules) in a gas: (Figure 29.2c)

- are very spread out
- have no fixed positions and the forces between them are very weak
- move with a rapid, random motion.

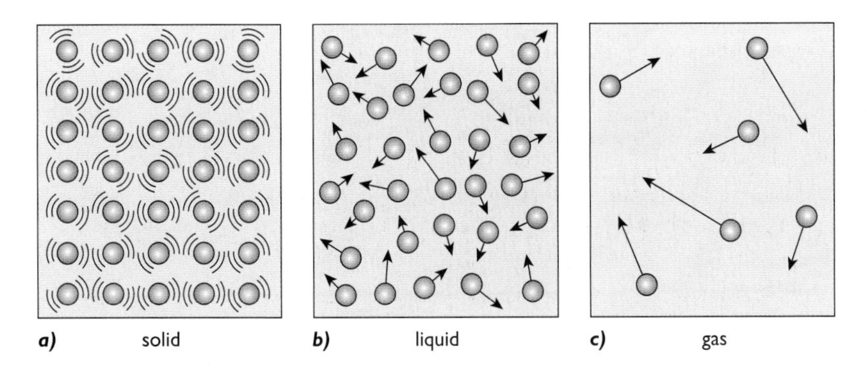

Figure 29.2 *The different arrangements of the particles in solids, liquids and gases.*

The gas laws

We are now going to focus our attention on the properties of gases. We shall explain the different properties in terms of the **kinetic theory**.

We have already said that gases are made up of particles that are moving. We believe that the particles in gases are spread out and constantly moving in a random, haphazard way.

Boyle's Law

The scientist Robert Boyle discovered something that you have probably noticed if you have ever used a bicycle pump: air is squashy! He noticed that you can squeeze air in a cylinder and that it springs back to its original volume when you release it. You can try this for yourself with a plastic syringe (Figure 29.4).

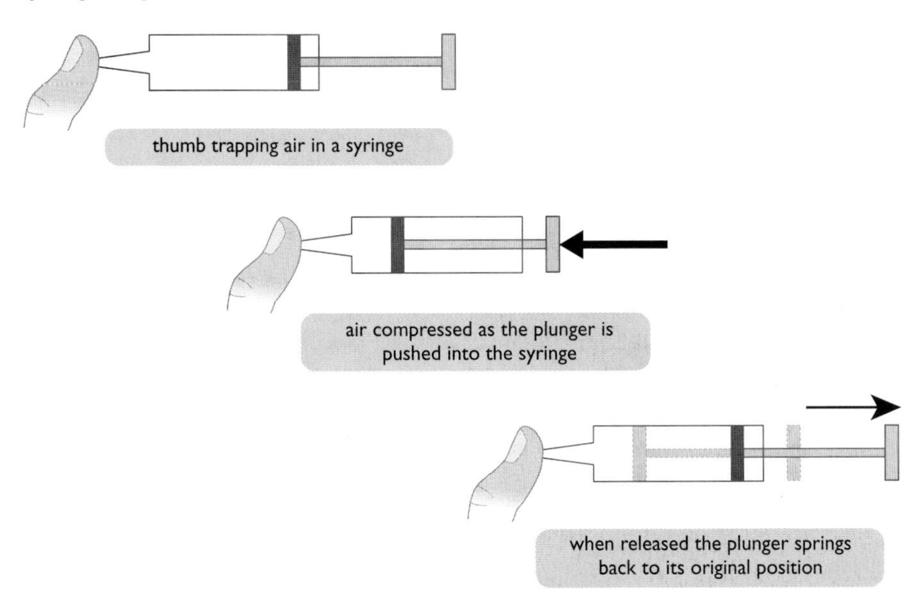

Figure 29.4 *Air is squashy!*

Figure 29.3 *Robert Boyle (1627–1691) was an Anglo-Irish chemist and philosopher. As well as discovering the law that bears his name, he worked with Robert Hooke at Oxford developing the air pump.*

Boyle devised an experiment to see how the volume occupied by a gas depends on the pressure exerted on it. **Pressure** is the force acting per unit area. This is measured in N/m^2. One N/m^2 is called a pascal (Pa) after the French mathematician and philosopher, Blaise Pascal (1623–1662). A version of Boyle's experiment is shown in Figure 29.5.

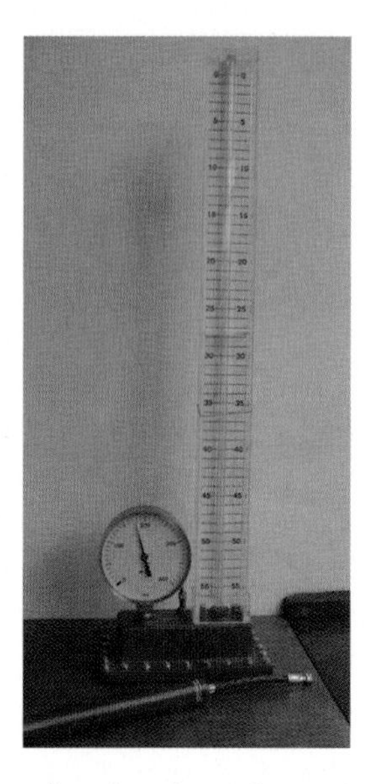

Figure 29.5 *Boyle's experiment to see how the volume of a gas depends on the pressure exerted on it.*

Boyle took care to make sure that the trapped gas he was studying stayed at the same temperature. He increased the pressure on the gas and made a note of the new volume. His results looked like the graph shown in Figure 29.6a.

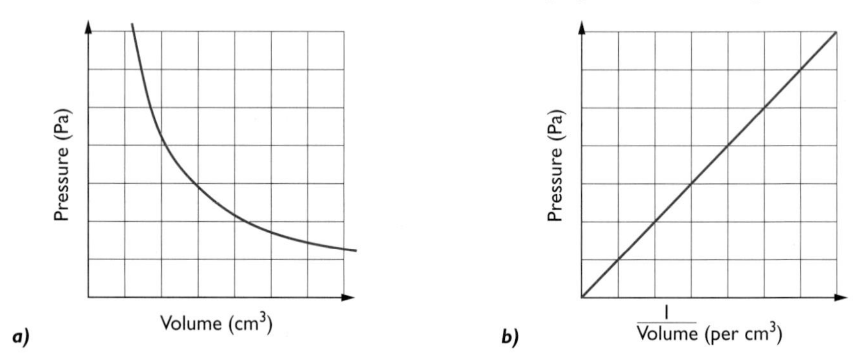

Figure 29.6 a) *Graph to show how the pressure of a gas at constant temperature varies with the volume,* b) *graph of pressure against 1/volume.*

Boyle noticed that when he doubled the pressure, the volume of the gas halved. If we plot pressure (p), against 1/volume (1/V), as in Figure 29.6b, we can see from the straight line passing through the origin that p is proportional to 1/V.

This discovery, called Boyle's Law, is expressed in the equation:

pressure × volume = a constant

$$pV = c$$

where c is a constant

Figure 29.7 shows a simple memory aid for Boyle's Law.

The constant depends on the particular gas, how much of it there is, and what the temperature is. For a fixed amount of a particular gas at constant temperature:

Pressure is proportional to 1/volume.

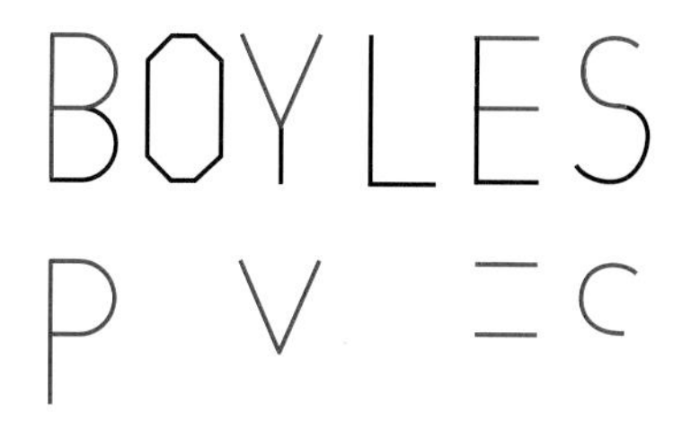

Figure 29.7 *Boyle's Law may be remembered from his own name.*

Gases can be compressed because the particles are very spread out. When a gas is squashed into a smaller container it presses on the walls of the container with a greater pressure. This is explained in terms of the kinetic theory as follows.

If the gas is kept at the same temperature, the average speed of the particles stays the same. (Remember that temperature is an indication of the kinetic energy of the particles.) If the same number of particles is squeezed into a smaller volume, they will hit the container walls more often. Each particle exerts a tiny force on the wall with which it collides. More collisions per second means a greater average force on the wall and, therefore, a greater pressure.

Absolute zero

Boyle took care to conduct his experiment at constant temperature. He was aware that temperature also had an effect on the pressure of a gas. Figure 29.8 shows an experiment to investigate how the pressure of a gas depends on its temperature.

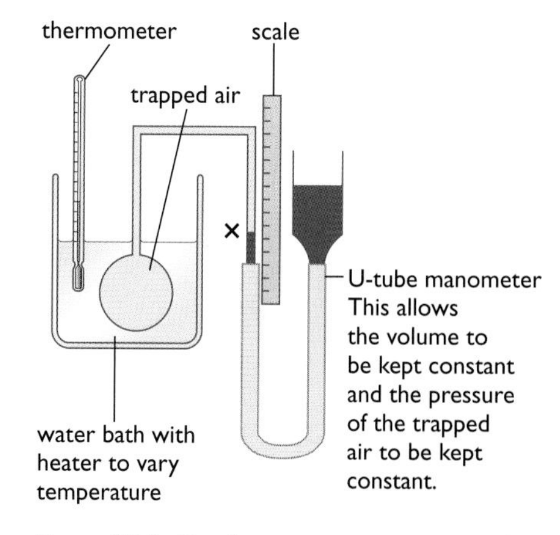

Figure 29.8 *Simple apparatus to measure the pressure of a fixed volume of gas at a range of temperatures.*

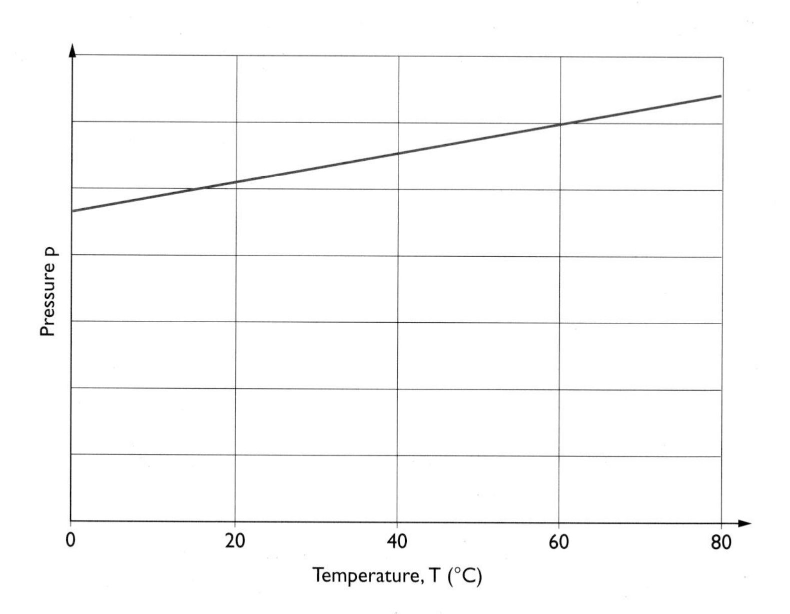

Figure 29.9 *Graph to show how the pressure of a fixed volume of gas varies with temperature.*

The gas is kept at constant volume and its pressure is measured at a range of temperatures. The graph in Figure 29.9 shows typical results for this experiment.

The graph shows that the pressure of the gas increases as the temperature increases. We could also say that the gas pressure gets smaller as the gas is cooled.

What happens if we keep cooling the gas? The graph in Figure 29.10 shows this.

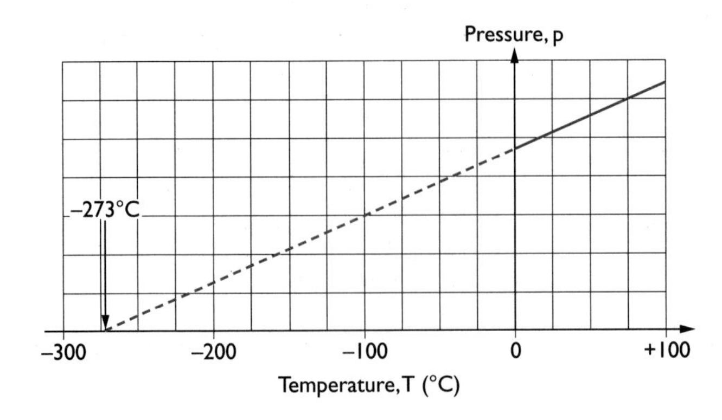

Figure 29.10 *At a temperature of –273°C, the pressure of the gas would be zero. This temperature is known as "absolute zero".*

As we cool the gas, the pressure keeps decreasing. The pressure of the gas cannot become less than zero. This suggests that there is a temperature below which it is not possible to cool the gas further. This temperature is called **absolute zero**. Experiments show that absolute zero is approximately –273°C.

The **kelvin** temperature scale starts from absolute zero. To convert a temperature on the Celsius scale (in °C) to a kelvin scale temperature (in K), add 273 to the Celsius scale temperature:

temperature in K = temperature in °C + 273

temperature in °C = temperature in K – 273

worked example

Example 1

a) At what temperature does water freeze, in kelvin?

Water freezes at 0°C. To convert 0°C to kelvin:

$T = (0°C + 273)$ K

$= 273$ K

b) What is room temperature, in kelvin?

Typical room temperature is 20°C, so:

$T = (20°C + 273)$ K

$= 293$ K

c) At what temperature does water boil, in kelvin?

Water boils at 100°C, so:

$T = (100°C + 273)$ K

$= 373$ K.

If we redraw the graph of pressure against temperature using the absolute or kelvin temperature scale, we get a graph that is a straight line passing through the origin, as shown in Figure 29.11. This shows that the pressure of the gas is proportional to its kelvin temperature. For example, if you heat a gas from 200 K (–73°C) to 400 K (127°C) its pressure will double.

Pressure, p
0
+100
+200
+300
+400
absolute temperature, T (K, kelvin)

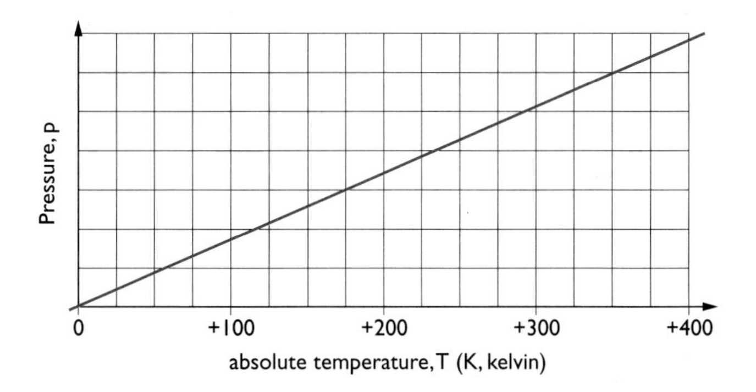

Figure 29.11 *Graph of pressure against absolute temperature for a fixed amount of gas at constant volume.*

Pressure is proportional to kelvin temperature for a fixed amount of gas at constant volume.

The relationship can be explained as follows.

The number of gas particles and the space, or volume, they occupy remain constant. When we heat the gas the particles continue to move randomly, but with a higher average speed. This means that their collisions with the walls of the container are harder and happen more often. This results in the average pressure exerted by the particles increasing.

When we cool a gas the kinetic energy of its particles decreases. The lower the temperature of a gas the less kinetic energy its particles have – they move more slowly. At absolute zero the particles have no thermal or movement energy, so they cannot exert a pressure.

Cooling things further becomes more and more difficult as they get closer to 0 K. Scientists have managed to cool gases down to within a few thousandths of a degree above absolute zero. At this temperature even gases like helium and hydrogen can be liquefied.

Charles' Law

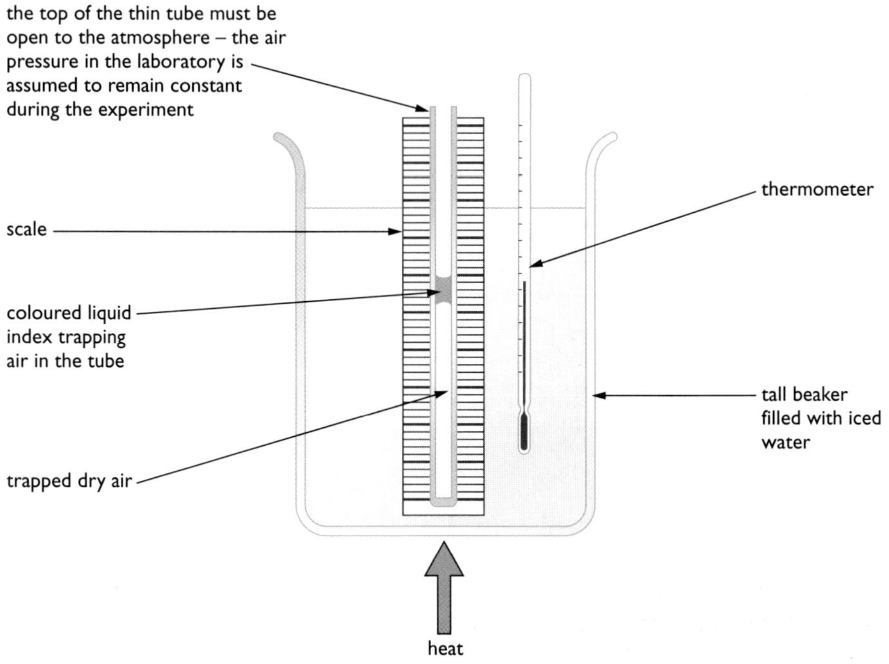

the top of the thin tube must be open to the atmosphere – the air pressure in the laboratory is assumed to remain constant during the experiment
thermometer
scale
coloured liquid index trapping air in the tube
tall beaker filled with iced water
trapped dry air
heat

Figure 29.12 *Apparatus used to investigate the effect of temperature on the volume of a fixed amount of gas kept at constant pressure.*

Boyle kept temperature constant and investigated the effect of pressure on the volume of a gas. We have also seen the effect of temperature on the pressure of gas kept at constant volume. The French scientist and aeronaut Jacques Charles (1746–1823) conducted an experiment to investigate the effect of temperature on the volume of a fixed amount of gas kept at constant pressure. A version of his experiment is shown in Figure 29.12 (on page 277), and Figure 29.13 shows typical results.

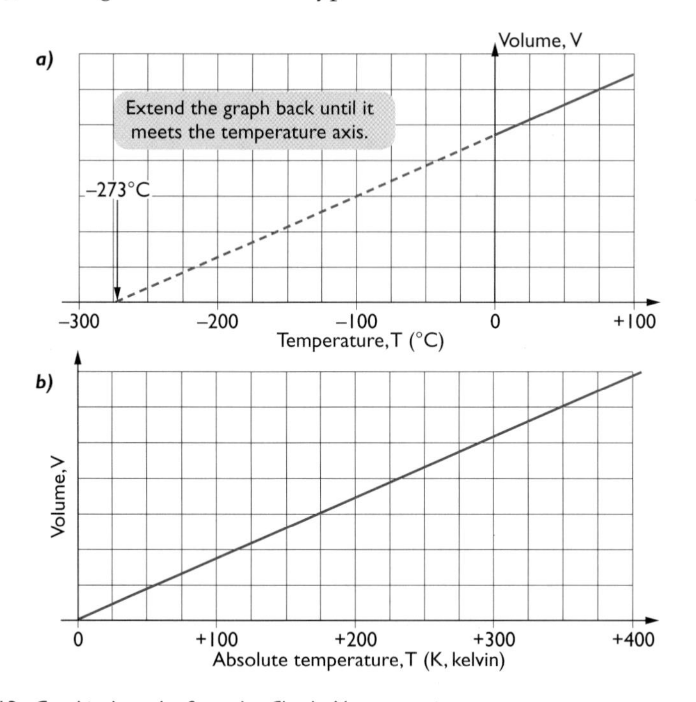

Figure 29.13 *Graphical results from the Charles' Law experiment.*

The graph in Figure 29.13a shows a similar trend to the pressure–temperature graph. As the gas is heated it expands, and as it is cooled its volume decreases. If we keep on cooling the gas (and it remains a gas) its volume keeps getting smaller. If we extend the graph back, it also cuts the V = 0 axis at –273°C. This provides further evidence for the absolute zero of temperature.

If we plot volume against kelvin temperature, as shown in Figure 29.13b, we get a straight line passing through the origin of the axes. This shows that:

The volume of a fixed amount of gas at constant pressure is proportional to its absolute or kelvin temperature.

This is Charles' Law, and it means that if we cool a gas from 300 K to 150 K (a chilly –123°C), for example, the volume it occupies will halve, provided we keep its pressure constant.

This happens because the particles in a gas move more slowly when the gas is cooled. As the particles move more slowly they do not put as much pressure on the walls of the container, so the gas gets squeezed into a smaller volume. This in turn means that the particles hit the walls more often because the space they are moving around in is smaller. The volume that the gas occupies will decrease until the gas exerts the same pressure as before (Figure 29.14).

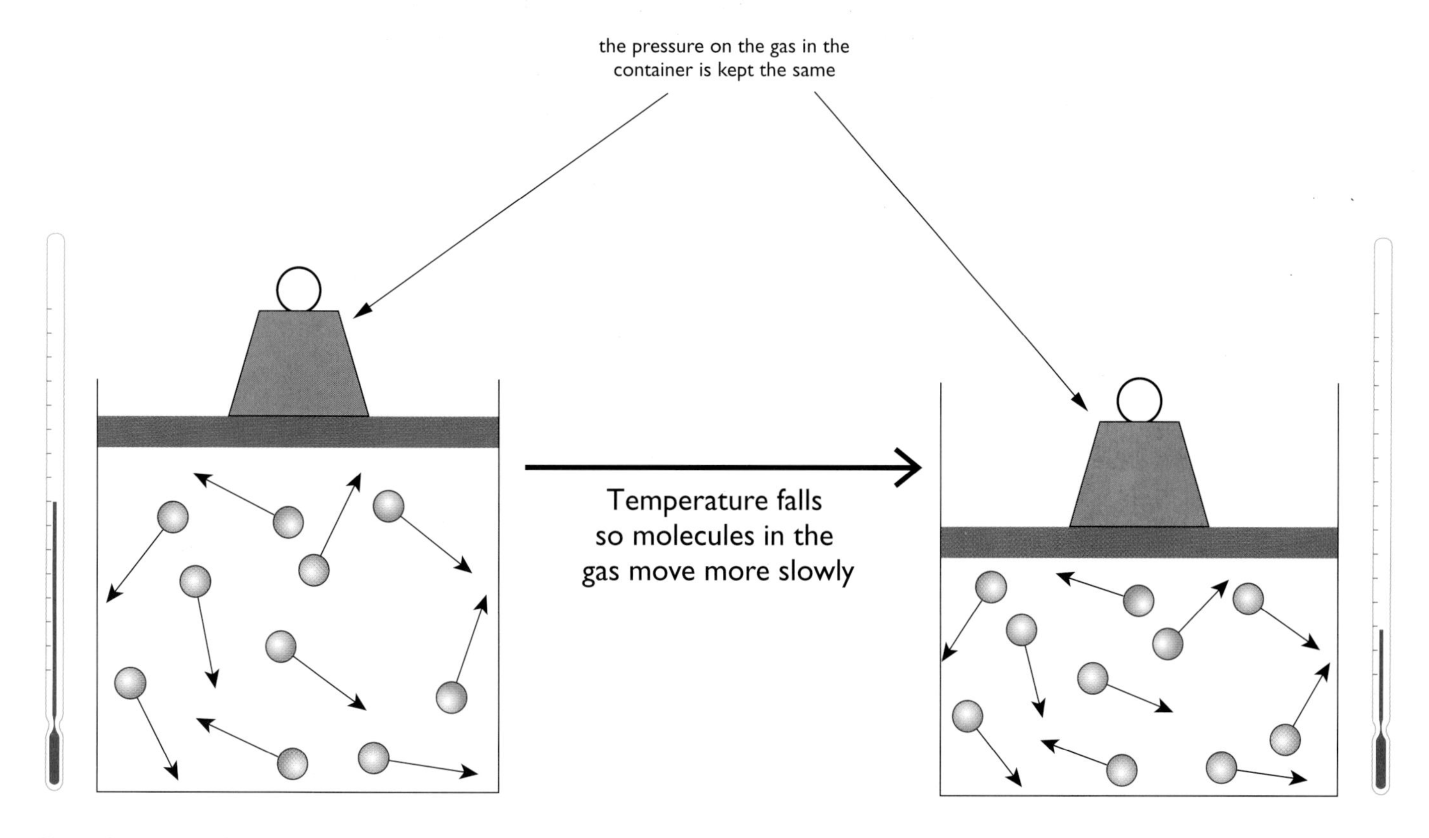

Figure 29.14 *The effect of cooling a fixed amount of gas keeping its pressure the same.*

The general gas equation

All the effects we have noted can be combined in a single equation, called the general gas equation:

(pressure × volume)/kelvin temperature = constant for a fixed amount of gas

$$\frac{pV}{T} = c$$

Remember that T must be measured in kelvin.

The value of the constant depends on the amount of gas involved. For a certain amount of gas, a more useful equation is:

$$\frac{p_1V_1}{T_1} = \frac{p_2V_2}{T_2}$$

p_1,V_1 and T_1 are the initial values of pressure, volume and temperature of the gas and p_2,V_2 and T_2 are the new conditions of pressure, volume and temperature of the gas after they have been changed. As usual, temperature must be measured in kelvin, and pressure and volume must be measured in *the same units on both sides of the equation.*

So, if we know the initial conditions of a gas and we change two of the factors we can predict the effect on the third.

This is best shown with an example.

***worked* example**

Example 2

A balloon contains some helium at a pressure of 120 kPa. (Atmospheric pressure is typically 100 kPa.) It has a volume of 2000 cm^3 at ground level where the temperature is 27°C. It floats up into the sky. At a certain height the pressure of the gas has fallen to 80 kPa and the temperature has fallen to –3°C. What is the volume of the balloon under these new conditions?

First write down what we know:

p_1 = 120 kPa

V_1 = 2000 cm^3

T_1 = 27°C but we *must convert this to kelvin*, so T_1 = 27 + 273 K = 300 K

p_2 = 80 kPa

T_2 = –3 +273 K = 270 K

Now we use the equation:

$$\frac{p_1V_1}{T_1} = \frac{p_2V_2}{T_2}$$

Since we want to find V_2, the new volume, we might rearrange the equation to make V_2 the subject:

$$V_2 = \frac{p_1V_1}{T_1} \times \frac{T_2}{p_2}$$

$$V_2 = \frac{120\text{ kPa} \times 2000\text{ cm}^3}{300\text{ K}} \times \frac{270\text{ K}}{80\text{ kPa}}$$

$$= 2700\text{ cm}^3$$

The balloon has expanded. Although the temperature has dropped to 0.9 of its original value (which, on its own, would make the volume decrease) the pressure has fallen by a greater proportion (a decrease in pressure tends to cause the gas to expand).

End of Chapter Checklist

If you haven't got a copy of your specification, read the introduction on page iv.

You will need to be able to do some or all of the following. Check your Awarding Body's specification (syllabus) to find out exactly what you need to know.

- Understand that matter is made up of particles (molecules or atoms) that are in continuous motion.
- Appreciate that the average energy of the motion of particles in a substance increases with temperature.
- Account for the different properties of solids, liquids and gases in terms of the arrangement and motion of the particles in the substance.
- Know that gases exert a pressure because of the collisions that gas particles make with other matter.
- Understand that when you heat a gas with a fixed volume its pressure will increase because the particles hit the walls of the container more frequently and more energetically.
- Explain the changes in pressure and volume of a gas as the gas temperature changes in terms of the motion of the particles of the gas.
- Describe how to investigate the relationship between pressure, volume and temperature of gases keeping one factor constant.
- Understand the reasons we believe that there is an absolute zero of temperature and know how to convert Celsius temperature to absolute or kelvin temperature.
- Recall and apply the gas law equations.

Questions

More questions on the kinetic theory of matter can be found at the end of Section E on page 291.

1 ***a)*** Convert the following Celsius temperatures to absolute (kelvin) temperatures.

i) 0°C

ii) 100°C

iii) 20°C

b) Convert the following absolute temperatures to Celsius temperatures.

i) 250 K

ii) 269 K

iii) 305 K

2 Copy and complete the following paragraph about the kinetic theory of matter.

Matter is made up of ____________ that are in continuous ____________ . When we supply heat energy to matter the ____________ move ____________ . Gases exert a pressure on their containers because the ____________ are continually ____________ with the walls. The pressure will ____________ when the gas is heated in a container of fixed volume because the ____________ are moving ____________ .

3 Explain how the kinetic theory of matter accounts for the absolute zero of temperature.

4 State what happens in the situations shown in the diagram. Explain your answers using the kinetic theory.

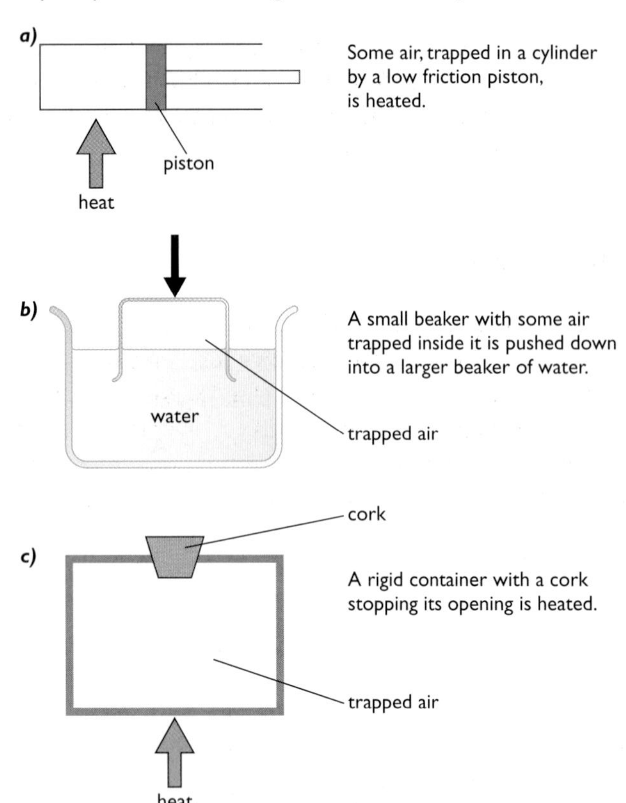

5 The table below gives some figures for a Charles' Law experiment in which a fixed amount of air is heated in such a way as to keep the pressure of the air constant.

Temperature (°C)	0	12	25	35	45	52	60	70
Volume (cm^3)	6.0	6.3	6.6	6.8	7.0	7.2	7.4	7.6

a) Draw a sketch diagram of a suitable arrangement of apparatus to carry out this investigation. State how your apparatus ensures that the trapped air is kept at constant pressure. Explain any other precautions that you would take to ensure the accuracy of your results.

b) Plot a graph using the given results, to obtain an estimate of the value of absolute zero in °C.

c) At what temperature would the volume of the gas be 9.0 m^3? Express your answer in *i)* °C, and *ii)* K.

6 A fixed amount of gas occupies a volume of 500 m^3 at a temperature of −23°C, when under a pressure of 100 kPa. Use the gas equation to calculate the volume that this amount of gas will occupy when its temperature is increased to 27°C and the pressure on the gas is increased to 125 kPa. You should show each stage of your calculation.

Chapter 30: Calorimetry

The heat energy required to raise the temperature of a substance by a certain amount depends of the specific heat capacity of the substance, and on the mass of substance to be heated. In this chapter, you will learn how to calculate the amount of energy transferred between bodies when temperature changes are involved.

Calorimetry is the measurement of heat. More specifically, it is the measurement of how much heat is transferred between objects at different temperatures. Before we can start measuring heat we need to be clear about the difference between heat and temperature.

Heat and temperature

We have already learned that heat is a form of energy (see page 227). **Temperature** is how hot something is. We experience temperature by touch and we can measure temperature using a thermometer. We saw, in Chapter 29, that the temperature of an object is a measure of the average speed of the particles of which it is made up. Heat and temperature are *not* different ways of saying the same thing. We can show this by a simple experiment, shown in Figure 30.1.

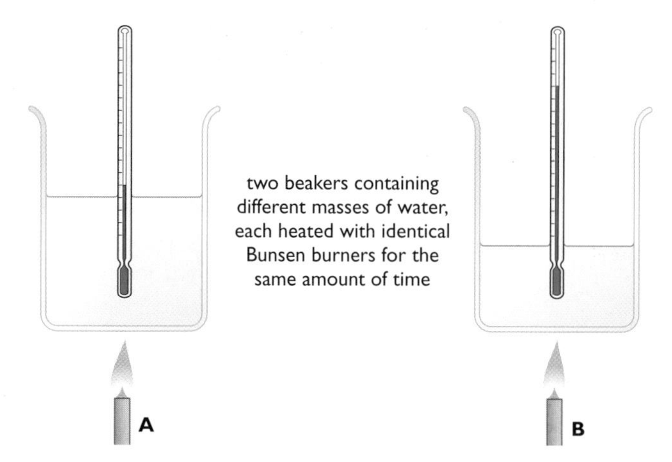

Figure 30.1 *An experiment to show that supplying the same amount of heat energy to two objects does not necessarily raise their temperatures by the same amount.*

Two identical beakers, A and B, are filled with different amounts of water and the temperature of each is measured. This is to check that they are both starting at the same temperature. They are both heated with identical Bunsen burners for the same amount of time. At the end of the heating process you will not be surprised to find that the temperature of the water in beaker B is greater than that in A.

Burning the same amount of gas in the Bunsen burner converts the same amount of chemical energy into heat. So we have given the same amount of heat energy to each beaker of water. The rise in temperature is not the same because there was more water to heat up in beaker A.

A higher temperature means that the average kinetic energy of the particles in the water is greater. If you supply the *same* amount of energy to a larger mass of substance – that is, with more particles – the average increase in kinetic energy of the particles will be smaller and so, too, will the temperature rise. If you want to raise the temperature of a large mass and a small mass by the same amount, you must supply *more* heat energy to the larger mass.

Heating different substances

We have just seen that more heat energy is needed to raise the temperature of a larger mass of water by a stated amount. We can do a similar experiment to see if different substances need different amounts of heat to raise their temperature by a given amount. Figure 30.2 shows an experiment to investigate this.

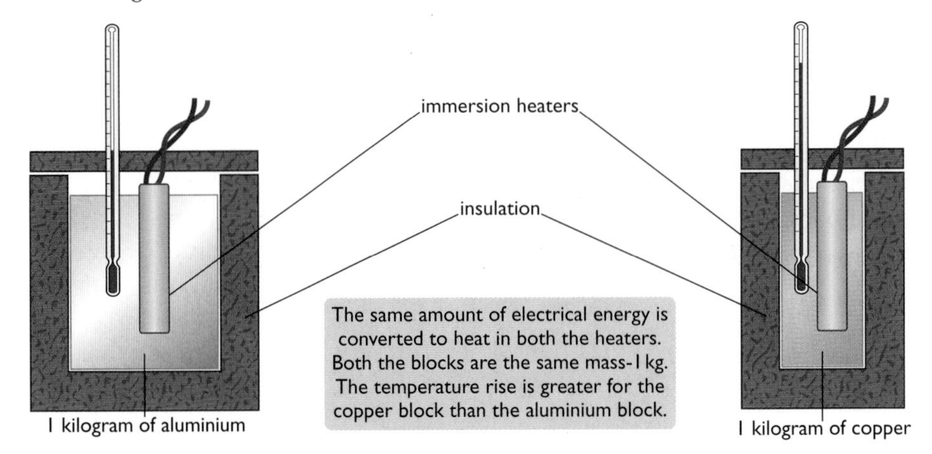

Figure 30.2 *Experiment to investigate whether different substances need different amounts of heat to raise their temperatures by the same amount.*

Two blocks of metal are heated electrically. The blocks have the same mass and are insulated to reduce heat loss to the surroundings. The amount of energy supplied to the blocks can be measured using an ammeter and a voltmeter. The thermometers are used to measure the temperature rise in each block. This experiment shows that it takes different amounts of energy to heat the same mass of different substances through the same rise in temperature.

We express this difference between different substances by defining the **specific heat capacity** or **shc** of a substance as follows:

A change in temperature of I kelvin is the same as a change of I°C.

The specific heat capacity of a substance is the amount of heat energy needed to raise the temperature of one kilogram of the substance by I°C or I K.

Specific heat capacity is measured in joules per kilogram per °C (J/kg/°C) or joules per kilogram per K (J/kg/K).

The table shows the specific heat capacities of some common substances.

Substance	Shc (J/kg/°C)
water	4200
copper	390
aluminium	900
glass	640
cooking oil	2000
air	1005

If we want to calculate how much heat is needed to raise the temperature of a known mass of a substance by a particular amount we use the equation:

$$\text{heat energy, E (in J)} = \text{mass, m (in kg)} \times \text{shc, c (in J/kg/°C)} \times \text{temperature rise, } \Delta T \text{ (in °C)}$$

$$E = mc\Delta T$$

ΔT is the symbol used for the rise in temperature. It is pronounced "delta tee". The Greek letter "delta" (Δ) is used in physics to mean "the change in" – so, for example, we could write Δv as mathematical shorthand for "the change in velocity" of an object.

Energy transfer calculations

Heating water for a bath

To find out how much energy is needed to heat the water for a bath, we need to know how much water is in the bath, and how much the temperature of the water has been raised.

worked example

Example 1

Suppose that a bath contains 120 kg of water and that the water has been heated from 18°C to 40°C (a rise of 22°C). The amount of heat needed to do this is calculated using the formula as follows:

heat energy = mass × shc × temperature rise

heat energy needed (in J) = 120 kg × 4200 J/kg/°C × 22°C

= 11 088 000 J or 11.1 MJ

(MJ stands for mega joule: 1 MJ = 10^6 J.)

If this water is heated using an electric immersion heater we can work out how long the heater needs to be switched on and how many units of electricity are used. If we can calculate the number of units or kWh, we can also work out the cost of taking the bath! A typical immersion heater is rated at 2.4 kW (it converts 2400 J of electrical energy into heat every second). The time taken in seconds to provide 11.1 MJ is calculated by rearranging the power formula:

$$\text{power (in watts)} = \frac{\text{energy (in joules)}}{\text{time (in seconds)}}$$

$$\text{time} = \frac{\text{energy}}{\text{power}}$$

$$= \frac{11\,088\,000\text{ J}}{2400\text{ W}}$$

= 4620 s or 77 min or 1.28h

If a 2.4 kW heater is on for 1.28 hours it consumes (2.4 × 1.28) kWh of electrical energy – that is, just over 3 kWh or units. If a unit costs 7 p then the cost of heating the bath is 21 p.

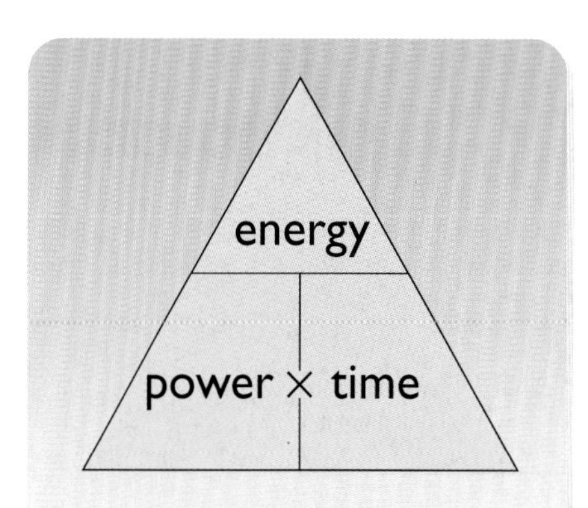

Figure 30.3 *You can use the triangle method for rearranging equations like power = energy/time.*

The original calculation assumes that all the electrical energy is converted to heat. In practice heat will be lost from the water as it is being heated and as the bath is run, so the total energy consumption and cost will be greater.

Cooling the bath!

The bath temperature in Example 1 is too high. Some cold water must be mixed with the bath to cool it to a more comfortable temperature, say 35°C. The heat lost by the water in the bath is transferred to the cold water run into the bath. Assuming no heat energy is lost to the surroundings while the cold water is being mixed into the bath, then:

$$\text{heat lost by the hot water } \textit{cooling} = \text{heat gained by the cold water } \textit{heating up}$$

As an equation this is written:

$$\text{mass of hot water} \times \text{shc} \times \text{temperature } \textit{fall} = \text{mass of cold water} \times \text{shc} \times \text{temperature } \textit{rise}$$

In this example we can see that the shc term can be cancelled since it appears on both sides of the equation.

***worked* example**

Example 2

If we substitute the figures from the description above, we can calculate how much cold water (mass m), initially at 18°C, must be run into the bath to cool it to 35°C:

$$120\,\text{kg} \times \cancel{4200\,\text{J/kg/°C}} \times (40°\text{C} - 35°\text{C}) = m \times \cancel{4200\,\text{J/kg/°C}} \times (35°\text{C} - 18°\text{C})$$

$$120\,\text{kg} \times 5°\text{C} = m \times 17°\text{C}$$

$$m = \frac{120\,\text{kg} \times 5°\text{C}}{17°\text{C}}$$

$$= 35.3\,\text{kg}$$

(Remember that, in this case, the heat losses that we have ignored will make the actual amount of water required smaller.)

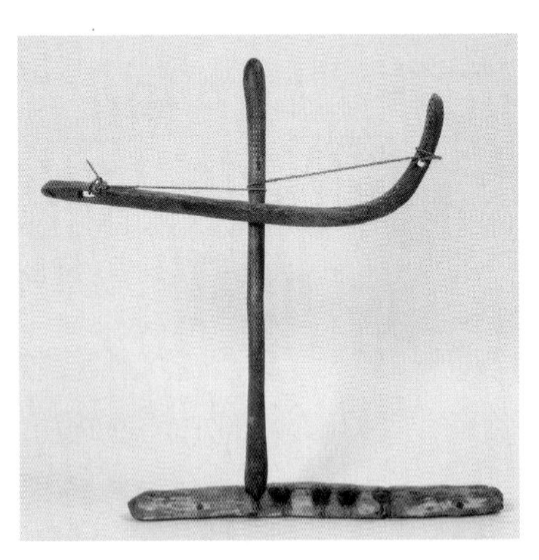

Figure 30.4 *The friction between the peg and the wood block generates enough heat energy to start a fire!*

Kinetic energy into heat

Figure 30.4 shows a fire being kindled by using a wood drill. A hard-wood peg is spun backwards and forwards with the bow in a notch cut into a block of softer wood. The kinetic energy of the spinning peg is converted into heat energy. Enough heat energy can be produced in this way to set light to dry kindling. Matches are a simpler option, but they, too, use heat converted from movement energy to start a chemical reaction. There are many other situations in which kinetic energy is converted into heat.

Suppose we fire an air gun pellet made of lead at a solid target. When it hits the target it is brought very rapidly to a halt. All its kinetic energy is converted to other forms. Some energy will be converted to sound, some is used in permanently deforming the air gun pellet and the target, but most of the energy will be converted into heat. This heat energy will raise the temperature of the air gun pellet and the target. If we can measure the temperature rise of the pellet we can work out how much heat energy is given to the pellet, using:

$$E = mc\Delta T$$

Figure 30.5 *When the pellet hits the target, all its kinetic energy is converted to other forms.*

worked example

Example 3

The mass, m, of an air gun pellet is 2 g (0.002 kg), and the shc, c, of the lead of which it is made is 120 J/kg/°C. If the temperature rise of the pellet, ΔT, after it hits the target is 10°C, calculate the minimum velocity of the pellet when it hit the target.

The heat energy, E, supplied to the pellet is given by:

$$E = m \times c \times \Delta T$$
$$= 0.002\,\text{kg} \times 120\,\text{J/kg/°C} \times 10\text{°C}$$
$$= 2.4\,\text{J}$$

We can use this figure to calculate the *minimum* velocity of the pellet when it hit the target. If *all* the KE of the pellet is converted to heat, and all this heat goes to raise the temperature of the pellet, then:

KE of the pellet = heat energy given to the pellet

$$\tfrac{1}{2}mv^2 = 2.4\,\text{J}$$
$$\tfrac{1}{2} \times 0.002 \times v^2 = 2.4$$
$$0.001 \times v^2 = 2.4$$
$$v^2 = 2400$$
$$v = \sqrt{(2400)}$$
$$v = 49\,\text{m/s}$$

Remember that this is the *minimum* velocity that the pellet must have. In practice, some of the kinetic energy is converted to sound and some of the heat produced is lost to the air or given to the target, so the pellet would have been moving faster than this.

Measuring temperature

You have probably used thermometers like those in Figure 30.6 in school. These use a liquid like mercury or alcohol (usually dyed red or blue) in a fine glass tube. The liquid expands as the temperature rises and contracts as the temperature falls. All thermometers rely on some property of a material that changes with temperature.

Data logging equipment uses electronic sensors to measure changes in the environment. Temperature sensors may use a number of properties of matter that change with their temperature. A suitable property for electronic sensors is resistance. The resistance of metals increases with temperature. Semiconductor materials are used to make components called **thermistors**, whose resistance can be made to fall as their temperature rises.

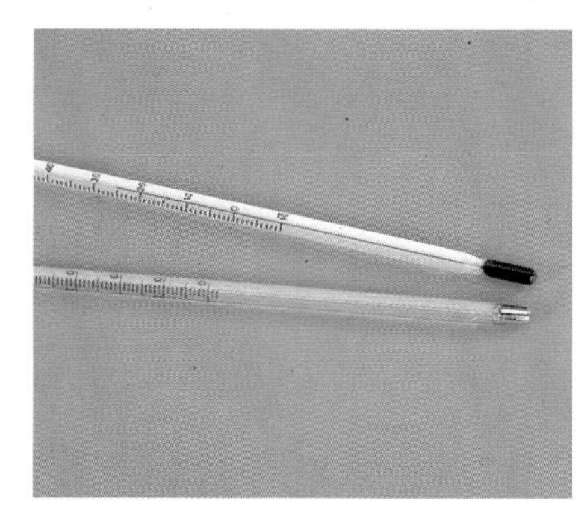

Figure 30.6 *Thermometers are used to measure temperature.*

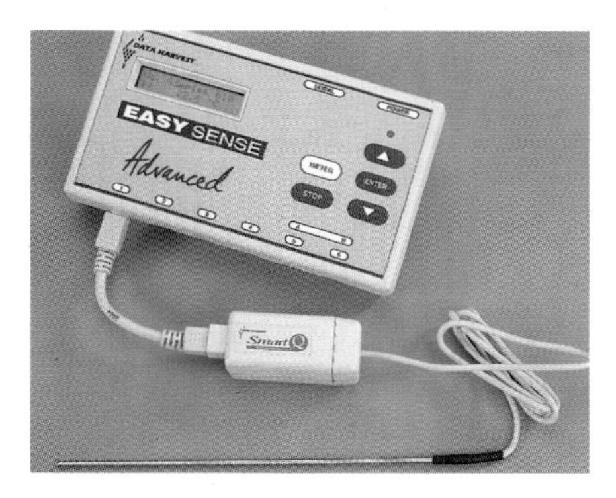

Figure 30.7 *Electronic sensors can be used to measure temperature and feed the data directly into data logging equipment.*

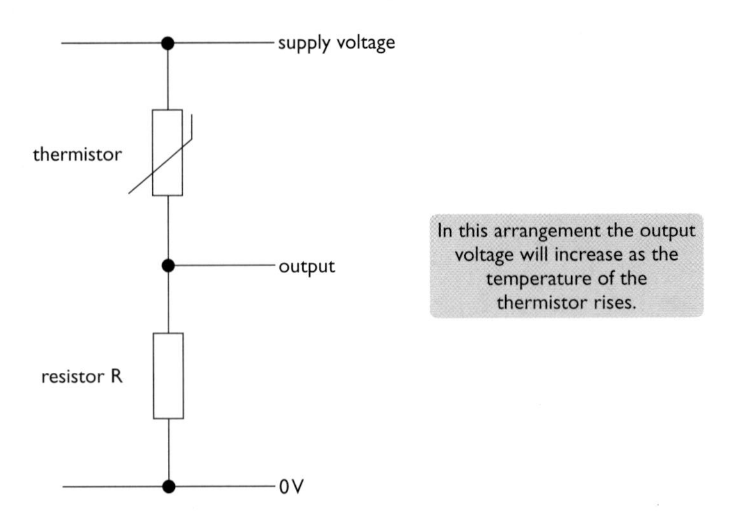

Figure 30.8 *This circuit can be used as a temperature sensor.*

We could calibrate the circuit without using another thermometer at all. All that is needed is to measure the value of the output voltage at two known temperatures, like the melting and boiling points of water.

The circuit shown in Figure 30.8 uses a potential divider circuit (see Chapter 7) to measure temperature. If the temperature rises the resistance of the thermistor decreases and this makes the output voltage rise. We could calibrate this thermometer circuit by plotting a graph of the output voltage from the circuit against temperature measured using a mercury-in-glass thermometer.

End of Chapter Checklist

If you haven't got a copy of your specification, read the introduction on page iv.

You will need to be able to do some or all of the following. Check your Awarding Body's specification (syllabus) to find out exactly what you need to know.

- Explain the difference between heat and temperature.
- Understand that it takes more heat energy to raise the temperature of a large mass of substance than a small mass of that substance.
- Understand that it takes different amounts of heat energy to make the temperature of the *same* mass of *different* substances increase by one degree.
- Recall and use the formula for the specific heat capacity (shc) of a substance.
- Calculate the heat transfer between bodies.
- Calculate energy conversions involving kinetic energy and heat.
- Understand that potential dividers using thermistors may be used to measure temperature.

Questions

More questions on calorimetry can be found at the end of Section E on page 291.

1 Copy and complete the following paragraphs about temperature and heat:

The ______________ of an object is a measure of how hot it is. The degree of hotness of an object depends on the average ______________ ______________ of the particles that it is made from. In a hot object the particles move around ______________ than in a cold object.
______________ is a form of energy that is transferred between places at different temperatures.

2 If it takes 8000 J of energy to raise the temperature of 2 kg of a metal by 10°C, how much energy does it take to heat:

a) 1 kg of the metal by 10°C

b) 1 kg of the metal by 1°C

c) 200 g of the metal by 1°C

d) 200 g of the metal by 25°C?

3 ***a)*** Copy and complete the following:

The specific heat capacity of water is 4200 J/kg/°C. This means that it takes ___________ J of heat ___________ to raise the temperature of ___________ ___________ of water by ______________________.

b) A hot water tank contains 80 kg of water. How much energy is needed to raise the temperature of the water in the tank by:

i) 1°C

ii) 25°C?

4 A hot water storage tank contains 75 kg of water and is heated by a 3 kW electric immersion heater. The cost of a unit of electricity (kWh) is 9 pence.

a) How much energy is needed to raise the temperature of the water in the tank from 15°C to 60°C?

b) How long does it take the immersion heater to supply this much energy?

c) How much does it cost to heat up the water?

d) What assumptions have you made in your calculations?

5 The diagram shows two blocks of metal, A and B. They are stuck together so that they are in thermal contact – that is, heat can be transferred easily from one block to the other. In the following questions, assume that the only heat transfer that takes place is between the two blocks.

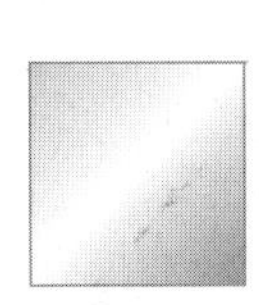

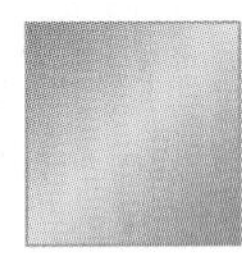

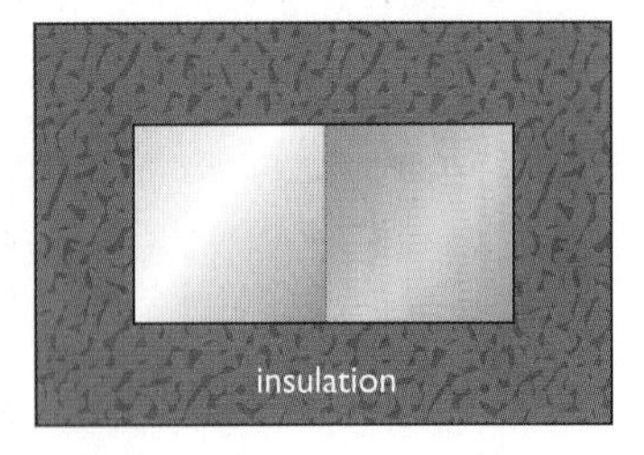

block **A** block **B** the two blocks brought into contact

a) The blocks are made of the same metal and have the same mass. Block A is initially at 100°C and block B is at 24°C. What is the final temperature of the two blocks when they are brought into contact? (Hint: Let the final temperature be X so that the temperature fall for block A is (100 – X)°C and the temperature rise for block B is (X – 24°C.)

b) Block A has twice the mass of block B. If block A is initially at 20°C and block B is at 35°C, what is the final temperature of the two blocks after they have been brought into contact?

c) Block A is made of a metal with half the shc of block B, but has three times the mass of block B. If block A is initially at 50°C and cools to 30°C when brought into contact with block B, what was the initial temperature of block B?

6 An insulated tube 1 m long contains 100 g of lead shot, initially at 20°C. The tube is swiftly turned upside down so that the lead shot falls to the bottom of the tube. This action is repeated a total of 50 times. Assume that no heat energy is lost from the lead shot and that the amount of energy converted to sound is very small.

a) Describe the energy conversions that take place in this experiment.

b) If lead has a shc of 120 J/kg/°C, what is the temperature of the lead shot at the end of the experiment?

End of Section Questions

1 Windmills and solar panels are both examples of renewable energy sources.

a) Explain what is meant by a *renewable energy source.* *(2 marks)*

b) What advantage do solar panels have over windmills as a source of energy? *(2 marks)*

Total 4 marks

2 A hydroelectric power station has turbines that are fed by a reservoir in the mountains, as shown in the diagram.

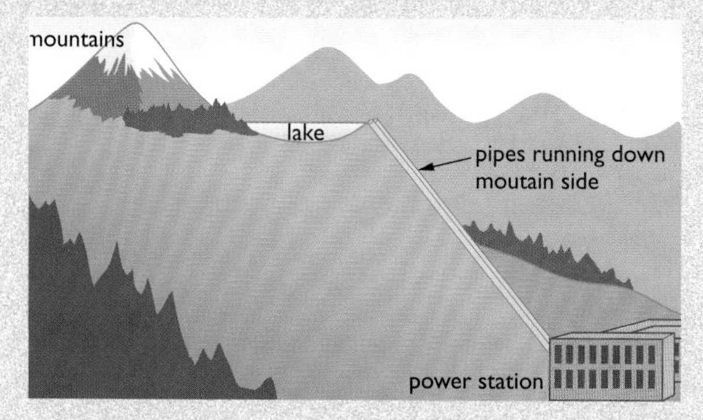

The reservoir is 400 metres above the power station. As water falls from the reservoir to the turbines in the power station it gains kinetic energy. The kinetic energy is converted to electrical energy by generators driven by the turbines.

a) How much kinetic energy does each kilogram of water gain as it falls to the turbines? (Take g as 10N/kg.) *(2 marks)*

b) The turbines are 60% efficient. What is the power output of the station if 300 kg of water passes through the turbines each second? *(4 marks)*

Total 6 marks

3 A power station produces 600 MW of electrical power. The output voltage of the generators is stepped up to 400 kV using a transformer. The power is then transmitted at this voltage on the National Grid.

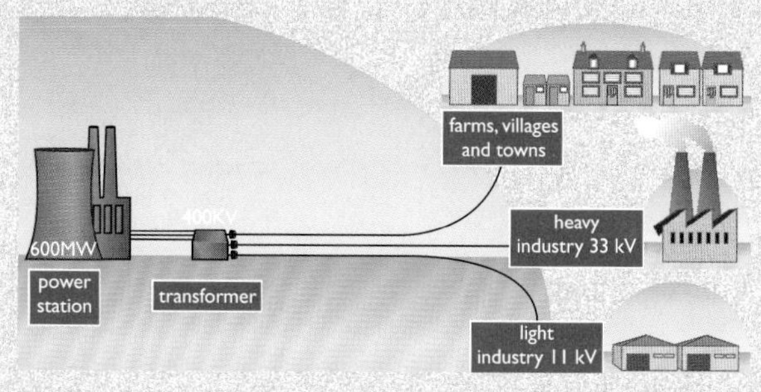

a) Describe, with the aid of a diagram, the basic structure of a transformer. *(4 marks)*

b) Write down the equation that allows you to calculate the output voltage of a transformer from its input voltage and a key feature of the design of the transformer. Define the terms in your equation. *(4 marks)*

c) Calculate the current that flows along the National Grid power lines. *(3 marks)*

d) What is the advantage of using a transformer to step up the voltage at which power is transmitted? *(3 marks)*

Total 14 marks

4 The diagram shows a lift powered by an electric motor.

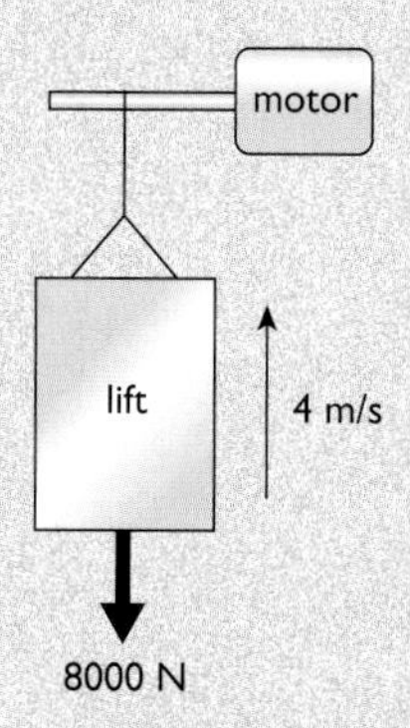

a) If the lift and its load of passengers have a total weight of 8000 N, and the lift is moving upwards at a steady rate of 4 m/s, what is the useful power output of the electric motor? *(3 marks)*

b) The system is not 100 % efficient. Describe the ways in which energy is wasted in this system. *(4 marks)*

Counterweights are used to reduce the energy needed to operate a lift. These are shown in the diagram below.

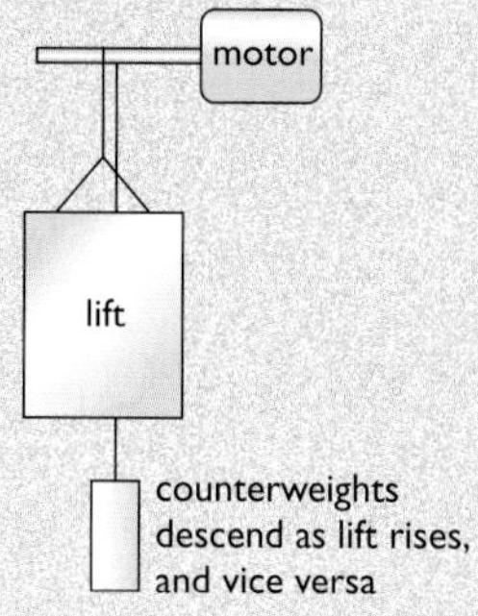

c) What weight should these counterweights have? *(2 marks)*

d) How do the counterweights save energy? *(3 marks)*

Total 12 marks

5 The diagram shows how an electric storage heater is constructed. The heater has a large mass of special thermal bricks that are heated up during the night, using electricity at a lower cost per unit. The mass of the bricks is 90 kg. The specific heat capacity of the brick used is 3400 J/kg/°C (or J/kg/K).

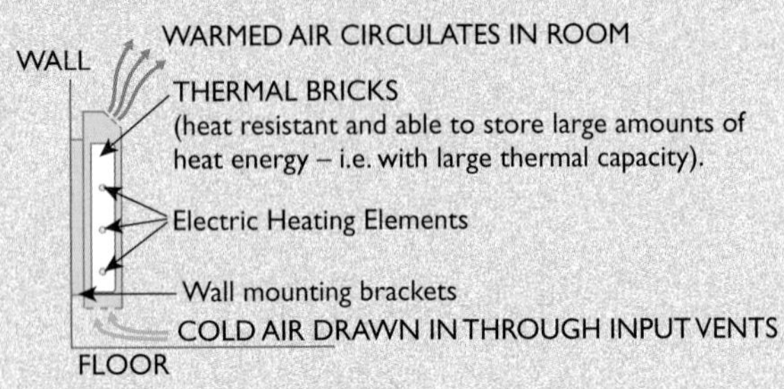

a) How much energy does it take to heat the bricks in the storage heater from 15°C to 60°C? *(3 marks)*

b) The heating element in the storage heater has a power rating of 1.2 kW. How long will it take to raise the bricks' temperature from 15°C to 60°C? *(2 marks)*

c) In practice it will take longer than the answer you have calculated in ***b)***. Explain why this is so. *(2 marks)*

The heating element is switched off at 6.00 am. By 4.00 pm the storage heater has cooled to a temperature of 25°C.

d) How much energy has the storage heater transferred to the room? *(3 marks)*

e) What is the average power output from the heater during this time? *(3 marks)*

f) When would you expect the power output of the storage heater to be a maximum? Give reasons for your answer. *(3 marks)*

Total 16 marks

6 The chart below shows the proportion of heat energy lost from a house by different methods. The annual heating cost for the house is £1000.

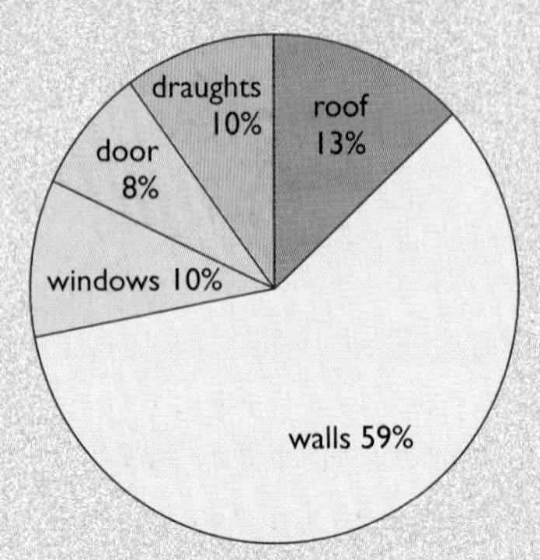

a) What is the annual cost of heat energy lost each year:

i) through the windows *(2 marks)*

ii) by draughts? *(2 marks)*

The householder is considering different ways of reducing the annual heating cost. The possibilities include double glazing, loft insulation and fitting draught excluders to all the doors and windows.

b) Explain how double glazing cuts down energy loss from a house. Your answer must consider all the mechanisms of heat transfer. *(6 marks)*

Double glazing the house reduces heat loss through the windows by 40%. The cost of double glazing is £9000. Draft excluders reduce heat loss by 15% and cost £80 to buy and fit.

c) Calculate the annual saving that could be achieved by:

i) fitting double glazing *(2 marks)*

ii) installing draught excluders. *(2 marks)*

d) Explain what is meant by *cost effectiveness* in terms of measures that the householder could take to cut down energy loss from the house. *(4 marks)*

e) The householder decides to install double glazing. What other factors may have influenced this decision? *(3 marks)*

Total 21 marks

7 A fan heater is used to heat up an office. The energy transfers taking place are shown in the diagram. The 'useful' energy output is heat.

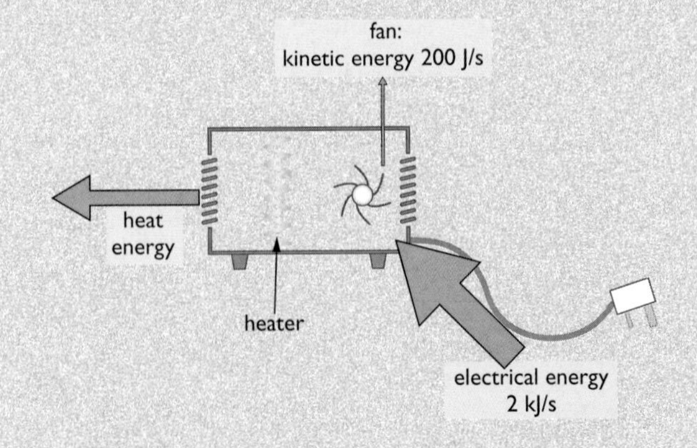

a) Draw a labelled Sankey (energy flow) diagram for this system. *(4 marks)*

b) Calculate the efficiency of the fan heater. *(3 marks)*

Total 7 marks

Chapter 31: Structure of the Atom

Atoms are made up of sub-atomic particles called neutrons, protons and electrons. It is the numbers of these particles that give each element its unique properties. In this chapter you will read about how our ideas about the structure of the atom have developed over centuries. You will see that the atom itself is made up of even smaller particles and learn about the properties of these basic bits.

Dalton's model

About 2500 years ago, a Greek philosopher called Demokritos suggested the idea that matter is made up of tiny, indivisible particles. The name for these particles – **atoms** – comes from a Greek word meaning "cannot be cut or divided".

John Dalton was a scientist who studied gases in the early part of the nineteenth century. He thought that Demokritos was right – that all matter is made up of atoms. He imagined these atoms to be tiny, hard spheres.

Dalton said that some substances were made up of identical atoms – these substances are called **elements**. An important discovery was that atoms of an element always had the same mass – different elements had atoms with different masses. Each different element has its own type of atom. Today we know of the existence of more than 110 different elements. About 90 of these occur naturally on the Earth.

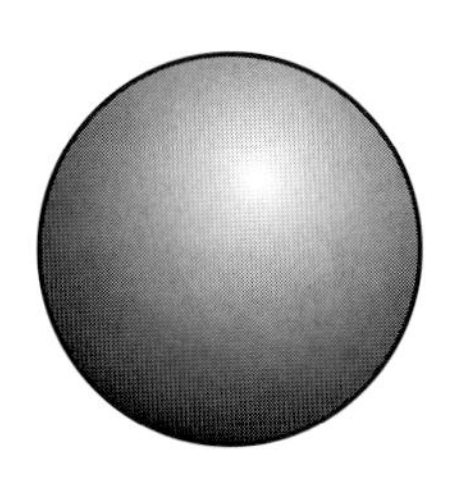

Figure 31.2 *Dalton's model of the atom as a solid sphere.*

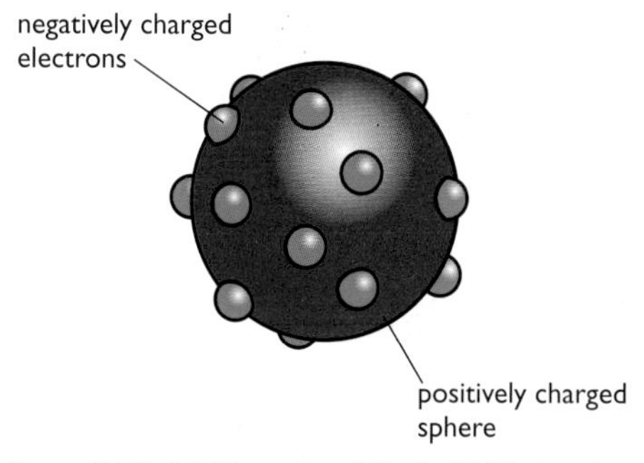

Figure 31.3 J. J. *Thompson (1865–1940) thought that atoms were positively charged spheres with negatively charged electrons embedded in them.*

Figure 31.1 *John Dalton (1766–1844) was a meteorologist who studied the properties of gases. He proposed his atomic theory in 1803. Though some of his ideas have since been discarded, his work marked the start of the development of modern atomic theory.*

The plum pudding model

In 1897, J. J. Thomson discovered the electron while studying the properties of cathode rays. He found that the atom itself was made up of smaller particles that we now call sub-atomic particles. The electron was the first sub-atomic particle to be discovered. Thomson showed that it was a negatively charged particle of very light weight.

He thought that the atom was a ball of positive charge with electrons dotted through it, rather like currants in a bun, or plums in a pudding. For this reason, Thomson's idea is called the "plum pudding" model of the atom.

J. J. Thomson (1856–1940), speaking of the electron in 1934 when he was 78 years old, said, "its mass is an insignificant fraction of the mass of a hydrogen atom …"

Evidence for the existence of the nucleus

ideas evidence

The Rutherford model of the atom

Ernest Rutherford (1871–1937) was a scientist who studied the atom at the start of the twentieth century. He worked with Hans Geiger, who is probably best known for developing the Geiger counter for measuring radiation. Geiger also worked with an undergraduate student called Ernest Marsden.

Together they carried out a series of experiments that involved firing **alpha particles** at very thin gold foil. Alpha particles are positively charged particles given off by some radioactive substances. (We shall learn more about alpha particles in Chapter 32.) Geiger and Marsden's experiment is shown in Figure 31.4.

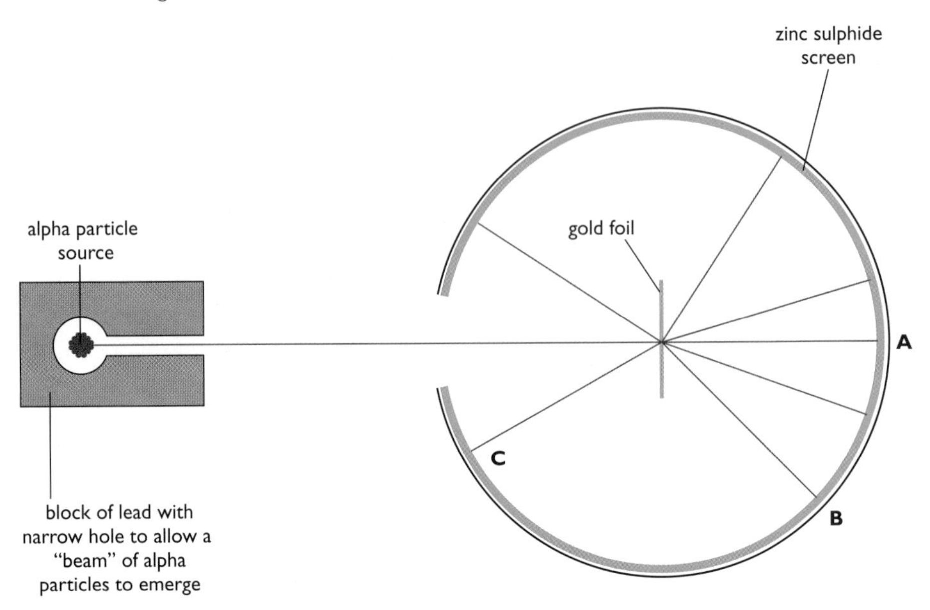

Figure 31.4 *Geiger and Marsden fired alpha particles at thin gold foil to see how they would be deflected.*

The zinc sulphide screen gives out tiny sparks of light, called **scintillations**, when hit by an alpha particle. Geiger and Marsden were able to see what happened to the alpha particles as they passed through the gold foil by noting where the sparks occurred on the screen.

Most of the alpha particles passed straight through the gold foil as if it wasn't there at all A. Once in a while an alpha particle was knocked off course B. A very small proportion (about 1 in 8000) actually seemed to bounce off the gold foil C.

Rutherford studied these results. He realised that, as most of the alpha particles passed straight through the gold foil, most of each gold atom must be empty space. However, Rutherford knew that atoms *did* have mass, so they could not be *just* empty space. The rare event of a rebound meant that an alpha particle had run into something very massive. Rutherford realised that the mass must be concentrated in a very tiny volume at the centre of the atom, which he called the **nucleus** (the plural is **nuclei**). The deflections and rebounds were because the positive charges on the alpha particles were repelled by positive charges in the nuclei. The amount of deflection depends on a number of factors:

1 the *speed* of the alpha particle – the alpha particle is deflected less if it is travelling faster, because it has more momentum

2 the *nuclear charge* – if the nucleus is strongly positive, then the alpha particle will be more strongly repelled away from it

3 *how close* the alpha particle gets to the positively charged nucleus (this is shown in Figure 31.5).

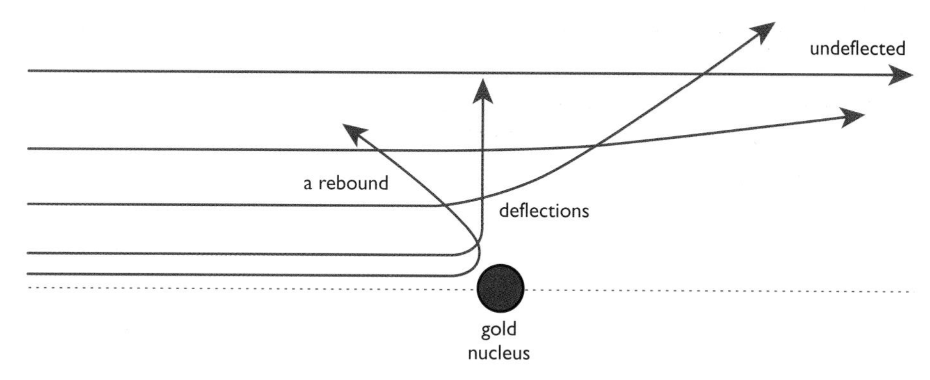

Figure 31.5 *Alpha particles passing very close to a gold atom nucleus are deflected more than those passing at a greater distance.*

Analysis of the results of Geiger and Marsden's experiment not only gave evidence for the existence of the nucleus but also allowed Rutherford to estimate the size of the nucleus (Figure 31.6).

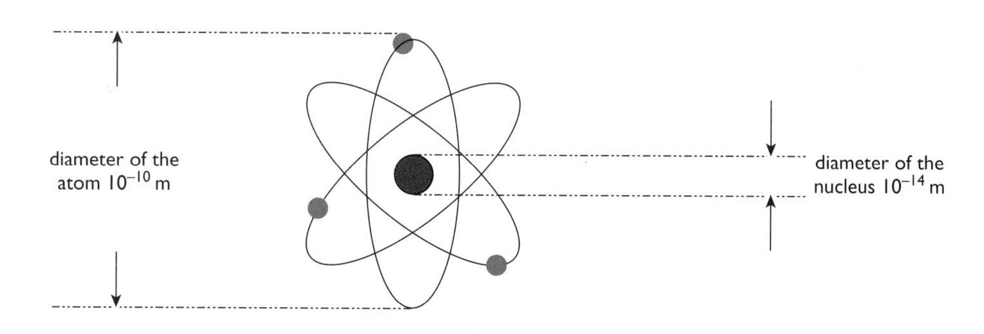

Figure 31.6 *Rutherford's model of the atom had a tiny, dense nucleus and electrons around the outside.*

Electrons, protons and neutrons

ideas
evidence

We now know that the atom is made up of **electrons**, **protons** and **neutrons**. Figure 31.7 shows the planetary model proposed by Nils Bohr in 1913.

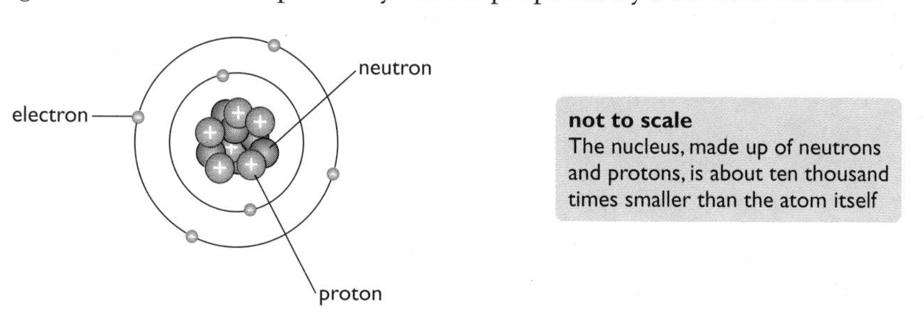

not to scale
The nucleus, made up of neutrons and protons, is about ten thousand times smaller than the atom itself

Figure 31.7 *Bohr's model proposed that there are protons and neutrons in the nucleus of the atom and electrons in orbits around the outside.*

The electron is a very light particle with very little mass. It has a *negative* electric charge. Electrons orbit the nucleus of the atom. Rutherford showed that the nucleus is very small compared to the size of the atom itself. The diameter of the nucleus is about 10 000 times smaller than the diameter of the atom. If the nucleus of an atom were enlarged to the size of a full stop on this page, the atom would have a diameter of around 2.5 metres.

The nucleus is made up of protons and neutrons. Protons and neutrons have almost exactly the same mass. Protons and neutrons are nearly 2000 times heavier than electrons. Protons carry *positive* electric charge but neutrons, as the name suggests, are electrically neutral or *uncharged*. The amount of charge on a proton is equal to that on an electron but opposite in sign.

The properties of these three atomic particles are summarised in the table below. Protons and neutrons are also called **nucleons** because they are found in the nucleus of the atom.

The *actual* mass of an electron is 9.1×10^{-31} kg and its *actual* charge is -1.6×10^{-19} C. The table shows the approximate relative masses of the particles and the relative amount of charge they carry. The mass in atomic mass units is discussed below.

Atomic particle	Relative mass of particle	Mass of particle in atomic mass units	Relative charge of particle
electron	1	0	−1
proton	2000	1	+1
neutron	2000	1	0

The atom

The nucleus of an atom is surrounded by electrons. We sometimes think of electrons as orbiting the nucleus in a way similar to the planets orbiting the Sun. It is more accurate to think of the electrons as moving rapidly around the nucleus in a cloud or shell. An atom is electrically neutral. This is because the number of positive charges carried by the protons in its nucleus is balanced by the number of negative charges on the electrons in the electron "cloud" around the nucleus.

Atomic number

The chemical behaviour and properties of a particular element depend upon how the atoms combine with other atoms. This is determined by the number of electrons in the atom. Although atoms may gain or lose electrons, sometimes quite easily, the number of protons in atoms of a particular element is always the same. The **atomic number** of an element tells us how many protons each of its atoms contains. For example, carbon has *six* protons in its nucleus – the atomic number of carbon is, therefore, 6. The symbol we use for atomic number is Z. Each element has its own unique atomic number.

Atomic mass

The total number of protons and neutrons in the nucleus of an atom determines its **atomic mass**. The mass of the electrons that make up an atom is tiny and can usually be ignored. The mass of a proton is approximately 1.7×10^{-27} kg. To save writing this down we usually measure the mass of an atom in **atomic mass units** (amu) – 1 amu is equal to the mass of one proton (or one neutron). The atomic mass of an element is given the symbol A.

Atomic notation – the recipe for an atom

Each particular type of atom will have its own atomic number, which identifies the element, and an atomic mass that depends on the total number of nucleons, or particles, in the nucleus. Figure 31.8 shows the way we represent an atom of an element whose chemical symbol is X, showing the atomic number and the atomic mass.

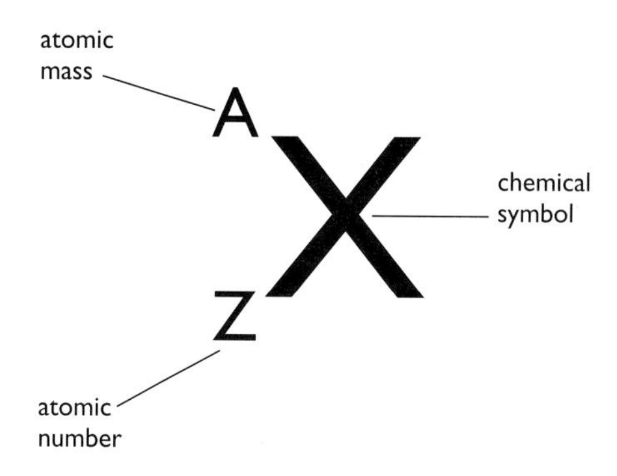

Figure 31.8 *Atomic notation*

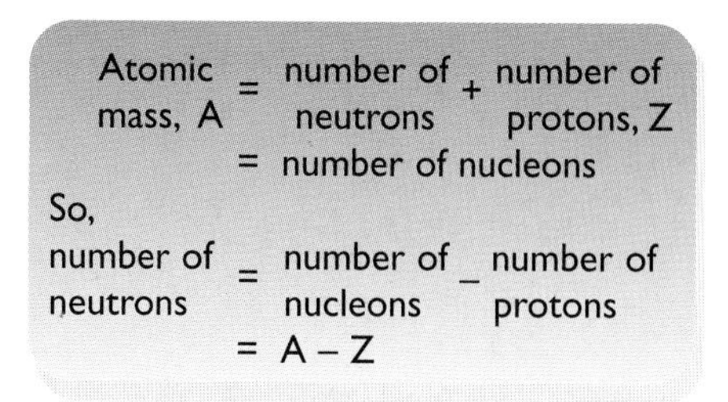

So, using this notation, an atom of oxygen is represented by:

$^{16}_{8}O$

The chemical symbol for oxygen is O. The atomic number is 8 – this tells us that the nucleus contains eight protons. The atomic mass is 16, so there are a total of 16 nucleons (protons and neutrons) in the nucleus. Since eight of these are protons, the remaining eight must be neutrons. The atom is electrically neutral overall, so the +8 charge of the nucleus is balanced by the eight orbiting electrons, each with charge –1. The "recipe" for the oxygen atom is complete.

Figure 31.9 shows some examples of the use of this notation for hydrogen, helium and carbon, together with a simple indication of the structure of an atom of each of these elements. In each case the number of orbiting electrons is equal to the number of protons in the nucleus, so the atoms are electrically neutral.

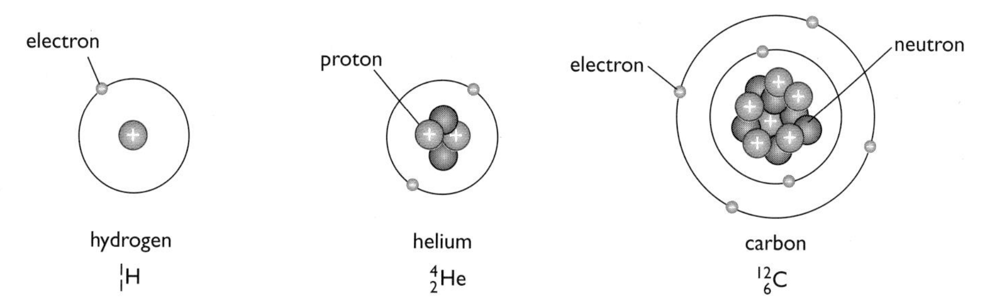

Figure 31.9 *The hydrogen atom has one proton in its nucleus and no neutrons, so the atomic mass, A = 1 + 0 = 1. As it has one proton, its atomic number, Z = 1. For helium, A = 4 (2 protons + 2 neutrons) and Z = 2 (2 protons). For carbon, A = 12 (6 protons + 6 neutrons) while Z = 6 (6 protons).*

Isotopes

The number of protons in an atom identifies the element. The chemical behaviour of an element depends on the number of electrons it has and, as we have seen, this always balances the number of protons in the nucleus. However, the number of neutrons in the nucleus can vary slightly. Atoms of an element with different numbers of neutrons are called **isotopes** of the element. The number of neutrons in a nucleus affects the mass of the atom. Different isotopes of an element will all have the *same* atomic number, Z, but *different* atomic mass numbers, A. Figures 31.10 and 31.11 show some examples of isotopes.

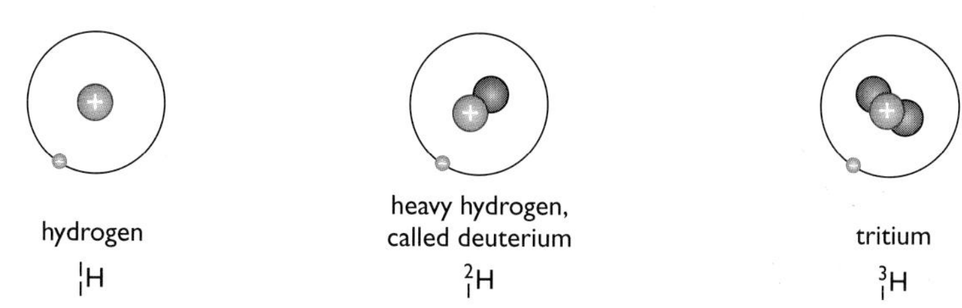

Figure 31.10 *Isotopes of hydrogen – they all have the same atomic number, 1, and the same chemical symbol, H.*

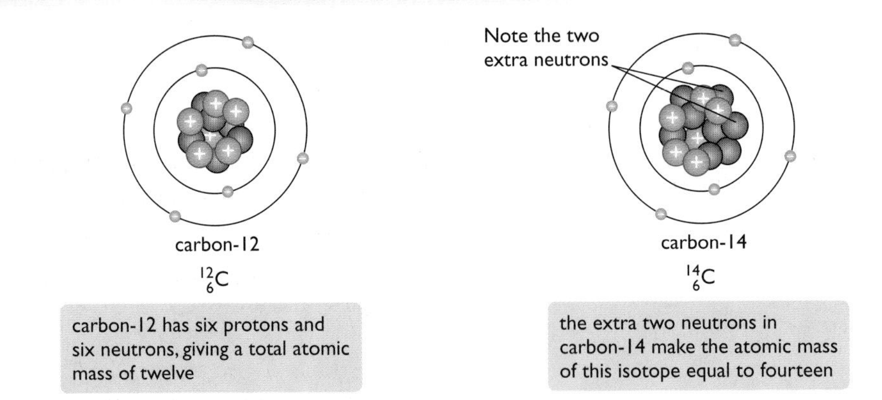

Figure 31.11 *Two isotopes of carbon – they are referred to as carbon-12 and carbon-14 to distinguish between them.*

The stability of isotopes

Isotopes of an element have different physical properties from other isotopes of the same element. One obvious difference is the mass! Another difference is the **stability** of the nucleus.

The protons are held in the nucleus by the **nuclear force**. This force is very strong and acts over a very small distance. It is strong enough to hold the nucleus together against the **electric force** repelling the protons away from each other. (Remember that protons carry positive charge and like charges repel.) The presence of neutrons in the nucleus affects the balance between these forces. Too many or too few neutrons will make the nucleus **unstable**. An unstable nucleus will eventually **decay**. When the nucleus of an atom decays, it splits apart giving out energy, radiation and fragments of the original nucleus.

Stable or unstable?

The stability of a nucleus depends on how many protons and neutrons it contains. Some elements have no stable isotope, others have several. Figure 31.12 shows a graph with isotopes plotted according to the number of neutrons and the number of protons in their nuclei. This is sometimes called a **nuclide chart**.

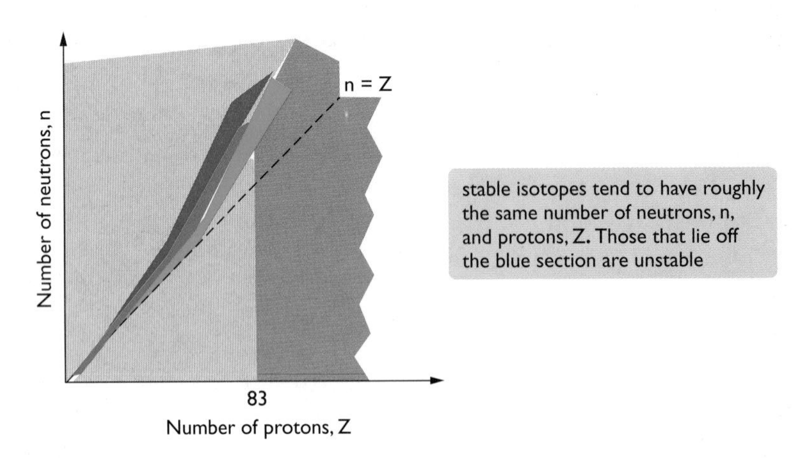

Figure 31.12 *You will notice that the stable isotopes lie along the curved central part of the graph, shaded in blue. Outside this curve lie isotopes that are unstable (green and pink areas) and these will spontaneously decay. (The red and grey shading indicates that the unstable isotopes in these areas decay in different ways.)*

End of Chapter Checklist

If you haven't got a copy of your specification, read the introduction on page iv.

You will need to be able to do some or all of the following. Check your Awarding Body's specification (syllabus) to find out exactly what you need to know.

- Understand the difference between the Thomson and Rutherford models of the atom.
- Know that most of the matter of an atom is concentrated at its centre in a very tiny volume called the nucleus.
- Describe the alpha particle scattering experiment carried out by Geiger and Marsden.
- Understand the factors that determine the amount of deflection of an alpha particle as it passes through gold foil.
- Compare the mass and charge of electrons, protons and neutrons.
- Understand the terms atomic number and atomic mass, and be able to use and interpret notation for a nucleus using these numbers.
- Know that the same element may exist in the form of several different isotopes.
- Appreciate that some isotopes are unstable because of the proportion of protons to neutrons in their nuclei and that the way they decay depends on this.

Questions

More questions on the structure of the atom can be found at the end of Section F on page 339.

1 Give one difference between Thomson's model of the atom and the Rutherford model.

2 ***a)*** Look at this diagram of Geiger and Marsden's apparatus. Write out the correct words for the labels A to E.

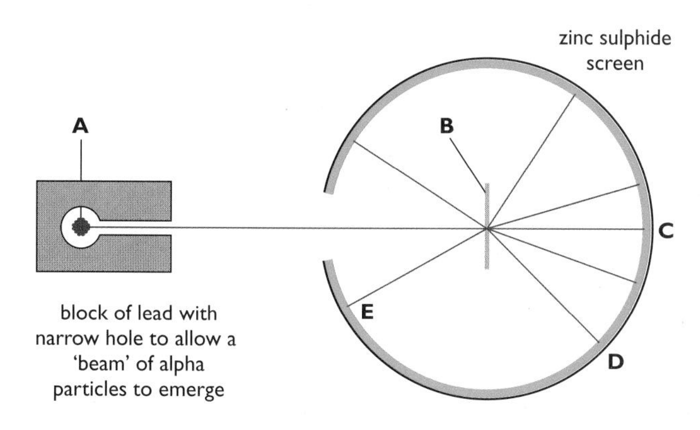

b) What did the results of Geiger and Marsden's experiment reveal about the atom?

3 The diagram below shows the paths of several alpha particles travelling towards a gold nucleus.

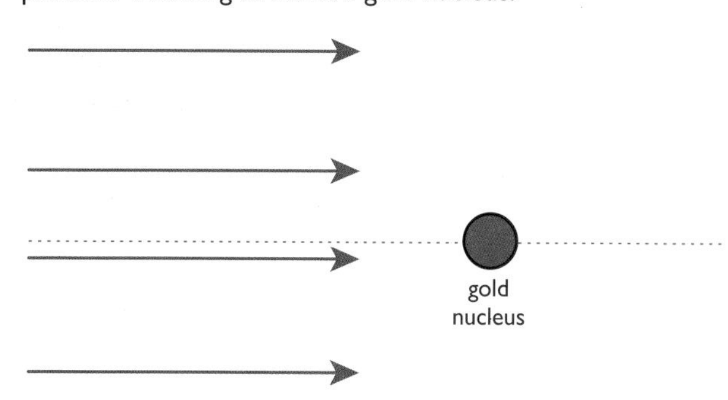

a) Copy and complete the diagram showing likely paths for the alpha particles as they continue towards the nucleus.

b) How would your diagram have differed if the particles had:

i) been travelling faster

ii) had only half as much electric charge?

4 Copy and complete the table below showing the relative charge and mass of each of the particles that make up atoms.

Atomic particle	Relative mass of particle	Mass of particle in atomic mass units	Relative charge of particle
	1		−1
		1	+1
	2000		0

5 Identify the following nuclear particles from their descriptions:

a) an uncharged nucleon

b) the particle with the least mass

c) the particle with the same mass as a neutron

d) the particle with the same amount of charge as an electron

e) a particle that is negatively charged.

6 Explain the following statements about the atom.

a) An atom is mainly made up of empty space.

b) Almost all the mass of an atom is in the nucleus.

c) Atoms are electrically neutral.

7 Explain the following terms used to describe the structure of an atom:

a) atomic number

b) atomic mass.

8 Copy and complete the table below, describing the structures of the different atoms in terms of numbers of protons, neutrons and electrons.

	$^{3}_{2}He$	$^{13}_{6}C$	$^{23}_{11}Na$
protons			
neutrons			
electrons			

Chapter 32: Radioactive Decay

There are three types of ionising radiation – alpha, beta and gamma. In this chapter, you will learn about their different properties, and about how each type behaves in a magnetic field. You will find out that when a nuclear decay involves the emission of a particle, the atomic number of the nucleus changes – the element transforms into a new element. You will also learn about ways in which we can detect radiation, and about the background level of radiation we experience all the time.

Ionising radiation

ideas
evidence

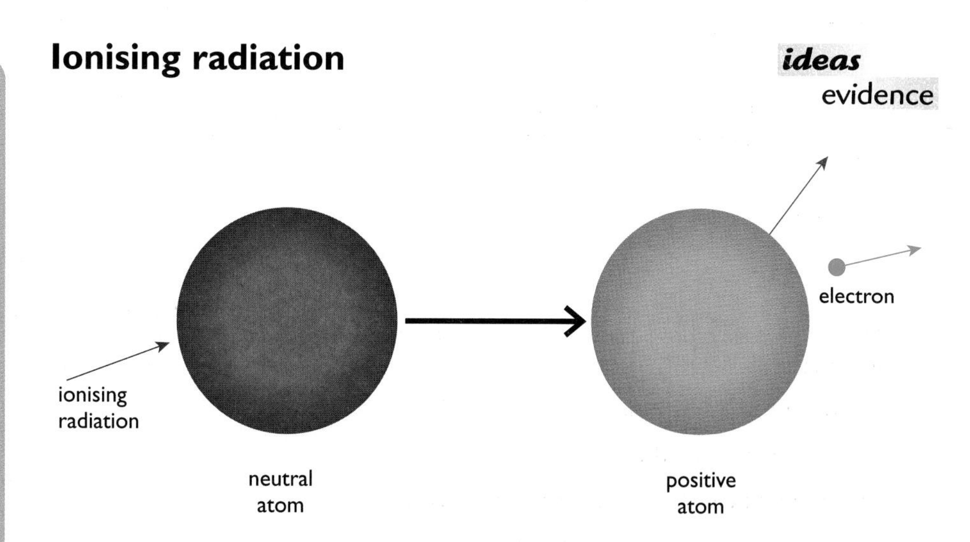

Figure 32.1 *When a neutral atom (or molecule) is hit by ionising radiation it loses an electron and becomes a positively charged ion.*

When unstable nuclei decay they give out **ionising radiation**. Ionising radiation causes atoms to gain or lose electric charge, forming **ions**. Unstable nuclei decay at **random**. This means that it is not possible to predict which unstable nucleus in a piece of radioactive material will decay, or when a decay will happen. We shall see that we can make measurements that will enable us to predict the probability that a certain proportion of a radioactive material will decay in a given time.

There are three basic types of ionising radiation: they are **alpha** (α), **beta** (β) and **gamma** (γ) radiation.

Alpha radiation

Alpha radiation consists of fast-moving particles that are thrown out of unstable nuclei when they decay. These are called **alpha particles**. Alpha particles are helium nuclei – helium atoms without their orbiting electrons. Figure 32.2 shows an alpha particle and the notation that is used to denote it in equations.

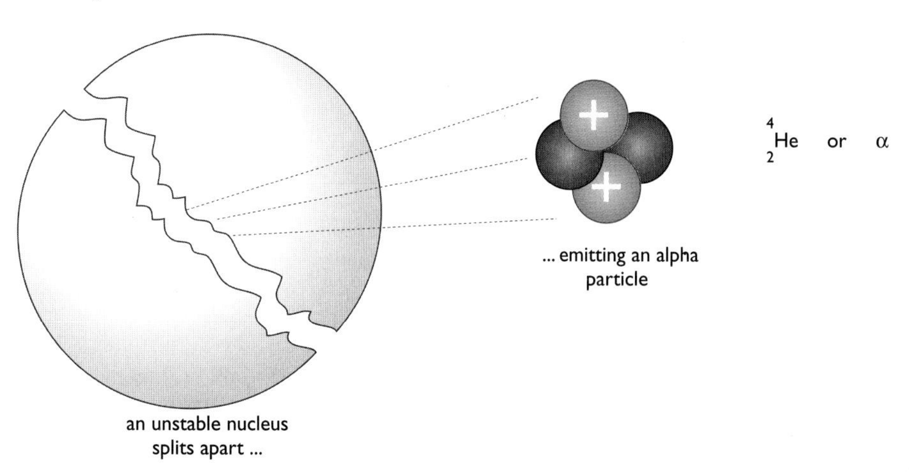

Figure 32.2 *An unstable nucleus splits, emitting an alpha particle.*

Alpha particles have a relatively large mass. They are made up of four nucleons and so have an atomic mass of 4. They are also charged because of the two protons that they carry. The relative charge of an alpha particle is +2.

Alpha particles have a short **range**. The range of ionising radiation is the distance it can travel through matter. Alpha particles can only travel a few centimetres in air and cannot penetrate more than a few millimetres of paper. They have a limited range because they interact with atoms along their paths, causing ions to form. This means that they rapidly give up the energy that they had when they were ejected from the unstable nucleus.

Beta radiation

Beta particles are very fast-moving electrons that are ejected by a decaying nucleus. The nucleus of an atom contains protons and neutrons, so where does the electron come from? The stability of a nucleus depends on the proportion of protons and neutrons it contains. The result of radioactive decay is to change the balance of protons and neutrons in the nucleus to make it more stable. Beta decay involves a neutron in the nucleus splitting into a proton and an electron. The proton remains in the nucleus and the electron is ejected at high speed as a beta particle.

Nucleons have roughly 2000 times the mass of an electron, and alpha particles are made up of 4 nucleons, so an alpha particle has 8000 times the mass of a beta particle.

Beta particles are very light – they have only 0.000125 times the mass of an alpha particle. The relative charge of a beta particle is –1.

Beta particles interact with matter in their paths less frequently than alpha particles. This is because they are smaller and carry less charge. This means that beta particles have a greater range than alpha particles. Beta particles can travel long distances through air, pass through paper easily and are only absorbed by denser materials like aluminium. A millimetre or two of aluminium foil will stop all but the most energetic beta particles.

Gamma rays

Gamma rays are electromagnetic waves (see page 187) with very short wavelengths. As they are waves, they have no mass and no charge. They are weakly ionising and interact only occasionally with atoms in their paths. They are extremely penetrating and pass through all but the very densest materials with ease. It takes several centimetres thickness of lead, or a metre or so of concrete, to stop gamma radiation.

Summary of the properties of ionising radiation

We have said that ionising radiation causes uncharged atoms to lose electrons. An atom that has lost (or gained) electrons has an overall charge. It is called an **ion**. The three types of radioactive emission all form ions.

As ionising radiation passes through matter, its energy is absorbed. This means that radiation can only penetrate matter up to a certain thickness. This depends on the type of radiation and the density of the material that it is passing through.

The ionising and penetrating powers of alpha, beta and gamma radiation are compared in the table on page 303. Note that the ranges given in the table are typical but they do depend on the energy of the radiation. More energetic alpha particles will have a greater range than those with lower energy, for example.

Radiation	Ionising power	Penetrating power	Example of range in air	Radiation stopped by
alpha, α	strong	weak	5–8 cm	paper
beta, β	medium	medium	500–1000 cm	thin aluminium
gamma, γ	weak	strong	virtually infinite	thick lead sheet

The effect of a magnetic field on radioactive emissions

Alpha and beta particles are charged. Magnetic fields produce a force on moving charged particles. The direction of this force depends on the direction of the current produced by the moving charge and the direction of the magnetic field. The effect of a magnetic field on the three types of radioactive emission is shown in Figure 32.3.

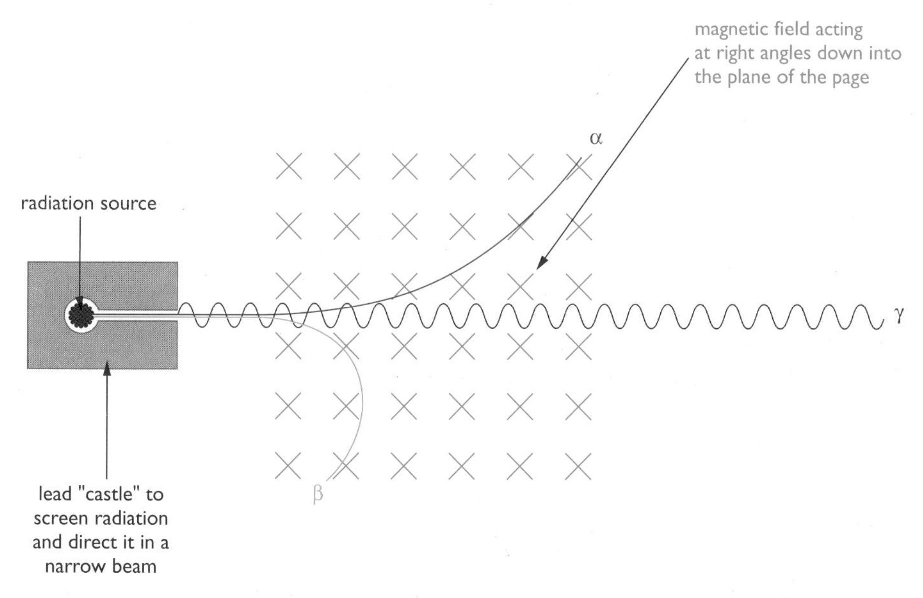

Figure 32.3 *The effect of a magnetic field on alpha, beta and gamma radiation.*

Figure 32.4 *This photograph shows alpha particle tracks in a school-type cloud chamber. The alpha particles leave dense tracks because they produce lots of ions along their path. Notice that the paths are not very long. The tracks are all of the same length because the source is emitting alpha particles all with the same energy.*

The gamma rays, which are uncharged, are not affected by the magnetic field and so travel in a straight line. Notice that the alpha and beta particles, which are oppositely charged, follow paths that curve in opposite directions. The curved path for the beta particles has a smaller radius. This is because beta particles travel much faster than alpha particles, so the force acting on them is greater. Beta particles also have a much lower mass (about 0.000125 of the mass of the alpha particle), so the acceleration produced by the force is greater. The charged particles follow circular paths because the force acting on them is at right angles to their motion.

Nuclear scientists investigate the behaviour of different types of ionising radiation by looking at the paths they leave as they pass through layers of photographic film. They can also show up the paths of some particles in devices called cloud chambers. Bubble chambers are also used for this purpose. The curved paths followed by radioactive particles are used to identify their relative charge and mass.

Figure 32.5 *This photograph shows the tracks of particles in a bubble chamber in a strong magnetic field. Notice that the paths are curving in opposite directions. This tells us that the particles carry charge of opposite signs. The amount of curvature of the paths also gives physicists information about the type of ionising radiation that is making the tracks.*

Nuclear transformations

As we said earlier, an unstable atom, or strictly speaking its nucleus, will decay by splitting and emitting radiation. If the decay process involves the nucleus ejecting either an alpha or a beta particle, the atomic number will change. This means that alpha or beta decay causes the original element to transform into a different element.

Alpha (α) decay

Here is an example of alpha decay:

$$^{222}_{88}\text{Ra} \rightarrow \ ^{218}_{86}\text{Rn} + \ ^{4}_{2}\text{He} + \text{energy}$$

radium atom → radon atom + alpha particle + energy

The radioactive isotope radium-222 decays to the element radon by the emission of an alpha particle. The alpha particle is sometimes represented by the Greek letter α. Radon is a radioactive gas that also decays by emitting an alpha particle. Note that the atomic number for radon, 86, is two *less* than the atomic number for radium.

The general form of the alpha decay equation is:

$$^{A}_{Z}\text{Y} \rightarrow \ ^{A-4}_{Z-2}\text{W} + \ ^{4}_{2}\text{He} + \text{energy}$$

(↑ $^{4}_{2}\text{He}$: **alpha particle, α**)

Reminder: A is the atomic mass of the element and Z is the atomic number. The letters W and Y are not the symbols of any particular elements as this is the *general equation*.

In alpha decay, element Y is transformed into element W by the emission of an alpha particle. Element W is two places before element Y in the Periodic Table. The alpha particle, a helium nucleus, carries away four nucleons, which reduces the atomic mass (A) by four. Two of these nucleons are protons so the atomic number of the new element is two less than the original element, Z – 2. Notice that the atomic mass and the atomic number are conserved through this equation – that is, the *total* numbers of nucleons and protons on each side of the equation are the same.

It is worth pointing out that the atomic mass number refers to the number of nuclear particles, or nucleons, involved in the transformation – not the exact mass. Mass is *not* conserved in nuclear transformations as some of it is transformed into energy.

Beta (β) decay

Here is an example of beta decay:

$$^{14}_{6}\text{C} \rightarrow \ ^{14}_{7}\text{N} + \ ^{0}_{-1}\text{e} + \text{energy}$$

carbon atom → nitrogen atom + beta particle + energy

The radioactive isotope of carbon, carbon-14, decays to form the stable isotope of the gas nitrogen, by emitting a beta particle. Remember that the beta particle is formed when a neutron splits to form a proton and an electron. The beta particle is sometimes represented by the symbol β. Figure 32.6 shows the standard atomic notation for a beta particle.

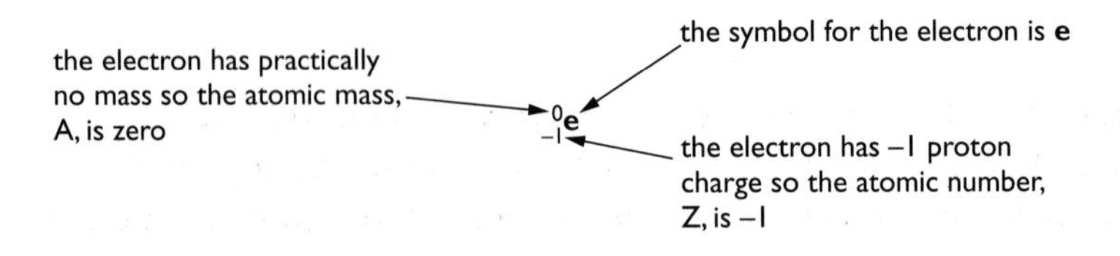

Figure 32.6 *A beta particle.*

The general form of the beta decay equation is:

$$^{A}_{Z}X \rightarrow {}^{A}_{Z+1}Y + {}^{0}_{-1}e + \text{energy}$$

↑ **beta particle, β^-**

In beta decay, element X is transformed into element Y by the emission of a beta particle. Element Y is the next element in the Periodic Table after element X. The beta particle, an electron, has practically no mass so the atomic mass, A, is the same in X and Y. As the beta particle has a charge of –1, the atomic number of the new element is increased to Z + 1. Again the atomic mass and the atomic number are conserved through this equation.

Other types of decay

Each particle that makes up the atom has an equivalent **antiparticle**. Antiparticles make up what we call **antimatter**. Antiparticles that are charged carry charge of the opposite sign to that on the equivalent particle that makes up ordinary matter.

The antiparticle of the electron is called a **positron**. The positron is very much like an electron except that it is positively charged. In nuclear equations, the positron has the symbol:

$$^{0}_{+1}e$$

sometimes written simply as e^+ or β^+.

For every particle that exists in our world of ordinary matter there is an equivalent antiparticle. Particles and antiparticles differ in other ways than having opposite charge but this is beyond the scope of GCSE physics courses.

β^+ decay

Here is an example of β^+ (positron) decay:

$$^{11}_{6}C \rightarrow {}^{11}_{5}B + {}^{0}_{+1}e + \text{energy}$$

carbon atom → boron atom + positron + energy

In this decay, the radioactive isotope carbon-11 emits a positron and transforms into boron. Boron has an atomic number of 5.

The positron equivalent of beta decay follows the following equation:

$$^{A}_{Z}Y \rightarrow {}^{A}_{Z-1}X + {}^{0}_{+1}e + \text{energy}$$

↑ **positron, β^+**

Here, a proton within the unstable nucleus has decayed to form a positron and a neutron. When the positron is ejected from the parent isotope, Y, it carries away negligible mass and a charge of +1. This means that the daughter isotope, X, has the same atomic mass, A.

What becomes of the positron?

Antimatter particles do not have a long life in our world of ordinary matter. As soon as a positron meets an electron, they mutually annihilate – that is, disappear as their mass is converted into energy. This energy is given out in a burst of gamma rays.

Electron capture

Another process, called **electron capture**, involves an electron being *absorbed* by a proton in the nucleus. This proton transforms to a neutron. This causes a new isotope with an atomic number decreased by 1. There is no change in the atomic mass.

ideas evidence

Detecting ionising radiation

Figure 32.7 *Henri Becquerel (1852–1908) studied X-rays.*

Using photographic film

Henri Becquerel was studying X-rays using uranium ore in 1896. He believed that the uranium emitted X-rays after being exposed to sunlight. To test this idea he placed some wrapped, unused photographic plates in a drawer with some samples of the uranium ore on top of them. He found a strong image of the ore on the plates when he developed them. He had discovered radioactivity.

The unit of radioactivity is named after Becquerel. The **becquerel** (Bq) is a measure of how many unstable nuclei are disintegrating per second – one becquerel means a rate of one disintegration per second. The becquerel is a very small unit. More practical units are the kBq (an average of 1 000 disintegrations per second) and the MBq (an average of 1 000 000 disintegrations per second).

Photographic film is still used to detect radioactivity. Scientists who work with radioactive materials wear a strip of photographic film in a badge. If the film becomes fogged it shows that the scientist has been exposed to a certain amount of radiation. These badges are checked regularly to ensure that the safety limit for exposure to ionising radiation is not exceeded.

The Geiger–Müller tube

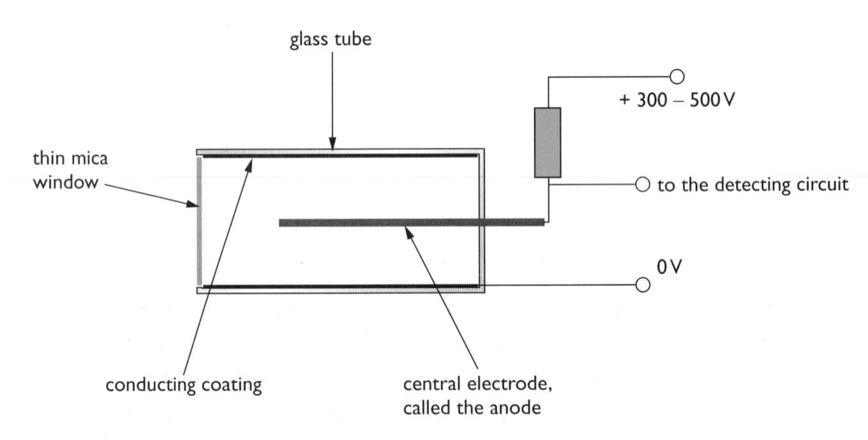

Figure 32.8 *A Geiger–Müller tube is used to measure the level of radiation.*

Figure 32.8 shows the basic construction of a Geiger–Müller (GM) tube. It is a glass tube with an electrically conducting coating on the inside surface. The tube has a thin window made of mica (a naturally occurring mineral that can be split into thin sheets). The tube contains a special mixture of gases at very low pressure. In the middle of the tube, electrically insulated from the conducting coating, there is an electrode. This electrode is connected, via a high value resistor, to a high voltage supply, typically 300–500 V.

Remember that γ radiation is in the form of electromagnetic waves. Energy is emitted in the form of waves as "packets" of energy, called **photons**.

When ionising radiation enters the tube it causes the low pressure gas inside to form ions. The ions allow a pulse of current to flow from the electrode to the conducting layer. This is detected by an electronic circuit.

The GM tube is usually linked up to a counting circuit. This keeps a count of how many ionising particles (or photons in the case of γ radiation) have entered the GM tube. Sometimes GM tubes are connected to rate meters.

These measure the number of ionising events per second, and so give a measure of the radioactivity in becquerels. Rate meters usually have a loudspeaker output so the level of radioactivity is indicated by the rate of "clicks" produced.

Background radiation

Background radiation is low-level ionising radiation that is produced all the time. This background radiation has a number of sources. Some of these are natural and some are artificial.

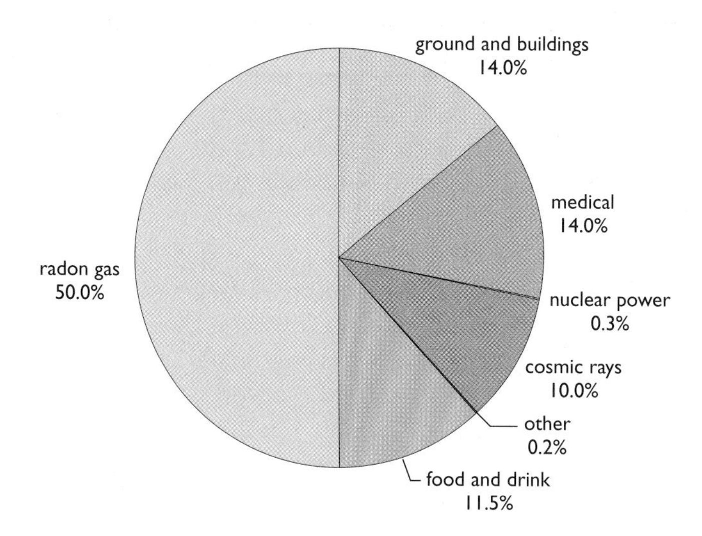

Figure 32.9 *Sources of background radiation in the UK.*

Natural background radiation from the Earth

Some of the radiation we receive comes from rocks in the Earth's crust. When the Earth was formed, around 4.5 billion years ago, it contained many radioactive isotopes. Some decayed very quickly but others are still producing radiation. Some of the **decay products** of these long lived radioactive materials are also radioactive, so there are radioactive isotopes with much shorter half-lives still present in the Earth's crust. (We will learn about half-life in Chapter 33.)

When an atom of a radioactive element decays it gives out radiation and changes to an atom of another element. This may also be radioactive, and decay to form an atom of yet another element. The elements formed as a result of a radioactive element undergoing a series of decays are called decay products.

One form of uranium is a radioactive element that decays very slowly. Two of its decay products are gases. These are the radioactive gases **radon** and **thoron**. Radon-222 is a highly radioactive gas produced by the decay of radium-226. Thoron, or radium-220, is an isotope of radium formed by the decay of a radioactive isotope of thorium (thorium-232).

As these decay products are gases, they seep out of radioactive rocks. They are dense gases so they build up in the basements and foundations of buildings. Some parts of the Earth's crust have higher amounts of radioactive material so the amount of background radiation produced in this way varies from place to place. In Cornwall, for example, where the granite rock contains traces of uranium, the risk of exposure to radiation from radon gas is greater than in some other parts of the UK.

Natural background radiation from space

Violent nuclear reactions in stars and exploding stars called **supernovae** produce very energetic particles and **cosmic rays** that continuously bombard the Earth. These pass through the atmosphere and produce lower energy particle showers as they interact with the atoms in the air. The atmosphere thus acts as a shield so, at ground level, cosmic radiation is mainly composed of particles called muons (not on GCSE syllabuses), with some gamma rays, beta particles and neutrons.

Internal radiation

The atoms that make up our bodies were formed in the violent reactions that took place in stars created at the beginning of the Universe. Some of these atoms are radioactive so we carry our own personal source of radiation around with us. Also, as we breathe we take in tiny amounts of the radioactive isotope of carbon, carbon-14. Because carbon-14 behaves chemically just like the stable isotope, carbon-12, we continuously renew the amount of the radioactive carbon in our bodies (see page 318).

Artificial radiation

We use radioactive materials for many purposes. Generating electricity in nuclear power stations has been responsible for the leaking of radioactive material into the environment. The levels are usually small, but there have been a number of major incidents around the world, notably at Three Mile Island in the USA in 1979 and at Chernobyl in the Ukraine in 1986. Testing nuclear weapons in the atmosphere has also increased the amounts of radioactive isotopes on the Earth.

Radioactive tracers are used in industry and medicine. Radioactive materials are also used to treat certain forms of cancer. However the majority of background radiation is natural – the amount produced from medical and civil use in industry is very small indeed.

End of Chapter Checklist

If you haven't got a copy of your specification, read the introduction on page iv.

You will need to be able to do some or all of the following. Check your Awarding Body's specification (syllabus) to find out exactly what you need to know.

- Know that radioactive nuclei may emit three different types of ionising radiation – alpha, beta and gamma.
- Know that ionising radiation can cause atoms to lose electrons and become charged ions.
- Describe the different properties of alpha, beta and gamma radiation considering ionising and penetrating power.
- Understand the effect of a magnetic field on different types of ionising radiation.
- Balance nuclear transformation equations with respect to mass number and atomic number.
- Understand that alpha and beta decay result in a nucleus of one element transforming into another.
- Be aware of positron decay and electron capture as possible ways that elements can transform.
- Know that the becquerel is a measure of the rate of nuclear decay.
- Understand different methods of detecting ionising radiation.
- Be aware of background radiation and describe its different sources.

Questions

More questions on radioactive decay can be found at the end of Section F on page 339.

1 Copy and complete the following sentences:

a) An alpha particle consists of four ________ . Two of these are ________ and two are ________ . An alpha particle carries a charge of ________ .

b) A beta particle is a fast-moving ________ that is emitted from the nucleus. It is created when a ________ in the nucleus decays to form a ________ and the beta particle.

c) The third type of ionising radiation has no mass. It is called ________ radiation. This type of radiation is a type of wave with a very ________ wavelength.

d) Gamma radiation is part of the ________ spectrum.

2 ***a)*** Explain what is meant by *background radiation*.

b) Explain the difference between natural background radiation and artificial background radiation.

c) Give three different sources of background radiation. Say whether your examples are natural or artificial sources.

3 ***a)*** Explain, simply, the principle of the Geiger–Müller tube.

b) The Geiger–Müller tube is often connected to a rate meter. Explain what this instrument measures.

c) The rate meter is calibrated in kBq. How is this unit defined?

4 A certain radioactive source emits different types of radiation. The sample is tested using a Geiger counter. When a piece of card is placed between the source and the counter, there is a noticeable drop in the radiation. When a thin sheet of aluminium is added to the card between the source and the counter, the count rate is unchanged. A thick block of lead, however, causes the count to fall to the background level.

What type (or types) of ionising radiation is the source emitting? Explain your answer carefully.

5 ***a)*** The nuclear equation below shows the decay of thorium. Copy and complete the equation by providing the missing numbers.

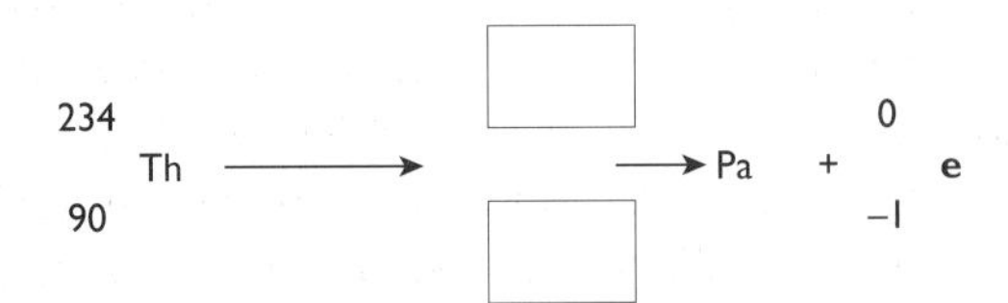

b) What type of decay is taking place in this transformation?

6 ***a)*** The nuclear equation below shows the decay of polonium. Copy and complete the equation by providing the missing numbers.

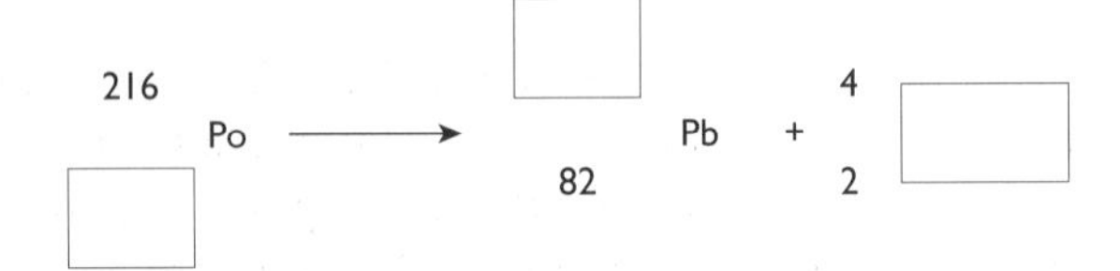

b) What type of decay is taking place in this transformation?

7 A bubble chamber is used by nuclear scientists to study different sub-atomic particles. Ionising particles leave a visible trail of minute bubbles as they pass through a bubble chamber. The bubble chamber is placed in a strong magnetic field.

The diagram shows an event sometimes observed in a bubble chamber. Two particles have been spontaneously created by a high-energy gamma photon.

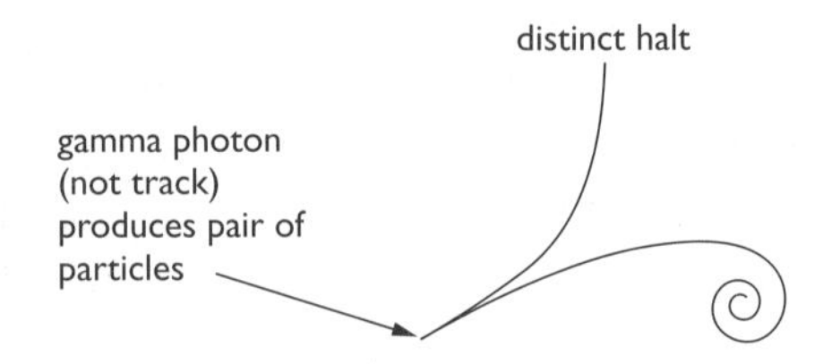

a) Can you suggest what types of particle have been created? Explain your answer in as much detail as possible.

b) One of the tracks suddenly stops. What do you think is the explanation for this?

Section F: Radioactivity

Chapter 33: Radioactive Decay and Half-life

We use a value called half-life to measure the activity of radioactive isotopes. In this chapter you will learn why half-life is a useful way of measuring radioactivity, and how the half-life of a radioisotope can be found.

Radioactive decay is a random process, just like tossing a coin. If we toss a coin we cannot say with certainty whether it will come down "heads" or "tails". If we toss a thousand coins we cannot predict which will land "heads" and which will land "tails". The same is true for radioactive nuclei. It is impossible to tell which nuclei will disintegrate at any particular time. However, if we tossed a thousand coins we would be surprised if the number that landed as "heads" was not around 500. We know that a fair coin has an equal chance of landing as a "head" or a "tail", so if we got 600 "heads" we would think it was unusual. If the proportion of "heads" were much greater than this we would be right to think that the coin was not fair.

Experimental demonstration of nuclear decay

We could, if we had the time, take 1000 coins and toss them. We could then remove all the coins that came down "heads", note the number of coins remaining and then repeat the process. If we did this for, say, six trials we would begin to see the trend. A set of typical results is shown in the table below and in Figure 33.1.

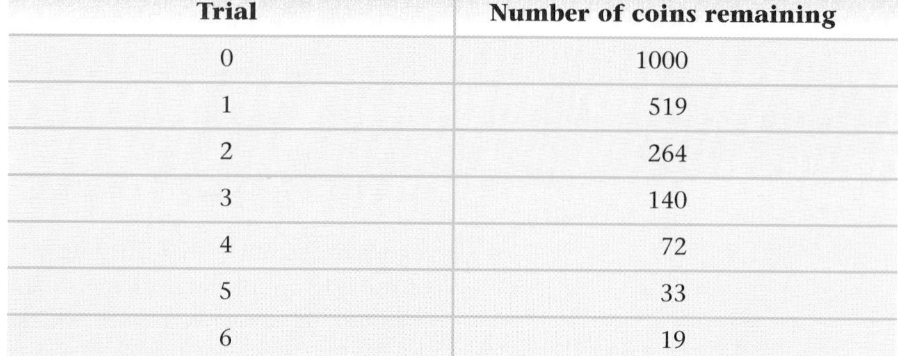

Trial	Number of coins remaining
0	1000
1	519
2	264
3	140
4	72
5	33
6	19

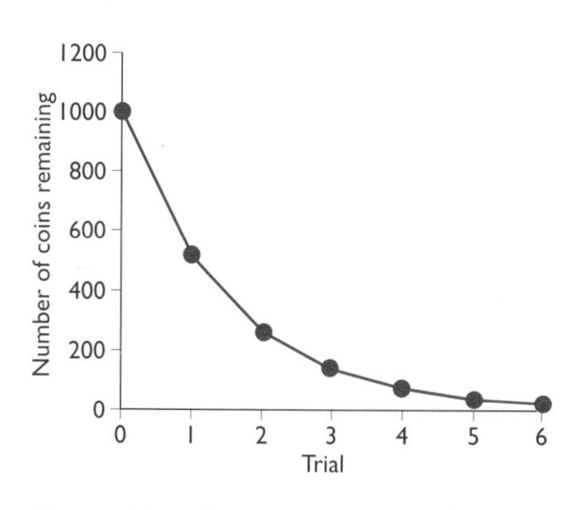

Figure 33.1 *Coin-tossing experiment. Each time the coins are tossed about 50% of them land as "heads" and are removed from the pile. The graph decreases steeply at first but then does so more and more slowly.*

Notice that the graph in Figure 33.1 falls steeply at first and more slowly after each toss. How quickly the graph falls depends on how many "heads" occur each toss. But as the number of coins decreases, the number of coins that come up "heads" also gets smaller. This graph follows a rule: "the smaller the quantity, the more slowly the quantity decreases". The "quantity" here is the number of coins still in the experiment. The name for this kind of decrease proportional to size is called **exponential decay**.

If we have a sample of a radioactive material, it will contain millions of atoms. The process of decay is random, so we don't know when an atom will decay but there will be a probability that a certain fraction of them will disintegrate in a particular time. This is the same as in the coin toss – there was a 50% probability that the coins would land "heads" each time we conducted a trial. Once an unstable nucleus has disintegrated, it is out of the "game" – it won't be around to disintegrate during the next period of time. If we plot a graph of number of disintegrations per second against time for a radioactive isotope we would, therefore, expect the rate of decay to fall as time passes because there are fewer nuclei to decay.

The coin-tossing experiment is a **model** of radioactive decay. A good model will show the features of the real process. We must remember that models have limitations, however, and do not perfectly represent the actual process. One limitation of the experiment is that of scale. In just one gram of uranium-235 there are millions of nuclei, and our model uses just 1000 coins. The model would be better if we used, say, 1 000 000 coins, but would take too long to perform. If we use a computer model to toss the coins we *could* deal with more realistic numbers.

Half-life

Our coin-tossing model of radioactive decay shows a graph that approaches the horizontal axis more and more slowly as time passes. The model will produce a number of throws after which all the coins have been taken out of the game. The number is likely to vary from trial to trial because the model becomes less and less reliable as the number of coins becomes smaller. With real radioactive decay we use a measure of activity called the **half-life**. This is defined as follows.

The half-life of a radioactive sample is the average time taken for half the original mass of the sample to decay.

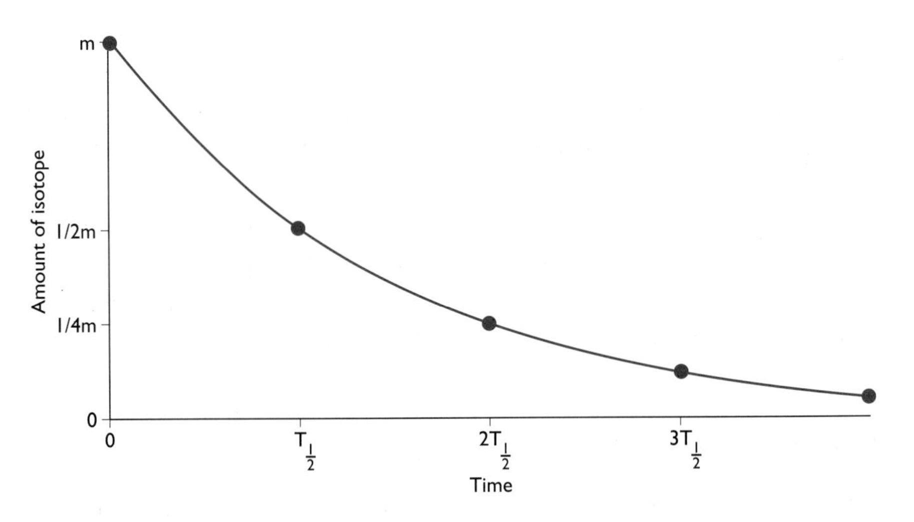

Figure 33.2 *Graph showing the half-life period for a radioactive isotope.*

Figure 33.2 shows what this means. After one half-life period, $t_{\frac{1}{2}}$, the amount of the original unstable element has halved. After a second period of time, $t_{\frac{1}{2}}$, the amount has halved again, and so on.

Measuring the half-life of a radioactive isotope

To measure the half-life of a radioactive material (radioisotope) we must measure the activity of the sample at regular times. This is done using a Geiger–Müller (GM) tube linked to a rate meter. Before taking measurements from the sample, we must measure the local background radiation. We must subtract the background radiation measurement from measurements taken from the sample so we know the radiation produced by the sample itself. We then measure the rate of decay of the sample at regular time intervals. The rate of decay is shown by the count rate on the rate meter. The results should be recorded in a table like the one shown below.

Average background radiation measure over 5 min = x Bq		
Time, t (min)	**Count rate (Bq)**	**Corrected count rate, C (Bq)**
0	y_0	$y_0 - x$
5	y_5	$y_5 - x$

The rate of decay, C, corrected for background radiation, is proportional to the amount of radioactive isotope present. If we plot a graph of C against time, t, we can measure the half-life from the graph, as shown in Figure 33.3.

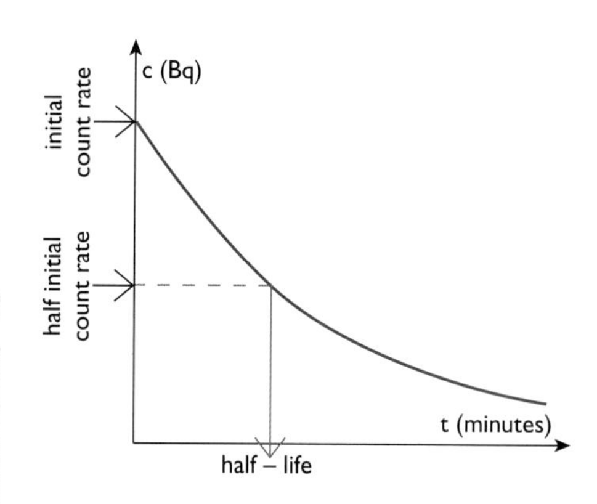

Figure 33.3 *You can find the half-life by reading from the graph the time taken for the count rate to halve.*

As we have already mentioned, different isotopes have widely differing half-lives. Some examples of different half-lives are shown in the table below.

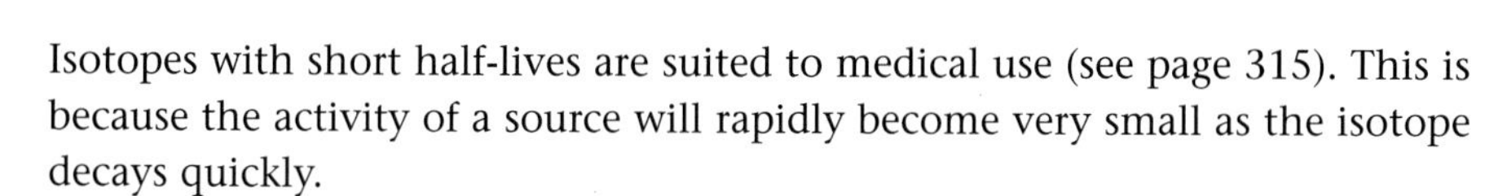

Isotope	Half-life	Decay process
uranium-238	4.5 billion years	α particle emission
radium-226	1590 years	α particle emission, γ ray emission
radon-222	3.825 days	α particle emission
polonium-218	3.05 minutes	α particle emission

Isotopes with short half-lives are suited to medical use (see page 315). This is because the activity of a source will rapidly become very small as the isotope decays quickly.

Isotopes used for dating samples of organic material need to have very long half-lives. This is because the activity will become difficult to measure accurately if it drops below a certain level. In Chapter 34, we shall see that there are suitable isotopes for these different applications.

Half-life calculations

Graphs of activity, in becquerels, against time can be used to find the half-life of an isotope, and this half-life information can be used to make predictions of the activity of the radioisotope at a later time.

worked example

Example 1

The activity of a sample of a certain isotope is found to be 200 Bq.

a) If the isotope has a half-life of 20 minutes, what will the activity of the sample be after one hour?

After 20 minutes the activity will have halved to 100 Bq.

After 40 minutes (2 half-lives) it will have halved again to 50 Bq.

After 60 minutes it will have halved again, so the activity will be 25 Bq.

b) What is the level of activity of this sample after 3 hours?

Three hours = 9 × 20 minutes – that is, nine half-life periods. This means the activity will have halved nine times. The level of activity (and the amount of the radioisotope remaining) will be:

$\frac{1}{2} \times \frac{1}{2} \times \frac{1}{2} \times \frac{1}{2} \times \frac{1}{2} \times \frac{1}{2} \times \frac{1}{2} \times \frac{1}{2} \times \frac{1}{2}$

or $\frac{1}{2^9}$ of the original value, and so $\frac{1}{512}$ of the original activity or amount.

End of Chapter Checklist

If you haven't got a copy of your specification, read the introduction on page iv.

You will need to be able to do some or all of the following. Check your Awarding Body's specification (syllabus) to find out exactly what you need to know.

- Understand that radioactive decay is a random process.
- Model radioactive decay processes and understand the limitations of such models.
- Understand and be able to measure the half-life of a radioisotope.
- Carry out calculations involving half-life.

Questions

More questions on half-life can be found at the end of Section F on page 339.

1 Two students decide to use some six-sided dice to model radioactive decay. They throw the dice and remove all those that come up "6" after each throw.

a) One student decides to use 40 dice and the other uses 200. Which student is likely to produce the best model? Explain your answer.

b) What proportion of the dice are, on average, likely to be removed after each throw?

c) How could the model be altered to simulate a rapid decay process?

(Assume that the dice are fair – that is, equally likely to come up with any number.)

2 Define what is meant by the *half-life* of a radioactive material.

3 The activity of a radioactive sample is measured. The activity, corrected for background radiation, is found to be 240 Bq. The activity is measured again after 1 hour 30 minutes and is now 30 Bq. What is the half-life of the sample?

4 In another model of radioactive decay, a student fills a burette with water as shown in the diagram, and starts a timer at the instant the valve at the bottom is opened. She notes the height of the column of water at regular intervals. It takes 35 seconds to empty from 50 ml to 25 ml. Assuming that the arrangement provides a good model of radioactive decay:

a) How long will it take for three quarters of the water in the burette to drain away?

b) How much water should remain in the burette after $1\frac{3}{4}$ minutes?

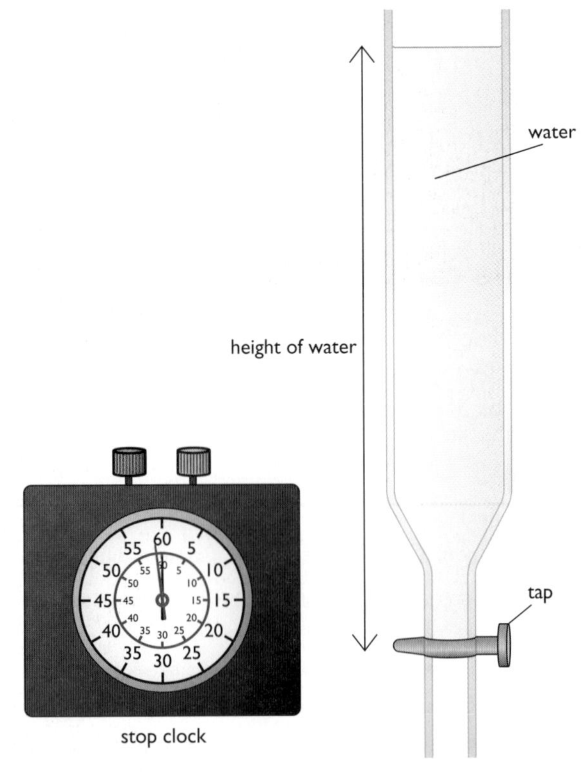

5 Radioactive decay is a *random* process. Explain what this means.

6 A student wants to measure the half-life of a radioactive isotope. He is told the isotope has a half-life of between 10 and 20 minutes. Illustrating your answers as appropriate, describe:

a) the measurements that he should take

b) how he should use the measurements to arrive at an estimate of the half-life for the isotope.

Chapter 34: Applications of Radioactivity

Radioactivity can be used in a wide variety of useful applications, including medicine, industry, power generation and archeology. In this chapter, you will read about these applications, and also learn about the hazards associated with the use of radioactivity.

The use of radioactivity in medicine

Using tracers in diagnosis

Radioactive isotopes are used as **tracers** to help doctors identify diseased organs. A radioactive tracer is a chemical compound that emits gamma radiation. The tracer is taken orally by the patient (swallowed) or injected. Its passage around the body can then be traced using a gamma ray camera.

Different compounds are chosen for different diagnostic tasks. For example, the isotope iodine-123 is absorbed by the thyroid gland in the same way as the stable form of iodine. The isotope decays and emits gamma radiation. A gamma ray camera can then be used to form a clear image of the thyroid gland.

The half-life of iodine-123 is about 13 hours. A short half-life is important as this means that the activity of the tracer decreases to a very low level in a few days.

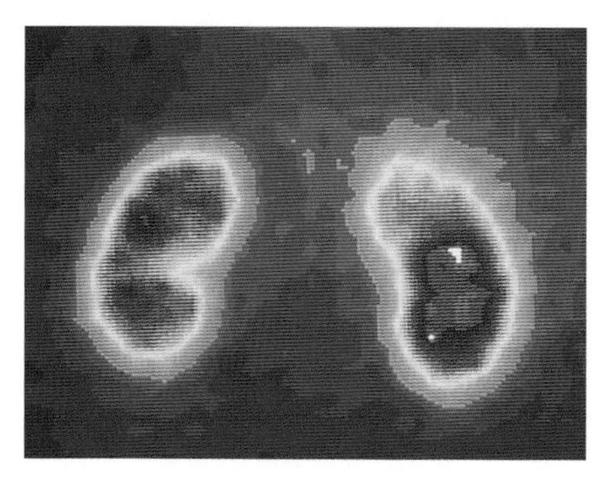

Figure 34.1 *This scan shows the kidneys in a patients body.*

Other isotopes are used to image specific parts of the body. Technetium-99 is the most widely used isotope in medical imaging. It is used to help identify medical problems that affect many parts of the body. Figure 34.1 shows a scan of a patient's kidneys. It shows clearly that one of the kidneys is not working properly.

Imaging techniques enable doctors to produce three-dimensional computer images of parts of a patient's body. These are of great value in diagnosis. Figure 34.2 shows the kind of equipment used for three-dimensional imaging.

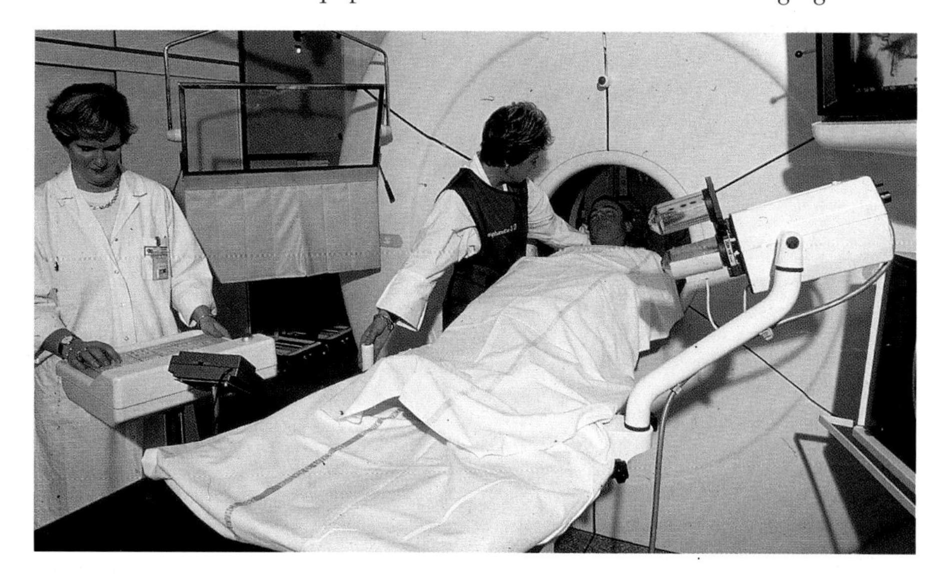

Figure 34.2 *Scanner used to provide 3D images of a patient's body.*

Treatment

Radiation from isotopes can have various effects on the cells that make up our bodies. Low doses of radiation may have no lasting effect. Higher doses may cause the normal function of cells to be changed. This can lead to abnormal growth and cancer. Very high doses will kill living cells.

Cancer can be treated by surgery that involves cutting out cancerous cells. Another way of treating cancer is to kill the cancer cells inside the body. This can be done with chemicals containing radioactive isotopes. Unfortunately, the radiation kills healthy cells as well as diseased ones. To reduce the damage to healthy tissue, chemicals are used to target the location of the cancer in the body. They may emit either alpha or beta radiation. Both these types of radiation have a short range in the body, so they will affect only a small volume of tissue close to the "target".

The radioisotope iodine-131 is used in the treatment of various diseases of the thyroid gland. It has a half-life of about eight days and decays by beta particle emission.

Sterilisation using radiation

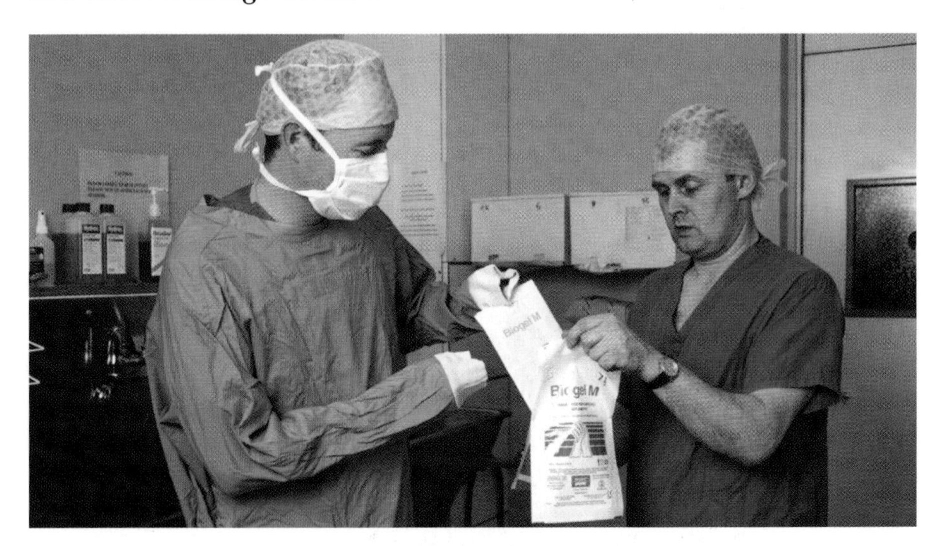

Figure 34.3 *Gamma radiation is used to sterilise medical equipment.*

Ionising radiation can kill living cells. It is therefore used to kill micro-organisms on surgical instruments and other medical equipment. The technique is called **irradiation**. The items to be sterilised are placed close to strongly ionising radiation sources. The items can be packaged in airtight bags to ensure that they cannot be re-contaminated before use. The radiation will penetrate the packaging and destroy bacteria without damaging the item.

Irradiation will not destroy any poisons that bacteria may have already produced in the food before it is treated.

Some food products are treated in a similar way to make sure that they are free from any bacteria that will cause the food to rot or will cause food poisoning. The irradiation of food is an issue that causes concern amongst the public, and is not a widely used procedure at the present time.

Irradiation does not destroy vitamins in the food like other means of killing bacteria, such as high temperature treatment.

The use of radioactivity in industry

Gamma radiography

A gamma ray camera is like the X-ray cameras used to examine the contents of your luggage at airports. A source of gamma radiation is placed on one side of the object to be scanned and a gamma camera is placed on the other. Gamma rays are more penetrating than X-rays. They can be used to check for imperfections in welded joints and for flaws in metal castings. Without this technique of gamma radiography, neither problem could be detected unless the welding or casting were cut through. An additional advantage of **gamma radiography** over the use of X-rays for this purpose is that gamma sources can be small and do not require a power source or large cumbersome equipment.

Figure 34.4 *A gamma camera image used to view inside a valve.*

Gauging

In industrial processes, raw materials and fuel are stored in large tanks or hoppers. Figure 34.5 shows how radioactive isotopes are used to **gauge**, or measure, how much material there is in a storage vessel.

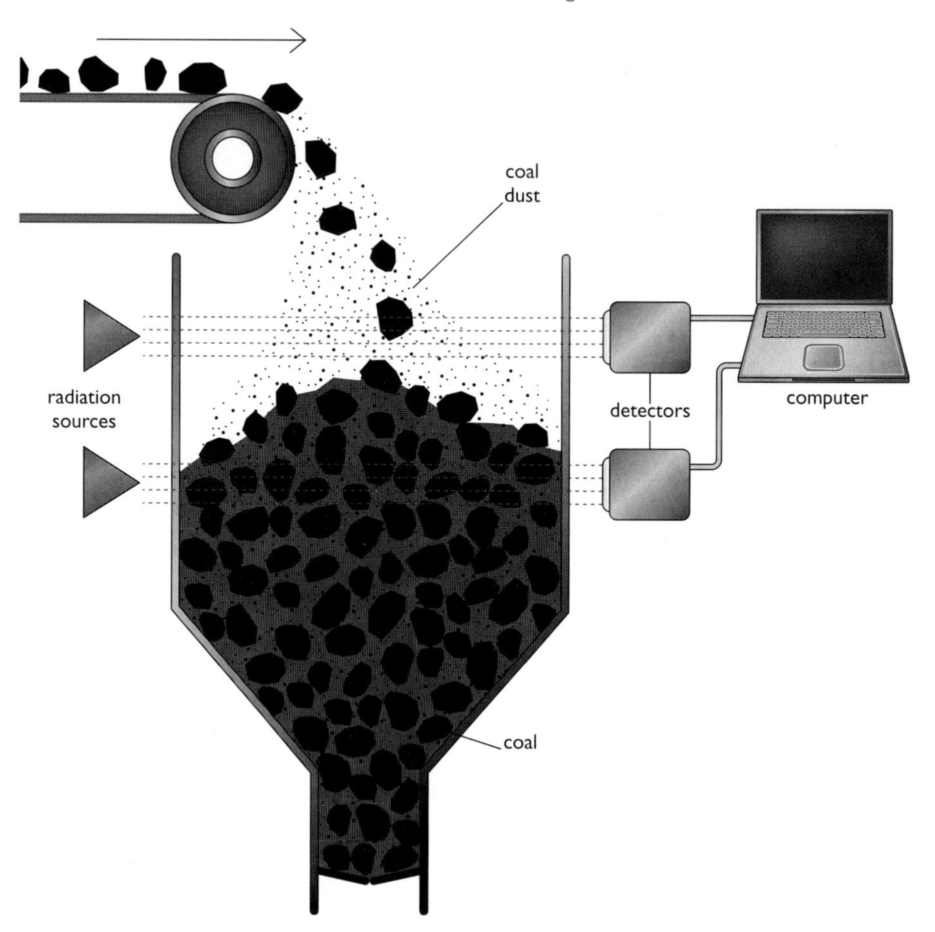

Figure 34.5 *The amount of coal in the hopper can be measured using gamma radiation.*

In Figure 34.5, the coal absorbs a large amount of the radiation so the reading on the lower detector will be small. As the upper part of the hopper is empty the upper detector will have a high reading.

This method of gauging has several advantages over other methods. There is no contact with the material being gauged. Also, coal dust might cause false readings with an optical gauging system (one using light beams). Coal dust is much less dense than coal so the gamma ray system still works properly.

Another example of gauging uses a similar process to monitor the thickness of plastic sheeting and film. The thicker the sheet, the greater the amount of radiation it absorbs. By monitoring the amount of radiation, the thickness of the sheeting can be closely controlled during manufacture.

Tracing and measuring the flow of liquids and gases

Radioisotopes are used to track the flow of liquids in industrial processes. Very tiny amounts of radiation can easily be detected. Complex piping systems, like heat exchangers in power stations, can be monitored for leaks. Radioactive tracers are even used to measure the rate of dispersal of sewage (Figure 34.6)!

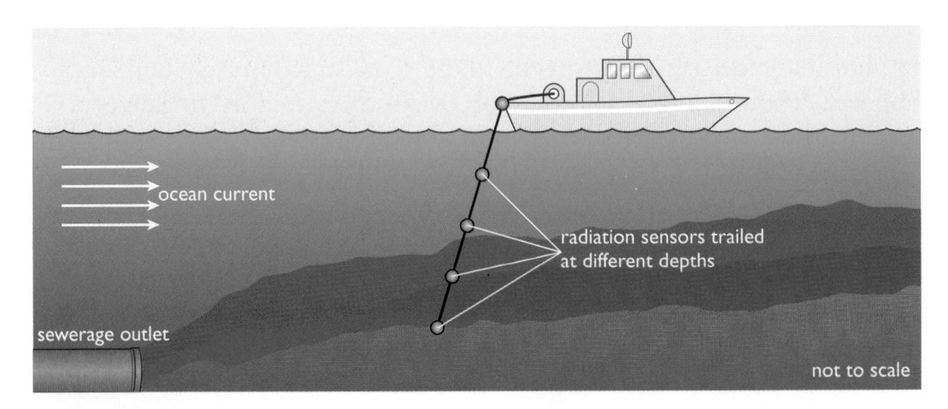

Figure 34.6 *Radioactive tracers released with the sewage allow its dispersal to be monitored, to make sure concentration does not reach harmful levels in any area.*

Radioactive dating

A variety of different methods involving radioisotopes are used to date minerals and organic matter. The most widely known method is **radiocarbon dating**. This is used to find the age of organic matter – for example, from trees and animals – that was once living. We shall also look at techniques that are used to find the age of inorganic material like rocks and minerals.

Radiocarbon dating

Radiocarbon dating measures the level of an isotope called carbon-14 (C-14). This is made in the atmosphere. Cosmic rays from space are continually raining down upon the Earth. These have a lot of energy. When they strike atoms of gas in the upper layers of the atmosphere, the nuclei of the atoms break apart. The parts fly off at high speed. If they strike other atoms they can cause nuclear transformations to take place. These transformations turn the elements in the air into different isotopes. One such collision involves a fast-moving neutron striking an atom of nitrogen. (Nitrogen forms nearly 80% of our atmosphere.) The nuclear equation for this process is:

$${}^{14}_{7}\mathrm{N} + {}^{1}_{0}\mathrm{n} \rightarrow {}^{14}_{6}\mathrm{C} + {}^{1}_{1}\mathrm{p}$$

Notice that, as in the other nuclear equations we have seen, the top numbers – which show the number of nucleons – add up to the same total on each side of the equation. This is because the mass number is conserved. The bottom numbers – which show the amount of charge on the particles – are also conserved.

In this equation, the neutron is represented by:

$${}^{1}_{0}\mathrm{n}$$

Its notation is explained in Figure 34.7.

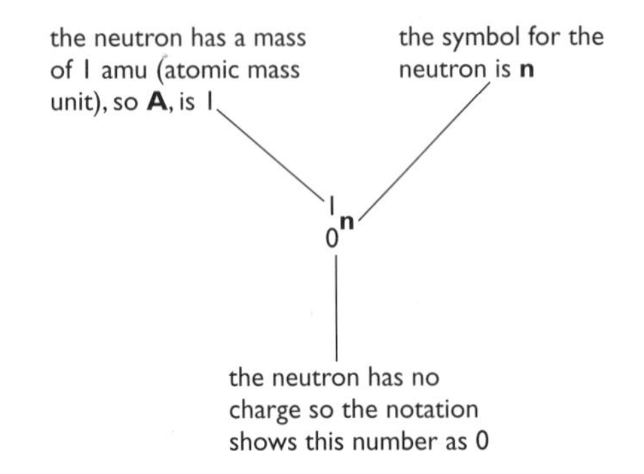

Figure 34.7 *The neutron.*

The proton formed as a result of the collision is represented by:

^1_1p

The notation for this is explained in Figure 34.8.

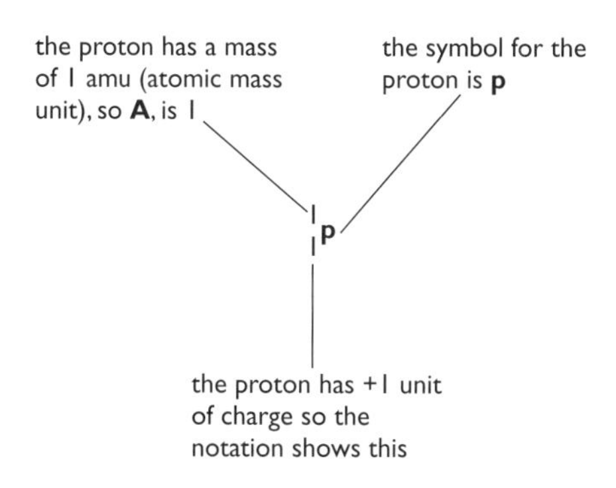

Figure 34.8 *The proton.*

The result of the collision of a neutron with a nitrogen atom is a nuclear transformation. The nitrogen atom is transformed into an atom of the radioactive isotope of carbon, carbon-14.

As we have already mentioned, isotopes of an element have the same chemical behaviour. This means that the carbon-14 atoms react with oxygen in our atmosphere to form carbon dioxide, just like the much more common and stable isotope, carbon-12. The carbon dioxide is then absorbed by plants in the process of photosynthesis. As a result, a proportion of the carbon that makes up any plant will be the radioactive form, carbon-14. Incorporated in plant material, the radioactive isotope carbon-14 enters the food chain, which means that animals and humans will also have a proportion of carbon-14 in their bodies. These carbon-14 atoms will decay but, in living plants and animals, they are continuously replaced by new ones.

When a living organism dies the replacement process stops. As time passes, the radioactive carbon decays and the proportion of radioactive carbon in the remains of the plant or animal, compared with the stable carbon isotope, decreases.

The half-life for the decay of carbon-14 is approximately 5600 years. This means that every 5600 years the proportion of carbon-14 in dead plant and animal material will halve. The amount of carbon-14 present in a sample of dead plant or animal material is found by measuring the activity of the sample. This is compared with the amount of carbon-14 that would have been present when the sample was part of a living organism. From this, it is possible to estimate when the source of the sample died.

***worked* example**

Example 1

120 g of living wood has a radioactive activity (corrected for background radiation) of 24 Bq. A 120 g sample of wood from an archaeological site is found to have an activity of 6 Bq. If the half-life of carbon-14 is 5600 years, estimate the age of the wood from the site.

$6\text{ Bq} = 24\text{ Bq} \times \frac{1}{2} \times \frac{1}{2}$

Since the activity of the sample has halved twice from that expected in living wood, two half-lives must have passed.

The age of the sample is therefore around 2 × 5600 years = 11 200 years.

There are limitations to the method of radiocarbon dating. It assumes the level of cosmic radiation reaching the Earth is constant, which is not necessarily an accurate assumption. Fortunately, the technique has been calibrated to take the variations of cosmic ray activity into account. This is done by testing samples of a known age, like material from the mummies of Egyptian Pharaohs and from very ancient living trees.

The radiocarbon method is not used to date samples older than 50 000 – 60 000 years because, after 10 half-lives, the amount of carbon-14 remaining in samples is too small to measure accurately.

Dating rocks

Inorganic, non-living matter does not absorb carbon-14, so different techniques must be used for finding out the age of rocks and minerals.

When a radioactive substance decays it transforms into a different isotope, sometimes of the same element, sometimes of a different element. The original radioisotope is called the **parent** nuclide (unstable nucleus) and the product is called the **daughter** nuclide. Many of the products of decay, the daughter isotopes, are also unstable and these too decay, in turn. This means that as the parent isotope decays it breeds a whole family of elements in what is called a **decay series**. The end of the decay series is a **stable** isotope – one that does not decay further.

The table shows some radioactive parent isotopes with the stable daughters formed at the end of their particular decay series. The half-life quoted is the time for half the original number of parent nuclei to decay to the stable daughter element.

Radioactive parent isotope	Stable daughter element	Half-life (years)
potassium-40	argon-40	1.25 billion
thorium-232	lead-208	14 billion
uranium-235	lead-207	704 billion
uranium-238	lead-206	4.47 billion
carbon-14	nitrogen-14	5568

For rocks containing such radioactive isotopes, the proportion of parent to stable daughter nuclide gives a measure of the age of the rock. Notice that the half-lives of most of the radioactive parent isotopes are extremely long, in some cases greater than the lifetime of the Earth.

The decay series of potassium-40 ends with argon gas. As potassium-40 decays in igneous rock, the argon produced remains trapped in the rock. Igneous rocks are formed when molten rock becomes solid. Igneous rocks are non-porous. The proportion of argon to potassium-40 again gives a measure of the age of the rock.

ideas
evidence

The health hazards of ionising radiation

Ionising radiation can damage the molecules that make up the cells of living tissue. Cells suffer this kind of damage all the time for many different reasons. Fortunately, cells can repair or replace themselves given time so, usually, no permanent damage results. However, if cells suffer repeated damage because of ionising radiation, the cell may be killed. Alternatively the cell may start to behave in an unexpected way because it has been damaged. We call this effect cell **mutation**. Some types of cancer happen because damaged cells start to divide uncontrollably.

Different types of ionising radiation present different risks. Alpha particles have the greatest ionising effect, but they have little penetrating power.

This means that an alpha source presents little risk, as alpha particles do not penetrate the skin. The problem of alpha radiation is much greater if the source of alpha particles is taken into the body. Here the radiation will be very close to many different types of cells and they may be damaged if the exposure is prolonged. Alpha emitters can be breathed in or taken in through eating food. Radon gas is a decay product of radium and is an alpha emitter. It therefore presents a real risk to health. Smokers greatly increase their exposure to this kind of damage as they draw the radiation source right into their lungs (cigarette smoke contains radon).

Beta and gamma radiation do provide a serious health risk when outside the body. Both can penetrate skin and flesh and can cause cell damage by ionisation. Gamma radiation, as we have mentioned earlier, is the most penetrating. The damage caused by gamma rays will depend on how much of their energy is absorbed by ionising atoms along their path. Beta and gamma emitters that are absorbed by the body present less risk than alpha emitters, because of their lower ionising power.

In all cases, the longer the period of exposure to radiation the greater the risk of serious cell damage. Workers in the nuclear industry wear badges to indicate their level of exposure. Some are strips of photographic film that become increasingly "foggy" as the radiation exposure increases. Another type of badge uses a property called **thermoluminescence**. Thermoluminescence means that the exposed material will give out light when it is warmed. The radiation releases energy to make heat so the thermoluminescent badges give out more light when exposed to higher levels of radiation. Workers have their badges checked regularly and this gives a measure of their overall exposure to radiation.

Safe handling of radioactive materials

Samples of radioactive isotopes used in schools and colleges are very small. This is to limit the risk to users, particularly those who use them regularly – the teachers! Although the risk is small, certain precautions must be followed. The samples are stored in lead containers to block even the most penetrating form of radiation, gamma rays. The containers are clearly labelled as a radiation hazard and must be stored in a locked metal cabinet. The samples are handled using tongs and are kept as far from the body as possible.

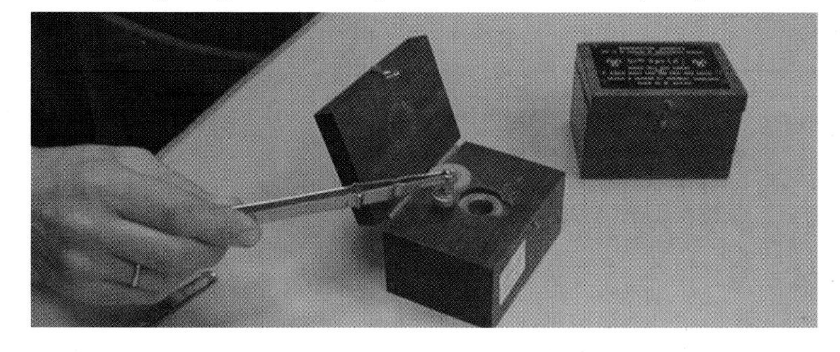

Figure 34.9 *Radioactive samples are stored in lead-lined containers and are handled with tongs.*

In the nuclear industry and research laboratories, much larger amounts of radioactive material are used. These have to be handled with great care. Very energetic sources will be handled remotely by operators who are shielded by lead, concrete and thick glass viewing panels.

Figure 34.10 Industrial sources of radioactivity must be handed with a lot of care.

The major problem with nuclear materials is long-term storage. Some materials have extremely long half-lives so they remain active for thousands and sometimes tens of thousands of years. Nuclear waste must be stored in sealed containers that must be capable of containing the radioactivity for enormously long periods of time.

ideas
evidence

Generating electricity using nuclear fuels

Uranium-235 is used as fuel in a nuclear reactor. It is used because its nuclei can be split by a neutron. The process of splitting an atom is called **fission.** Uranium-235 is called a **fissile** material because it goes through the splitting process easily. The fission process is shown in Figure 34.11.

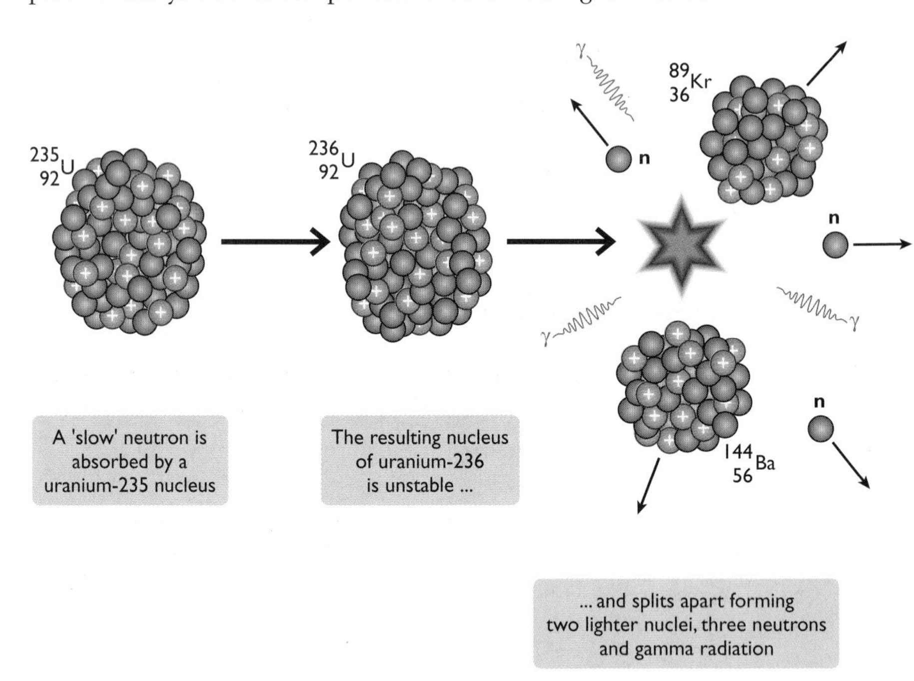

Figure 34.11 *Fission of uranium-235.*

A "slow" neutron is a low-energy neutron produced by a nuclear decay. Faster moving, more energetic neutrons do not cause fission.

In the fission reaction, a "slow" neutron is absorbed by a nucleus of uranium-235.

The resulting nucleus of uranium-236 is unstable and splits apart. The fragments of this decay are the two lighter nuclei of barium-144 and krypton-89. The decay also produces gamma radiation and three more neutrons. The equation for this decay is:

$$^{236}_{92}\text{U} \rightarrow {}^{144}_{56}\text{Ba} + {}^{89}_{36}\text{Kr} + 3\,{}^{1}_{0}\text{n} + \gamma \text{ radiation}$$

The fission reaction produces a huge amount of energy. This is because some of the mass of the original uranium-236 nuclei is converted to energy. Most of the energy is carried away as the kinetic energy of the two lighter nuclei. Some is emitted as gamma radiation. The three neutrons produced by the fission may hit other nuclei of uranium-235, so causing the process to repeat, as shown in Figure 34.12. If one neutron from each fission causes one nearby uranium-235 to split, then the fission reaction will keep going. If more than one neutron from each fission causes fission in surrounding nuclei, then the reaction escalates – a bit like an avalanche.

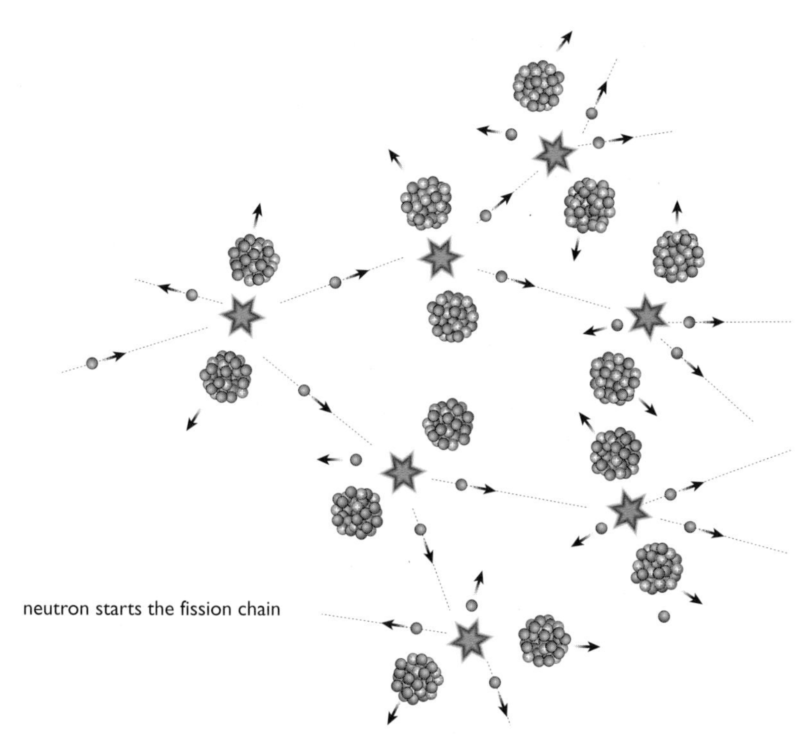

Figure 34.12 *The chain reaction in uranium-235.*

This is called a **chain reaction**. If this reaction is allowed to take place in an uncontrolled way, the result is a nuclear explosion. This involves the sudden release of enormous amounts of heat energy and radiation. In a nuclear reactor the process is controlled so that the heat energy is released over a longer period of time. The heat produced in the **core** or heart of the reactor is used to heat water. The steam produced then drives turbines to turn generators. The basic parts of a nuclear reactor are shown in Figure 34.13.

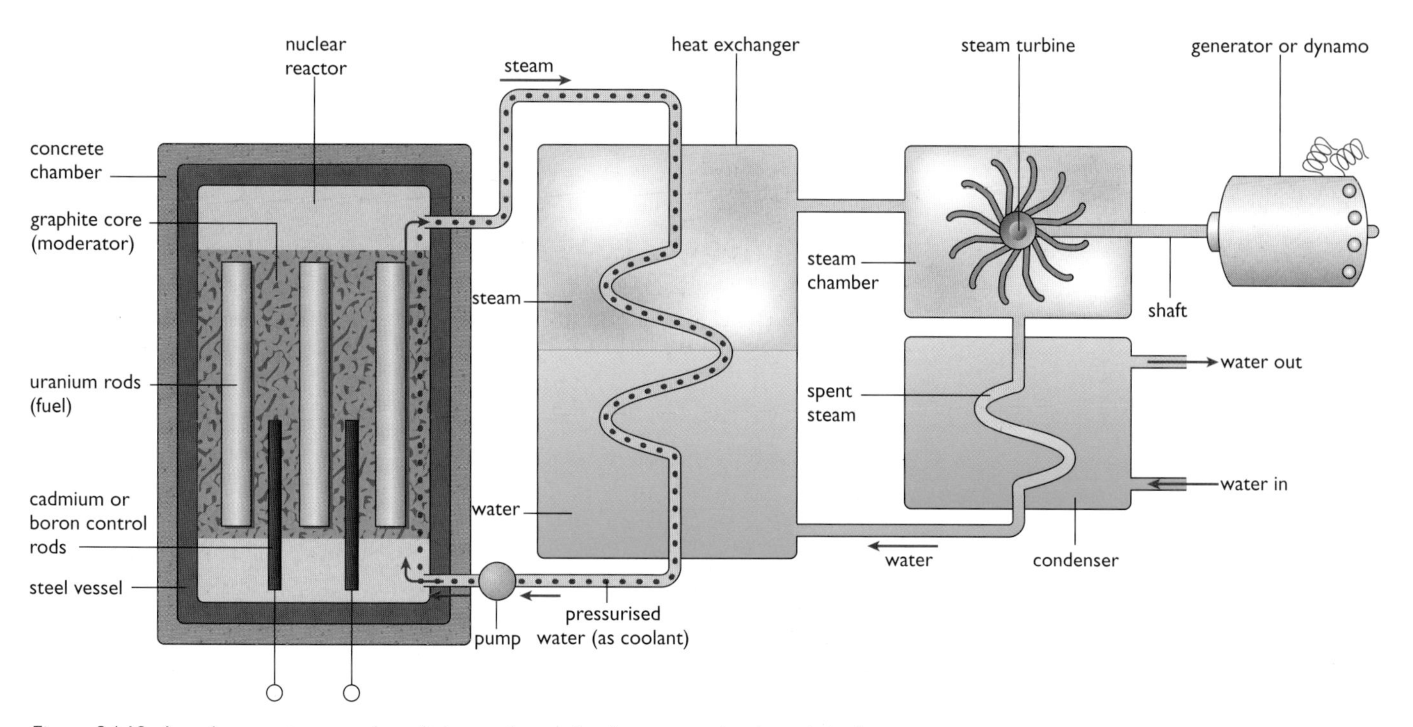

Figure 34.13 *A nuclear reactor controls a chain reaction so that heat energy is released slowly.*

The reactor core contains fuel rods of enriched uranium. Enriched uranium is uranium-238 with a higher proportion of uranium-235 than is found in natural reserves of uranium. Graphite is used as a moderator. The job of the **moderator** is to absorb some of the kinetic energy of the neutrons to slow them down. This is because slow neutrons are more easily absorbed by uranium-235. A neutron slowed in this way can start the fission process. In the nuclear reactor there are also **control rods**, made of boron or cadmium. These absorb the neutrons and take them out of the fission process completely. When the control rods are fully lowered into the core, the chain reaction is almost completely stopped and the rate of production of heat is low. As the control rods are raised, the rate of fission increases producing heat at a greater rate.

ideas evidence

Nuclear power stations do not produce carbon dioxide or acidic gases as fossil fuel power stations do. This means that nuclear power does not contribute to global warming or acid rain. Only small amounts of uranium are needed for a chain reaction and the supply of nuclear fuel will last many hundreds of years – unlike some fossil fuels that could run out in the next fifty years.

The nuclear process in a reactor produces a variety of different types of radioactive material. Some have relatively short half-lives and decay rapidly. These soon become safe to handle and do not present problems of long-term storage. Other materials have extremely long half-lives. These will continue to produce dangerous levels of ionising radiation for thousands of years. These waste products present a serious problem for long-term storage. They are usually sealed in containers that are then buried deep underground. The sites for underground storage have to be carefully selected. The rock must be impermeable to water and the geology of the site must be stable – storing waste in earthquake zones or areas of volcanic activity would not be sensible!

Some reactors are designed to produce **plutonium**. Plutonium is a very radioactive artificial element. Small amounts of plutonium represent a serious health hazard. Plutonium is another fissile material. If a large enough mass of plutonium is brought together a chain reaction will start. Plutonium can be used in the production of nuclear weapons.

End of Chapter Checklist

If you haven't got a copy of your specification, read the introduction on page iv.

You will need to be able to do some or all of the following. Check your Awarding Body's specification (syllabus) to find out exactly what you need to know.

- Be aware of the various medical applications of radioactive isotopes.
- Know that some isotopes are used in diagnosis, either by imaging particular parts of the body, or by showing whether an organ is working properly by the rate at which it processes particular chemicals.
- Know that some isotopes are used to destroy diseased cells in the treatment of illness.
- Select appropriate isotopes for specific medical tasks.
- Be aware of the industrial uses of radioactive isotopes in tracing the movement of liquids and gases, identifying leaks and in gauging.
- Understand the principles of dating organic materials and minerals using radioisotopes.
- Know the health hazards presented by different types of radiation from radioisotopes and the need to monitor exposure.
- Appreciate the need for appropriate handling techniques.
- Understand the chain reaction process that may occur in fissile material.

Questions

More questions on applications of radioactivity can be found at the end of Section F on page 339.

1 The most widely used isotope in medicine is technetium-99. It has a half-life of 6 hours and decays by the emission of low-energy gamma rays and beta particles.

a) Explain why the characteristics of technetium-99 make it suitable for diagnostic use in medicine.

b) Technetium-99 can be chemically attached to a wide variety of pharmaceutical products so that it can be targeted at particular tissues or organs. How can its progress through the body be measured and monitored?

2 Technetium-99 is produced from molybdenum-99 in a device called a technetium generator. Which decay process – α, β or γ – could cause molybdenum-99 to decay to technetium-99? Explain your answer. (**Hint**: Look at Chapter 33.)

3 A radioactive isotope of iodine is used in both the diagnosis and treatment of a condition of the thyroid gland. This gland naturally takes up ordinary iodine as part of its function. If a patient has an overactive thyroid it concentrates too much iodine in the gland and this has serious effects on the patient's health.

How might the radioisotope iodine-131 be used to:

a) identify an overactive thyroid gland

b) treat the overactive thyroid?

(Iodine-131 has a half-life of 8 days and is a high-energy beta emitter.)

4 Explain the advantages of using ionising radiation for sterilising surgical equipment.

5 *a)* Paper is made in a variety of different "weights", with different thicknesses. How could ionising radiation be used to check the thickness of paper during production? You should consider the following:

- the type of radiation to be used
- how it will be used to measure the paper thickness
- what checks should be made to ensure that the measurements are accurate
- safety procedures.

b) Find out how radioactivity is used in measuring or gauging.

6 Radiocarbon dating is used to estimate the age of organic (once-living) materials. It uses a radioisotope of carbon, carbon-14 (C-14).

a) How is carbon-14 formed?

b) Why does all living matter contain a proportion of C-14?

c) What happens to the proportion of C-14 in an organism once it has died?

d) What assumptions are made in the process of radiocarbon dating?

e) Why is this method unsuitable for accurately dating material that is more than 50 000 years old?

7 Most radioactive isotopes of elements have half-lives that are extremely short compared to the age of the Earth. The Earth is about 4.5 billion years old. Radium has a half-life of only about 1600 years.

a) How many half-lives of radium have there been since the Earth formed?

Student A says all the radium formed when the Earth condensed out of the Sun's atmosphere should have decayed away to an unmeasurably small amount by now.

Student B says that it depends on how much there was to start with.

Student C says that there is still a significant amount of radium on the Earth.

b) Discuss these statements. Who do you think is right?

c) What other information would you need to make a better decision?

8 An isotope that decays by alpha emission is relatively safe when outside the body but very dangerous if absorbed by the body, either through breathing or eating.

a) Explain why this is so.

b) Why is radon-220 a particularly dangerous isotope?

9 *a)* Uranium-235 is a *fissile* material. What does this mean?

b) If there is a large enough mass of uranium-235, it may cause a chain reaction. (This is called the *critical mass* for the isotope.)

i) What is a chain reaction?

ii) Why does a chain reaction depend on how much of the fissile material there is in one piece?

10 List two advantages and two disadvantages of nuclear fission as a way of producing energy.

Chapter 35: Electrons and Other Particles

Atoms are made up of electrons, protons and neutrons. Electrons are indivisible, but protons and neutrons are themselves made up of smaller particles called quarks. Quarks, electrons, and the antimatter equivalents of electrons – positrons – are what all matter is made up of. In this chapter, you will learn what happens to these particles during some nuclear transformations. You will also find out how beams of electrons are used to create the pictures you see on television and computer screens, and to generate X-rays.

Fundamental particles are the basic building blocks of all matter, and have no internal structure. The atom was once thought to be a fundamental particle, but we now know that it is made up of electrons, protons and neutrons. More recently scientists have found that protons and neutrons are not fundamental, but are made up of smaller particles called **quarks**. Scientists continue to search for more fundamental particles.

Fundamental particles

Electrons and positrons

The **positron** is an example of an **antiparticle** – that is, one of the particles that make up **antimatter**. Antimatter is similar in many ways to the ordinary matter that makes up our world and, as far as we can tell, the whole of our Universe but antimatter and ordinary matter cannot exist together. This makes antimatter very scarce indeed.

The particles that make up antimatter can be made using **particle accelerators**. A particle accelerator is a piece of apparatus used by scientists studying high-energy physics. It is used to accelerate charged particles like protons and electrons to very high speeds – indeed, close to the speed of light (300 000 km/s). These high-speed, high-energy particles are fired into atoms to break them apart so that physicists can work out just how they are made. Figure 35.1 shows the particle accelerator used in high-energy research at CERN in Geneva.

Figure 35.1 *Particle accelerator at CERN, Geneva.*

If just 1 g of antimatter entered our world it would very quickly be annihilated by 1 g of ordinary matter. The resulting explosion would release an enormous amount of energy – 9×10^{13} J. This amount of energy would run a 3 kW electric fire continuously for more than 950 years. It is equivalent to 25 million kWh of energy.

The electron and its antiparticle, the positron, are considered to be fundamental particles. They have no internal structure – that is, they are not made from simpler, smaller particles. The electron and the positron have the same mass. The electron is negatively charged. The positron carries an equal amount of positive charge. When a positron and an electron meet they cancel each other out and their mass is completely converted into energy. This process is called **mutual annihilation**.

The name "quark" was given by a particle physicist called Murray Gell-Mann. He took the word from a novel by the famous Irish writer, James Joyce.

Quarks

Protons and neutrons are made of particles called **quarks**. There are a number of different types of quarks and they have rather odd names, like "strange" quarks and "charmed" quarks.

Protons and neutrons are made up of two types of quark. These are called the **up** quark and the **down** quark. The up and down quarks are the only stable quarks in ordinary matter. These two quarks have different amounts of electric charge (Figure 35.2).

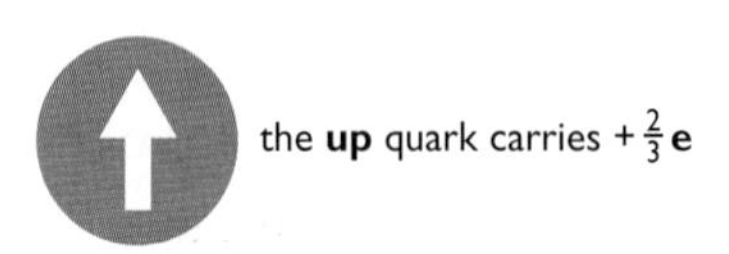

e is the charge on an electron, -1.6×10^{-19}C

Figure 35.2 *The up and down quarks carry different amounts of electric charge.*

A proton is made up of two up quarks and one down quark, and this is written as (uud), as shown in Figure 35.3.

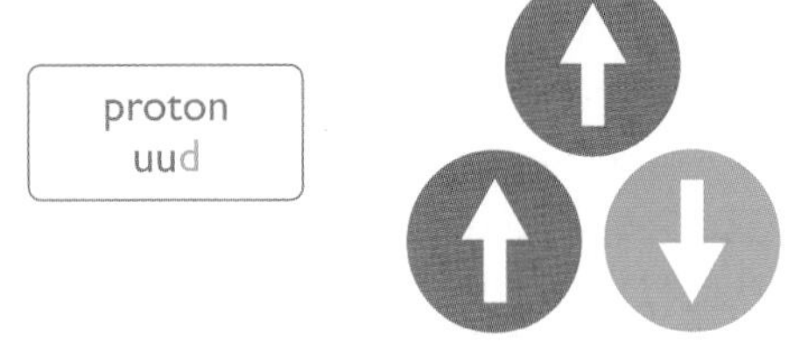

Figure 35.3 *The quark structure of a proton.*

This gives the proton a charge of:

$$+\frac{2}{3}e + \frac{2}{3}e - \frac{1}{3}e = +e$$

A neutron is made up of two down quarks and one up quark, written as (udd), as in Figure 35.4.

So the total charge of a neutron is:

$$+\frac{2}{3}e - \frac{1}{3}e - \frac{1}{3}e = 0$$

ideas evidence

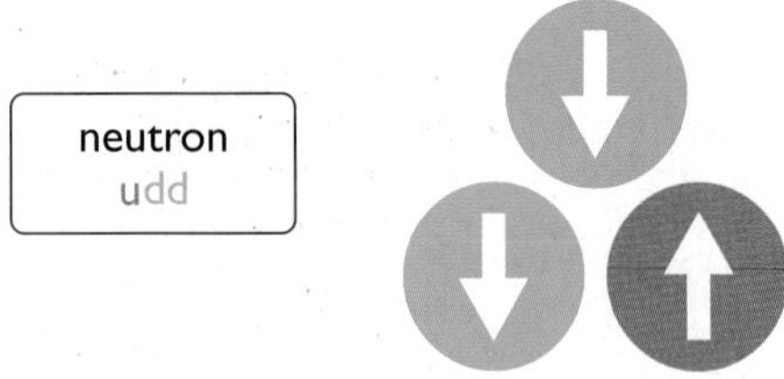

Figure 35.4 *The quark structure of a neutron.*

Quarks and beta decay

Beta⁻ decay involves a neutron in the nucleus splitting into a proton and a high-speed electron (beta particle, β^-), which leaves the nucleus. When beta decay occurs we believe a down quark has turned into an up quark. This is shown in Figure 35.5.

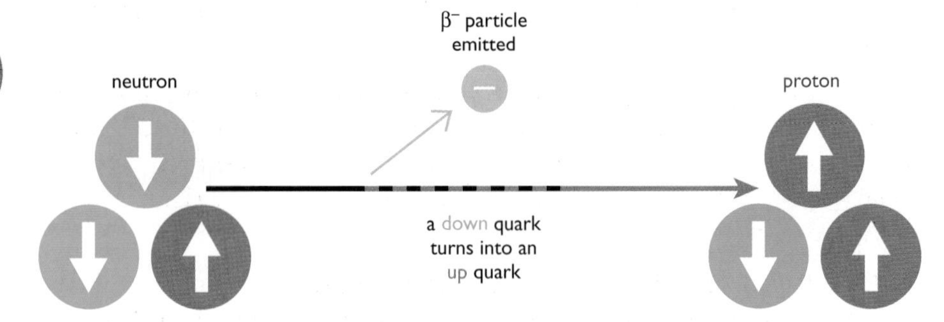

Figure 35.5 *β^- decay.*

Electric charge is conserved during beta–decay:

$$+\tfrac{2}{3}e - \tfrac{1}{3}e - \tfrac{1}{3}e \rightarrow +\tfrac{2}{3}e + \tfrac{2}{3}e - \tfrac{1}{3}e - e$$

$${}^{1}_{0}n \rightarrow {}^{1}_{1}p + {}^{0}_{-1}e$$

↑ emitted beta particle, β^-

Beta+ (positron) decay involves an up quark in a proton turning into a down quark, with a positron (β^+) emitted to conserve the electric charge. This is shown in Figure 35.6.

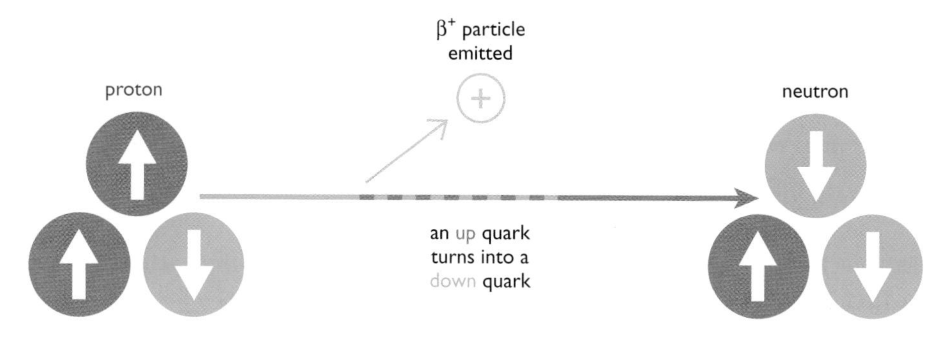

Figure 35.6 β^+ *decay.*

Electric charge is conserved during beta+ decay:

$$+\tfrac{2}{3}e + \tfrac{2}{3}e - \tfrac{1}{3}e \rightarrow +\tfrac{2}{3}e - \tfrac{1}{3}e - \tfrac{1}{3}e + e$$

$${}^{1}_{1}p \rightarrow {}^{1}_{0}n + {}^{0}_{-1}e$$

↑ emitted positron, β^+

Electrons, thermionic emission and the cathode ray oscilloscope

Electrons are used to produce the pictures we see on television screens and computer monitors. They are also used to produce the graphs we see on cathode ray oscilloscopes (CROs). When fast-moving electrons strike certain materials, some of their energy is converted into light. These materials are called **phosphors**.

The electrons needed for this purpose are produced by heating metal to high temperatures. Metals contain lots of free electrons and those near the surface of hot metals have enough energy to escape from the metal surface. You could say that they were "boiled off" the metal surface. This is called **thermionic emission**.

To "paint" pictures with electrons, we must produce a fine beam of them. This is done with an **electron gun**. The electron gun is made up of shaped metal plates called **electrodes**. These electrodes are connected to different voltage power supplies in the CRO. The electron gun is at the back of the cathode ray tube (CRT), which is completely emptied of air to create a vacuum inside it. The basic structure of a cathode ray tube and electron gun is shown in Figure 35.7.

Before the electron was discovered and identified as a minuscule negatively charged particle, some of the effects it produced were explained as being the result of an unknown kind of ray produced by the cathode in a discharge tube. These rays were named "cathode rays".

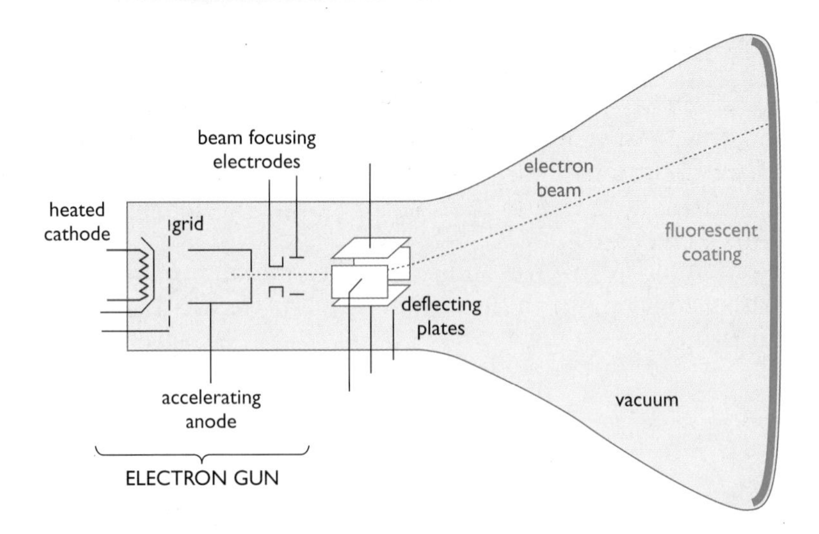

Figure 35.7 *The cathode ray tube.*

The electron gun consists of several electrodes. One, called the **cathode**, is heated to make it produce electrons by thermionic emission. A large voltage is connected between the cathode and a specially shaped electrode called the **accelerating anode**. The anode is positive with respect to the cathode. The anode attracts the negatively charged electrons. When they reach the anode, the electrons are travelling very fast. Some electrons will shoot straight through the hole in the anode. These emerge as a beam of high-speed electrons. Further electrodes are used to focus this beam to a fine point on the inside of the CRT, which is coated with a phosphor. When the beam of electrons hits the phosphor coating the phosphor gives out light.

The energy of the electron beam

The voltage between the cathode and the accelerating anode in an electron gun creates and **electric field**. This field makes a force act on the electrons, which causes them to accelerate. As the electrons accelerate, their kinetic energy increases. The kinetic energy of an electron in an electron beam is given by:

kinetic energy, KE = electronic charge × accelerating voltage

$$KE = eV$$

worked example

Example 1

Work out the kinetic energy of an electron (charge -1.6×10^{-19} C) after it has been accelerated through a voltage of 5000 V.

Use: $KE = eV$

Substitute $e = 1.6 \times 10^{-19}$ C* and $V = 5000$ V:

$KE = 1.6 \times 10^{-19} \text{C} \times 5000 \text{V}$

$= 8.0 \times 10^{-15} \text{J}$

(*Notice that we do not need to include the minus sign for the electronic charge in this equation.)

This is a tiny amount of energy, but, because its mass is also extremely small, the elctron is travelling very fast indeed (at approximately 130 million m/s).

Current in an electron beam

An electron beam is made up of moving electrically charged particles (electrons). In Chapter 1 we saw that an electric current is a flow of electric charge. The beam therefore carries an electric current. We can work out the size of the current by calculating how much charge, in coulombs, passes any point in the path of the beam in one second.

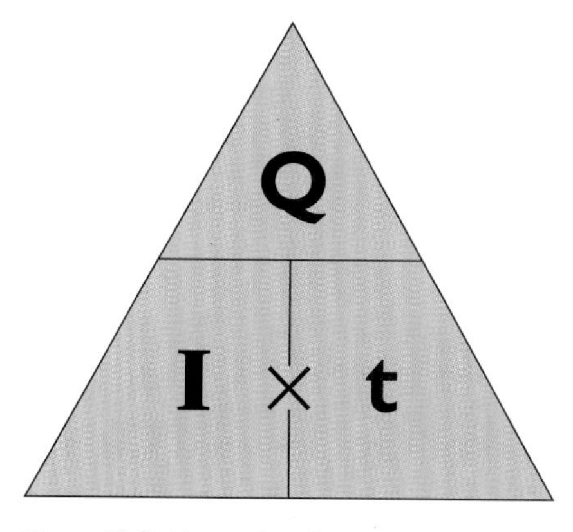

Figure 35.8 *Remember that you can use the triangle method for rearranging equations like $I = Q/t$.*

$$\text{current, } \mathbf{I} = \frac{\text{charge, Q}}{\text{time, t}}$$

$$\mathbf{I} = \frac{Q}{t}$$

worked
example

Example 2

An electron gun fires a beam of 625 000 000 electrons per second. How much current does the beam carry?

Each electron carries a charge of -1.6×10^{-19} C so the total charge leaving the electron gun per second is given by:

$$\text{current, I} = \frac{625\,000\,000 \times -1.6 \times 10^{-19}\,\text{C}}{1\,\text{s}}$$

$$\text{I} = 10^{-10}\,\text{A}$$

Electron beam deflection

Using an electric field

Electrons are negatively charged. If they are passed between charged metal plates, they will be attracted towards the positive plate and repelled by the negative plate. The voltage connected across the plates creates an electric field between them. This is shown in Figure 35.9.

As the electrons pass between the deflecting plates, the beam curves downwards. If the voltage is reversed the beam can be bent upwards. By varying the size of the voltage, the degree of bending of the beam can be altered. This system of beam deflection is used in an oscilloscope to control the vertical position of the beam on the screen. Another pair of plates is used to deflect the beam horizontally. By controlling the size and direction of the electric field between both pairs of plates it is possible to deflect the beam to any part of the screen.

The cathode ray oscilloscope produces a graph on its screen that shows how a voltage signal is changing with time. The signal, after being amplified, is applied to the vertical deflecting plates to make the beam move up and down on the screen. At the same time the beam is deflected by the horizontal deflecting plates so that it sweeps across the screen at a known rate.

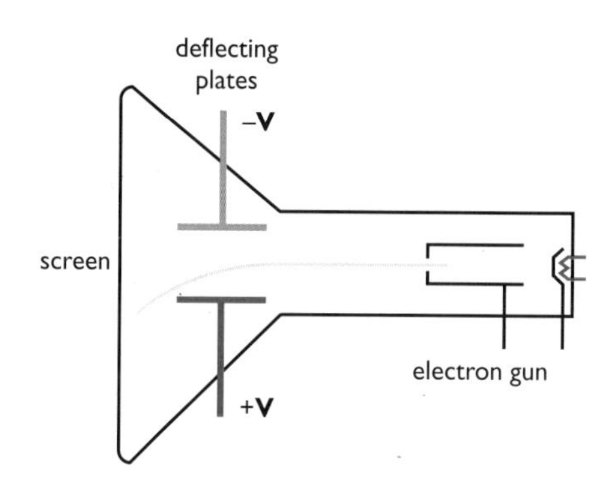

Figure 35.9 *This diagram shows the pair of plates in an oscilloscope that deflect the beam up and down (in the vertical direction). The pair of plates that deflect the beam from side to side have been left out to keep the diagram clear, but they operate in the same way.*

Deflecting ink droplets

Inkjet printers use an electric field to deflect ink droplets to hit the paper at the correct spot. The print head of the printer produces tiny droplets of ink that are charged by static electricity. The drops pass through an electric field in a similar way to the electron beam in an oscilloscope. By controlling the electric field, the printer guides each ink droplet to the correct point on the paper.

Using a magnetic field

As we have seen, electron beams carry current. Currents produce a magnetic effect – when a current passes through an external magnetic field, a magnetic force acts on the charge carriers. So when an electron beam is fired through a magnetic field it will experience a magnetic force. This force will cause the beam to bend.

In televisions and computer monitors, the electron beam that "paints" the picture on the screen is deflected using a magnetic field. The magnetic field is produced by pairs of coils positioned above and below the cathode ray tube. A current in these coils produces the magnetic field that deflects the electron beams.

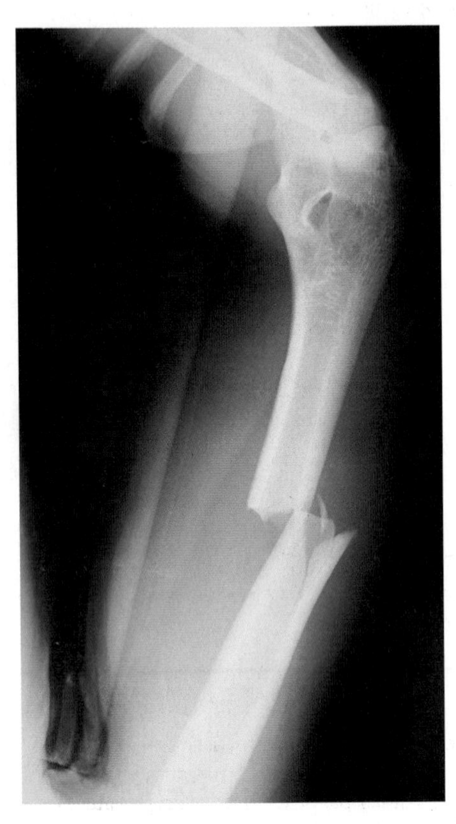

Figure 35.10 *This patient's fracture can be seen clearly using X-rays.*

Using electron beams to create X-rays

X-rays are a part of the electromagnetic spectrum (see page 184). They are electromagnetic waves with a very short wavelength. They can pass easily through low-density materials, like soft body tissue. Denser materials, like bones, absorb the rays so that fewer can penetrate. X-rays leave a shadow on photographic plates, which means that they can be used to examine the internal structure of a patient's body without the need to operate on the patient.

X-rays are created when very energetic (fast-moving) electrons are caused to decelerate very quickly. This is done by using an electron gun with a very high accelerating voltage. The high-speed electrons produced are fired at a target. The target is a dense metal with a high melting point, like tungsten. As the electrons hit the target their energy is absorbed and some is re-radiated as X-rays. The process also produces heat. The X-rays produced in this way are filtered before being used, to form clearer images of a patient's internal organs and bones.

End of Chapter Checklist

If you haven't got a copy of your specification, read the introduction on page iv.

You will need to be able to do some or all of the following. Check your Awarding Body's specification (syllabus) to find out exactly what you need to know.

- Know that electrons, positrons and quarks are considered to be fundamental particles.
- Know that protons and neutrons are composed of two different types of quark.
- Explain beta decay in terms of quark transformation.
- Describe the process of thermionic emission.
- Draw the basic structure of the electron gun used in cathode ray tubes.
- Calculate the energy gained by an electron when it is accelerated by a voltage.
- Calculate the current in an electron beam.
- Describe how charged particles can be deflected using either electric or magnetic fields.
- Describe how X-rays are produced by making electrons decelerate from high speeds.

Questions

More questions on electrons and other particles can be found at the end of Section F on page 339.

1 State the quark structure of:

a) a proton

b) a neutron.

2 In β^- decay, a neutron splits to form a proton and the electron that is ejected is a beta particle.

a) Explain this process in terms of the quarks that make up a neutron.

b) How does β^+ decay differ from β^- decay?

3 Explain what is meant by *thermionic emission.*

4 ***a)*** Draw a simple labelled diagram showing the main parts of an electron gun.

b) What is the energy gained by an electron when it is accelerated through a voltage of 7.5 kV?

c) In the Geiger and Marsden alpha particle scattering experiment described on page 293–4, an alpha particle source emitted alpha particles at an average rate of 1000 per second. What electric current is carried by this beam of alpha particles?

5 The electron beam in a cathode ray tube (CRT) consists of moving charged particles. Describe two methods of deflecting the beam to make it strike a different part of the screen.

Chapter 36: The Cathode Ray Oscilloscope

In this chapter, you will learn how to set up and use a cathode ray oscilloscope to make measurements during experiments.

Oscilloscopes are used to make measurements of voltage and time. They are voltmeters that can measure rapidly changing voltages and show how the voltage varies with time. They do this by "drawing" a graph on the screen. The vertical scale on this graph is in volts and the horizontal scale is in units of time.

We have already discussed some of the principles of operation of an oscilloscope in Chapter 35. The main part of an oscilloscope is the cathode ray tube. This is a glass tube that has had all the air removed from it. The tube has a flat fluorescent screen at one end. Inside the tube at the other end is the electron gun and other components that control the focus and the brightness of the line drawn on the screen by the electron beam. The beam is deflected using two pairs of plates to create horizontal and vertical electric fields.

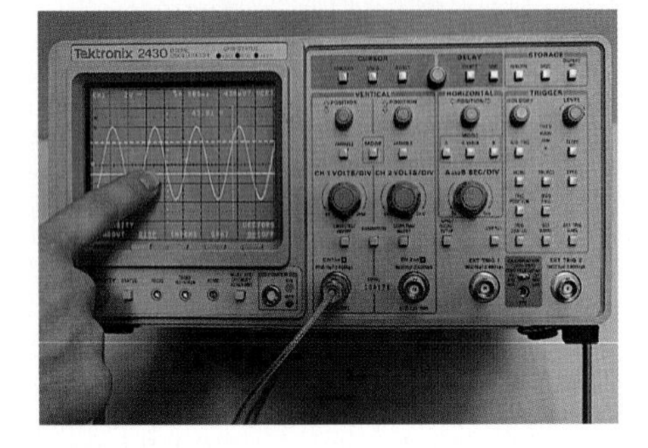

Figure 36.1 *An oscilloscope is used to measure voltage and time.*

The controls of a typical oscilloscope are shown in Figure 36.2.

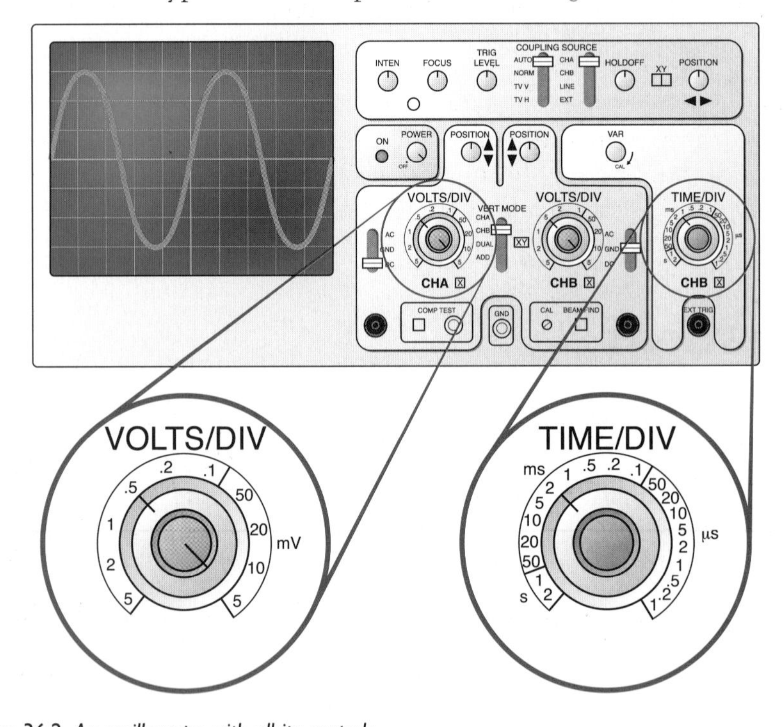

Figure 36.2 *An oscilloscope with all its controls.*

The time scale

Reminder: 1 ms is one thousandth of a second (10^{-3} s). 1 μs is one millionth of a second (10^{-6} s).

The electron beam is produced by the electron gun and sweeps across the screen horizontally. (This is done by an internal circuit, called the **timebase**. The timebase changes the voltage across the horizontal deflecting plates. This makes the electron beam move across the screen horizontally.) The time taken for the beam to travel across one of the divisions on the screen can be controlled. Figure 36.2 shows the control that sets the horizontal time scale. It is marked in time/division (time/div). Notice that some sections of the scale around this control are in seconds per division (s/div), others in milliseconds per division (ms/div) and some in microseconds per division (μs/div).

This setting tells you the horizontal time scale of the graph that the oscilloscope draws on the screen. In Figure 36.2, the timebase is set to 2 ms/div so each horizontal division on the screen represents a time of 2 ms.

The voltage scale

The varying voltage signal that you want to see drawn on the oscilloscope screen is connected to the **Y input**. The signal is amplified and then connected to the vertical deflecting plates. As the voltage changes, the electron beam will be deflected up and down the screen. Figure 36.2 shows the Y-gain control of the amplifier. This control sets the vertical voltage scale of the oscilloscope screen. It is marked out in volts per division (volts/div or V/div) and millivolts per division (mV/div). This setting tells you the vertical voltage scale of the oscilloscope graph – in Figure 36.2 it is 0.5 V/div.

Setting up an oscilloscope for use

First turn the oscilloscope on. Usually this will result in some kind of trace (line) appearing on the screen. If not, look for the horizontal and vertical position controls. These controls will have arrows like those shown in Figure 36.3a, indicating either horizontal or vertical motion. When you have found them, set them to the middle of their range of movement.

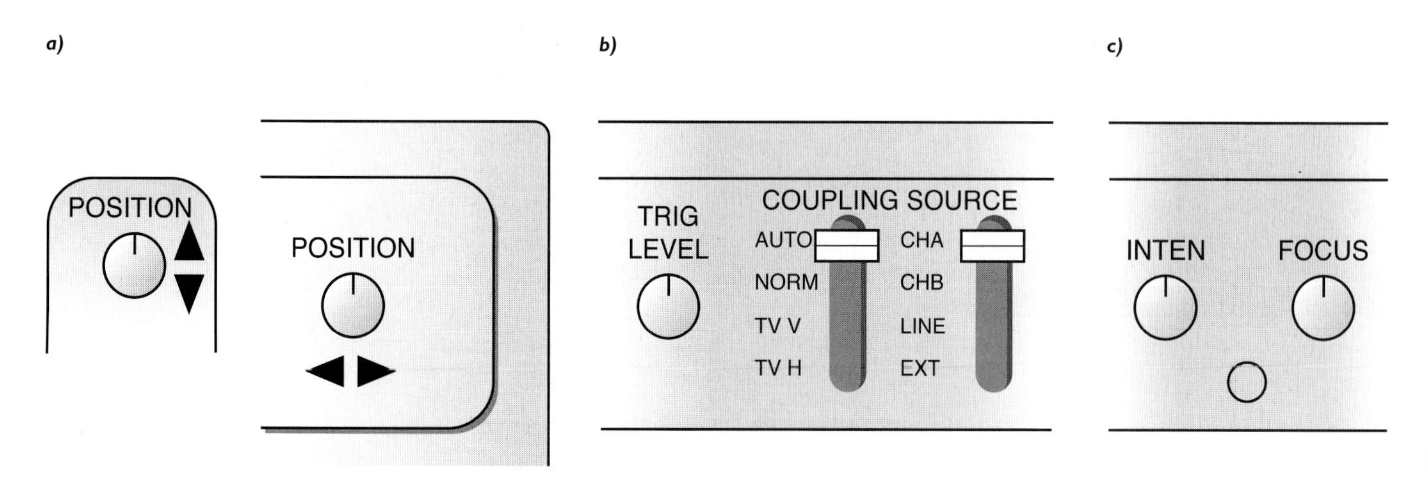

Figure 36.3 a) *The vertical and horizontal position controls – each should be set to the middle of its range of movement,* b) *trigger control – set to "auto" and the channel in use, if the oscilloscope is dual trace,* c) *brightness and focus controls.*

As you do this, you may notice the trace moving into view on the screen. If the trace is still not visible, make sure that the trigger control (Figure 36.3b) is set to "auto". One further reason that the trace may not be visible is if the brightness (sometimes called intensity) control has been turned down. A standard setting for the brightness control is around two thirds of its maximum range. If you have done these things, you should now have a horizontal line on the screen. If the line appears blurred you can focus it with the focus control, shown in Figure 36.3c. (The line will also appear blurred if the brightness is set too high.)

Using the oscilloscope

Suppose you want to find out the frequency and amplitude of a signal produced by a signal generator. First you must connect the signal generator up to the oscilloscope, as shown in Figure 36.4.

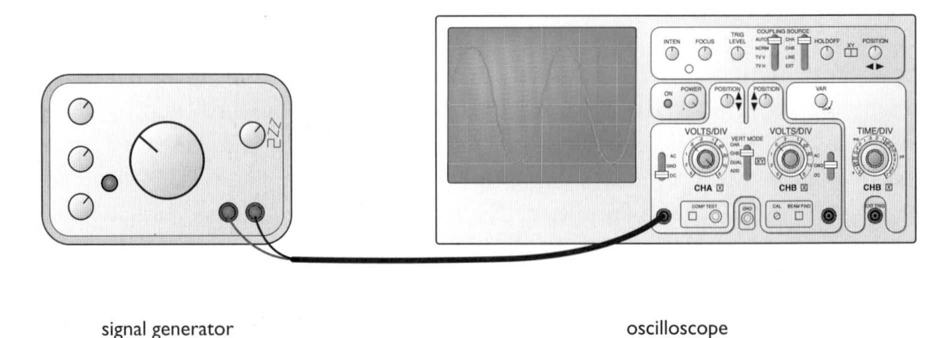

Figure 36.4 *Connecting the CRO to a signal generator.*

Next you must adjust the time and voltage scales on the oscilloscope so you can see the voltage waveform clearly. Figure 36.5a shows what you should see with suitable settings of the controls. Figures 36.5b and 36.5c show incorrect settings and how to correct them.

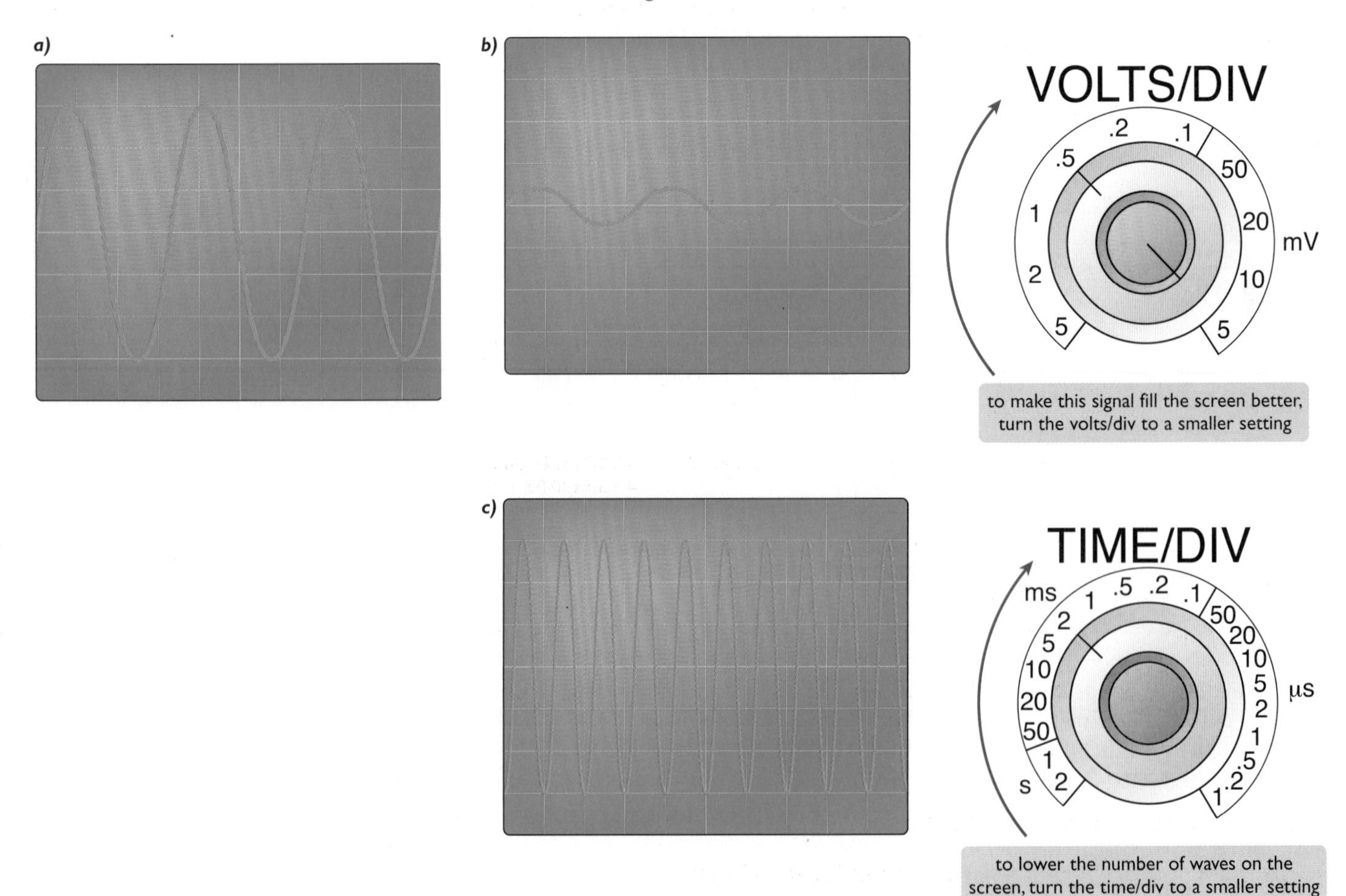

Figure 36.5 *Adjusting the time and voltage scales on the oscilloscope so that you can see the voltage waveform clearly.*

Measuring the amplitude

The **amplitude** of a wave is its height from the mid position (see page 142). Measure the height of the voltage signal against the scale marked on the CRO screen. Multiply the height in screen divisions by the setting on the volts/div (V/div) control to find the amplitude (Figure 36.6).

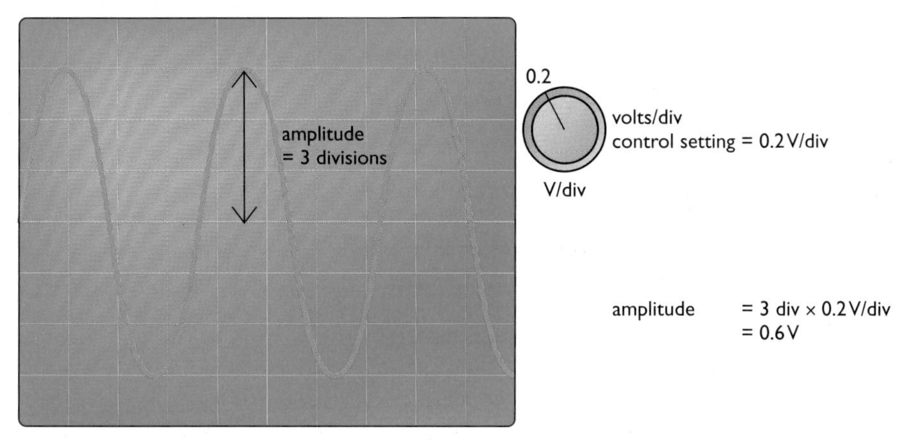

Figure 36.6 *Measuring the amplitude of a signal.*

Measuring the frequency

The **frequency** of a wave is the number of cycles of the wave there are in one second (see page 142). To measure the frequency of a voltage signal, you first need to measure the **period**, T, of the signal. To do this, count the number of divisions across the screen that a whole cycle of the wave takes up. Multiply the number of divisions by the setting on the time/div control (Figure 36.7).

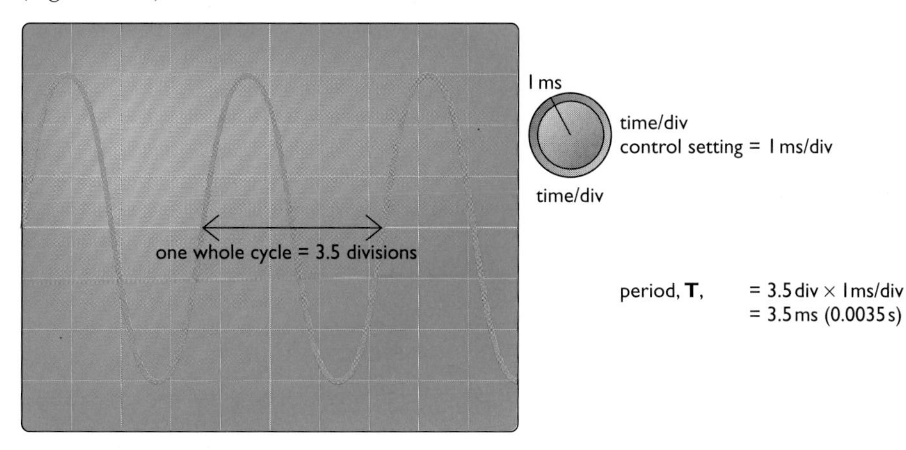

Figure 36.7 *Finding the period of a waveform.*

The frequency, f, of a waveform is given by:

$$\text{frequency, f} = \frac{1}{\text{period, T}}$$

where f is measured in hertz (Hz) if T is measured in seconds (s).

So the frequency of the waveform shown in Figure 36.7 is:

$$\text{frequency, f} = \frac{1}{0.0035\,\text{s}} = 286\,\text{Hz}$$

End of Chapter Checklist

If you haven't got a copy of your specification, read the introduction on page viii.

You will need to be able to do some or all of the following. Check your Awarding Body's specification (syllabus) to find out exactly what you need to know.

- Understand the basic controls of a cathode ray oscilloscope (CRO).
- Use the volts/div (V/div) and time/div controls to produce a clear graph of a waveform.
- Make voltage and time measurements using a CRO.
- Define the terms amplitude and period for a waveform.
- Calculate the frequency of a waveform using the CRO to measure its period.

Questions

More questions on the cathode ray oscilloscope can be found at the end of Section F on page 339.

1 The oscilloscope can be used to measure two quantities to a high degree of accuracy. What are these two quantities?

2 By using a flow diagram or a series of numbered instructions, outline how you would set up a CRO for use.

3 The diagram below shows a trace on an oscilloscope. A student wants to measure its amplitude and period as accurately as possible. Is the trace suitable for these measurements to be made? If not, describe the changes you would make to the CRO settings, with your reasons for making them.

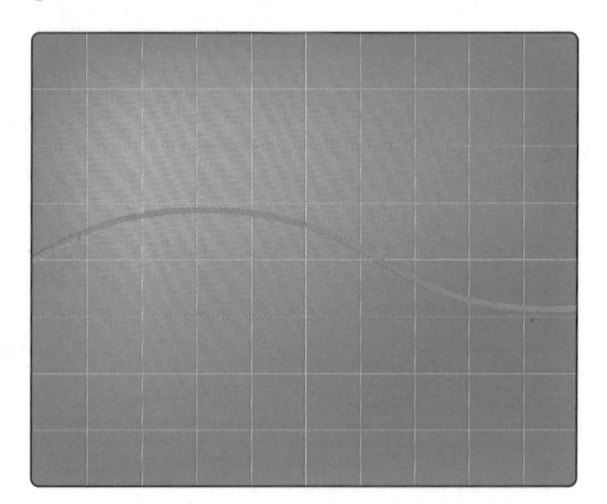

4 The diagram below shows a triangular waveform on a CRO, together with the settings of the volts/div and time/div controls.

a) Find the amplitude of the waveform.

b) Find the time period of the waveform.

c) Use your answer to ***b)*** to calculate the frequency of the waveform.

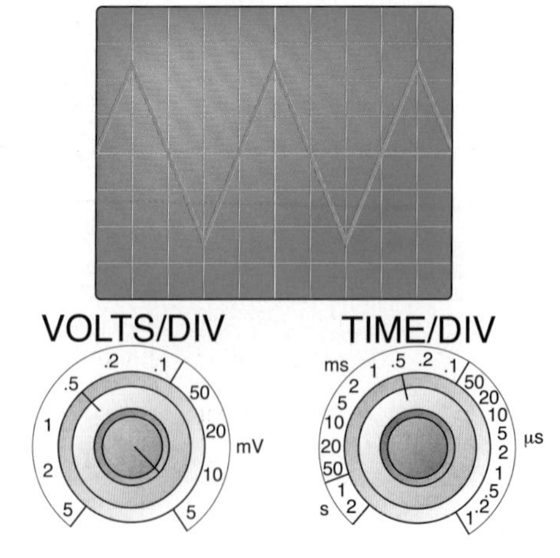

5 Draw a grid measuring 10 cm across by 8 cm high on a piece of graph paper. This represents the screen of a CRO. Draw a triangular waveform of frequency 500 Hz and amplitude 1.5 V on this "screen". State the settings for the volts/div and time/div controls that you have chosen. Remember that you are trying to get a clear graph of the waveform – filling the screen as far as possible with no more than a few cycles of the wave shown.

End of Section Questions

1 Copy and complete the following passage about the structure of the atom, filling in the spaces.

The atom is made up of three basic particles: ___________ , which orbit the central core of the atom, and ___________ and ___________ , which together make up the central core. The central core is called the ___________ of the atom, and the two types of particles in the central core are collectively called ___________ .

The atom is electrically neutral because it has equal numbers of ___________ and ___________ .

Total 7 marks

2 Geiger and Marsden conducted an experiment involving firing alpha particles at a thin sheet of gold foil.

a) What are alpha particles? *(2 marks)*

b) Describe, with the aid of a labelled diagram, the apparatus that Geiger and Marsden used in their experiment. *(4 marks)*

c) Describe the results that they obtained from their experiment and the conclusions that Rutherford drew from them. *(4 marks)*

d) Before this experiment, scientists had a different model of the atom. It was sometimes referred to as the "plum pudding" model. Describe this model. *(2 marks)*

Total 12 marks

3 For many years physicists believed that electrons, neutrons and protons were fundamental particles. In the latter part of the twentieth century some of these were found not to be fundamental.

a) What do physicists mean by *fundamental particles*? *(2 marks)*

b) Which of the particles mentioned above are no longer considered to be fundamental? *(2 marks)*

c) What are the new fundamental particles called? *(1 mark)*

Total 5 marks

4 The notation for the most abundant isotope of carbon is:

$^{12}_{6}C$

a) Explain what the numbers 12 and 6 in this notation mean. *(4 marks)*

b) Explain the term *isotope*. *(4 marks)*

c) Give an example of another isotope of carbon. You should write down the notation for the isotope and explain how it differs from the usual or most abundant isotope. *(4 marks)*

d) Explain the process of radiocarbon dating, used to estimate the age of organic material. *(4 marks)*

Total 16 marks

5 ***a)*** Describe how you would measure the half-life of a radioactive isotope, assuming the half-life can be measured in minutes (rather than hours or years, for example). You should mention the apparatus you would use, the measurements that you would need to take and any safety precautions you must follow. *(6 marks)*

b) The results of such an experiment are shown in the graph below.

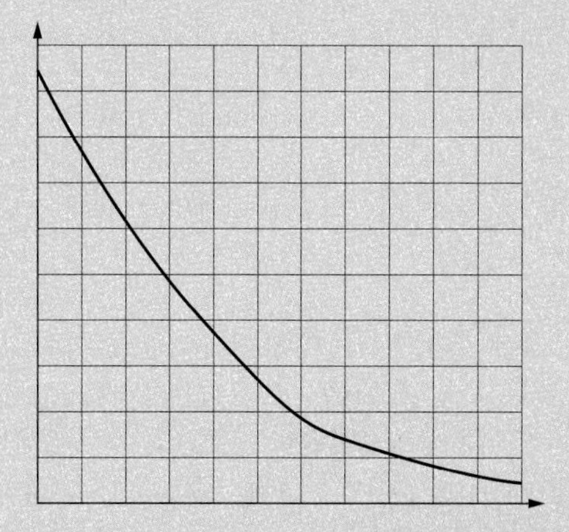

i) What units should be marked on the vertical scale? *(1 mark)*

ii) Copy the graph, and show how you would use it to measure the half-life of the isotope. *(3 marks)*

iii) What is the half-life of the isotope in minutes? *(2 marks)*

c) In another experiment a student has a 200 μg sample of a radioactive isotope that has a half-life of 10 minutes. How much of the sample will remain after half an hour? (Show how you calculated your answer.) *(3 marks)*

Total 15 marks

6 Copy and complete the following paragraph about different types of ionising radiation emitted from radioactive isotopes.

____________ particles are high-speed, negatively charged particles emitted by an unstable nucleus that has too many ____________ . No stable elements exist with atomic numbers greater than ____________ ; elements with higher atomic numbers decay by emitting ____________ particles. Nuclear decay is frequently accompanied by very high frequency electromagnetic radiation called ____________ ____________ .

Total 5 marks

7 Radioactive isotopes are used in the treatment and diagnosis of certain illnesses. The type of isotope chosen will depend on whether the purpose is diagnosis or treatment of the illness. The atomic number of the isotope is also a factor that influences the choice.

a) What type of isotope is likely to be used for treatment of a cancer? Give reasons for your answer. *(3 marks)*

b) What type of isotope is likely to be chosen for diagnosis? Give reasons for your choice. *(4 marks)*

c) A radioisotope of iodine is sometimes used in the treatment of thyroid cancer. Why is iodine used in preference to other isotopes with similar qualities? *(3 marks)*

Total 10 marks

8 Copy and complete the following nuclear equations by filling in the boxes.

a) $^{220}_{86}\text{Rn} \longrightarrow {}^{\square}_{84}\text{Po} + {}^{4}_{\square}\text{He}$

(2 marks)

b) $^{\square}_{91}\text{Pa} \longrightarrow {}^{234}_{92}\text{U} + {}^{0}_{\square}\text{e}$

(2 marks)

c) $^{131}_{53}\text{I} \longrightarrow {}^{131}_{54}\text{Xe} + {}^{\square}_{\square}\square$

(3 marks)

d) $^{22}_{11}\text{Na} \longrightarrow {}^{22}_{10}\text{Ne} + {}^{\square}_{\square}\square$

(3 marks)

Total 10 marks

Appendix A: Practical Investigations

The practical investigation counts 20% towards your final GCSE grade. It isn't difficult to get full (or very close to full) marks, but it does take time and patience. If you are prepared to spend that time, you will establish a really good foundation on which to build in your end-of-course or end-of-module exams.

The next few pages show you how to gain a high score on an investigation. It is important to listen to your teacher's advice on exactly what you need to do in order to get *full* marks. After all, he or she will be marking your piece of work, and will have up-to-date knowledge of what your examiners want.

The example investigation is:

Investigate why white light is dispersed when it travels through a prism and why the colours of the spectrum are always in the same order.

You must realise that the version of the investigation given here in this book is *incomplete*. You will find it in full on the website supporting the book at www.longman.co.uk/gcsephysics.

The text here gives ideas about finding information for this particular investigation, but what you need to do will be just the same whatever investigation you are given. Use this and other GCSE books to get you started. Don't forget the possibility of using A-level text books or Data Books to fill in some details.

It might also be worth seeing what you can find on the Awarding Bodies' websites (try all the Awarding Bodies – not just your own!). There may be material designed for teachers that you could make use of. You will find web addresses in the introduction on page iv.

Be very wary of any examples of coursework provided by other students on the Internet. They aren't difficult to find, but just because they are on the Internet, it doesn't mean that they are any good!

How to start

Your starting point is to gain a thorough understanding of the words "dispersion" and "spectrum". What is dispersion? What causes it? What causes the colours to behave differently? What measurements could you take to confirm your ideas? You'll find the answers to these questions, and many others, on pages 343–349 in this book.

When you design and carry out your experiments, remember that there are several factors that might affect your results. Although you should be changing only one variable at a time, it is important to be aware of the other variables that might affect your experiment. These must be controlled to ensure that your experiment will be a fair test.

For example, in the case of this investigation, you must use a transparent block made from the same material as your prism (see "What I already know" on page 342), and use the same block each time so that you can be certain that no variations in the properties of the block will affect the experiment.

Don't restrict yourself to this book. Use other GCSE and even A-level books. Nobody expects you to produce an A-level answer, but you should be pushing at the limits of GCSE if you want an A*.

Planning

Your teacher will be marking the planning part of your work by matching it to this checklist. The important terms are explained in the next page or two.

If you can	Mark awarded
outline a simple procedure	2
plan to collect evidence which will be valid plan the use of suitable equipment or sources of evidence	4
use scientific knowledge and understanding to plan and communicate a procedure, to identify key factors to vary, control or take into account, and to make a prediction where appropriate decide a suitable extent and range of evidence to be collected	6
use detailed scientific knowledge and understanding to plan and communicate an appropriate strategy, taking into account the need to produce precise and reliable evidence, and to justify a prediction, when one has been made use relevant information from preliminary work, where appropriate, to inform the plan	8

To score 8 marks, your work must match both the statements in the last box, above, but must also fulfil all the other statements for 6, 4 and 2 in the previous boxes. In other words, for full marks, you need to do everything in the table. It is important to aim for the highest possible mark. Even if you miss it, you can still score well.

You can score odd-numbered marks if your work falls just short of a level. For example, you might score 7 marks if you satisfied the first statement needed for 8 marks, but didn't do any preliminary experiments when some would have been helpful.

You could use the headings over the next few pages to help you to get everything in a logical order during the planning stage.

What I am going to do

It's a good idea at the start of these investigations to have a mini brainstorming session where you write down absolutely everything that you feel might affect the quantity you are measuring. Having done this, you can then identify all those variables you need to control. Telling the examiner that you have done this – better still, handing in a neat version of your brainstorm – will indicate to the examiner that you have given this problem some thought, rather than rushing in with little idea of how to tackle it.

What I already know

To score 8 marks for planning, everything must be based on "detailed scientific knowledge and understanding" – and you have to make it clear how you are going to use that knowledge and understanding. List all the *relevant* things you have found out from books or other sources, and say why you think they might be useful to you.

So, now you have done your research you should know in very broad terms what is happening when white light is dispersed by a prism, and what you are going to do in your experiments. You should have read that light slows down when it travels from air into glass and that this change in speed causes the light to change direction or **refract**. You may also have discovered that there is a law called Snell's Law, which links the amount by which a ray of light is refracted and the speed at which the light is travelling. So you have the idea that the reason light is dispersed into its different colours is because each colour of light is slowed down by a different amount as it enters the glass prism – that is, the different coloured lights are travelling at different speeds.

To test this, you are going to pass a ray of coloured light through a glass block and measure its angle of incidence and its angle of reflection. You are going to repeat this six times, put your results in a table, and then draw a graph to find the refractive index of the block for that colour of light. Finally, you are going to work out the speed of that colour of light while it is travelling in the block. You are then going to repeat this experiment for at least six different colours.

In this investigation, it will be extremely useful to draw some ray diagrams to show dispersion by a prism and refraction of light by the glass block.

Warning! None of the examples given are complete. You will need to decide for yourself if other details should be included.

Preliminary work

Preliminary work involves doing rough experiments to find out the best conditions for carrying out your investigation. It is essential if you are going to score 8 marks for planning. Describe your preliminary work carefully, and say exactly how it helped you to decide your final plan.

Again, wherever possible, explain your choices in terms of "detailed scientific knowledge and understanding".

For example, in this case, it is important to discover what black-out conditions are necessary to allow the rays of light to be seen clearly and yet also provide enough light in the room for accurate measurements to be made. Does the size or the position of the slit which produces the ray of light from the ray box affect how accurately its path can be marked? How are you going to mark the direction of the ray while it is inside the block? What range of angles of incidence of the ray are you going to use? Does the place where the ray strikes the block affect the experiment?

> I placed a rectangular glass block in the centre of a plain piece of paper and drew around it. I then shone a ray of white light onto the centre of the top surface of the block. I marked the direction of the ray of light as it entered and left the block with two crosses. I removed the glass block, drew in the path of the ray and a normal where the ray entered the block. I measured the angle of incidence and the angle of refraction. As I did this experiment, I altered the distance between the ray box and the block to see if it affected the brightness of the ray as it travelled into and out of the block. I discovered that the ray box should be no more than 5 cm from the block, otherwise the ray emerging from the block becomes dim and more difficult to mark. I also discovered that as long as the lights were out and the blinds on my side of the lab were down I could carry out all parts of the experiment.

Again, remember that this is not a complete version of what you would have to write in your account of your preliminary work.

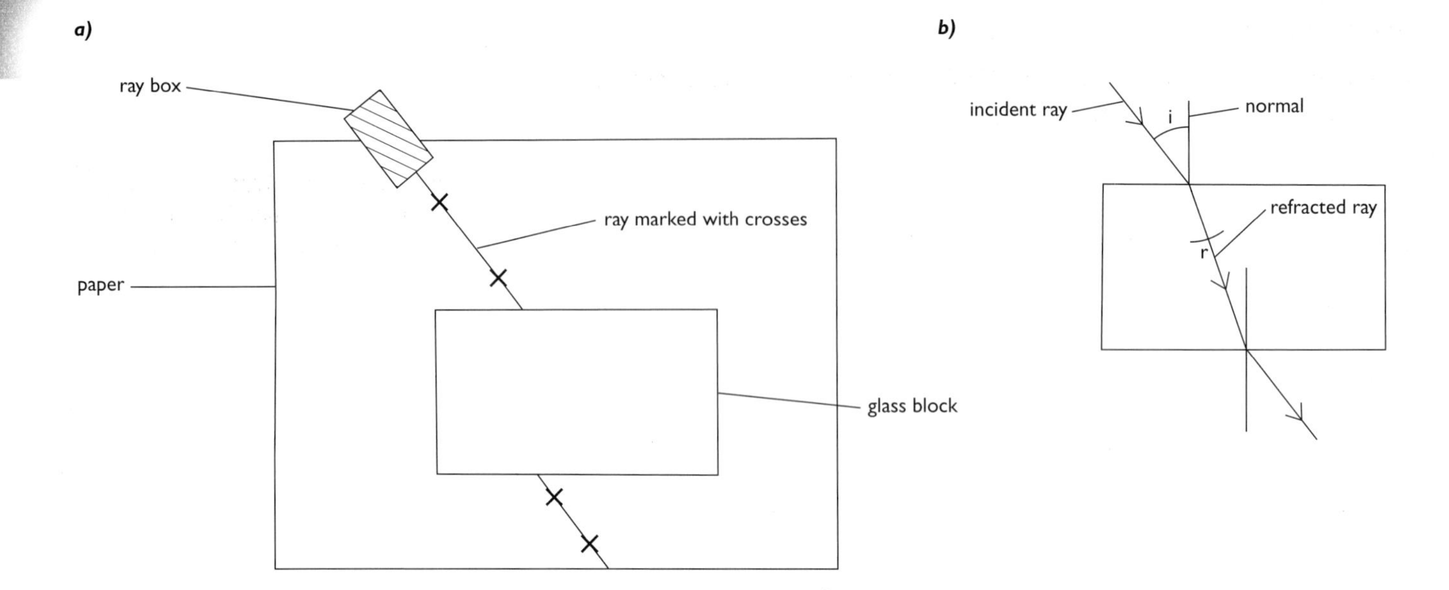

Producing precise and reliable evidence

"Precise" means that you are measuring things as accurately as possible. Particularly where the quantity you are measuring is small, you should try to measure it using the most accurate equipment you have access to. "Reliable" means that if you repeat the measurements, you will get the same results.

> I realised when I did my preliminary experiment that because I was using a rectangular glass block the angles of incidence and refraction on the upper and lower sides of the block should be the same. I am therefore going to measure both angles of incidence and both angles of refraction and then work out their averages. This should improve the accuracy of my measurements.

Take advice from your teacher about how many repeat experiments you are expected to do in order to get full marks. Remember that your teacher is going to mark your work, although it might also be checked by someone from the Awarding Body. You also need to know how many sets of results you are expected to produce in order to earn full marks. For example, in this experiment by taking six pairs of readings for each colour of light you will have enough results to draw a graph from which you can determine the refractive index of the glass for that colour of light.

Safety

In some experiments, where perhaps you are heating things or are using chemicals, it is important to demonstrate to your examiner that you have taken steps to ensure that the experiment is carried out safely, and have explained why you need to take each precaution. There are some simple safety precautions you could take with this experiment – for instance, making sure that bags and unused stools are moved out of the way so that people do not trip over them if they are moving around the lab when it is dark.

Doing the experiment

Describe what you are going to do in detail, listing all the apparatus you need. Draw diagrams to show exactly how the apparatus is used so that there is no uncertainty about what you mean.

When you have finished describing your method, read it critically and ask yourself whether someone else could carry it out successfully if they did it *exactly* as you have written it. You can assume that they know how to use standard apparatus like ray boxes and power packs, but you could stress any points that are particularly important for accuracy.

My prediction

Again, your predictions must be justified in terms of "detailed scientific knowledge and understanding".

I predict that the different colours of light travel through glass at different speeds. I believe this to be the case because the amount by which a ray of light is refracted depends upon the refractive index of the medium it has entered and the refractive index of a material is related to the speed at which light travels through it, by the equation:

$$\text{refractive index of medium} = \frac{\text{speed of light in a vacuum}}{\text{speed of light in the medium}}$$

I know that the speed of light in air is 3×10^8 m/s. If I can calculate the refractive index of the glass for each of the colours, I can calculate their speeds in the glass. I expect that the order of their speeds is the same as the order in which they are found in the spectrum.

I can calculate the refractive index of the glass for each colour of light using Snell's Law. This states that:

$$\text{refractive index of medium} = \frac{\sin i}{\sin r}$$

where i is the angle of incidence and r is the corresponding angle of refraction.

Warning! Don't forget that these are not complete predictions.

Obtaining evidence

Your teacher will be using the checklist below.

If you can	Mark awarded
collect some evidence using a simple and safe procedure	2
collect appropriate evidence which is adequate for the activity record the evidence	4
collect sufficient systematic and accurate evidence and repeat or check when appropriate record clearly and accurately the evidence collected	6
use a procedure with precision and skill to obtain and record an appropriate range of reliable evidence	8

Draw up a table of results

To get full marks, you must record everything in a logical and clear manner. In this particular case, that means clear tables of results for the angle of incidence, i, and the angle of reflection, r, and for the sine values of these angles.

Label the columns and rows in your table, *and include the correct units*. Then write down the numbers to reflect their accuracy.

For red light:

i	r	sin i	sin r
20°	13°	0.342	0.228
30°	20°	0.500	0.333
40°	25°	0.643	0.429
50°			
60°			
70°			
80°			

Analysing your evidence and drawing conclusions

Your teacher will be using the checklist below.

If you can	Mark awarded
state simply what is shown by the evidence	2
use simple diagrams, charts or graphs as a basis for explaining the evidence identify patterns and trends in the evidence	4
construct and use suitable diagrams, charts, graphs (with lines of best fit where appropriate), or use numerical methods, to process evidence for a conclusion draw a conclusion consistent with the evidence and explain it using scientific knowledge and understanding	6
use detailed scientific knowledge and understanding to explain a valid conclusion drawn from processed evidence. explain the extent to which the conclusion supports the prediction, if one has been made	8

Notice that to gain 6 or 8 marks you don't necessarily have to draw graphs. You can use "numerical methods" instead. That means doing some reasonably complicated calculations. Although working out a simple average is a "numerical method", it isn't complicated enough to earn you 6 (or 8) marks.

Calculations

To determine the refractive index of the glass block for a particular colour, you could use the equation:

$$\text{refractive index of medium} = \frac{\sin i}{\sin r}$$

for each pair of readings. Adding these together and dividing the total by six will then give you an average value, which is likely to be more accurate than just one pair of readings. The examiner will also see that you have taken steps to improve the accuracy of your experiment – so be sure to show all your working out.

Alternately you could draw a graph of sin i against sin r. By drawing a line of best fit through your results and measuring its gradient you are again finding an averaged value. If any results do not fit the pattern or trend shown by the rest, they should be ignored when the line of best fit is drawn but mentioned as anomalous results that have been discarded.

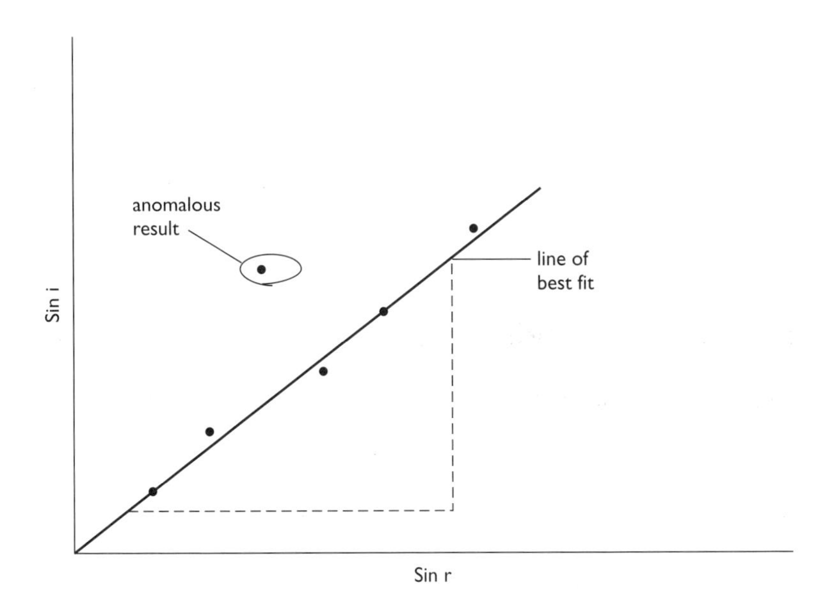

Displaying your final results

Having found a refractive index for each of the colours, you can calculate their speeds using the equation:

$$\text{refractive index of medium} = \frac{\text{speed of light in a vacuum}}{\text{speed of light in the medium}}$$

For example:

$$\text{speed of red light in glass} = \frac{\text{speed of light in a vacuum}}{\text{refractive index of glass for red light}}$$

The results of these calculations should be presented together in a final table.

Drawing your conclusions

You have to decide whether your results fit your original prediction. Remind the person marking your work exactly what your prediction was. Don't just refer back to your original scientific explanation. Give it again, together with any modifications or extras that your results show are necessary.

> *My results clearly show that the refractive index of glass is different for each of the different colours of light and as a result would bend the colours through different angles causing them to disperse as they pass through a prism. The results also show that I was correct in predicting that the order of the spectrum is determined by the different speeds with which the different colours travel.*

It doesn't matter if your prediction doesn't work out as you expected, provided you can use your knowledge and understanding to come up with a convincing explanation.

Evaluating your investigation

Your teacher will be using the checklist below.

If you can	Mark awarded
make a relevant comment about the procedure used or the evidence obtained	2
comment on the quality of the evidence, identifying any anomalies comment on the suitability of the procedure and, where appropriate, suggest changes to improve it	4
consider critically the reliability of the evidence and whether it is sufficient to support the conclusion, accounting for any anomalies describe, in detail, further work to provide additional relevant evidence	6

It is easy to throw marks away in this last stage. You must remember that this part is worth almost as many marks as each of the other three stages, and so you must take just as much care over it. Students frequently get bored at this point, and hope to get full marks with a paragraph of generalised waffle. No chance!

Evaluating the experiment

You need to point out any results that appear to be wrong, and try to account for them.

> I carried out the experiment over two lessons. In the second lesson I wasn't sure whether I had the same glass block as the first lesson, as the label had fallen off. Although this didn't affect the overall pattern, it might explain why the last three sets of readings for green, blue and indigo light seemed to form their own trend.

If you are doing an investigation that produces a definite numerical answer, always try to find out what the accepted answer is. That will give you an immediate idea of how accurate your experiment was. You can then look for reasons to account for your degree of inaccuracy.

Improving the experiment

Make sure that the improvements you suggest are detailed and specific to your investigation. You won't get much credit for general comments like "use more accurate equipment" or "take more care with measurements". There is something to be said for leaving some flaws in your original experiment so that you have something to talk about at this evaluation stage!

> If I were doing the experiment again, I would make sure I used the same glass block for the whole experiment.
>
> If I had had time, I would have liked to have repeated the readings that gave us our two anomalous results.

Extending the experiment

This is to fulfil the 6-mark statement, "describe, in detail, further work to provide additional relevant evidence". In the case we're looking at, there are several areas of further interest.

> I could repeat the experiment with blocks made from different materials to confirm that the relationship is true for all media.
>
> I could extend the experiment to discover if the same effect is present in transparent media that are liquids.

Appendix B: Electrical Circuit Symbols

Description	Symbol
conductors crossing with no connection	
junction of conductors	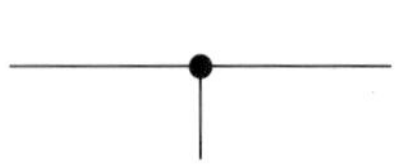
open switch	
closed switch	
open push switch	
closed push switch	
cell	
battery of cells	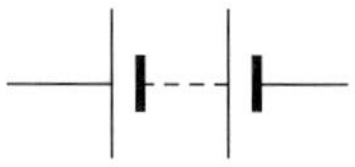
power supply	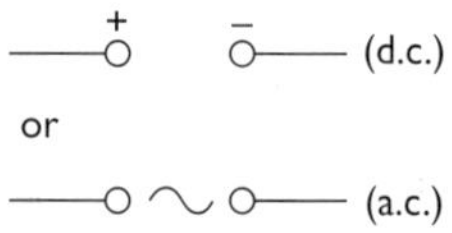 + − (d.c.) or (a.c.)
transformer	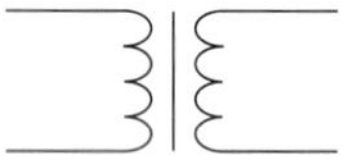
ammeter	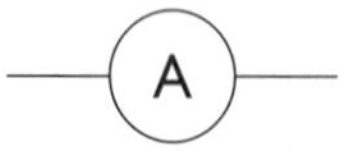 A
milliammeter	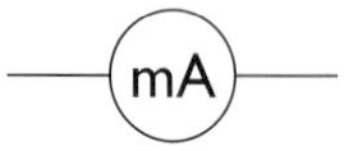 mA
voltmeter	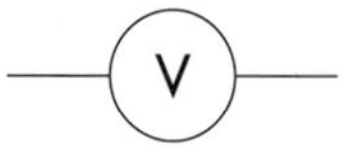 V
fixed resistor	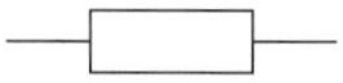
variable resistor	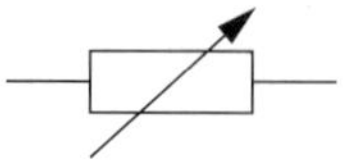
NPN transistor	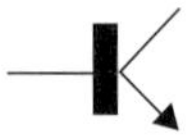
heater	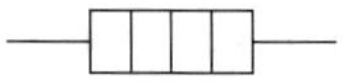
thermistor	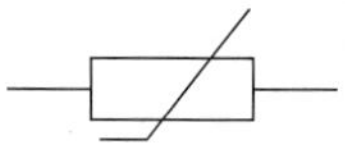

Description	Symbol
light-dependent resistor (LDR)	
relay	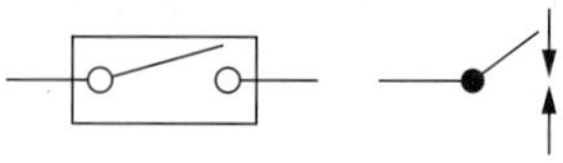
diode	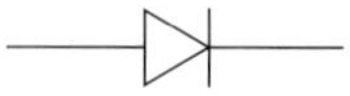
light-emitting diode (LED)	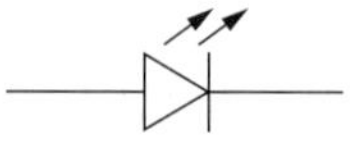
lamp	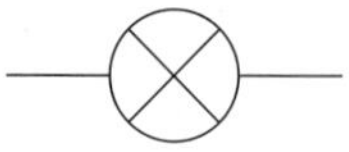
loudspeaker	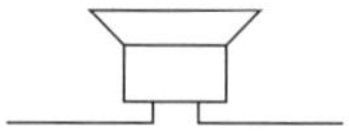
microphone	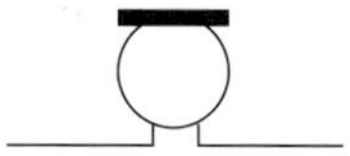
electric bell	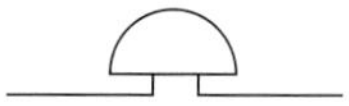
earth or ground	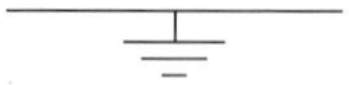
motor	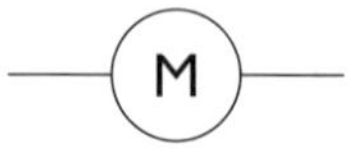 M
generator	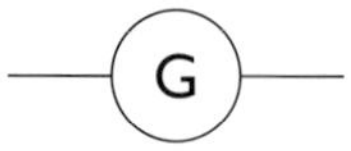 G
fuse/circuit breaker	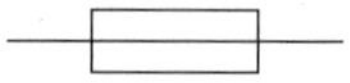
NOT gate (inverter)	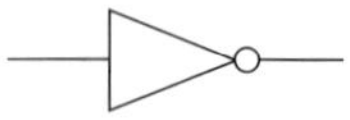
AND gate	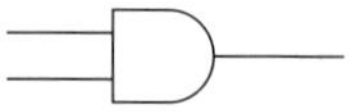
NAND gate	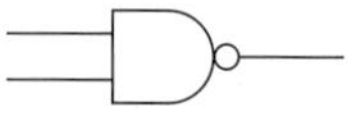
OR gate	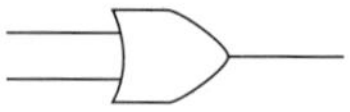
NOR gate	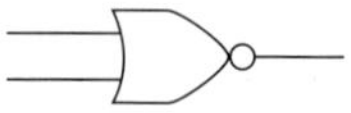
EOR gate	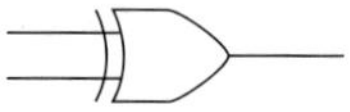

The resistor colour and preferred values

The resistor colour code

Resistors usually have four coloured bands on them. This is a code that allows you to work out their value.

The first two coloured bands represent the numbers 0 to 9. The third band, sometimes called the multiplier, tells you what you have to multiply the two digit number shown by the first and second band to get the value of the resistor in ohms. The fourth band tells you the tolerance of the resistor – this is how accurately the resistor is made – the *actual* value of the resistor may vary from its *nominal* or stated value because of the manufacturing process.

Below is a table showing the colour code for resistors, with two worked examples of how to use the colour code. You are not expected to remember the colour code but you should know how to use it if you are given it in a question.

	First band	Second band	Multiplier	Tolerance
Black	0	0	× 1	
Brown	1	1	× 10	± 1%
Red	2	2	× 100	± 2%
Orange	3	3	× 1000	
Yellow	4	4	× 10 0000	
Green	5	5	× 100 000	Gold ± 5%
Blue	6	6	× 1000 000	Silver ± 10%
Violet	7	7		
Grey	8	8		
White	9	9		

***worked* example**

1. A resistor has a colour code of brown black red silver. What is its nominal value? What are the maximum and minimum values that the resistor may have?

Look at the first three coloured bands to find the nominal value.

brown black red

1 0 × 100

This gives the nominal value of the resistor as 1000 Ω or 1 kΩ.

The fourth band tells us that these resistors are made to a tolerance of 10%. This means that the actual value of a particular resistor with this colour code can vary by ± 10% of the nominal (stated) value. In this example 10% of 1000 Ω is 100 Ω, so the actual value can be anything from (1000 – 100) Ω to (1000 + 100) Ω. So the maximum value should be 1100 Ω and the minimum value should be no less than 900 Ω.

2. Another resistor has a colour code of yellow violet black gold. Work out its nominal value and the range of values it may actually have.

   Remember, the first three coloured bands give the nominal value of the resistor.

   yellow violet black

   4 7 ×1

   So the nominal value is 47 Ω.

   The gold fourth band tells us that these resistors are made to a tolerance of 5%, so the actual value of the resistor can vary by ±5% of 47 Ω. 5% of 47 is 2.35 so the range of values you would expect for a resistor with this colour code is:

   44.65 Ω to 49.35 Ω

Preferred values

Resistors are not made with every possible value of resistance. It would be impractical and also pointless because of the tolerance on the values. A 10% tolerance resistor with a nominal value of 10 Ω might have any value from 9 Ω to 11 Ω, so there is little point in having separate resistor values of 9, 10 and 11 Ω. The E12 preferred value series has just 12 different values of resistance:

10, 12, 15, 18, 22, 27, 33, 39, 47, 56, 68, 82

and the multiples of these, such as 100, 1000, 10 000, 100 000 1 000 000 and 120, 1200, 12 000, 120 000, 1 200 000 etc.

You will not be expected to remember the preferred value series but you may be asked why resistors are only available with certain values.

Appendix C: Formulae and Relationships

The formulae and relationships listed below may not be provided for you in exams, so you will need to remember them.

1 The relationships between speed, distance and time:

$$\text{speed} = \frac{\text{distance}}{\text{time}}$$

2 The relationship between force, mass and acceleration:

$$\text{force} = \text{mass} \times \text{acceleration}$$

$$\text{acceleration} = \frac{\text{change in velocity}}{\text{time taken}}$$

3 The relationship between density, mass and volume:

$$\text{density} = \frac{\text{mass}}{\text{volume}}$$

4 The relationship between force, distance and work:

$$\text{work done} = \text{force} \times \text{distance moved in direction of force}$$

5 The energy relationships:

$$\text{energy transferred} = \text{work done}$$

$$\text{kinetic energy} = \tfrac{1}{2} \times \text{mass} \times \text{speed}^2$$

$$\text{change in potential energy} = \text{mass} \times \text{gravitational field strength} \times \text{change in height}$$

$$\text{power} = \frac{\text{work done}}{\text{time taken}}$$

6 The relationship between mass, weight and gravitational field strength:

$$\text{weight} = \text{mass} \times \text{gravitational field strength}$$

7 The relationship between an applied force, the area over which it acts and the resulting pressure:

$$\text{pressure} = \frac{\text{force}}{\text{area}}$$

8 The relationship between the moment of a force and its distance from the pivot:

$$\text{moment} = \text{force} \times \text{perpendicular distance from pivot}$$

9 The relationships between charge, current, voltage (potential difference), resistance, electrical power and energy:

charge = current × time

voltage = current × resistance

electrical power = voltage × current

energy transferred = power × time

energy transferred = voltage × charge

10 The relationship between efficiency, energy input and useful energy output:

$$\text{efficiency} = \frac{\text{useful energy transferred by device}}{\text{total energy supplied to device}}$$

11 The relationship between speed, frequency and wavelength:

wave speed = frequency × wavelength

12 The relationship between the voltage across the coils in a transformer and the number of turns in them:

$$\frac{\text{voltage across secondary coil}}{\text{voltage across primary coil}} = \frac{\text{number of turns in secondary coil}}{\text{number of turns in primary coil}}$$

Appendix D: Physical Quantities and Units

Fundamental physical quantities

Physical quantity	Unit(s)
length	metre (m) kilometre (km) centimetre (cm) millimetre (mm)
mass	kilogram (kg) gram (g) milligram (mg)
time	second (s) millisecond (ms)
temperature	degrees Celsius (°C) kelvin (K)
current	ampere or amp (A) milliampere or milliamp (mA)

Derived quantities and units

Physical quantity	Unit(s)
area	cm^2 m^2
volume	cm^3 dm^3 m^3 litre (l) millilitre (ml)
density	kg/m^3 g/cm^3
force	newton (N)
pressure	pascal (Pa or N/m^2) N/cm^2
speed	m/s km/h
acceleration	m/s^2
energy	joule (J) kilojoule (kJ) megajoule (MJ)
power	watt (W) kilowatt (kW) megawatt (MW)
frequency	hertz (Hz) kilohertz (kHz)
electrical charge	coulomb (C)
potential difference (voltage)	volt (V) millivolt (mV)
resistance	ohm (Ω)
gravitational field strength	N/kg
radioactivity	Becquerel (Bq)
sound intensity	decibel (dB)

Index